Molecular and Integrative Toxicology

Series Editors
Jamie C. DeWitt, East Carolina University, Greenville, NC, USA
Sarah Blossom, Department of Pharmaceutical Sciences, UNM College of
Pharmacy, University of New Mexico, Albuquerque, NM, USA

Molecular and Integrative Toxicology presents state-of-the-art toxicology in a useful context. Volumes emphasize the presentation of cellular and molecular information aimed toward the protection of human or animal health or the sustainability of environmental systems. This book series is committed to maintaining the highest level of integrity in the content published. It has a Conflict of Interest policy in place and complies with international, national and/or institutional standards on research involving Human Participants and/or Animals and Informed Consent. The Book series is a member of the Committee on Publication Ethics (COPE) and subscribes to its principles on how to deal with acts of misconduct thereby committing to investigate allegations of misconduct in order to ensure the integrity of research. This book series may use plagiarism detection software to screen the submissions. If plagiarism is identified, the COPE guidelines on plagiarism will be followed.

Content published in this book series is peer reviewed. It solicits reviews from the senior editors of the volume since they are the experts in the field, and they are already known to the authors. In cases, however, where there are particular concerns, a single-blind review system is used where independent external experts are asked for advice. These reviewers know the names of the authors, but the identities of the reviewers are unknown to the authors. Submitted manuscripts are reviewed by at least two experts (the senior editors of the volume and or the consulting editors) as well as the series editor. The reviews evaluate whether the manuscript is scientifically sound and coherent, whether it meets the guidelines regarding style, content, and level of writing. The volume editors decide on the suitability of the manuscript based on their reviews.

Titles in this series are indexed in SCOPUS.

More information about this series at http://www.springer.com/series/8792

Kavindra Kumar Kesari • Niraj Kumar Jha
Editors

Free Radical Biology and Environmental Toxicity

 Springer

Editors
Kavindra Kumar Kesari
Department of Applied Physics and
Department of Bioproducts and Biosystems
School of Science
Aalto University
Espoo, Finland

Niraj Kumar Jha
Department of Biotechnology
School of Engineering and Technology
Sharda University
Greater Noida, Uttar Pradesh, India

ISSN 2168-4219   ISSN 2168-4235   (electronic)
Molecular and Integrative Toxicology
ISBN 978-3-030-83445-6   ISBN 978-3-030-83446-3   (eBook)
https://doi.org/10.1007/978-3-030-83446-3

This Springer imprint is published by the registered company Springer Nature Switzerland AG
The registered company address is: Gewerbestrasse 11, 6330 Cham, Switzerland

Foreword

The edited book, *Free Radical Biology and Environmental Toxicity*, by Kavindra Kumar Kesari and Niraj Kumar Jha represents an interdisciplinary literature by integrating the basic principles of physics, chemistry, and biology. The title itself defines the theme of the book, wherein the authors have considered relevant environmental issues impacting human and animal health. This book contains 17 chapters which cover most of the environmental issues. The arrangement of the chapters is orchestrated, synchronized, and coherently connect to each other with suitable schematic presentation of mechanisms and pathways to comprehend the subject in a better way. It would be appropriate to say that this book is a good recipe for the readers and researchers in the field of environmental toxicology. Though the major focus of the book is on nanotoxicity, fine particle contaminants, heavy metals, and radiation toxicity, including a chapter on COVID-19 makes it more pertinent with the impact of environmental toxicity on the worsening of the COVID-19 pandemic. One of the most important components of this book is elucidating mechanisms and presenting evidence-based studies, which is an occasional combination in available environmental toxicology books. In the majority of the chapters, the connection of free radical biology, a complement of oxidative stress with environmental toxicity, represents the mechanistic clues of environmental toxicants induced oxidative damage. The chapters in this book are relevant with the theme of the book as it begins with introductory chapters on the environment persistent free radicals, followed by causative factors, and ends with protective measures including numerous therapeutic strategies. To discuss a few of the chapters, Chapter "Reactive Oxygen Species Producing Photoactivatable Molecules and Their Biological Application" showcases the free radicals–induced photoactivated molecules that may work as photosensitizers and pave the way for therapeutic opportunities in biomedical sciences. Arsenic is one of the carcinogens which may cause severe health issues, and Chapter "Role of Arsenic in Carcinogenesis" presents the health hazards of arsenic and provides evidence-based information for better understanding. Chapter "Regulation of Glucose Transporters in Cancer Progression" deals with glucose transporters in cancer progression that open new doors for future research. Several other chapters are dedicated to toxic metal contaminations and radiation-induced toxicity, which are leading health hazards. The role of environmental toxicants in neurological diseases is well covered in the book.

This book may serve as a reference book for academic programs and represents an update for environmental toxicology researchers. The book will also be useful for general readers as it covers many public issues. The editors of this book deserve special thanks for editing such an important book on an important topical issue.

Professor, Department of Applied Physics,   Janne Ruokolainen
School of Science, Aalto University,
Espoo, Finland
e-mail: janne.ruokolainen@aalto.fi

Associate Professor of Pharmacology and   Shreesh Ojha
Therapeutics, College of Medicine and Health Sciences,
United Arab Emirates University,
Al Ain, United Arab Emirates
e-mail: shreeshojha@uaeu.ac.ae

Preface and Acknowledgements

In the present scenario, the major concerns to our environment and its toxicity have raised a great interest globally, especially for the younger generation. With spreading networking of mutagenic factors in the environment, every individual is exposed to different chemicals and fine particles, which may react with our genetic material and may cause myriad health issues. The fundamental questions on the effects of environmental toxicants and pollutants mainly focus on the multidisciplinary field of science by applying the basic principles of physics, chemistry, and biology. We are all surrounded with various natural and man-made environmental toxicants, mutagens, and infectants, which may present significant risk to human health. The global industrialization together with the development of our health system has led to an exponential growth in the number of toxic chemicals used for different purposes. There are varieties of chemical, physical, and biological contaminants that are present in the environment. These toxicants are mainly heavy metals, toxic chemicals, compounds, tobacco smoke, nanoparticles, pesticides, viruses, radiations (ionizing and non-ionizing), and several others (fine particles, pollutions, climate change). These toxicants are highly associated with our daily life and affect our health. In principle, any exposure to these toxicants may induce oxidative stress and start producing free radicals within the cells and may eventually cause severe diseases such as neurological disorders and cancer. It would not be entirely incorrect to assume that the oxidative stress is the root cause of most of the severe health conditions.

This book is a value-added collection of 17 chapters which discuss the multidisciplinary approaches with a focus on the environmental toxicity and free radical formation. Chapter "Environment Persistent Free Radicals – Long Lived Particles" discusses the significant role of long-lasting environment persistent free radicals (EPFRs) on human health. Study also highlights a set of potential risk factors, which are associated with the EPFRs in bioremediation of organic pollutants and human health. Although, in Chapter "Reactive Oxygen Species Producing Photoactivatable Molecules and Their Biological Application", photosensitizers have been reported as reactive oxygen species (ROS) producing molecules and found exclusive for site-specific targeting. Photosensitizers prove to be the emerging therapeutic molecules for tumor treatment and cancer cell destruction with high specificity. This chapter describes the mechanisms, types, and the therapeutic applications of photosensitizers. Chapter "Effects of Transformation of Metallic Nanoparticles in the Environment and its Toxicity on the Aquatic and Terrestrial Life Forms" discusses the toxicity of nanomaterials that may raise serious concern to human health, as these are potentially used in several research fields. Similar to EPFRs (Chapter "Environment Persistent Free Radicals – Long Lived Particles"), living organisms in the environment may experience a long-lasting unwanted effect due to the release of nanomaterials to the environment. Therefore, this chapter describes the various sources of the unwanted release of nanomaterials in the environment, their mechanism of actions, and associated side-effects on the living system. Along with nanomaterials, nanodrugs or drug compounds may also induce oxidative stress which may induce cellular toxicity as reported in Chapter "Drug Induced Oxidative Stress and Cellular Toxicity". In this chapter, the role of ROS and reactive nitrogen species (RNS) in response to various drugs and further retrieval effects after co-supplementation of the antioxidants have been discussed. Evidences show that any kind of drug supplementation either nanoparticle and/or drug compounds may affect majorly the elderly populations. Therefore, Chapter "Adversities of Nanoparticles in Elderly Populations" discusses the pros and cons of the nanoparticles in the elderly populations. Study highlights the association of nanotoxicity with the phenomenon of ageing and possible link of various organs-related diseases. The major sources of nanoparticles or nanodrugs are being used in food and other commonly used products at the large scale. Therefore, Chapter "Toxicity of Titanium Dioxide Nanoparticles and Oxidative Stress" discusses the role of titanium dioxide (TiO_2) nanoparticles-induced oxidative stress which may cause oxidative damage and genotoxicity in the human system. Study mainly focuses on exploring the role of nanotoxicity based on the nanoparticle size, shape, route of administration, and various other sources. In this chapter, study highlights the applications of TiO_2 nanoparticles and their toxic potential to cause severe diseases in humans. The sources of contamination through the consumption of various modes such as food, air, water, and waste irrigated crops may play an important role in cell cycle progression and severe disease development, especially water has raised serious health issues due to chemical and heavy metal contaminations. Chapter "Role of Arsenic in Carcinogenesis" discusses the arsenic contamination in drinking water leading to poisoning in the healthy living population. Evidences reported in this chapter suggest that the arsenic has potential to lead many conditions such as diabetes, hearing loss, portal fibrosis, cardiovascular and peripheral vascular diseases, as well as various cancer types, such as skin, bladder, kidney, prostate, liver, and lung cancer. These diseases may be caused due to oxidative stress induced by arsenic, which may alter the DNA structure by insertion, deletion, and by several other mutations like gene amplification and suppression of p53 leading to the manifestation of cancer. One of the important mechanisms of arsenic poisoning-induced oxidative stress is lipid peroxidation, which may increase the arsenic-induced mutations up to five-fold. Cancer is one of the leading diseases globally. Until now, from first chapter to seventh chapter, there is one thing in common, i.e., nanotoxicity and chemical toxicity raising serious health concerns. The reported toxicants mostly alter mitochondrial functions, where mitochondria have been found a key player in maintaining the metabolism, bioenergetics, biosynthesis, and cell. In connection to this, Chapter "Role of Mitochondrial Oxidative Stress in Pathophysiology of Lung Cancer" discusses the role of ROS in regulating the molecular signaling pathway alterations, which may lead to lung cancer progression. Besides the mitochondrial metabolic changes, mitochondrial dynamics such as fission and fusion events are also affected in cancer cells. Study also focuses on the various redox-sensitive miRNAs and their roles in pathophysiology of lung cancer progression. After detailed discussions on the mutagenic factors and cancer, Chapter "Regulation of Glucose Transporters in Cancer Progression" shows great impact towards the regulation of glucose transporters in cancer progression. Study discusses glycolysis, which is often the preferred metabolic pathway for cancer cells that enables them to acquire energy and other metabolites for their growth and survival. After going through Chapter "Role of Mitochondrial Oxidative Stress in Pathophysiology of Lung Cancer", it provides a better understanding of regulatory pathways where glucose transporters are rate-limiting checkpoints, abnormally regulated in cancer. The regulation of glucose transporters (GLUTs) by key proliferation and pro-survival pathways including the phosphatidylinositol 3-kinase (PI3K)-Akt, hypoxia-inducible factor-1, and HIF-1 pathways provide pioneering ways for future research. GLUTs can be targeted in a relatively tumor cell-specific manner to block glucose-regulated processes more comprehensively for cancer therapy which has been discussed in detail in this chapter.

In the follow-up studies, chemical contamination has raised severe health concerns since past few decades. The previous chapters of this book have already discussed the pros and cons after consuming contaminated crops, food, and water, which has causative impact on our biological system. In the continuation, Chapter "Oxidative Stress: A Potential Link Between Pesticide Exposure and Early-Life Neurological Disorders" elaborates different classes of pesticides and their effects on the oxidative stress parameters. This chapter highlights the association between pesticide exposure and the development of different neurological disorders. In Chapter "Sleep Disturbance Induced Free Radical Formation in Gut May Blocked by Melatonin", the oxidative stress has been recognized as a responsible factor to cause sleep disturbances and produces free radicals in the gut. The inhibitory effect of oxidative stress by melatonin has been reported to improve the sleep/wake cycle. Melatonin is an antioxidant which has the potential to cure the sleep-related complications. Therefore, this chapter introduces the natural products and their ability to produce novel secondary metabolites which may act as potential drug candidate in future research. In the follow-up, Chapter "Initiation of Neurodegenerative Disorders (NDDs) Through Metal Toxicity Generated Oxidative Stress" discusses the metal-induced toxicity which may initiate the neurodegeneration. Heavy metal toxicity is the most common contamination in our daily life and could be consumed easily through various sources. In this chapter, the authors depict various unwanted effects of metal toxicity in human brain and its favorable antidotal strategies in treating the intoxication. In this chapter, study mainly aims to explore the pros and cons of metal toxicity in the brain and associated neurodegenerative diseases, i.e., Alzheimer's and Parkinson's diseases.

Until now, after reviewing first chapter to twelfth chapter, it could be concluded that oxidative stress is the factor responsible for increase in the level of ROS, and as resultant, it has been considered to be a hallmark for diabetes-induced pressure ulcers, which has been reported in Chapter "Reactive Oxygen Species and Oxidative Stress on the Formation of Diabetic Ulcer". For better understanding of the underlying signaling cascades, which may lead to diabetic wounds could facilitate pharmacological intervention and thereby helping to combat with this debilitating disease conditions. Diabetes-induced pressure ulcers are a growing burden for any healthcare system, and it has been observed that wound healing is delayed by hyperglycemia, excess levels of ROS, and the resulting imbalance at the wound site. In Chapter "Role of Mitochondrial Oxidative Stress in Pathophysiology of Lung Cancer", it has been reported that mitochondria are major targets to ROS, which is also reported in Chapter "Chronic Oxidative Stress Leads to Genomic Instability in the Pathogenesis of Fanconi Anemia". In this chapter, role of mitochondria in cellular oxidative stress regulation along with altered mitochondrial dynamics in relation with the clinical symptoms of Fanconi Anemia (FA) has been discussed. Study also reports the first human disease evidence where mitochondrial role in genomic instability is depicted. Studying mitochondrial dysfunction in this chapter would offer deeper insights as well as diagnostic and therapeutic solutions to FA.

In the environment, both ionizing and non-ionizing radiations play a major role in biological toxicity, where these radiations produce the free radicals and cause severe health concerns, i.e., cancer and neurological disorders. In this connection, Chapter "Toxicity with Waste Generated Ionizing Radiations: Blunders Behind the Scenes" discusses the various sources of ionizing radiations and their effects on human health. Evidences suggest that the radiation may affect the cellular system by producing higher number of free radicals which may start inducing health detriments as discussed in this chapter. Furthermore, provoking biotechnological implications from radiation-tolerant life forms have been sufficed and discussed in detail. The networking of environmental toxicants may also responsible to cause infectious diseases, e.g., COVID-19 detected as biggest tragedy in the twenty-first century. Therefore, Chapter "The Antioxidant Arsenal Against COVID-19" addresses the role of free radicals in post-coronavirus infections in the human biological system. In this chapter, a compelling view of the various mechanistic networks within which COVID-19-led oxidative stress could deter the homeostatic harmony and how antioxidants would emerge as a highly sought therapy is discussed. Chapter "Synergistic Effects of Heavy Water in Health Prospects", the final chapter, provides an insight of heavy water/deuterium use in the medicinal chemistry and biotechnology, due to its physico-chemical characteristics that are reported to have potential applications in applied research. Study discussed in this chapter is on the synergistic effects of heavy water, and health prospects are vividly sketched to understand its explorable properties. Further, few case studies on various pharmaceuticals employing the active isotopes in the drugs that are under clinical trials have been discussed in this chapter. This is a very interesting chapter which addresses environmental toxicity perspectives.

All the chapters presented in this book have follow-up links from each other and conclude that oxidative stress-induced free radical formation is the responsible factor for all types of serious health concerns. This book focuses on molecular and integrative toxicology in understanding the mechanisms of toxicity associated with free radical generation. One who can read all the chapters, from first chapter to seventeenth chapter, will inculcate a better understanding of environmental challenges and be able to measure the mutagenic factors in the environment and understand how they are impacting human and animal health. Moreover, a brief discussion on the other challenges such as changing environment, climate, and lifestyle factors are growing concerns in the twenty-first century. I hope this book will serve as both an excellent review and a valuable reference for formulating suitable measures against environmental toxicology and for promoting the science involved in this area of research.

Finally, we would like to dedicate this book to all the frontline workers globally, who are fighting 24/7 against COVID-19 to save our lives, and also COVID-19 warriors who lost their lives in saving humanity. We would like to thank all authors who have contributed to this book. Last but not least, our special thanks go to series editors (Jamie C. Dewitt and Sarah Blossom), publisher, and entire Springer editorial team for their sincere assistance and support. Our special thanks go to Carolyn Spence (Senior Publishing Editor) for her continuous support and suggestions throughout the book editing.

Espoo, Finland   Kavindra Kumar Kesari
Greater Noida, India   Niraj Kumar Jha

Contents

Environment Persistent Free Radicals: Long-Lived Particles

Ankita Vinayak, Gaurav Mudgal, and Gajendra B. Singh

Contents

Abstract Environment persistent free radicals (EPFRs) are emerging environmental pollutants that have longer lifetime than other free radicals. EPFRs are more stable in ambient environment because they are protected by surrogate particles like particulate matter. They are primarily emitted during thermal processes such as pyrolysis and organic material combustion. The byproducts formed from these processes provide breeding ground to the radicals for EPFR generation. Such surface bound free radicals become more persistent and resistant to atmospheric oxidation. They are also found in contaminated soil, tar ball, and cigarette smoke. The stabilization and formation of EPFR are significantly affected by elemental precursor composition and environmental factors. EPFRs are capable of producing reactive oxygen species like hydroxyl radicals, responsible for inducing oxidative stress in living organisms thereby posing adverse environmental and human health effects.

A. Vinayak · G. Mudgal · G. B. Singh (✉)
University Institute of Biotechnology, Chandigarh University, Mohali, Punjab, India

© The Author(s), under exclusive license to Springer Nature Switzerland AG 2021
K. K. Kesari, N. K. Jha (eds.), *Free Radical Biology and Environmental Toxicity*,
Molecular and Integrative Toxicology, https://doi.org/10.1007/978-3-030-83446-3_1

Despite acting as persistent environment pollutant, recent research investigations have shown EPFRs to be potentially active candidate for degradation of organic environmental contaminants. Consequently, research studies regarding occurrence, formation, toxicity, and environmental application of EPFRs have attracted global attention in the past few years. However, the existing data and investigations in this field are discrete and limited. Hence, this chapter is organized to discuss research studies and important finding on persistent free radicals, highlighting sources, occurrences, generation mechanism, exogenous factors for formation, as well as environmental implication. Moreover, role of EPFRs in the degradation of organic pollutants for environmental remediation is also being discussed.

Keywords Environment persistent free radical · Organic pollutants · Pyrolysis · Particulate matter · Degradation

1 Introduction

Free radicals are simply defined as atoms or molecules having unpaired electron and is very unstable. It includes alkoxyl radicals, hydroxyl radicals, intermediate molecules, transition metal species, and also excited state molecules. The formation of radical species occurs through redox reactions and ionizing radiations (Chaudhuri et al. 2010). The multiplicity of chemical reactions increases the chances of generation of free radical species. As redox reactions occur normally in all biological systems, likely is the generation of free radicals. They are basically involved in providing immunity and in eliminating pathogenic viruses. However, introduction of new chemicals into environment through industrial, medicinal, or agricultural use has added excessive number of radicals into the surrounding. The presence of unpaired electron makes them highly reactive. They have the tendency to react with proteins, lipids, and nucleic acids of living cells, thereby causing oxidative stress and ultimately cell damage (Alkadi 2020; Bartolomei et al. 2015).

Free radicals are also emitted in thermal treatments such as waste incineration and combustion systems, but those produced species are more stable and persistent. They are also known as "long-lived particles" because they have longer life span ranging from hours to months in comparison to free radicals which have lifetime of picoseconds only (Lomnicki et al. 2008; Pan et al. 2019). In the combustion system like in pyrolysis they are primarily formed in cooling zones of processors (Wang et al. 2019). When particles emitted in the system leave the high temperature flame, they tend to attach to the surface of fine particles. The combination or chemisorption of radical species onto the surface of some other particle like metal oxide and coarse particle imparts extra stability to radicals. It results in the generation of bio-stable free radical, which is a newly formed particle pollutant system, highly stable and persistent termed as – "environment persistent free radicals" (EPRFs) (Dugas et al. 2016). These resonance and surface stabilized radicals include cyclopentadienyls, phenoxyls, and semiquinones formed during thermal decomposition of catechol, phenols, and hydroquinone. They are mainly detected in particulate matter, combustion system, organic compound contaminated soil, plastic generation, during pyrolysis (biochar/biodiesel), coal combustion, and cigarette tar (Gao et al. 2018). EPFRs have the tendency to carry out prolonged oxidation–reduction reactions, thus forming large number of radical species. This property confers high level of toxicity to EPFRs which is more pronounced than either pollutant or particle. EPFRs are considered as emerging pollutants because they display toxic effects not only on plant and human health but also on ecological health. The EPRFs require special chemical reactions for their removal and treatment (Xu et al. 2020). In spite of being a new bio-stable environmental pollutant posing adverse toxic effects, EPFRs mediate degradation of organic contaminants also. They have the tendency to activate persulfate for the formation of reactive oxygen species like superoxide and hydroxyl ion. The formation and activation of intrinsic radicals induce the transformation and degradation of organic pollutants without catalyst and additional energy source (Qin et al. 2018). Therefore, we aim to discuss knowledge and research on EPFRs to date, focusing particularly on possible sources, environmental occurrences, formation, lifetime, toxicity, and their environmental implications as well as potential application of EPFRs for the wide-ranging remediation application of EPFRs.

2   Occurrence of EPFRs

EFRFs are ubiquitous environmental pollutant as it is easily generated in combustion systems and in cooling regions of thermal processes such as pyrolysis and waste incineration (Fig. 1). Recently, EPFRs are also found in the matrix of ultrafine and airborne fine particulate matter. Particulate matter (PM) is emitted into the environment by both natural and anthropogenic processes. The five major source PMs in the atmosphere are coal combustion emissions, (16.8%), vehicular emission (32.1%), industrial processes (11.7%), dust storm (27.2%), and nitrates (3.4%). It is a mixture of organic, inorganic species, solid and liquid components of metals which have the tendency to form radicals (Saravia et al. 2013). Based on aerodynamic size, PM is classified into three categories: ultrafine particles (<0.1 μm), smaller fine particles (0.1–2.5 μm), and coarse particles (2.5–10 μm). The EPRFs are primarily associated with ultrafine and coarse airborne PM particles (Runberg et al. 2020). Studies have demonstrated that organic pollutants like aromatic hydrocarbons bind and chemisorb onto the surface of PM, resulting in surface stabilized radical species. The resultant complex formed are weakly reactive toward atmospheric oxygen, enhancing its persistence in air for hours to months. The strong chemical bond between them makes their removal difficult. The persistence of radical species is due to cyclic regeneration of ROS species whereas stability is because of surface bound particles (Gehling and Dellinger 2013). Mostly semiquinones and its similar type radicals are found to be chemisorbed on PM surface. The studies have reported the ambient 24 hrs concentration of EPFRs in PM ranging from 10^{19} to 10^{22} spins g^{-1}. The environmental samples collected from densely populated and industrialized sites have been reported with approximately 10-fold concentrations

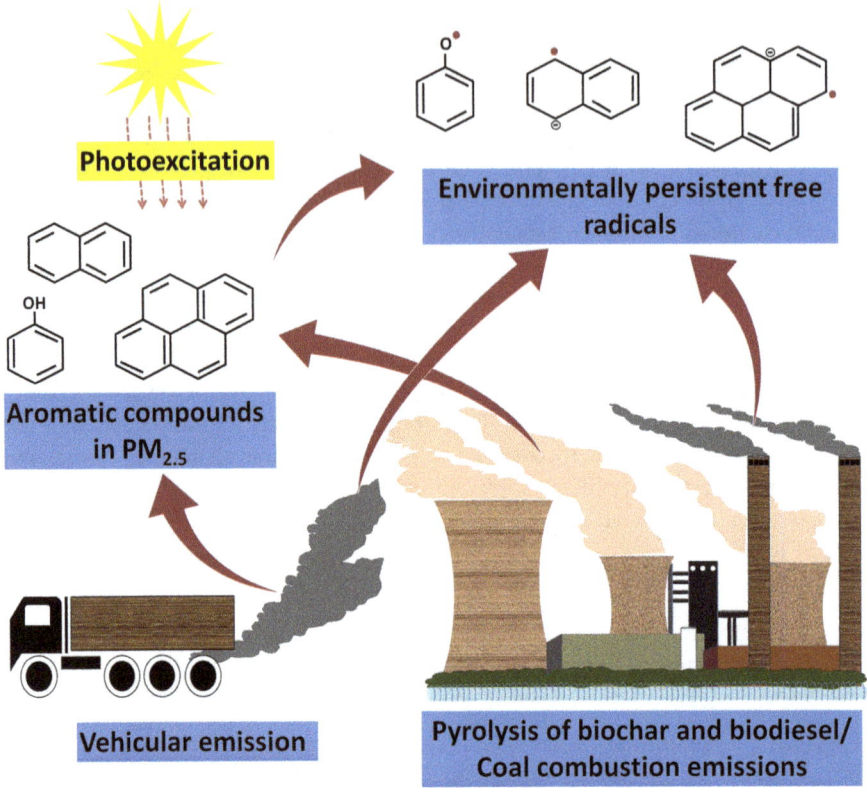

Fig. 1 Potential sources, occurrence, and transport of EPFRs

of EPFRs compared to less populated areas (Yang et al. 2017; Guo et al. 2020). The inhalation of EPFR-PM complex is implicated in various human health-related complications like cardiovascular, neurodegenerative, and respiratory system.

EPFRs are also widely detected in soil contaminated with organic compounds. Various studies have demonstrated the presence of free radicals in soil since soil is a reservoir of various complex molecules such as industrial organic pollutants and soil organics. The animal and plant residues in soil form antioxidant phenolics like tannins and vitamin C. These phenolics get adsorbed on the surface of metals in soil and form the aromatic radical cations. Organic pollutants are also formed during coal mining, coke production, and in soil contaminated with petrochemical and crude oil (Singh et al. 2013). Being hydrophobic in nature, organic pollutants like polycyclic aromatic hydrocarbons (PAHs) get deposited in soil sediment. The chemical stability and soil conditions inhibiting physical sequestration of PAHs makes them stable and slows down their degradation and leads to sorption of PAHs onto surface of clay minerals forming undesirable dioxins (Eom et al. 2007). The various remediation methods like photodegradation, chemical oxidation, and thermal processing are employed for their removal. Among all, volatilization of organic pollutants by thermal processing is a major phenomenon. The products, intermediates, and byproducts of thermal treatment of PAHs are detrimental to environment. During such remediation processes, they get transformed into surface stabilized radical species. Jia et al. reported that during the treatment of soil contaminated with anthracene, it gets prone to mineral transformation reactions and readily generates EPFRs (Jia et al. 2018). Similarly, thermal oxidation of catechol contaminated soil favors persistent radical formation. The studies have further demonstrated that soil contaminated with phenolics compounds and oxygen rich environment have higher concentration of EPFRs. The superfund soil containing polychlorinated substituted compound, pentachlorophenol, and PAHs contains 2–30 times higher concentrations of EPFRs as compared to original soil (dela Cruz et al. 2014). In addition to this, atmospheric oxidation of PAHs also forms persistent radicals. The photoexcitation of organic matter and humus like substances during visible light illumination favored the formation of PAHs like EPFRs (Chen et al. 2019). The results suggest that EPFRs' formation occurs via electron transfer mechanism in mixture matrix of biological, organic, and inorganic constituents.

3   Sources of EPFRs

3.1   Pyrolysis of Biochar and Biodiesel

Biochar is a stable, carboniferous byproduct derived from pyrolysis of low-cost biomass such as animal or plant at temperatures over 250 °C. Recently, biochar has captured attention because of its promising application in environment remediation, agricultural crop production, and bio-refinery activities. Biochar is considered as organic substituent in environment and agricultural areas because of its potential for contaminant adsorbent, pH buffer, water and nutrient retention, catalyst, and energy production. It can also increase soil fertility through stimulation of microbial community, retention of fertilizers, and reduction of greenhouse gases (Beesley et al. 2011). Being formed from thermal processing of organic biomass residues in oxygen limited conditions, the biochar synthesis forms bio-stable radicals in environment. The thermal decomposition of organic biomass under carbonization conditions at charring temperature facilitates EPFRs' formation. The organic molecules of biomass residues like cellulose, lignin, and hemicellulose are major precursors for EPFRs generation. During decomposition process they form oligosaccharides which is further cleaved at glycosidic linkage to form monomers of radical species (Liao et al. 2014). The various abiotic factors such as pH, temperature, electrical conductivity, elemental composition, calcium carbonate content, electrical charge density, and volatile/nonvolatile molecules influence the EPFRs' formation in pyrolysis process. Among all, the high temperature (250 °C) is a crucial determinant for EPFR generation in biochar pyrolysis. For example, EPFRs' generation was reported in biochar residues produced during thermal processing of maize straw, pine needle, and wheat straw at temperature range of 300–500 °C (Fang et al. 2014). In the biochar residues, the commonly found stable radicals are phenyl radical, hydrogen peroxide, superoxide, hydroxyl, and catechol type radicals and semiquinones types. In terms of stability, semiquinones type radicals viz. benzo-semiquinone and o-semiquinones which are oxygen centered are most stable species, whereas carbon centered radicals like cyclopentadienyls are least stable and are vulnerable to atmospheric oxidation (Khachatryan et al. 2006; Tian et al. 2009). Similarly, the pyrolysis of vegetable oil and animal fats biomass at high temperature for biodiesel formation favors the formation of stable and transient radical species which get adsorbed on particulate matter (Kibet et al. 2018). In 2018, Mosonik and coworkers reported the formation of EPFRs during thermal processing of *Croton megalocarpus* for biodiesel. The pyrolysis of *Croton megalocarpus* at 600 °C facilitated emission of particulate matter and free radical generation having half-life of 431 days, making them EPFR. Further characterization of free radicals revealed that species were carbon-centered with a low decay constant (1.86×10^{-8} s^{-1}) and resembled radical species of that in coal (Mosonik et al. 2018). Thus, the formation and persistence of EPFRs formed during pyrolysis of biochar and biodiesel contaminate the environment and have detrimental implications on the environment and human health, such as oxidative stress, cancer, and cardiac abnormalities.

3.2 Transition Metals–Mediated Generation of EPFRs

Transition metals are considered as key precursors favoring EPFRs' formation under various environmental conditions. The zinc, vanadium, iron, nickel, and copper are considered as most favorable metals for EPFRs' generation. The hydroxyl- and chlorine-substituted benzenes when decomposed at 150–400 °C chemisorb onto the surface of metal oxide with subsequent electron transfer. It results in metal reduction and simultaneous persistent free radical generation (Yang et al. 2017). Various metal oxides like Al_2O_3 and ZnO are emitted during combustion processes. They enhance the stabilization and formation of EPFR. The catalytic ability of metal oxide–generated EPFR generally depends on the standard reduction potential of transition metals. Studies indicated that series of chemical reaction, physisorption, chemisorption, and finally electron transfer from metal surface to adsorbed organic molecules lead to stable radical formation. The EPFRs formed from different metal oxides such as Fe_2O_3, NiO, Cu_2O, and ZnO were compared, and it was observed that ZnO EPFRs were most stable and long lasting (Vejerano et al. 2012a).

3.3 Other Sources of EPFRs

Plastics are used for various purposes in day-to-day life. Huge amounts of plastics are produced, utilized, and disposed into environment. Approximately 10% of plastics used are dumped into surrounding environment. With time, effects from photoaging, chemicals, and microbial actions corroborate the break down of macroplastics into much smaller versions (less than 5 mm) known as microplastics (MPs). MPs are widespread in natural environment, found in soil, sewage, sediments, and rivers (Andrady 2011). They may induce adverse effects on living forms such as significantly affecting feeding capacity, reproductivity and other system also. Most importantly, MPs can form reactive oxygen species leading to oxidative stress in living organisms. When MPs are released into the environment, they undergo both biotic and abiotic transformations (Zhu et al. 2020a). During these transformations, organic pollutants lead to formation of free radical species. Many MPs such as phenol formaldehyde resin and polystyrene contain substituted benzene rings. During photoaging they undergo oxidation leading to cleavage of C–H and C–C bond and ultimately forming stable radical species (Zhu et al. 2020b). Similarly, other plastic materials like polyvinyl chloride and polyethylene are also subjected to photooxidation in natural environment and form reactive oxygen species (Cai et al. 2018).

Tar balls and oil mineral aggregates are formed from weathering of spilled crude oil. These aromatic molecules tend to chemisorb on the surface of transition metals and suspended minerals in sediments. The metal and crude chemical interactions result in the formation of stable free radicals. Kiruri et al. reported the presence of two semiquinone type organic radicals in the tar ball collected samples. One is asphaltene radical species (g value – 2.0035) occurring in crude oil and the other one is new radical species formed from partial oxidation and interaction of crude components (g value 2.0041–2.0047) (Kiruri et al. 2013).

4 Mechanism of EPFR Formation

The resonance stabilized EPFRs are formed by association of substituted aromatic compounds with fine particle or transition metal center. Metals and organic compounds act as model particles for persistent free radical generation. With the adsorption of organic compounds on metals, EPFFRs are formed at transition metal center through metal reduction. The general mechanism for the formation of EPFRs involves initial physisorption, involving weak bonding between organic molecule and transition metal. It is followed by chemisorption of organic compound via removal of water and subsequent electron transfer from adsorbed species to the metal oxide center, resulting in metal reduction and EPFRs' formation (Fig. 2) (Qin et al. 2018). The metal center interaction imparts stability to the formed radical, providing it a resistance to degradation, oxidation, and recombination. The resonance of formed organic radicals allows formation of either carbon centered or oxygen centered EPFRs. Generation of such discreet EPFRs is mediated by transition metals or organic moieties such as those formed after incomplete combustion events like pyrolysis of biochar (Gehling and Dellinger 2013). The possible mechanism of EPFRs' formation in both includes these mentioned steps, while only the precursors such as aromatics, metal oxides, and exogenous factors differ. Various model studies have demonstrated that thermal processes such as pyrolysis and metal smelting processes generate abundant transition metal oxides which mediate the formation of bio-stable free radicals in the cooling zone of smelter. Vejerano et al. investigated

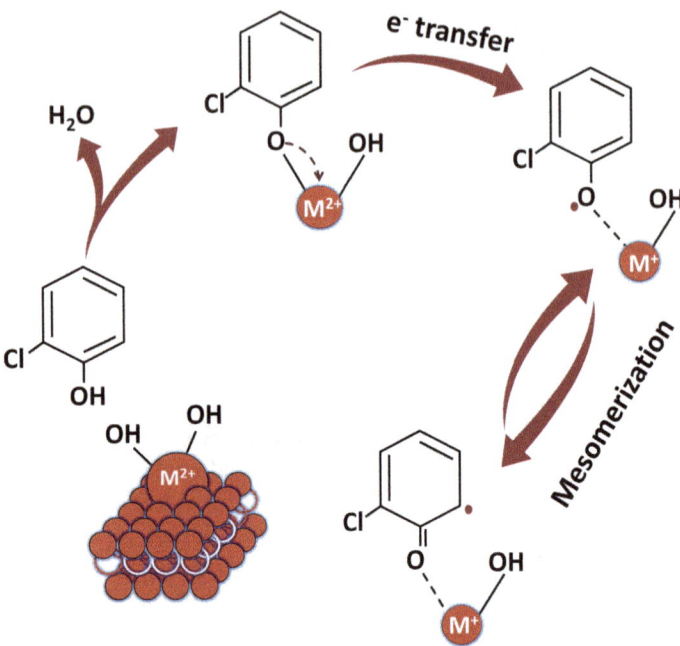

Fig. 2 Diagrammatic representation of M(II)O mediated EPFR generation from chlorinated benzene (M-Cu, Ni, Zn, Fe, Ti)

the formation and stabilization of persistent free radicals on NiO surface. The results showed that aromatic molecular adsorbates like monochlorobenzene, catechol, and 1,2-dichloronezene gets adsorbed on the surface of NiO/silica particles (5%) at 150–400 °C. The electron transfer occurs from the adsorbate to the center of metal, thereby reducing NiO and resulting in various types of formation of EPFRs. Depending on exogenous conditions such as nature of adsorbate and temperature, different types of radicals were formed, such as semiquinone type radical (g value – 2.0050–2.0080) and phenoxyl type radical (2.0029–2.0044) (Vejerano et al. 2012b). EPFRs formed from different metal oxides such as ZnO, Fe_2O_3, Ti_2O, and NiO were studied and compared for concentration and half-life. It was found that lifetime of EPFR formed strongly depend upon the standard reduction potential and catalytic oxidizing strength of metal (Yang et al. 2017). The amount and type of EPFR formed depend on different factors such as type of precursors, process time, temperature, and concentration of metal oxides.

The pyrolysis of biomass constituents such as lignin, cellulose, and hemicellulose occurs at high temperature (300–400 °C) for synthesis of biochar (Nzihou et al. 2013). The homolytic cleavage of C–O, C=O, and alpha/beta alkyl bond instigates formation of monomer and free radical. The formed radical then couples with the incomplete combustion by products like hydroquinone, phenol, and polyaromatic compounds for the generation of surface bound EPFR. The extended pyrolysis time can even increase the accumulation of free radical in residue (Collard and Blin 2014). The cooling zones of thermal processors favor more formation of EPFR. The radical species formed in the matrix of organic moieties are protected from reacting with chemicals and with each other, providing them extra stability. Biochar synthesized from different biomass feedstock showed electron paramagnetic resonance spectra with varying intensity and g factor. The chemical composition of biomass determines the decomposition temperature and also the energy requirement for bond cleavage (Fang et al. 2015a). Thus, EPFR generation mediated by organic molecules significantly depends on carbon content of feedstock and poorly on metal element concentration. In 2014, Liao and colleagues studied the formation of EPFRs during the pyrolysis of biochar from wheat straw, rice, and corns at 200 °C. The results showed that cleavage of bond and decomposition of biomass structure promoted the formation of stable free radicals (Liao et al. 2014). Studies have shown that biochar residues mainly form oxygen centered, carbon centered, and oxygenated carbon centered radicals. Among all, oxygen centered (semiquinone type species) are more stable while carbon centered (cyclopentadienyls) are least stable as they are prone to atmospheric oxidation (Khachatryan et al. 2006; Tian et al. 2009).

5 Factors Affecting EPFRs' Formation

Although EPFRs are long-lived and stable radical species, its lifetime and formation are affected by various abiotic factors like light irradiation, precursors, presence of metals, temperature, pH, humidity, thermal processing time, and oxygen (Gao et al. 2018; Liu et al. 2015). The compounds that can form stable radicals through electron transfer or thermal decomposition are known as precursors. They are considered as important factors influencing the EPFRs' formation. The organic components of precursor molecules undergo different reactions thereby producing different radical species under same process conditions. The investigations have also reported that different types of raw material for biochar synthesis under same conditions form EPFRs that vary in their concentration and g value. Liao and his coworkers compared the concentration and nature of radicals in biochar synthesized from cellulose and lignin at 200 °C. It was found that radical species formed from lignin was more stable and five times higher in concentration as compared to cellulose. The loading of biomass with organic components also rapidly increases the EPFRs' concentration in biochar because of significant effect on g value of EPFRs (Liao et al. 2014; Fang et al. 2015a). Also, the presence of functional groups such as chlorine and hydroxyl groups in benzene ring favors the formation of stable radical species. The relationship between the nature of precursors and radical species will greatly determine the occurrence and behavior of EPFRs (Nwosu et al. 2016a). The various types of EPFRs commonly found in environment with their g values are summarized in Table 1.

The presence of metals is an influential parameter controlling the stable free radical formation. The transition metals significantly affect the yield and type of EPFRs

Table 1 Common types of EPFRs with g values

Type of EPFRs	Characteristics	Examples	g value	References
Oxygen centered radicals	Includes semiquinone and quinone type radicals, Formed from reaction of radical and phenolic compound, More stable in environment	1,4-Benzosemiquinone	2.0046– 2.0049	Xu et al. (2020), Segal et al. (1965), Chen et al. (2018), Pryor et al. (1983), and Neta and Fessenden (1974)
		1,4-Benzoquinone	2.0050	
		Benzoquinone	2.0039	
		1,4-Naphthoquinone	2.0044	
		Poly-1,4-Naphthoquinone	2.0035	
		Poly-1,7-Naphthoquinone	2.0035	
		1,4-Benzosemiquinone anion	2.0046	
		Phenanthrene quinone	2.0039	
Carbon centered radicals	Consists of aromatic, alkyl and aryl radicals, Formed at high temperature of pyrolysis, vulnerable to atmospheric oxidation, induce formation of reactive oxygen species	Pyrene anion	2.0027	Xu et al. (2020), Segal et al. (1965), and Odinga et al. (2020)
		Benzene anion	2.0029	
		Perylene anion	2.0027	
		Perylene cation	2.0026	
		Graphitic carbon	2.0028	
		Anthracene anion	2.0027	
		Anthracene cation	2.0026	
		Naphthalene anion	2.0028	
		Polyaromatic hydrocarbon radicals	2.0026	
		Fluoranthene anion	2.0027	
Oxygenated carbon centered radicals	Stable radicals, inhibits autooxidation of organic substances	Phenoxyl	2.0046– 2.0053	Xu et al. (2020), Neta and Fessenden (1974), Odinga et al. (2020), Graf et al. (1977), and Barclay et al. (1994)
		2-Chloro-phenoxyl	2.0062	
		4-Chloro-phenoxyl	2.0063	
		2-Carboxyphenoxyl	2.0046	
		2,6-Dichloro-phenoxyl	2.0065	
		3,5-Dichloro-phenoxyl	2.0049	
		Phenyl radical	2.0030– 2.0040	
		3,4-Dimethoxylphenacyl	2.0036	
		2,4,6-Trichloro-phenoxyl	2.0076	

formed. The oxidizing ability of metal oxides largely affects the formation of EPFRs. The copper and zinc catalyze the formation of semiquinone and phenoxyl type radicals only, whereas iron and nickel have potential to form mixture of radicals with different g factor (Yang et al. 2017). g factor is basically the ratio of magnetic moment of electron to its angular moment, considered as unique property of electron in specific environment. For instance, NiO formed two types of radical species: semiquinone type (g value – 2.0050–2.0081) and phenoxyl type (g value – 2.0029–2.0044) radicals. Similarly, Fe_2O_3 produces phenoxyl type and semiquinone type radicals but with slightly different g values (Vejerano et al. 2011). The oxidation potential of metal oxides also determines the yield of EPFRs. Among metal oxides of Ni, Fe, Cu, and Zn, NiO has the highest oxidizing strength and thus maximum EPFRs' formation ability. The amount of metal ions also affects the formation of EPFRs. The increased concentration of metals has negative impact on the yield of radical formation. Fang and coworkers assessed the effect of increasing number of metals on EPFRs' formation in biochar. It was found that EPFRs' concentration was higher at 0.1 mmol L^{-1} of metal as compared to 0.2 mmol L^{-1} of metal. It is due to the fact that excess amount of metal ions consumes free radicals because EPFRs can mediate electron transfer thereby reducing metals ions such as Cr^{6+} and Fe^{3+} (Fang et al. 2015a).

Sunlight represents the only exogenous light source for the formation of radicals in environment. Light energy enhances the electron transfer process generating degradation products and free radicals. Chen et al. studied the EPFRs' formation by visible light illumination of PM of organic matter. The formed radicals were similar to PAHs and phenol derived radicals and have lifetime of 30 min to 1 day (Chen et al. 2019). However, direct photo-irradiation of radicals facilitates their decay. It is found that light irradiation promotes transformation of radical cations into oxygenic radicals such as O^{2-} and OH (Jia et al. 2019). Therefore, in-depth investigations are required for photosensitive organic molecules. Oxygen content is another significant factor catalyzing EPFRs generation and stability. Oxygen may reduce the transition metal EPFRs and can also directly interact with less stable radical species like cyclopentadienyl radicals (Vejerano et al. 2011). The oxygen centered radicals are less stable and persistent as compared to carbon centered radicals. For instance, in air EPFRs have lifetime of only 12.4 days whereas under vacuum conditions EPFRs have 151.8 days of lifetime (Nwosu et al. 2016b). The EPFRs' formation is also controlled by water content during process. A general hypothesis is that humidity has determinantal negative effect on EPFR concentration (Pan et al. 2019). Jia et al. studied the effect of increased humidity and decay values of EPFRs. It was found that radicals were stable for 2 months at 7% humidity whereas they exhibit fast decay at nearly 100% humid conditions. This is because of hydration of metal oxides and also the competition of water molecules with radicals for Lewis acid site, thereby decreasing the formation of radicals or elimination of already existing stable radicals (Jia et al. 2017).

Other abiotic factors like temperature and time of process also determine the type and amount of EPFRS formed. For example, the concentration of EPFRs increased rapidly when pyrolysis temperature is increased from 300 to 600 °C. But any further increase in pyrolysis temperature decreased markedly the formation of free radicals because of reorganization and breakdown of organic structures resulting in decay of free radicals. Similarly, the time of different stages of pyrolysis process influences the formation and type of EPFRs (Gao et al. 2018; Qiu et al. 2007).

The different trends on free radical concentration were observed with the increase in time of pyrolysis process. For example, Fang reported that EPFRs' concentration significantly increased when pyrolysis time prolonged from 1 to 12 h (Fang et al. 2015b).

6 Lifetime and Persistence of EPFRs

The lifetime of persistence free radicals is generally longer in comparison to free radicals, as EPFRs are surface stabilized by particles making them more stable. Their half-lives range from hours to months depending on environmental conditions. The half-lives of EPFRs are usually associated with redox potential of metals and also the chemical reaction of radical with metal oxides. Thus, the EPFRs associated metals determine the persistence as well as stability of radical species. For example, the half-lives of EPFRs formed on Cu_2O are less than that of EPFRs formed on Fe_2O_3. The half-lives of EPFRs generated on Fe_2O_3 and NiO are same nearly 1–5 days. However, out of all transition metal studied, the EPFRs formed on ZnO have longest half-life ranging from nearly 3 to 73 days. The EPFRs associated with ZnO can even persist for a year. It is due to increased interaction of metals with free radicals owing to the closed shell structure, small ionic size, and lack of crystal field lattice energy (Vejerano et al. 2012a) . Thus, the stability of radical species, redox potential, and ability to accept electrons significantly affect the lifetime of EPFRs. In addition to reducibility of metal, the concentration of metal also significantly affects the lifetime of radical species. Kiruri et al. in 2014 reported that chlorophenoxyl radical on CuO/silica at 0.75% concentration and phenoxyl radicals on CuO/silica at 0.5% concentration have longer time span (Kiruri et al. 2014).

Being stable molecule, EPFRs are less prone to chemical reactions. The atmospheric oxygen is the only sink for stable free radicals converting them into molecular species. The EPFRs decay in atmosphere depends on reaction of EPFRs with molecular oxygen. Depending on reaction, EPFRs display no decay, low decay, and fast decay. The *no decay* character of EFFRs is attributed to restriction or entrapment of radical species in the bulk of particulate matter. In the matrix of PM, the unpaired electrons of EPFRs are delocalized over aromatic bonds. On the other hand, phenoxyl radicals undergo fast decay and semiquinone radicals are subjected to slow decay (Xu et al. 2020).

7 Potential Risks of EPFRs

The stability, persistence, and redox recycling of EPFRs confer them biological toxic properties. The prolonged cycling of persistent free radicals forms large number of radical species which enhances their toxicity. The potential risks of EPFRs are more pronounced than either of pollutant or particle because of its surrogate association. The adverse effects of EPFRs are due to oxidative stress induced through the generation of reactive oxygen species (ROS) such as hydroxyl radicals and superoxide anion radicals, thereby inducing cellular oxidative stress, cytotoxicity, and DNA damage (Dugas et al. 2016). They lower the levels of cellular antioxidants and ultimately lead to cell death. The EPFRs have shown to affect the biotic and abiotic components of environment, viz. humans, animals, plants, microbes, and soil (Fig. 3). The natural and industrial process tends to transport EPFRs into

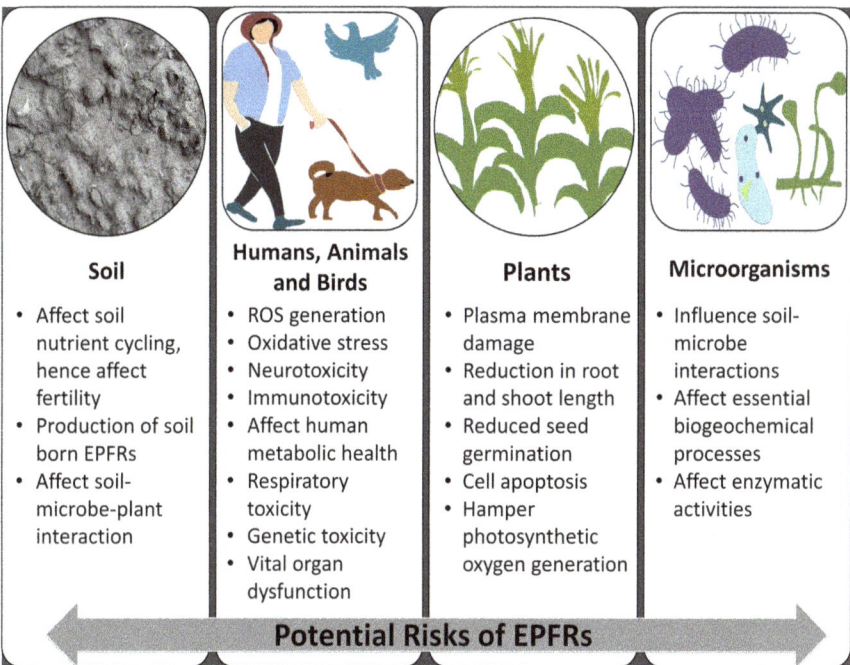

Fig. 3 Toxic effects of EPFRs on biotic and abiotic components of environment

the soil matrix such as through weathering, runoff, and decomposition. Once in the soil and rhizosphere, free radicals take part in metal complexation and chemical redox reactions. They influence the soil–microbe interaction and essential biogeochemical processes. For example, the biochar derived EPFRs alter the soil microbe community and microbial and enzymatic activity. It directly affects the nutrient cycling (N, P, S) soil–microbe–plant relationship, soil organic matter turnover, and radical scavenging (Spokas et al. 2011). Other than these, pyrolysis and remediation processes are known to induce EPFRs' formation in soil, which might confer internalization led toxicity to soil microbes. Similarly, EPFRs may much easily and variously disseminate into the atmosphere resulting in their easy exposure to human and other lifeforms.

The stable free radicals are widespread in environment and are associated with various human diseases. The presence of phenoxyl type and semiquinone type species in PM induces the generation of activated species which cause human health risks through oxidative stress. Continued aerial persistence of EPFRs may variously lead to serious health complications. Numerous investigation have demonstrated that combustion derived stable radicals are responsible for declining human metabolic health (Xu et al. 2020). EPFRs are correlated with acute infections like respiratory toxicity and influenza virus infection and also chronic risks like pneumonia, DNA damage, and liver dysfunction. They are also reported to be causative agents of cardiac arrest, cancer, aging, and escalated pulmonary pressure. The exposure of PM related radicals to humans have showed cardiovascular disease, respiratory dysfunction, and type II diabetes (Lord et al. 2011; Mahne et al. 2012). Chuang et al. observed the effect of bio-stable free radicals on mouse cardiomyocytes and found that radicals cause cell mortality by inducing oxidative stress and resulted in cell death. It showed that radical species in experimental animals leads to cellular stress and cardiac toxicity (Chuang et al. 2017). In another experiment, combustion derived radicals led to mitochondrial breakdown of muscles and reduced the energy expenditure. EPFRs also restrict the activity of native proteins in cells, cytochrome 450 – well known for breakdown of foreign bodies. The inhibition of cytochrome 450 will hinder the metabolism of extracellular substances (Reed et al. 2015). Like combustion-derived stable radicals, biochar-associated EPFRs produce the ROS species and reduce the levels of cellular enzymes such as superoxide dismutase and glutathione peroxidase, ultimately decreasing the cell membrane integrity.

The bio-stable radicals are not only toxic to humans and animals but also significantly affect the plants. They are linked with inhibitive traits like plasma membrane damage, reduction in root and shoot length, and reduced germination of seeds. Liao et al. performed experimental study to assess the effect of free radicals produced during biochar pyrolysis of rice straw, wheat, and corns on plant growth and germination. The electromagnetic paramagnetic resonance (EPR) spectra were obtained for original raw material and biomass formed biochar. EPR signal and intensity were only observed in biochar biomass, which even increased by increasing pyrolysis temperature. The root length, shoot length, and growth of test plants were evaluated on corn seedling. Significant retardation in root and shoot length, reduced seed germination, and plant growth was observed in plants exposed to biochar free radicals. In addition, ROS generated from EPFRs also reacted with membrane glycoproteins causing destabilization of cellular integrity and membrane thereby inducing cell apoptosis (Liao et al. 2014). The semiquinone type radical species were also reported to inhibit photosynthesis in plants. The mechanism of photosynthesis inhibition by EPFRs is mainly observed in major components of biochar, such as humic acid and lignin. They act as electron scavengers, altering electron transfer process and hampering photosynthetic oxygen generation in plants (Pflugmacher et al. 2006).

8   Role of EPFRs for Remediation of Organic Contaminants

The particle associated EPFRs are active oxidizing agents catalyzing various oxidants. They have potential to generate reactive oxygen species, which are active against organic contaminants. The free radicals effectively activate hydrogen peroxide or persulfate to produce ROS species such as superoxide, singlet oxygen, or OH^-. The activation of small intrinsic radical species induces the degradation of environmental organic pollutants (Fang et al. 2014; Yi et al. 2019). Fang and his colleagues studied the role of hydroxyl radical in biochar suspension for the degradation of diethyl phthalate in the presence of oxygen. The persistent free radicals catalyze the electron transfer to oxygen producing O^{2-} and OH^-, which further induced the formation of hydroxyl radical for diethyl phthalate degradation. The results showed that in the presence of oxygen nearly 89–100% of diethyl phthalate (5.0 mg L^{-1}) was degraded in 24 h. Moreover, ROS species generated from EPFRs under UV radiation also exhibited potential to degrade diethyl phthalate. It was observed that 52.3–72.3% of diethyl phthalate was degraded in 2 hrs reaction time at pH 7.0 (Fang et al. 2015b). It was concluded that biochar derived EPFRs produces OH^- and singlet oxygen for the diethyl phthalate degradation.

Following the same hypothesis, Yang studied the important role of OH^- in the removal of organic contaminant p-nitrophenol (Yang et al. 2016). In another study, He and coworkers investigated the potential of activated persulfate in iron-porphyrin biochar for the degradation of perfluorooctanoic acid (PFOA). It was found that bioactive free radicals transfer electrons from persulfate to iron-porphyrin to speed up the formation of hydroxyl and sulfate radical. The addition of electronic circulation agent ascorbic acid increases the further rate of reaction. The results showed that in iron-porphyrins biochar-ascorbic acid system, 75.90% of PFOA was degraded in 30 min and reaching up to 90.88% when reaction time was increased (He et al. 2020). For the enhanced activation of persistent free radicals of biochar transition metals are added to further increase the removal of organic pollutants. Fang et al. assessed the effect of phenolic compound and transition metals addition for the formation of EPFRs. Results showed that loaded phenolic compounds and metals not only changed the type of biologically stable radicals but also elevated the concentration of EPFRs in biochar. Manipulation of biochar in this way by metals and phenolic compounds may interpret effective management of EPFRs and so can cater the transformation of organic contaminants (Fang et al. 2015a). In yet another study, a potential role of EPFRs in the remediation of soils contaminated with PAHs and coking byproducts is also documented. The potential role of EPFRs for the remediation of soil contaminated with PAHs and coking process byproduct was also investigated. The results suggested that EPFRs act as an electron shuttle for persulfate and induced the formation of ROS. The generated ROS back reacts with anthracene resulting in its oxidation and complete mineralization (Jia et al. 2018). The semiquinone type persistent radicals also play an important role in the removal of inorganic contaminants from environment. For instance, carbon centered radicals have potency to reduce heavy metal chromium. Similarly, Zhao et al. reported that modified biochar generates EPFRs, which then acts as electron donor for transformation of chromium (Zhao et al. 2018). The observed studies suggested the development of EPFRs-based remediation technologies for environmental pollutants.

9 Conclusions

EPFRs are ubiquitous pollutants of emerging concerns, formed of coupled particle pollutants system exhibiting toxicological implications. EPFRs' formation is primarily mediated by transition metals and organic compounds formed during incomplete combustion or thermal processes. They are capable of inducing oxidative stress through ROS formation. In the past few decades, research investigation for the environmental occurrence and associated health risks of EPFRs has attracted increased research attention. Still the information and research regarding the persistent free radicals are at infant stage. This work provides detailed insight into sources, lifetime, types, decay, as well as toxicity of EPFRs. Based on available literature, we tried to fill the knowledge gaps about the environmental occurrence of EPFRs in various sources. The original data of published work on EPFRs is used to discuss and compare the formation mechanism of EPFRs mediated by different agents. The detailed outlook on the environmental factors affecting the EPFRs' formation such as precursors, process time, temperature, and light radiation provides a way to fully understand generation mechanism. The redox recycling of EPFRs imparts toxicity to surface-bound EPFRs, and considerable focus has been given to discuss the toxic effects of EPFRs on living forms including humans and plants. Recent research studies have provided insights into role of EPFRs for environment remediation. EPFRs possess property for the catalysis and activation of persulfate for producing ROS for the degradation of organic contaminants. Although EPFRs are long-lived environmental pollutants, the tendency to activate free radical generation can make them wide range environmental remediation tool.

Acknowledgments The authors thank the fraternity of University Institute of Biotechnology and also the University Center for Research and Development (UCRD) at Chandigarh University (CU) for support. All the authors have equally contributed to the manuscript and revised the manuscript. The institute to which the authors are currently affiliated, however, has no role in shaping the manuscript. All the authors declare no conflict of interest.

References

Alkadi H. A review on free radicals and antioxidants. Infect Disord-Drug Targets (Formerly Curr Drug Targets-Infect Disord). 2020;20(1):16–26.

Andrady AL. Microplastics in the marine environment. Mar Pollut Bull. 2011;62(8):1596–605.

Barclay LRC, Cromwell GR, Hilborn J. Photochemistry of a model lignin compound. Spin trapping of primary products and properties of an oligomer. Can J Chem. 1994;72(1):35–41.

Bartolomei V, Gomez Alvarez E, Wittmer J, Tlili S, Strekowski R, Temime-Roussel B, et al. Combustion processes as a source of high levels of indoor hydroxyl radicals through the photolysis of nitrous acid. Environ Sci Technol. 2015;49(11):6599–607.

Beesley L, Moreno-Jiménez E, Gomez-Eyles JL, Harris E, Robinson B, Sizmur T. A review of biochars' potential role in the remediation, revegetation and restoration of contaminated soils. Environ Pollut. 2011;159(12):3269–82.

Cai L, Wang J, Peng J, Wu Z, Tan X. Observation of the degradation of three types of plastic pellets exposed to UV irradiation in three different environments. Sci Total Environ. 2018;628:740–7.

Chaudhuri L, Sarsour EH, Goswami PC. 2-(4-Chlorophenyl) benzo-1, 4-quinone induced ROS-signaling inhibits proliferation in human non-malignant prostate epithelial cells. Environ Int. 2010;36(8):924–30.

Chen Q, Sun H, Wang M, Mu Z, Wang Y, Li Y, et al. Dominant fraction of EPFRs from nonsolvent-extractable organic matter in fine particulates over Xi'an, China. Environ Sci Technol. 2018;52(17):9646–55.

Chen Q, Sun H, Wang M, Wang Y, Zhang L, Han Y. Environmentally persistent free radical (EPFR) formation by visible-light illumination of the organic matter in atmospheric particles. Environ Sci Technol. 2019;53(17):10053–61.

Chuang GC, Xia H, Mahne SE, Varner KJ. Environmentally persistent free radicals cause apoptosis in HL-1 cardiomyocytes. Cardiovasc Toxicol. 2017;17(2):140–9.

Collard F-X, Blin J. A review on pyrolysis of biomass constituents: mechanisms and composition of the products obtained from the conversion of cellulose, hemicelluloses and lignin. Renew Sust Energ Rev. 2014;38:594–608.

dela Cruz ALN, Cook RL, Dellinger B, Lomnicki SM, Donnelly KC, Kelley MA, et al. Assessment of environmentally persistent free radicals in soils and sediments from three Superfund sites. Environ Sci Process Impacts. 2014;16(1):44–52.

Dugas TR, Lomnicki S, Cormier SA, Dellinger B, Reams M. Addressing emerging risks: scientific and regulatory challenges associated with environmentally persistent free radicals. Int J Environ Res Public Health. 2016;13(6):573.

Eom I, Rast C, Veber A, Vasseur P. Ecotoxicity of a polycyclic aromatic hydrocarbon (PAH)-contaminated soil. Ecotoxicol Environ Saf. 2007;67(2):190–205.

Fang G, Gao J, Liu C, Dionysiou DD, Wang Y, Zhou D. Key role of persistent free radicals in hydrogen peroxide activation by biochar: implications to organic contaminant degradation. Environ Sci Technol. 2014;48(3):1902–10.

Fang G, Liu C, Gao J, Dionysiou DD, Zhou D. Manipulation of persistent free radicals in biochar to activate persulfate for contaminant degradation. Environ Sci Technol. 2015a;49(9):5645–53.

Fang G, Zhu C, Dionysiou DD, Gao J, Zhou D. Mechanism of hydroxyl radical generation from biochar suspensions: Implications to diethyl phthalate degradation. Bioresour Technol. 2015b;176:210–7.

Gao P, Yao D, Qian Y, Zhong S, Zhang L, Xue G, et al. Factors controlling the formation of persistent free radicals in hydrochar during hydrothermal conversion of rice straw. Environ Chem Lett. 2018;16(4):1463–8.

Gehling W, Dellinger B. Environmentally persistent free radicals and their lifetimes in PM2. 5. Environ Sci Technol. 2013;47(15):8172–8.

Graf F, Loth K, Günthard HH. Chlorine hyperfine splittings and spin density distributions of phenoxy radicals. An ESR and quantum chemical study. Helvetica Chim Acta. 1977;60(3):710–21.

Guo X, Zhang N, Hu X, Huang Y, Ding Z, Chen Y, et al. Characteristics and potential inhalation exposure risks of PM2. 5–bound environmental persistent free radicals in Nanjing, a mega–city in China. Atmos Environ. 2020;224:117355.

He W, Zhu Y, Zeng G, Zhang Y, Wang Y, Zhang M, et al. Efficient removal of perfluorooctanoic acid by persulfate advanced oxidative degradation: inherent roles of iron-porphyrin and persistent free radicals. Chem Eng J. 2020;392:123640.

Jia H, Zhao S, Nulaji G, Tao K, Wang F, Sharma VK, et al. Environmentally persistent free radicals in soils of past coking sites: distribution and stabilization. Environ Sci Technol. 2017;51(11):6000–8.

Jia H, Zhao S, Zhu K, Huang D, Wu L, Guo X. Activate persulfate for catalytic degradation of adsorbed anthracene on coking residues: role of persistent free radicals. Chem Eng J. 2018;351:631–40.

Jia H, Zhao S, Shi Y, Zhu K, Gao P, Zhu L. Mechanisms for light-driven evolution of environmentally persistent free radicals and photolytic degradation of PAHs on Fe (III)-montmorillonite surface. J Hazard Mater. 2019;362:92–8.

Khachatryan L, Adounkpe J, Maskos Z, Dellinger B. Formation of cyclopentadienyl radical from the gas-phase pyrolysis of hydroquinone, catechol, and phenol. Environ Sci Technol. 2006;40(16):5071–6.

Kibet JK, Mosonik BC, Nyamori VO, Ngari SM. Free radicals and ultrafine particulate emissions from the co-pyrolysis of Croton megalocarpus biodiesel and fossil diesel. Chem Central J. 2018;12(1):1–9.

Kiruri LW, Dellinger B, Lomnicki S. Tar balls from Deep Water Horizon oil spill: environmentally persistent free radicals (EPFR) formation during crude weathering. Environ Sci Technol. 2013;47(9):4220–6.

Kiruri LW, Khachatryan L, Dellinger B, Lomnicki S. Effect of copper oxide concentration on the formation and persistency of environmentally persistent free radicals (EPFRs) in particulates. Environ Sci Technol. 2014;48(4):2212–7.

Liao S, Pan B, Li H, Zhang D, Xing B. Detecting free radicals in biochars and determining their ability to inhibit the germination and growth of corn, wheat and rice seedlings. Environ Sci Technol. 2014;48(15):8581–7.

Liu J, Jiang X, Shen J, Zhang H. Influences of particle size, ultraviolet irradiation and pyrolysis temperature on stable free radicals in coal. Powder Technol. 2015;272:64–74.

Lomnicki S, Truong H, Vejerano E, Dellinger B. Copper oxide-based model of persistent free radical formation on combustion-derived particulate matter. Environ Sci Technol. 2008;42(13):4982–8.

Lord K, Moll D, Lindsey JK, Mahne S, Raman G, Dugas T, et al. Environmentally persistent free radicals decrease cardiac function before and after ischemia/reperfusion injury in vivo. J Recept Signal Transduct. 2011;31(2):157–67.

Mahne S, Chuang GC, Pankey E, Kiruri L, Kadowitz PJ, Dellinger B, et al. Environmentally persistent free radicals decrease cardiac function and increase pulmonary artery pressure. Am J Phys Heart Circ Phys. 2012;303(9):H1135–H42.

Mosonik BC, Kibet JK, Ngari SM, Nyamori VO. Environmentally persistent free radicals and particulate emissions from the thermal degradation of Croton megalocarpus biodiesel. Environ Sci Pollut Res. 2018;25(25):24807–17.

Neta P, Fessenden RW. Hydroxyl radical reactions with phenols and anilines as studied by electron spin resonance. J Phys Chem. 1974;78(5):523–9.

Nwosu UG, Khachatryan L, Youm SG, Roy A, dela Cruz ALN, Nesterov EE, et al. Model system study of environmentally persistent free radicals formation in a semiconducting polymer modified copper clay system at ambient temperature. RSC Adv. 2016a;6(49):43453–62.

Nwosu UG, Roy A, dela Cruz ALN, Dellinger B, Cook R. Formation of environmentally persistent free radical (EPFR) in iron (III) cation-exchanged smectite clay. Environ Sci Process Impacts. 2016b;18(1):42–50.

Nzihou A, Stanmore B, Sharrock P. A review of catalysts for the gasification of biomass char, with some reference to coal. Energy. 2013;58:305–17.

Odinga ES, Waigi MG, Gudda FO, Wang J, Yang B, Hu X, et al. Occurrence, formation, environmental fate and risks of environmentally persistent free radicals in biochars. Environ Int. 2020;134:105172.

Pan B, Li H, Lang D, Xing B. Environmentally persistent free radicals: occurrence, formation mechanisms and implications. Environ Pollut. 2019;248:320–31.

Pflugmacher S, Pietsch C, Rieger W, Steinberg CE. Dissolved natural organic matter (NOM) impacts photosynthetic oxygen production and electron transport in coontail Ceratophyllum demersum. Sci Total Environ. 2006;357(1-3):169–75.

Pryor WA, Hales BJ, Premovic PI, Church DF. The radicals in cigarette tar: their nature and suggested physiological implications. Science. 1983;220(4595):425–7.

Qin Y, Li G, Gao Y, Zhang L, Ok YS, An T. Persistent free radicals in carbon-based materials on transformation of refractory organic contaminants (ROCs) in water: a critical review. Water Res. 2018;137:130–43.

Qiu N, Li H, Jin Z, Zhu Y. Temperature and time effect on the concentrations of free radicals in coal: evidence from laboratory pyrolysis experiments. Int J Coal Geol. 2007;69(3):220–8.

Reed JR, dela Cruz ALN, Lomnicki SM, Backes WL. Environmentally persistent free radical-containing particulate matter competitively inhibits metabolism by cytochrome P450 1A2. Toxicol Appl Pharmacol. 2015;289(2):223–30.

Runberg HL, Mitchell DG, Eaton SS, Eaton GR, Majestic BJ. Stability of environmentally persistent free radicals (EPFR) in atmospheric particulate matter and combustion particles. Atmos Environ. 2020;240:117809.

Saravia J, Lee GI, Lomnicki S, Dellinger B, Cormier SA. Particulate matter containing environmentally persistent free radicals and adverse infant respiratory health effects: a review. J Biochem Mol Toxicol. 2013;27(1):56–68.

Segal BG, Kaplan M, Fraenkel GK. Measurement of g values in the electron spin resonance spectra of free radicals. J Chem Phys. 1965;43(12):4191–200.

Singh GB, Gupta S, Gupta N. Carbazole degradation and biosurfactant production by newly isolated Pseudomonas sp. strain GBS. 5. Int Biodeterior Biodegradation. 2013;84:35–43.

Spokas KA, Novak JM, Stewart CE, Cantrell KB, Uchimiya M, DuSaire MG, et al. Qualitative analysis of volatile organic compounds on biochar. Chemosphere. 2011;85(5):869–82.

Tian L, Koshland CP, Yano J, Yachandra VK, Yu IT, Lee S, et al. Carbon-centered free radicals in particulate matter emissions from wood and coal combustion. Energy Fuel. 2009;23(5):2523–6.

Vejerano E, Lomnicki S, Dellinger B. Formation and stabilization of combustion-generated environmentally persistent free radicals on an Fe $(III)_2O_3$/silica surface. Environ Sci Technol. 2011;45(2):589–94.

Vejerano E, Lomnicki S, Dellinger B. Lifetime of combustion-generated environmentally persistent free radicals on Zn (II) O and other transition metal oxides. J Environ Monit. 2012a;14(10):2803–6.

Vejerano E, Lomnicki SM, Dellinger B. Formation and stabilization of combustion-generated, environmentally persistent radicals on Ni (II) O supported on a silica surface. Environ Sci Technol. 2012b;46(17):9406–11.

Wang Y, Li S, Wang M, Sun H, Mu Z, Zhang L, et al. Source apportionment of environmentally persistent free radicals (EPFRs) in PM2. 5 over Xi'an, China. Sci Total Environ. 2019;689:193–202.

Xu Y, Yang L, Wang X, Zheng M, Li C, Zhang A, et al. Risk evaluation of environmentally persistent free radicals in airborne particulate matter and influence of atmospheric factors. Ecotoxicol Environ Saf. 2020;196:110571.

Yang J, Pan B, Li H, Liao S, Zhang D, Wu M, et al. Degradation of p-nitrophenol on biochars: role of persistent free radicals. Environ Sci Technol. 2016;50(2):694–700.

Yang L, Liu G, Zheng M, Jin R, Zhao Y, Wu X, et al. Pivotal roles of metal oxides in the formation of environmentally persistent free radicals. Environ Sci Technol. 2017;51(21):12329–36.

Yi P, Chen Q, Li H, Lang D, Zhao Q, Pan B, et al. A comparative study on the formation of environmentally persistent free radicals (EPFRs) on hematite and goethite: contribution of various catechol degradation byproducts. Environ Sci Technol. 2019;53(23):13713–9.

Zhao N, Yin Z, Liu F, Zhang M, Lv Y, Hao Z, et al. Environmentally persistent free radicals mediated removal of Cr (VI) from highly saline water by corn straw biochars. Bioresour Technol. 2018;260:294–301.

Zhu K, Jia H, Sun Y, Dai Y, Zhang C, Guo X, ct al. Long-tcrm phototransformation of microplastics under simulated sunlight irradiation in aquatic environments: roles of reactive oxygen species. Water Res. 2020a;173:115564.

Zhu K, Jia H, Sun Y, Dai Y, Zhang C, Guo X, et al. Enhanced cytotoxicity of photoaged phenol-formaldehyde resins microplastics: combined effects of environmentally persistent free radicals, reactive oxygen species, and conjugated carbonyls. Environ Int. 2020b;145:106137.

Reactive Oxygen Species Producing Photoactivatable Molecules and Their Biological Applications

Suman Das and Dhermendra K. Tiwari

Contents

Abstract Many photoactivatable molecules generate reactive oxygen species (ROS) upon light irradiation. If these molecules are present in a living system, the generated ROS will deactivate surrounding proteins, generate localized cytotoxicity, and kill nearby cell upon light irradiation. The duration and effect of generated free radicals are short duration and localized. Therefore, these photosensitizers are used for various research and medical applications in chromophore-assisted light inactivation, photoablation, and photodynamic therapy. In this chapter, we summarize the various photosensitizers, their biological applications, and various other uses.

1 Introduction

Reactive oxygen species comprises both non-free radical and free radical oxygen intermediates including hydrogen peroxide (H_2O_2), hydroxyl radical ($^{\bullet}OH$), singlet oxygen ($^1O_2$), and superoxide ($O_2^{\bullet-}$) which are unstable in nature and can easily react with other molecules. An increased level of ROS in cellular systems may cause

S. Das · D. K. Tiwari (✉)
Department of Biotechnology, Faculty of Life Sciences and Environment, Goa University, Taleigao Plateau, Goa, India
e-mail: dktiwari@unigoa.ac.in

© The Author(s), under exclusive license to Springer Nature Switzerland AG 2021
K. K. Kesari, N. K. Jha (eds.), *Free Radical Biology and Environmental Toxicity*, Molecular and Integrative Toxicology, https://doi.org/10.1007/978-3-030-83446-3_2

damage to the cell by affecting macromolecules such as DNA, RNA, proteins, and lipids and may also induce cell death. This toxic property of ROS has therefore gained a worldwide acceptance in various biological field ranging from application in microscopy to cancer treatment (Allison et al. 2004; Shu et al. 2011). Photoactivator molecules commonly known as photosensitizers (PS) are molecules which upon irradiation with a particular wavelength of light undergo photochemical reactions and produce ROS (Liao et al. 1994). ROS performs diverse roles in an organism's physiology and pathophysiology. Higher ROS production is considered as toxic for the survival of living system. However, excessive levels of ROS have been intentionally generated at target sites in case of PSs-based therapy which leads to damage or destruction of the targeted cells (Pieczenik and Neustadt 2007).

Dougherty first conceives the idea of PS-based therapy in the late 1970 when he placed agents that are radiation sensitive, with cell cultures near the lab window and observed subsequent death of the cells (Dougherty et al. 1978). Later he concluded that this effect was because of porphyrin-type molecules. In the last one decade, the development of therapy based on PS has gained a very strong attention among scientific communities due to its easy and safe therapeutic practice. Due to its minimum side effects, photodynamic therapy (PDT) has been well recognized for the treatment of critical disease like cancer (Allison et al. 1998). Apart from cancer treatment, PSs have been used in various application such as photocatalysis, photodegradation, photopolymerization, and chromophore-assisted light inactivation (Liao et al. 1994; Calixto et al. 2016; Jewhurst et al. 2014).

Majority of the scientifically and medically proven PSs molecules are synthesized chemically known as chemical photosensitizers (CPS) which can be classified as porphyrin and non-porphyrin PS, based on their precursor molecule. Porphyrin-based PSs are further categorized as first, second, and third generation PSs depending on their target specificity (O'Connor et al. 2009). The first generation PSs involve hematoporphyrin derivative (HpD) and photofrin, which exhibited antitumorigenic activity against colon, breast, prostate, head, and neck cancer (Dougherty et al. 1978; Schuitmaker et al. 1996). Second generation PSs include Levulan®, Foscan®, Metvix®, and NPe6, which are approved for clinical use and exhibit promising results against colorectal cancer, head, and neck cancer, and actinic keratosis treatment (Kujundžić et al. 2007; Braathen et al. 2007; Markham and Collins 2001). Third generation PSs are designed by packaging or conjugating second generation PSs with some nanomaterials or carrier molecules that efficiently deliver them to target cells or tissues (Wöhrle et al. 1998). The non-porphyrin-based PSs include cationic dyes such as toluidine blue and methylene blue, which selectively aggregate in mitochondria of transformed cells (Leonard et al. 1999).

Various proteins isolated from living organisms have been known to show high ROS production when irradiated with light. These genetically encoded photosensitizers (GEPSs) have advantages over CPSs considering their intracellular localization and ease of application with other genetic elements or genes. The GEPS chromophore showed a high rate of intersystem crossing upon excitation in order to form the excited triplet state and subsequently produce ROS that can act as effective PS (Wojtovich and Foster 2014). Several studies suggested that the green fluorescent protein (GFP) based variants upon excitation with light can produce both fluorescence and sufficient ROS, resulting to toxic environment for the biological systems (Ansari et al. 2016; Liu et al. 1999). However, the production of ROS production by GFP variants is quite low as compared to commonly used PSs. Some fluorescent proteins which are photodynamically active have shown several fold higher ROS production as compared to GFP and examples of such proteins include KillerRed, miniSOG, and their variants. These GEPSs have been used as an optogenetic tool that allow the light-mediated inactivation of specific groups of neurons in transgenic zebrafish, mouse retina, *Caenorhabditis elegans,* and *Xenopus laevis* embryos (Jewhurst et al. 2014; Williams et al. 2013; Kobayashi et al. 2013; Byrne et al. 2013; Xu and Chisholm 2016). KillerRed has been used for diverse photosensitizer-based applications by targeting several sub-cellular organs such as lysosomes, cell membranes, nucleus, and mitochondria (Wang et al. 2013; Shibuya and Tsujimoto 2012; Berendzen et al. 2016). The smallest known GEPS called miniSOG is dependent on flavin mononucleotide (FMN) cofactor for its photosensitizing activity (Shu et al. 2011). It is very useful for the study of live cells because of its smaller size and ease of expression within the living system. However, the compromised redox potential, excitation limited in visible region, and limited availability restrict the use of GEPS.

This chapter summarizes various application of GEPS, CPS, and their combination with nanoparticle or nanomaterial in a biological perspective including, photoablation, spatio-temporal protein inactivation, CALI, correlation light-electron microscopy, and photodynamic therapy (PDT).

2 Mechanism of ROS Production by Photoactivator Molecules or Photosensitizers

PSs are specific group of molecules which upon irradiation with a specific wavelength of light undergo chemical reactions and produce ROS in a photochemical process (Liao et al. 1994). Upon light activation, PSs shifted from ground energy state (S_0) to first or second energy state (S_1 or S_2) and subsequently undergo long-lasting and more stable triplet excited states (T_1) through the process of intersystem crossing. In a biological system, PSs which are present in T_1 state upon reacting with molecular oxygen (O_2) release ROS (Kujundžić et al. 2007). There are two types of mechanisms proposed for the reactivity of PSs such as type I reaction and type II reaction (Fig. 1). In case of type I reaction, the energy of photons is transferred to cellular biomolecules via electron-hydrogen transfer and thereby damages the biomolecules and cells possessing such molecules. On the other hand, in case of type II reaction, when the energy of photon is transferred to O_2, highly electrophilic ROS such as singlet oxygen ($^1O_2$) is formed. This will further induce the destruction or damage of pathogenic microorganisms, targeted cellular structures, targeted protein deactivation, and carcinogenic tumors (de Freitas and Hamblin 2016). Apart from the characteristic of singlet oxygen production by PSs, their cytotoxic effect is

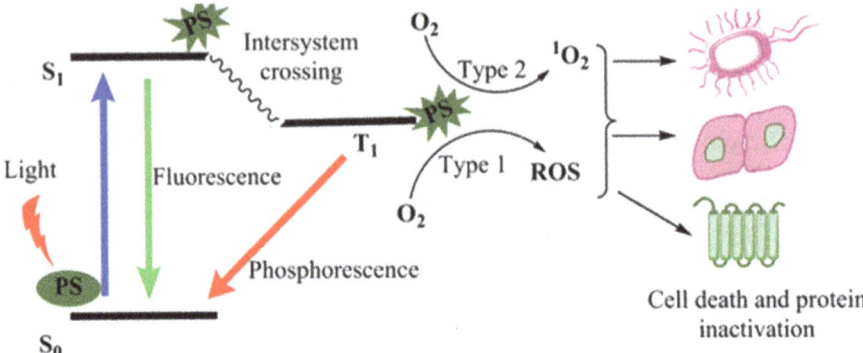

Fig. 1 ROS production and photosensitizing mechanism of photosensitive molecules. (With permission from Royal Society of Chemistry, Das et al. 2020)

highly dependent on accumulation efficiency and various cellular cues. The efficiency of photosensitizers in case of deep tissue *in vivo* applications depends on the radiation type that has been used for excitation leading to ROS production. PS which shows excitation in near-infrared light windows is very useful considering the effective penetration efficiency in deep-tissue regions by NIR light. This is due to the low scattering, least absorption, weak autofluorescence, and efficient tissue penetration of NIR light in NIR window in living systems.

2.1 Application of Photosensitizers in Cancer Treatment

In recent years, various types of photosensitizers have been developed that exhibit promising results in cancer treatment. Variety of immunotoxins like antibodies conjugated with chemicals, antibody fragments, PS, and toxins have been developed for high ROS generation and targeted specificity in a controlled manner. A highly stable single chain variable fragment antibody (scFv) of IgG1 called 4D5-single-chain-Fv fragment (4D5scFv) used for the treatment of tumors showing HER2/neu overexpression. Fusion of KillerRed with a 4D5scFv of an anti-p185HER-2-ECD antibody was the first genetically encoded immunophotosensitizer that has been developed to target HER2 positive breast cancer. The p185HER-2-ECD receptors which are overexpressed in cell line like SKOV-3 are the specific target of this PS construct (Serebrovskaya et al. 2009). This study has been successfully exploited the dimeric nature of KillerRed as the scFv fragment showed strong functionality and binding affinity in its multivalent form. The binding activity of KillerRed alone on SKOV-3 cells was insignificant. However, upon irradiation with light the fusion construct exhibited high ROS production and increased binding affinity, resulting in tumor cells death (Serebrovskaya et al. 2009). KillerRed also possesses the full potential of tumor killing in PDT. *In vitro* and *in vivo* studies on spheroid tumor and mouse xenograft tumor suggested that significant PDT effect was observed when

KillerRed expressing HeLa Kyoto cell lines were irradiated with 593 nm light (Shirmanova et al. 2013; Kuznetsova et al. 2015). *In vivo* studies showed KillerRed expressing non-pathogenic bacterial strains like *E. coli* guided PDT to treat tumor xenograft in mouse model. In this study, athymic nude mice containing CNE2 tumor were subcutaneously injected KillerRed-*E. coli* and maintained in a dark room for overnight so that the fluorescence area and intensity would be larger enough to occupy the whole tumor *in vivo* (Yan et al. 2015). After one week of post-irradiation, the tumors gradually disappeared followed by healing of the skin and tumor reoccurrence was not observed for more than two months in these PDT-treated mice (Yan et al. 2015).

In another study, Mironova and co-workers designed a fully genetically encoded immune-PS by fusing 4D5scFv as a targeting component with miniSOG. This recombinant immuno-PS-named 4D5scFv-miniSOG exhibited target specificity and highly specific phototriggered cytotoxic effect on HER2/neu-positive human breast adenocarcinoma SK-BR-3 cells with an improved IC_{50} value when compared to pyropheophorbide and verteporfin CPs (Mironova et al. 2013; Bhatti et al. 2008).

Apart from the GE-PSs, a large number of CPSs have been developed and successfully employed in cancer treatment. CPSs are commonly classified into porphyrins and non-porphyrins groups, among them porphyrins are further categorized into three categories considering their evolution, application modalities, functionalities, and limitations (O'Connor et al. 2009) (Fig. 2). Hematoporphyrin derivatives (HpD) and photofrin are considered as the first generation photosensitizers. The second generation photosensitizers are derivatives of porphyrin that are chemically pure from its antecedent generation and exhibit light absorption at higher wavelengths (Schuitmaker et al. 1996). Third generation photosensitizers are modified second generation molecules with various conjugates for target specific implications (Juzeniene et al. 2007). Non-porphyrins on the other hand involve mostly cationic dyes and their derivatives as photosensitizers.

The first generation PS such as hematoporphyrin was first developed by Scherer and co-workers in 1841 but its fluorescent property was explored much later in 1867 (Thudichum 1867). Initial study of hematoporphyrin suggested its potential in cancer detection, following which a key drawback was observed in the required dosage, resulting in improper photosensitization (Auler and Banzer 1942; Figge et al. 1948). Hematoporphyrin derivative (HpD), which was obtained through the manipulation of hematoporphyrin, comprises combination of different monomers, dimers, and oligomers showed enhanced properties when compared to its precursor molecule (Schwartz et al. 1955; Dougherty et al. 1984). HpD exhibits good PDT response against a wide variety of melanomas, malignant tumors, carcinomas of colon, breast, and prostate (Dougherty et al. 1978; O'Connor et al. 2009). Photofrin (Axcan Pharma Inc.) was derived by further purification of hematoporphyrin that exhibits an absorption maximum of 630 nm and stands as a "gold standard" in the PDT of non-skin derived cancers (Dougherty et al. 1984; Brown et al. 1993; Dolmans et al. 2003). Photofrin induces a cascade of chemical reactions during PDT through which free radicals are produced followed by cellular damage. The light in the PDT process initiates these reactions which causes the photofrin to absorb radiation and

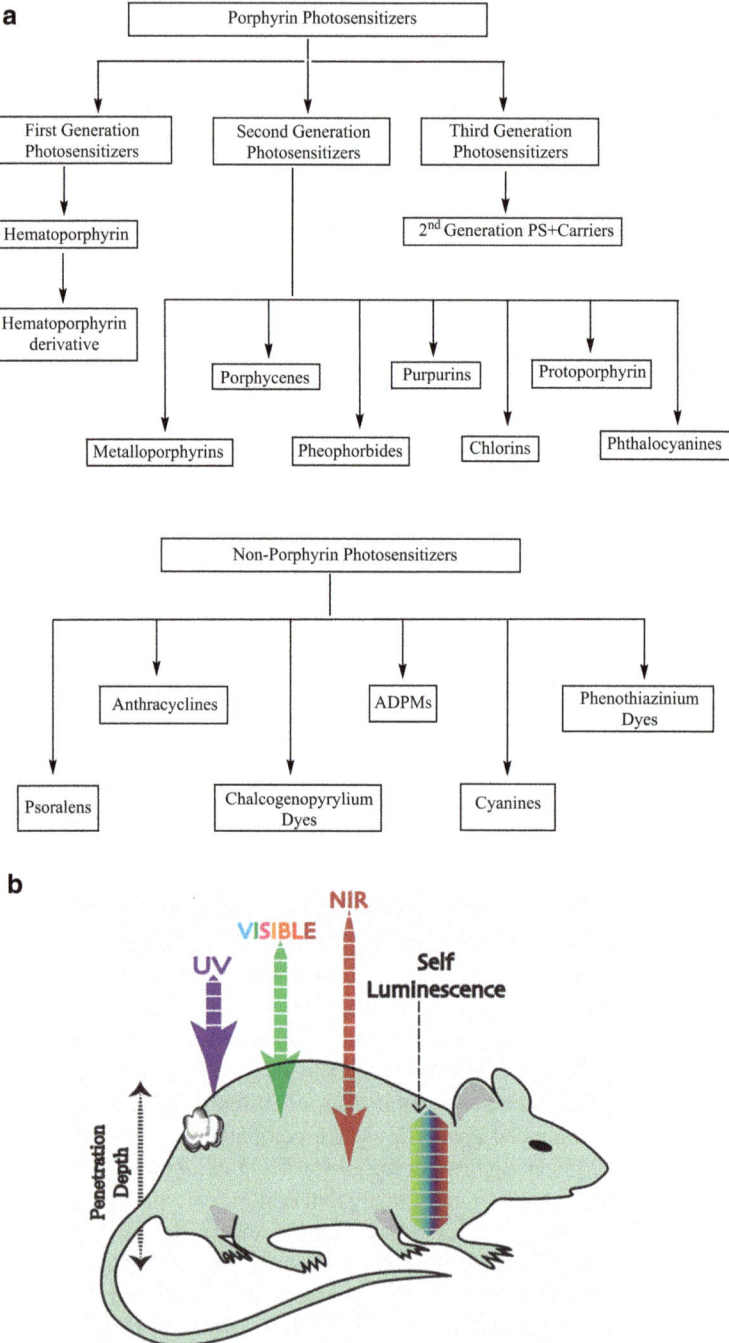

Fig. 2 (**a**) Classification of chemical photosensitizers and (**b**) suggestive penetration depth of the different wavelengths of the light for photoactivation of photosensitizers for in vivo applications. (With permission from Royal Society of Chemistry, Das et al. 2020)

thereby reach at a porphyrin excited state. The singlet oxygen state is generated by the spin transfer from the excited state to molecular oxygen which subsequently undergoes radical reactions to produce hydroxyl and superoxide radicals responsible for photosensitizing activity (Brown et al. 1993). Photofrin® is approved for medical use in Japan, Canada, and the USA for early- and late-stage lung cancer, esophageal cancer, bladder cancer, early-stage cervical cancer, and non-malignant and malignant skin diseases (Schuitmaker et al. 1996). It is also considered as a promising therapy against brain cancer, breast cancer, head and neck cancer, Kaposi's sarcoma, Barrett's esophagus dysplasia, and psoriasis (Sharman et al. 1999).

The development of second generation PS was based on the limitations of the first generation PSs (Sharman et al. 1999). Examples of second generation PSs include porphycenes, protoporphyrins, metalloporphyrins, pheophorbides, chlorins, purpurins, and phthalocyanines. Lutrin (motexafin lutetium or Lu-tex), which belongs to metalloporphyrin group exhibited photoactivation at 732 nm, has been tested under clinical trials for breast and prostate cancer (Sharman et al. 1999; Du et al. 2006; Wieman et al. 1999; Hsi et al. 2001). The only drawback toward this molecule is the pain experienced upon administration (Pandey 2012). Tookad® (or palladium bacterio-pheophorbide), which belongs to the group of pheophorbide, was obtained from the bacterial photosynthetic pigment BChl-a. This molecule exhibits near-infrared range absorption of 763 nm and a high molar extinction coefficient of nearly about 10.86×10^4 M^{-1} cm^{-1} (CHEN and HETZEL 1998). The drug has been reported to provide promising results when tested in *in vivo* conditions involving bone metastasis, human prostate cancer, and orthotopic prostatic models (Koudinova et al. 2003). Purlytin® (also called tin ethyl etiopurpurin or SnET2) which is a synthetic purpurin shows photoactivation at 664 nm and a molar extinction coefficient of 2.8×10^4 M^{-1} cm^{-1} had been employed in the treatment of cutaneous adenocarcinoma, basal cell carcinoma (BCC), and prostate cancer in canine models (Sharman et al. 1999; Josefsen and Boyle 2008; Rifkin et al. 1997; Kaplan et al. 1998; Selman et al. 2001). Foscan® (also known as temoporfin or meso-tetra-hydroxyphenyl-chlorine), which is a chlorin derivative, shows photoactivation at 652 nm and a molar extinction coefficient of 3×10^4 M^{-1} cm^{-1} was approved clinically by European Union (EU) in 2001. It was reported to exhibit promising results against cutaneous lesions, head and neck tumors, esophageal, pulmonary, and GI cancer (D'Cruz et al. 2004). Aptocine® (also known as talaporfin or mono-l-asoartyl chorine 6 or NPe6) is another derivative of chlorine showed promising results against various carcinomas under several clinical trials including squamous cell carcinoma (SCC) (Kato et al. 2003). In 2004, NPe6 was clinically approved in Japan for PDT of lung cancer and trademarked as Aptocine®. HPPH or Photochlor or 2(1-hexyloxyethly)-2-devinylpyropheophorbide-a, which belongs to the chlorine group, is hydrophilic in nature and has been employed successfully to treat a variety of cancers like skin cancer, lung cancer, basal cell carcinoma (BCC), head and neck cancer, and esophageal cancers (Bellnier et al. 2006; Bellnier et al. 2003; Dougherty et al. 2000; Anderson et al. 2003). Levulan® (also known as 5-Aminolevulinic acid or ALA), which belongs to the group of protoporphyrin, is a prodrug which activates after metabolization. This amino acid is available naturally and is enzymatically converted into protoporphyrin IX (PpIX) (Blume and Oseroff 2007). This PS shows photoactivation at 635 nm and a molar extinction coefficient of $<5 \times 10^3$ M^{-1} cm^{-1} (Sharman et al. 1999). PDT based on ALA has been used against squamous cell and basal cell cancers of the skin, for head and neck tumors, through invasive lesions but unsuccessful in achieving complete response (Blume and Oseroff 2007; Warloe et al. 1995). Phthalocyanines, which is a large organic macrocyclic compound, exhibits photoactivation at 670–770 nm and a molar extinction coefficient of 2.5×10^5 M^{-1} cm^{-1} shows promising results in the treatment of non-skin cancers (Miller et al. 2007; Lukyanets 2012).

The third generation PS were developed from second generation PS by conjugating them with certain carrier molecules to obtain high specificity to the target tissue. Liposomes, monoclonal antibodies (mABs), and polymers have been used as conjugating material to develop third generation PS for target specificity (Moser 1998). The use of third generation PS increases sensitivity toward a selected target, specifically a tumor and thereby greatly reduces the chance of any collateral damage to the healthy tissues or cells surrounding the target area (Wöhrle et al. 1998). Some example of such PS including amine-functionalized polyacrylamide nanoparticles possessing photofrin has been used against rat brain tumor model (Reddy et al. 2006). Recently, Huan Liang's group developed a ROS-sensitive amphipathic prodrug composed of hyaluronic acid (HA) grafted with gambogic acid (GA) in which PS Ce6 have been loaded into the hydrophobic core (HA-GA@Ce6) which upon laser activation showed site-specific cytotoxicity and efficient tumor cells killing both in vitro (4T1 murine breast cancer cells) and in vivo (4T1 tumor model) (Liang et al. 2018).

The non-porphyrin-based PSs include cationic molecules like hypericin, methylene blue, cyanines, toluidine blue, calcogenopyrilium dyes, tetra-aryl-azadipyrromethenes, etc., constituting the dye family of PS (Leonard et al. 1999). They exhibit activation at 650–850 nm range (Allison et al. 2004). Due to their hydrophobic nature, most of these dyes require liposomal delivery system. Phthalocyanines with an absorption maximum at 670 nm exhibit no absorption in 400–600 nm range and hence show strong photosensitization characteristics together with decreased photosensitivity to skin due to natural light. They also contain fluorescence and photoacoustic characteristics, broadening their functionalities to other imaging and diagnostic applications. Photocyanine, photosens, Pc 4, etc. are some of the examples of such photosensitizers, while photosens is only clinically approved and is available commercially in Russia. Photosens has been reported to show effective treatment in head and neck tumors, cutaneous and endobronchial lesions; aluminum phthalocyanine tetrasulfate showed promising results with tumors in cats (Peaston et al. 1993; Sokolov et al. 1995; Stranadko et al. 2001). Among other phthalocyanines, photocyanine has been reported to exhibit high PDT activity. It exhibits strong level of phototherapeutics and stability while showing lower dark toxicity. A dosage concentration of 2mg/kg was reported to show high inhibition of mice S180 and U14 tumors. Pc4, which is a silicon-based phthalocyanine PS, developed at the Case Western Reserve University, has shown to cause notable reductions in colon and ovarian cancers in animal models at a dosage concentrations of 1 mg/kg. These dyes aggregate specifically in transformed cell's mitochondria and cause photosensitization upon irradiation (Leonard et al. 1999).

3 Application of Photosensitizers as Immunomodulatory agent

The PDT-mediated destruction of tumor triggers acute inflammatory response with the infiltration of leukocytes into the tumor site followed by production of proinflammatory factors and cytokines (Krosl et al. 1995). The damaged or dying tumor cells will subsequently induce a systemic immune response against tumor, responsible for the secondary cause of tumor cell death. PDT is capable of triggering both the innate and adaptive arm of the immune system by presenting tumor cell undergoing PDT to correspondent immune cells (Wachowska et al. 2015; Reginato et al. 2014). Various cell types such as natural killer cells (NK cells), phagocytes (dendritic cells, macrophages, neutrophils), and the complement system which reacts to pathogenic invaders together comprise the innate arm of the immune system. The PDT-mediated acute inflammatory response further invoked the activation of innate immune system by activating complement system, releasing cytokines, and recruiting and activating the above-mentioned innate immune cells (Jalili et al. 2004; Preise et al. 2009). The expression of damage-associated molecular patterns (DAMPs) on the surface of the damaged cells occurs due to the oxidative stress generated by PDT. The immune phagocytes of the innate immune system concede those DAMPs as a danger signal and neutralize them (Garg et al. 2011). On the contrary, the adaptive arm of the immune system comprises antigen presenting cells (APCs) which trigger the naive T cell's maturation to produce tumor cell specific cytotoxic T lymphocytes (CTLs) and presenting antigens to B cells for antibody production. PDT-mediated apoptosis of cancer cells further triggers dendritic cell activation that subsequently activate specific T cell response and consequently generate an effective immune response against cancer (Galluzzi et al. 2012). Dendritic cells play a critical role by acting as a connector between innate and adaptive arm of the immune systems by subjecting tumor-associated antigens (TAAs) to immature T cells, resulting in the generation of CTLs, which further eliminate the remnant cancer cells. The mature dendritic cells, by expressing peptide-major histocompatibility complexes (MHC) on their cell surfaces, can also induce adaptive immune response by governing CD4+ and CD8+ T cells to CTLs (Reginato et al. 2014; Zheng et al. 2016; Vatansever and Hamblin 2015). However, in the majority of cases, immunosuppressive cytokines and pro-tumorigenic molecules secreted by the tumor cells through nonimmunogenic pathways make PDT insufficient in order to generate immune response that could lead to complete tumor killing. Therefore, to trigger an immune response that is strong enough to achieve higher tumor cures, researchers developed next generation PSs conjugated with immunostimulatory agents such as complement activators, hormones, exogenous immunoadjuvants, cytokines, microbial vaccines, and growth factors (Denis et al. 2011; Yang et al. 2016). The tumor antigens produced after PDT together with the co-existence of such immunostimulatory agents induced a comparatively enhanced antitumor immune response having a much better therapeutic outcome.

It was suggested that DAMPs expression by cancer cells is closely related to the type of photosensitizers that have been used for PDT and such photosensitizers include photofrin (Garg et al. 2011), 5-aminolevulinic acid (5-ALA), Rose Bengal acetate (Panzarini et al. 2014), foscan (Wang et al. 2015), and hypericin (Garg et al. 2012). Bastien's group reported that non-porphyrinic PS called OR141 triggered the expression of DAMPs such as Hsp90, annexin A1, and HMGB1 in SCC7 and A431 squamous cell carcinoma cells upon photoactivation and thereby inhibit the growth of SCC in mice. They further injected SCC7 cells killed with PDT treatment and dendritic cells primed with PDT-killed SCC7 cells in SCC7 tumor bearing mice and observed tumor growth inhibition in both the cases while the mice injected with dendritic cells showed more survival rate (Doix et al. 2019).

Differentiation of monocytes produces two different variety of macrophages such as M1 macrophages which are generally involved in the activation of invader attack and immune system, and the second one is M2 macrophages that are mainly involved in damage healing and promoting tumor progression also known as tumor-associated macrophage (TAM) (Lee et al. 2006; Shih et al. 2006; Solinas et al. 2009). Hamblin's group suggested that intratumorally injected maleylated albumin conjugated with chlorine (e6) can effectively target TAM residing within the tumor which is responsible for enhanced antitumor immune response and highly selective PDT killing (Anatelli et al. 2006; Korbelik and Hamblin 2015). Wen's group used hybrid nanoparticle composed of naturally occurring noninfectious cowpea mosaic virus (CPMV)/dendron as a carrier for the delivery of Zn-EpPor PS to M2 macrophages as well as cancer cells and studied the differences in uptake of photosensitizer between the populations of M2 and M1 macrophages and observed its efficiency in PDT of B16F10 melanoma cells (Wen et al. 2016).

Cheong's group by using hematoporphyrin manipulated the intracellular ROS generation and studied their specific role in the activation of dendritic cell. They observed that transient photogeneration of intracellular ROS can effectively prime dendritic cell maturation and increases its migration *in vitro* and *in vivo* with insignificant cellular death. They proposed that this technique which uses photodynamic reaction together with immature dendritic cells as vaccine adjuvants was sufficient enough to induce *in vivo* adaptive T cell responses (Cheong et al. 2015). Saji's group reported that combination therapy of dendritic cells injected intratumorally and ATXS10Na(II)-PDT in BALB/c mice possessing CT26 tumors exhibited greater tumor destruction than individual therapies (Saji et al. 2006). In another study, Jalili's group suggested that intratumoral injection of naive dendritic cells along with local PDT can prime cytotoxic T cells, which considerably inhibit subsequent tumor's growth and lead to regression of tumors at metastatic sites including multiple lung metastasis (Jalili et al. 2004). Blaudszun's group suggested an innovative drug delivery combination mediated by adoptive transferred T lymphocytes, cellular immunotherapy, and PDT. They employed PS mTHPP loaded within *ex vivo* activated human donor T cells as targeted delivery system for PDT as well as adoptive T cell therapy. Bispecific antibody redirected T lymphocytes contain dual specificity for recognizing a CD3 component present on T lymphocyte as well as surface marker like TAA on the target cell. When co-cultured with EpCAM expressing human carcinoma cells, the PS was transported to the target cells and consequently T lymphocytes induce cytolytic activity. As a result, upon irradiation increased cytotoxic ability of photosensitizer boosted T cells was observed (Blaudszun et al. 2015). Gao's group suggested that PDT mediated by phthalocyanine dye-labeled probe (DSAB-HK) not only inhibited subcutaneous tumor growth of 4T1 tumors, but also induced maturation of dendritic cell, resulting in CD8+ CTL recruitment for immune response. They also blocked the inhibitory receptors present on tumor cells which are known to evade the host immune response by using monoclonal antibodies programmed death-1 (PD-1), resulting in eradication of primary tumor and elimination of metastatic lesions due to the synergistic effect of PDT and immune checkpoint inhibition (Gao et al. 2016). Reginato's group showed that PDT after the administration of cyclophosphamide CY caused an absolute decline in the number of regulatory T cells (Treg) and subsequently strengthened PDT-mediated immunity that leads to complete tumor relapse and long-term survival (Reginato et al. 2013).

The use of nanomaterial-based photosensitizers has advanced the immunotherapy-based cancer treatment due to their multifunctional ability and improved target specificity (Zheng et al. 2020). In a study, Chunbai's group developed a multimodality nanoscale coordination polymer (NCP) NPs which carry the pyrolipid and oxaliplatin PS's (NCP@pyrolipid) to increase antitumor immunity. When NCP@pyrolipid was employed together with anti-PD-L1, it exhibited three treatment modalities such as chemotherapy by oxaliplatin, PDT by pyrolipid, and anti-programmed death-ligand 1 (PD-L1)-mediated checkpoint blockade therapy in mice bearing colorectal cancer. Oxaliplatin triggers apoptotic cell death and stimulates pre-apoptotic CRT expression on the cell surfaces, which behaves as an "eat me" signal to dendritic cells and macrophages. The PDT mediated by pyrolipid stimulates proinflammatory cytokines secretion like IL-6, TNF-α, and INF-γ, which further induced acute inflammation subsequently stimulating T-cell response specific for tumor. The activity of T cells against tumor cells was restored due to the use of anti-PD-L1, thereby reducing tumor volume and inhibiting cancer metastasis (He et al. 2016). Dangge's group reported that PDT-mediated cancer immunotherapy could be augmented by PD-L1 knockdown in tumor cells. They developed a micelleplex by covalently conjugating PS Pheophorbide A with an ultra pH-responsive diblock copolymer poly(ethylene glycol)-block-poly(di-isopropanol amino ethyl methacrylate-cohydroxyethyl methacrylate) (PEG-b-P(DPA-co-HEA), termed as PDPA along with small interfering RNA (siRNA) accountable for blocking of PD-1/PD-L1 interaction by PD-L1 knockdown. This micelleplex is activated only when acidic endocytic vesicles of tumor cells internalize them. The combination of PDT together with PD-L1 knockdown in a B16-F10 melanoma xenograft tumor model exhibited significantly increased level of tumor infiltrating CD8+ and CD4+ T cells which further inhibit distant metastasis and tumor growth when compared to PDT alone (Wang et al. 2016). Wen Song's group developed a chimeric peptide PpIX-1MT and reported that the peptide PpIX-1MT has the capability to form nanoparticles in PBS and via the enhanced effect of penetration and retention could aggregate in tumor areas. The PpIX-1MT NP when irradiated with 630 nm of light produce ROS which induce apoptosis in cancer cells by triggering caspase-3 expression and TAA production, which further generate an intense immune response. The caspase-3 cleavage stimulated the release of 1MT which further acted as an immune system booster to activate CD8+ T cells effectively (Song et al. 2018). Ce6 is one of the most studied photosensitizers for cancer PDT as mentioned above in detail (Anatelli et al. 2006; Korbelik and Hamblin 2015). A similar approach with some modifications was employed by Guangbao Yang's group who developed a smart NP comprised of Ce6 doped catalase (CAT) encapsulated by hollow silica nanoparticles (CAT@S/Ce6). The NP was further modified with (3-carboxypropyl)triphenylphos-phonium bromide (CTPP), which is a mitochondria targeting molecule and with a charge-convertible polymer that is highly responsive to acidic pH by electrostatic interaction (CAT@S/Ce6-CTPP/DPEG). These NPs exhibited charge conversion within the acidic microenvironment of tumor that would favor their tumor retention, their ability to target mitochondria, which are highly vulnerable to ROS, and their capability to overcome the tumor hypoxia by decomposing tumor endogenous H_2O_2 by enhancing PDT of solid tumors. When 4T1 tumors possessing Balb/c mice were treated with CAT@S/Ce6-CTPP/DPEG along with anti-PD-L1, it triggered CD8+ CTL infiltration in the primary tumor as well as in distant tumors which may efficiently inhibit primary tumor progression (Yang et al. 2018). Victoria's group employed photosens and photodithazine to trigger ICD during PDT and observed that both photosens and photodithazine significantly triggered cell death in glioma GL261 and fibrosarcoma MCA205 cells. They showed that the accumulation of photodithazine occurred in the endoplasmic reticulum and Golgi apparatus while localization of photosens predominantly occurred in the lysosomes. They also suggested that photodithazine or photosens PDT-induced dying cancer cells produce calreticulin, HMGB1, and ATP and were phagocytosed by bone-marrow-derived dendritic cells (BMDCs), which upon maturation and activation produced IL-6 (Turubanova et al. 2019) (Fig. 3).

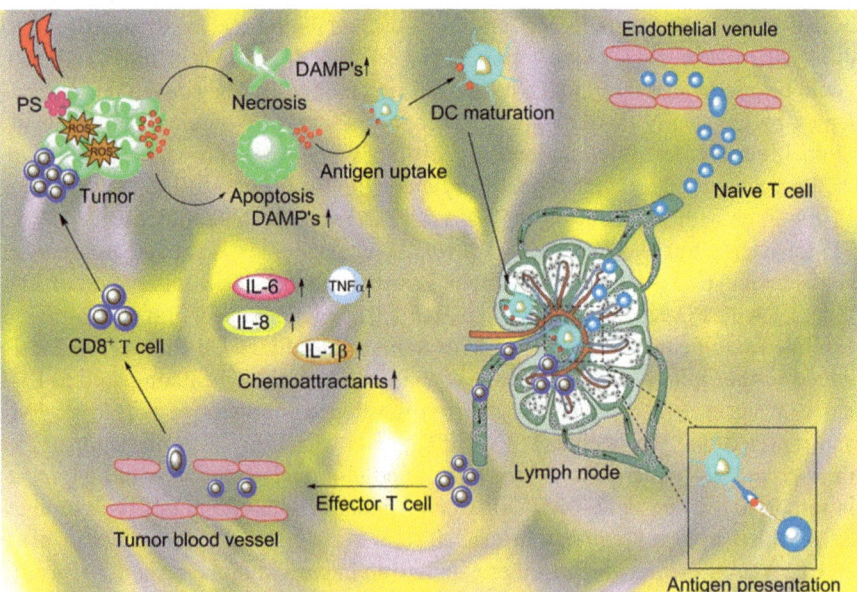

Fig. 3 Proposed mechanism of photosensitizer's interaction and activation of immune system in photodynamic therapy. (With permission from Royal Society of Chemistry, Das et al. 2020)

4   Miscellaneous Application of Photosensitizers

Miscellaneous application of PS mainly comprises CALI, photoablation, CLEM, etc. In these applications, GEPS-based approaches showed promising results compared to other classes of PSs. In CALI, a PS chromophore is generally fused with a target-specific protein by genetic engineering. Upon light activation, PS produces short-lived ROS, which deactivates the target protein and due to the short-life time of ROS, surrounding cells remain unaffected. CALI is a very useful tool to inactivate protein of interest which efficiently helps in the study of loss-of-function or gain-of-function of a target gene. GEPSs not only used to target proteins but also cell membrane, chromatins, and mitochondria to trigger cell death in order to analyze cell-specific functions among the mixed population. GEPSs such as SuperNova, KillerRed, miniSOG, and KillerOrange have been used for various CALI and photoablation studies. They are useful for CALI due to their suitability for live cell study, target specificity, least chemical use in cell, and low autofluorescence (Serebrovskaya et al. 2009; Mironova et al. 2013; Makhijani et al. 2017; Proshkina et al. 2015). GEPSs have been successfully employed in various *in vivo* and *in vitro* studies such as destruction of tumor cells, deactivation of target proteins, development study of transgenic population of worms, and *C. elegans* using CALI and photoablation (Shu et al. 2011; Williams et al. 2013; Qi et al. 2012; Takemoto et al. 2013; Sarkisyan et al. 2015).

KillerRed has been utilized to inactivate targeted cellular proteins in various organelles using CALI such as peroxisomes, mitochondria, nucleus, lysosomes (Bulina et al. 2006; Jarvela and Linstedt 2014; Destaing et al. 2010; Matsunaga et al. 2012), and photoablation for the study of developmental biology (Jewhurst et al. 2014; Williams et al. 2013; Kobayashi et al. 2013; Byrne et al. 2013; Shibuya and Tsujimoto 2012) and localized cellular oxidative stress study (Shibuya and Tsujimoto 2012). Jarvela's group studied KillerRed-based CALI in order to distinguish the role of two different variants of Golgi reassembly and staking protein (GRASPs) called GRASP65 and GRASP55 in cisternae-specific tethering and their compartmentalization in HeLa cells (Jarvela and Linstedt 2014). They observed that both the GRASPs were crucial to achieve correct cisternae-specific tethering to induce Golgi-ribbon formation and proper compartmentation. They also explained the intermixing of mis-compartmentalization and cisternae after inactivation of individual GRASP through CALI (Jarvela and Linstedt 2014). Destaing's group conducted a study to evaluate the function of β1 integrin in the organization and function of invadosome (Destaing et al. 2010). Invadosomes are adhesion structures that perform an important function in tissue invasion, where intensely active actin is present, and are thought to be involved in the degradation of extracellular matrix (ECM). They conducted a set experiments based on CALI with KillerRed and observed that the inactivation of cytoplasmic domain of KillerRed-β1A integrin immediately disassembles and disorganizes invadosome. This experiment clearly suggested that the β1A integrin performs an important function in ECM degradation and actin dynamics in invadosome (Destaing et al. 2010). KillerRed exhibited its potential as a ROS generator and arrest phase specific cell division due to phototoxicity to chromatin. CALI mediated by KillerRed identified the function of "RNA-binding motif protein encoded on X-chromosome (RBMX)" in the process of chromosome morphogenesis via binding with chromatin but not RNA (Matsunaga et al. 2012). In another study, KillerRed triggered the apoptosis of COS-7 and HeLa cells by targeting mitochondria and subsequent production of ROS further enhanced mitochondrial dysfunction and stimulates Cyt c/caspase-3 pathway (Li et al. 2019). Riani's group investigated whether SNG has the ability to drive selective cell death while retaining the viability of SNR expressing cells. They co-cultured HeLa cell lines stably expressing SNG or SNR in the mitochondrial matrix. They reported that upon 447 nm light irradiation, 94% of the cells expressing SNG were successfully ablated while almost all of the SNR expressing cells survived and upon 562 nm light irradiation reverse case happened (Riani et al. 2018).

GEPSs like KillerRed and MiniSOG have been utilized in several CALI studies in the animal model with a key focus on neuroscience. Lukyanov's group reported that the cells expressing H2B-tKR (tandem KillerRed) exhibit damage in the genome of mammalian cells and retardation of cell division and inhibition of neuronal development of *Xenopus laevis* tadpole forebrain when irradiated with green light (Serebrovskaya et al. 2011). Lin's group established the fact that upon blue light irradiation, miniSOG-VAMP2 caused paralysis and slow movement of *C. elegans* through the process of inhibition of synapses with CALI (InSynC) (Lin et al. 2013). In another study where minoSOG has been used for DNA CALI suggested that ROS generated by histone-miniSOG due to light induced extrachromosomal transgene integration and broad spectrum mutations in *C. elegans* (Noma and Jin 2015). Makhijani's group studied cell ablation by using miniSOG2 in *Drosophila*. They developed an upstream activating sequence (UAS) – miniSOG2 T2A Histone 3.3 (H3.3)-EGFP transgenic line and subsequently crossed this transgenic line with the *ppk*-GAL4 line to trigger the expression of miniSOG2 in the class IV dendritic arborization (da) neurons. Upon blue light irradiation coming from a xenon lamp, they found loss of orange fluorescence of tdTomato, which confirmed the successful ablation of dendrites and axons while leaving the unexposed neighboring neurons intact. They also analyze the neurons after blue-light irradiation for 2 hrs and found that the irradiated neuron appeared swelled, fragmented, and severed dendrites, which are the characteristics of class IV da neurons undergoing pruning (Makhijani et al. 2017).

MiniSOG has been successfully employed in correlative light electron microscopy without the need of any exogenous probes, ligands, or destructive permeabilizing detergents. When excited with blue light, miniSOG produces $^1O_2$, which locally triggers the polymerization and precipitation of diaminobenzidine that can be stained with osmium tetroxide, which then permitted to gain resolution well below 10 nm in CLEM imaging of stained cells (Shu et al. 2011). By using miniSOG, it is possible to distinguish syn-CAM1 and syn-CAM2 synaptic adhesion complex proteins present in mouse brain and neuronal cells in a closely related pre- and post-synaptic complex through CLEM imaging which was not achieved before by using antibody staining in conventional EM (Shu et al. 2011). The use of miniSOG in CLEM imaging provides a breakthrough in 3D imaging, study of protein degradation and trafficking of α-synuclein, and its organization in pre-synaptic terminal with tubulovesicular structure (Boassa et al. 2013). Another high-resolution CLEM study confirmed the association of plasma membrane with a small PKA-RI specific membrane kinase A-anchoring protein, known as smAKAP, at dense cell–cell junctions within the filopodia (Burgers et al. 2012). By using miniSOG in CLEM imaging, researchers also reported that the Mp1(receptor of thrombo-protein), in its immature form, transported and recruited in plasma membrane by using both ER-Golgi pathway and autolysosome secretory pathway (Cleyrat et al. 2014). In another interesting study, a new type of multifunctional polymer called E4-ORF3, produced by adeno viral protein was visualized by using miniSOG, and established the fact that the inactivation of this host protein triggers pathological viral replication (Ou et al. 2015).

5 Conclusion and Future Prospect

From the last several years, a variety of photosensitizers have been developed and used for different biological applications in redox biology. The use of photosensitizers has been considered to be the best and is easily executed for cancer treatment along with other commonly used therapeutics. GEPSs are rapidly broadening the horizons of redox biologists to study the specific function of gene without using molecular biology tools that are time consuming, exploring a particular cell type and their specific role in a mixed cell populations by photoablation, chromophore-assisted light inactivation to study variety of cell functions and producing high-resolution electron microscopy images without harsh chemical treatment. The capability of a photosensitizer to enter into the excited triplet state determines its efficiency of ROS production. The longer the triplet state, the better the production of ROS due to increasing chances of interaction with oxygen. Besides GEPS, chemical photosensitizers are also explored widely for PDT despite their limitations related to selectivity and phototoxicity. New approaches are emerging day by day in order to overcome such limitations that can allow them to specifically target a particular site. One such approach is nanoparticle-based drug delivery system which shows the path of their use in PDT.

Research based on photodynamic therapy is ongoing worldwide exploring various aspects of cancer treatment including PDT based on immunotherapy. Immuno-PDT is arising as a potential cancer treatment, where the ability of photosensitizers developed in the last few years is using to activate patient's own immune system to kill tumor cells. Researchers also developed some photosensitizers to not only induce immune system to kill tumor cells but also decrease the rate of tumor progression. These advanced photosensitizers utilize intelligent engineering by using previously existing PS with required modification by immune activator or antibody targeted and nanotechnologically driven design. Despite these developments, there are many obstacles related to effectiveness of PSs in both visible and near-infrared range with greater specificity, excitation in deep-tissue PDT, and high recycling capacity, which still require attention of scientific community in the coming future to develop advance PSs.

References

Allison RR, Mang TS, Wilson BD. Photodynamic therapy for the treatment of nonmelanomatous cutaneous malignancies. In: Seminars in cutaneous medicine and surgery, 1998, pp. 153–163

Allison RR, Downie GH, Cuenca R, Hu X-H, Childs CJ, Sibata CH. Photosensitizers in clinical PDT. Photodiagn Photodyn Ther. 2004;1(1):27–42.

Anatelli F, Mroz P, Liu Q, Yang C, Castano AP, Swietlik E, Hamblin MR. Macrophage-targeted photosensitizer conjugate delivered by intratumoral injection. Mol Pharm. 2006;3(6):654–64.

Anderson TM, Dougherty TJ, Tan D, Sumlin A, Schlossin JM, Kanter PM. Photodynamic therapy for sarcoma pulmonary metastases: a preclinical toxicity study. Anticancer Res. 2003;23(5A):3713–8.

Ansari AM, Ahmed AK, Matsangos AE, Lay F, Born LJ, Marti G, Harmon JW, Sun Z. Cellular GFP toxicity and immunogenicity: potential confounders in in vivo cell tracking experiments. Stem Cell Rev Rep. 2016;12(5):553–9.

Auler H, Banzer G. Untersuchungen über die Rolle der Porphyrine bei geschwulstkranken Menschen und Tieren. J Cancer Res Clin Oncol. 1942;53(2):65–8.

Bellnier DA, Greco WR, Loewen GM, Nava H, Oseroff AR, Pandey RK, Tsuchida T, Dougherty TJ. Population pharmacokinetics of the photodynamic therapy agent 2-[1-hexyloxyethyl]-2-devinyl pyropheophorbide-a in cancer patients. Cancer Res. 2003;63(8):1806–13.

Bellnier DA, Greco WR, Nava H, Loewen GM, Oseroff AR, Dougherty TJ. Mild skin photosensitivity in cancer patients following injection of Photochlor (2-[1-hexyloxyethyl]-2-devinyl pyropheophorbide-a; HPPH) for photodynamic therapy. Cancer Chemother Pharmacol. 2006;57(1):40–5.

Berendzen KM, Durieux J, Shao L-W, Tian Y, Kim H-e, Wolff S, Liu Y, Dillin A. Neuroendocrine coordination of mitochondrial stress signaling and proteostasis. Cell. 2016;166(6):1553–1563.e10.

Bhatti M, Yahioglu G, Milgrom LR, Garcia-Maya M, Chester KA, Deonarain MP. Targeted photodynamic therapy with multiply-loaded recombinant antibody fragments. Int J Cancer. 2008;122(5):1155–63.

Blaudszun A-R, Moldenhauer G, Schneider M, Philippi A. A photosensitizer delivered by bispecific antibody redirected T lymphocytes enhances cytotoxicity against EpCAM-expressing carcinoma cells upon light irradiation. J Control Release. 2015;197:58–68.

Blume JE, Oseroff AR. Aminolevulinic acid photodynamic therapy for skin cancers. Dermatol Clin. 2007;25(1):5–14.

Boassa D, Berlanga ML, Yang MA, Terada M, Hu J, Bushong EA, Hwang M, Masliah E, George JM, Ellisman MH. Mapping the subcellular distribution of α-synuclein in neurons using genetically encoded probes for correlated light and electron microscopy: implications for Parkinson's disease pathogenesis. J Neurosci. 2013;33(6):2605–15.

Braathen LR, Szeimies R-M, Basset-Seguin N, Bissonnette R, Foley P, Pariser D, Roelandts R, Wennberg A-M, Morton CA. Guidelines on the use of photodynamic therapy for nonmelanoma skin cancer: an international consensus. J Am Acad Dermatol. 2007;56(1):125–43.

Brown S, Vernon D, Holroyd J, Marcus S. Pharmacokinetics of photofrin in man, Excerpta medica international congress series. Elsevier; 1993. p. 475.

Bulina ME, Chudakov DM, Britanova OV, Yanushevich YG, Staroverov DB, Chepurnykh TV, Merzlyak EM, Shkrob MA, Lukyanov S, Lukyanov KA. A genetically encoded photosensitizer. Nat Biotechnol. 2006;24(1):95.

Burgers PP, Ma Y, Margarucci L, Mackey M, van der Heyden MA, Ellisman M, Scholten A, Taylor SS, Heck AJ. A small novel A-kinase anchoring protein (AKAP) that localizes specifically protein kinase A-regulatory subunit I (PKA-RI) to the plasma membrane. J Biol Chem. 2012;287(52):43789–97.

Byrne LC, Khalid F, Lee T, Zin EA, Greenberg KP, Visel M, Schaffer DV, Flannery JG. AAV-mediated, optogenetic ablation of Müller glia leads to structural and functional changes in the mouse retina. PLoS One. 2013;8(9):e76075.

Calixto GMF, Bernegossi J, De Freitas LM, Fontana CR, Chorilli M. Nanotechnology-based drug delivery systems for photodynamic therapy of cancer: a review. Molecules. 2016;21(3):342.

Chen Q, Hetzel FW. Laser dosimetry studies in the prostate. J Clin Laser Med Surg. 1998;16(1):9–12.

Cheong T-C, Shin EP, Kwon E-K, Choi J-H, Wang K-K, Sharma P, Choi KH, Lim J-M, Kim H-G, Oh K. Functional manipulation of dendritic cells by photoswitchable generation of intracellular reactive oxygen species. ACS Chem Biol. 2015;10(3):757–65.

Cleyrat C, Darehshouri A, Steinkamp MP, Vilaine M, Boassa D, Ellisman MH, Hermouet S, Wilson BS. Mpl traffics to the cell surface through conventional and unconventional routes. Traffic. 2014;15(9):961–82.

Das S, Tiwari M, Mondal D, Sahoo BR, Tiwari DK. Growing tool-kit of photosensitizers for clinical and non-clinical applications. J Mater Chem B. 2020;8(48):10897–940.

D'Cruz AK, Robinson MH, Biel MA. mTHPC-mediated photodynamic therapy in patients with advanced, incurable head and neck cancer: a multicenter study of 128 patients. Head Neck. 2004;26(3):232–40.

de Freitas LF, Hamblin MR. Antimocrobial photodynamic inactivation and antitumor photodynamic therapy with fullerenes. Morgan & Claypool Publishers; 2016.

Denis TGS, Aziz K, Waheed AA, Huang Y-Y, Sharma SK, Mroz P, Hamblin MR. Combination approaches to potentiate immune response after photodynamic therapy for cancer. Photochem Photobiol Sci. 2011;10(5):792–801.

Destaing O, Planus E, Bouvard D, Oddou C, Badowski C, Bossy V, Raducanu A, Fourcade B, Albiges-Rizo C, Block MR. β1A integrin is a master regulator of invadosome organization and function. Mol Biol Cell. 2010;21(23):4108–19.

Doix B, Trempolec N, Riant O, Feron O. Low photosensitizer dose and early radiotherapy enhance antitumor immune response of photodynamic therapy-based dendritic cell vaccination. Front Oncol. 2019;9:811.

Dolmans DE, Fukumura D, Jain RK. Photodynamic therapy for cancer. Nat Rev Cancer. 2003;3(5):380.

Dougherty TJ, Kaufman JE, Goldfarb A, Weishaupt KR, Boyle D, Mittleman A. Photoradiation therapy for the treatment of malignant tumors. Cancer Res. 1978;38(8):2628–35.

Dougherty T, Potter W, Weishaupt K. The structure of the active component of hematoporphyrin derivative, Porphyrins in tumor phototherapy. Springer; 1984. p. 23–35.

Dougherty T, Pandey R, Nava H, Smith J, Douglass H, Edge S, Bellnier D, O'Malley L, Cooper M. Preliminary clinical data on a new photodynamic therapy photosensitizer, HPPH for treatment of obstructive esophageal cancer. Proc SPIE. 2000:25–7.

Du K, Mick R, Busch T, Zhu T, Finlay J, Yu G, Yodh A, Malkowicz S, Smith D, Whittington R. Preliminary results of interstitial motexafin lutetium-mediated PDT for prostate cancer. Lasers Surg Med. 2006;38(5):427–34.

Figge FH, Weiland GS, Manganiello LO. Cancer detection and therapy. Affinity of neoplastic, embryonic, and traumatized tissues for porphyrins and metalloporphyrins. Proc Soc Exp Biol Med. 1948;68(3):640–1.

Galluzzi L, Kepp O, Kroemer G. Enlightening the impact of immunogenic cell death in photodynamic cancer therapy. EMBO J. 2012;31(5):1055–7.

Gao L, Zhang C, Gao D, Liu H, Yu X, Lai J, Wang F, Lin J, Liu Z. Enhanced anti-tumor efficacy through a combination of integrin αvβ6-targeted photodynamic therapy and immune checkpoint inhibition. Theranostics. 2016;6(5):627.

Garg AD, Krysko DV, Vandenabeele P, Agostinis P. DAMPs and PDT-mediated photo-oxidative stress: exploring the unknown. Photochem Photobiol Sci. 2011;10(5):670–80.

Garg AD, Krysko DV, Vandenabeele P, Agostinis P. Hypericin-based photodynamic therapy induces surface exposure of damage-associated molecular patterns like HSP70 and calreticulin. Cancer Immunol Immunother. 2012;61(2):215–21.

He C, Duan X, Guo N, Chan C, Poon C, Weichselbaum RR, Lin W. Core-shell nanoscale coordination polymers combine chemotherapy and photodynamic therapy to potentiate checkpoint blockade cancer immunotherapy. Nat Commun. 2016;7(1):1–12.

Hsi RA, Kapatkin A, Strandberg J, Zhu T, Vulcan T, Solonenko M, Rodriguez C, Chang J, Saunders M, Mason N. Photodynamic therapy in the canine prostate using motexafin lutetium. Clin Cancer Res. 2001;7(3):651–60.

Jalili A, Makowski M, Świtaj T, Nowis D, Wilczyński GM, Wilczek E, Chorąży-Massalska M, Radzikowska A, Maśliński W, Biały Ł. Effective photoimmunotherapy of murine colon carcinoma induced by the combination of photodynamic therapy and dendritic cells. Clin Cancer Res. 2004;10(13):4498–508.

Jarvela T, Linstedt AD. Isoform-specific tethering links the Golgi ribbon to maintain compartmentalization. Mol Biol Cell. 2014;25(1):133–44.

Jewhurst K, Levin M, Mclaughlin KA. Optogenetic control of apoptosis in targeted tissues of Xenopus laevis embryos. J Cell Death. 2014;7:JCD. S18368.

Josefsen LB, Boyle RW. Photodynamic therapy and the development of metal-based photosensitisers. Metal-based Drugs. 2008;

Juzeniene A, Peng Q, Moan J. Milestones in the development of photodynamic therapy and fluorescence diagnosis. Photochem Photobiol Sci. 2007;6(12):1234–45.

Kaplan MJ, Somers RG, Greenberg RH, Ackler J. Photodynamic therapy in the management of metastatic cutaneous adenocarcinomas: case reports from phase 1/2 studies using tin ethyl etiopurpurin (SnET2). J Surg Oncol. 1998;67(2):121–5.

Kato H, Furukawa K, Sato M, Okunaka T, Kusunoki Y, Kawahara M, Fukuoka M, Miyazawa T, Yana T, Matsui K. Phase II clinical study of photodynamic therapy using mono-L-aspartyl chlorin e6 and diode laser for early superficial squamous cell carcinoma of the lung. Lung Cancer. 2003;42(1):103–11.

Kobayashi J, Shidara H, Morisawa Y, Kawakami M, Tanahashi Y, Hotta K, Oka K. A method for selective ablation of neurons in C. elegans using the phototoxic fluorescent protein, KillerRed. Neurosci Lett. 2013;548:261–4.

Korbelik M, Hamblin MR. The impact of macrophage-cancer cell interaction on the efficacy of photodynamic therapy. Photochem Photobiol Sci. 2015;14(8):1403–9.

Koudinova NV, Pinthus JH, Brandis A, Brenner O, Bendel P, Ramon J, Eshhar Z, Scherz A, Salomon Y. Photodynamic therapy with Pd-bacteriopheophorbide (TOOKAD): Successful in vivo treatment of human prostatic small cell carcinoma xenografts. Int J Cancer. 2003;104(6):782–9.

Krosl G, Korbelik M, Dougherty G. Induction of immune cell infiltration into murine SCCVII tumour by photofrin-based photodynamic therapy. Br J Cancer. 1995;71(3):549–55.

Kujundžić M, Vogl T, Stimac D, Rustemović N, Hsi RA, Roh M, Katičić M, Cuenca R, Lustig R, Wang S. A phase II safety and effect on time to tumor progression study of intratumoral light infusion technology using talaporfin sodium in patients with metastatic colorectal cancer. J Surg Oncol. 2007;96(6):518–24.

Kuznetsova DS, Shirmanova MV, Dudenkova VV, Subochev PV, Turchin IV, Zagaynova EV, Lukyanov SA, Shakhov BE, Kamensky VA. Photobleaching and phototoxicity of KillerRed in tumor spheroids induced by continuous wave and pulsed laser illumination. J Biophotonics. 2015;8(11–12):952–60.

Lee C-C, Liu K-J, Huang T-S. Tumor-associated macrophage: its role in tumor angiogenesis. J Cancer Mol. 2006;2(4):135–40.

Leonard KA, Nelen MI, Simard TP, Davies SR, Gollnick SO, Oseroff AR, Gibson SL, Hilf R, Chen LB, Detty MR. Synthesis and evaluation of chalcogenopyrylium dyes as potential sensitizers for the photodynamic therapy of cancer. J Med Chem. 1999;42(19):3953–64.

Li X, Fang F, Gao Y, TangG XW, Wang Y, Kong R, Tuyihong A, Wang Z. ROS induced by KillerRed targeting mitochondria (mtKR) enhances apoptosis caused by radiation via Cyt c/ caspase-3 pathway. Oxid Med Cell Longev. 2019;

Liang H, Zhou Z, Luo R, Sang M, Liu B, Sun M, Qu W, Feng F, Liu W. Tumor-specific activated photodynamic therapy with an oxidation-regulated strategy for enhancing anti-tumor efficacy. Theranostics. 2018;8(18):5059.

Liao JC, Roider J, Jay DG. Chromophore-assisted laser inactivation of proteins is mediated by the photogeneration of free radicals. Proc Natl Acad Sci. 1994;91(7):2659–63.

Lin JY, Sann SB, Zhou K, Nabavi S, Proulx CD, Malinow R, Jin Y, Tsien RY. Optogenetic inhibition of synaptic release with chromophore-assisted light inactivation (CALI). Neuron. 2013;79(2):241–53.

Liu H-S, Jan M-S, Chou C-K, Chen P-H, Ke N-J. Is green fluorescent protein toxic to the living cells? Biochem Biophys Res Commun. 1999;260(3):712–7.

Lukyanets EA. Phthalocyanines as photosensitizers in the photodynamic therapy of cancer. J Porphyrins Phthalocyanines. 2012;03(06):424–32.

Makhijani K, To T-L, Ruiz-González R, Lafaye C, Royant A, Shu X. Precision optogenetic tool for selective single-and multiple-cell ablation in a live animal model system. Cell Chem Biol. 2017;24(1):110–9.

Markham T, Collins P. Topical 5-aminolaevulinic acid photodynamic therapy for extensive scalp actinic keratoses. Br J Dermatol. 2001;145(3):502–4.

Matsunaga S, Takata H, Morimoto A, Hayashihara K, Higashi T, Akatsuchi K, Mizusawa E, Yamakawa M, Ashida M, Matsunaga TM. RBMX: a regulator for maintenance and centromeric protection of sister chromatid cohesion. Cell Rep. 2012;1(4):299–308.

Miller JD, Baron ED, Scull H, Hsia A, Berlin JC, McCormick T, Colussi V, Kenney ME, Cooper KD, Oleinick NL. Photodynamic therapy with the phthalocyanine photosensitizer Pc 4: the case experience with preclinical mechanistic and early clinical–translational studies. Toxicol Appl Pharmacol. 2007;224(3):290–9.

Mironova KE, Proshkina GM, Ryabova AV, Stremovskiy OA, Lukyanov SA, Petrov RV, Deyev SM. Genetically encoded immunophotosensitizer 4D5scFv-miniSOG is a highly selective agent for targeted photokilling of tumor cells in vitro. Theranostics. 2013;3(11):831.

Moser JG. Photodynamic tumor therapy: 2nd and 3rd generation photosensitizers. Harwood; 1998.

Noma K, Jin Y. Optogenetic mutagenesis in Caenorhabditis elegans. Nat Commun. 2015;6:8868.

O'Connor AE, Gallagher WM, Byrne AT. Porphyrin and nonporphyrin photosensitizers in oncology: preclinical and clinical advances in photodynamic therapy. Photochem Photobiol. 2009;85(5):1053–74.

Ou HD, Deerinck TJ, Bushong E, Ellisman MH, O'Shea CC. Visualizing viral protein structures in cells using genetic probes for correlated light and electron microscopy. Methods. 2015;90:39–48.

Pandey RK. Recent advances in photodynamic therapy. J Porphyrins Phthalocyanines. 2012;4(4):368–73.

Panzarini E, Inguscio V, Fimia GM, Dini L. Rose Bengal acetate photodynamic therapy (RBAc-PDT) induces exposure and release of damage associated molecular patterns (DAMPs) in human HeLa cells. PLoS One. 2014;9(8).

Peaston A, Leach M, Higgins R. Photodynamic therapy for nasal and aural squamous cell carcinoma in cats. J Am Vet Med Assoc. 1993;202(8):1261–5.

Pieczenik SR, Neustadt J. Mitochondrial dysfunction and molecular pathways of disease. Exp Mol Pathol. 2007;83(1):84–92.

Preise D, Oren R, Glinert I, Kalchenko V, Jung S, Scherz A, Salomon Y. Systemic antitumor protection by vascular-targeted photodynamic therapy involves cellular and humoral immunity. Cancer Immunol Immunother. 2009;58(1):71–84.

Proshkina G, Shilova O, Ryabova A, Stremovskiy O, Deyev S. A new anticancer toxin based on HER2/neu-specific DARPin and photoactive flavoprotein miniSOG. Biochimie. 2015;118:116–22.

Qi YB, Garren EJ, Shu X, Tsien RY, Jin Y. Photo-inducible cell ablation in Caenorhabditis elegans using the genetically encoded singlet oxygen generating protein miniSOG. Proc Natl Acad Sci. 2012;109(19):7499–504.

Reddy GR, Bhojani MS, McConville P, Moody J, Moffat BA, Hall DE, Kim G, Koo Y-EL, Woolliscroft MJ, Sugai JV. Vascular targeted nanoparticles for imaging and treatment of brain tumors. Clin Cancer Res. 2006;12(22):6677–86.

Reginato E, Mroz P, Chung H, Kawakubo M, Wolf P, Hamblin MR. Photodynamic therapy plus regulatory T-cell depletion produces immunity against a mouse tumour that expresses a self-antigen. Br J Cancer. 2013;109(8):2167–74.

Reginato E, Wolf P, Hamblin MR. Immune response after photodynamic therapy increases anti-cancer and anti-bacterial effects. World J Immunol. 2014;4(1):1.

Riani YD, Matsuda T, Takemoto K, Nagai T. Green monomeric photosensitizing fluorescent protein for photo-inducible protein inactivation and cell ablation. BMC Biol. 2018;16(1):50.

Rifkin R, Reed B, Hetzel F, Chen K. Photodynamic therapy using SnET2 for basal cell nevus syndrome: a case report. Clin Ther. 1997;19(4):639–41.

Saji H, Song W, Furumoto K, Kato H, Engleman EG. Systemic antitumor effect of intratumoral injection of dendritic cells in combination with local photodynamic therapy. Clin Cancer Res. 2006;12(8):2568–74.

Sarkisyan KS, Zlobovskaya OA, Gorbachev DA, Bozhanova NG, Sharonov GV, Staroverov DB, Egorov ES, Ryabova AV, Solntsev KM, Mishin AS. KillerOrange, a genetically encoded photosensitizer activated by blue and green light. PLoS One. 2015;10(12):e0145287.

Schuitmaker J, Baas P, Van Leengoed H, Van Der Meulen F, Star W, van Zandwijk N. Photodynamic therapy: a promising new modality for the treatment of cancer. J Photochem Photobiol B Biol. 1996;34(1):3–12.

Schwartz S, Absolon K, Vermund H. Some relationships of porphyrins, X-rays and tumors. Univ Minn Med Bull. 1955;27:7–8.

Selman SH, Albrecht D, Keck RW, Brennan P, Kondo S. Studies of tin ethyl etiopurpurin photodynamic therapy of the canine prostate. J Urol. 2001;165(5):1795–801.

Serebrovskaya EO, Edelweiss EF, Stremovskiy OA, Lukyanov KA, Chudakov DM, Deyev SM. Targeting cancer cells by using an antireceptor antibody-photosensitizer fusion protein. Proc Natl Acad Sci. 2009;106(23):9221–5.

Serebrovskaya EO, Gorodnicheva TV, Ermakova GV, Solovieva EA, Sharonov GV, Zagaynova EV, Chudakov DM, Lukyanov S, Zaraisky AG, Lukyanov KA. Light-induced blockage of cell division with a chromatin-targeted phototoxic fluorescent protein. Biochem J. 2011;435(1):65–71.

Sharman WM, Allen CM, Van Lier JE. Photodynamic therapeutics: basic principles and clinical applications. Drug Discov Today. 1999;4(11):507–17.

Shibuya T, Tsujimoto Y. Deleterious effects of mitochondrial ROS generated by KillerRed photodynamic action in human cell lines and C. elegans. J Photochem Photobiol B Biol. 2012;117:1–12.

Shih J-Y, Yuan A, Chen JJ-W, Yang P-C. Tumor-associated macrophage: its role in cancer invasion and metastasis. J Cancer Mol. 2006;2(3):101–6.

Shirmanova MV, Serebrovskaya EO, Lukyanov KA, Snopova LB, Sirotkina MA, Prodanetz NN, Bugrova ML, Minakova EA, Turchin IV, Kamensky VA. Phototoxic effects of fluorescent protein KillerRed on tumor cells in mice. J Biophotonics. 2013;6(3):283–90.

Shu X, Lev-Ram V, Deerinck TJ, Qi Y, Ramko EB, Davidson MW, Jin Y, Ellisman MH, Tsien RY. A genetically encoded tag for correlated light and electron microscopy of intact cells, tissues, and organisms. PLoS Biol. 2011;9(4):e1001041.

Sokolov V, Stranadko E, Zharkova N, Iakubovskaia R, Filonenko E, Astrakhankina T. The photodynamic therapy of malignant tumors in basic sites with the preparations photohem and photosens (the results of 3 years of observations). Vopr Onkol. 1995;41(2):134–8.

Solinas G, Germano G, Mantovani A, Allavena P. Tumor-associated macrophages (TAM) as major players of the cancer-related inflammation. J Leukoc Biol. 2009;86(5):1065–73.

Song W, Kuang J, Li C-X, Zhang M, Zheng D, Zeng X, Liu C, Zhang X-Z. Enhanced immunotherapy based on photodynamic therapy for both primary and lung metastasis tumor eradication. ACS Nano. 2018;12(2):1978–89.

Stranadko E, Garbuzov M, Zenger V, Nasedkin A, Markichev N, Riabov M, Leskov I. Photodynamic therapy of recurrent and residual oropharyngeal and laryngeal tumors. Vestn Otorinolaringol. 2001;3:36–9.

Takemoto K, Matsuda T, Sakai N, Fu D, Noda M, Uchiyama S, Kotera I, Arai Y, Horiuchi M, Fukui K. SuperNova, a monomeric photosensitizing fluorescent protein for chromophore-assisted light inactivation. Sci Rep. 2013;3:2629.

Thudichum J. Tenth report of the medical officer of the privy council. London: HM Stationary Office; 1867.

Turubanova VD, Balalaeva IV, Mishchenko TA, Catanzaro E, Alzeibak R, Peskova NN, Efimova I, Bachert C, Mitroshina EV, Krysko O. Immunogenic cell death induced by a new photodynamic therapy based on photosens and photodithazine. J ImmunoTher Cancer. 2019;7(1):350.

Vatansever F, Hamblin MR. Photodynamic therapy and antitumor immune response. Cancer Immunol Springer. 2015:383–99.

Wachowska M, Muchowicz A, Demkow U. Immunological aspects of antitumor photodynamic therapy outcome. Central-Eur J Immunol. 2015;40(4):481.

Wang B, Van Veldhoven PP, Brees C, Rubio N, Nordgren M, Apanasets O, Kunze M, Baes M, Agostinis P, Fransen M. Mitochondria are targets for peroxisome-derived oxidative stress in cultured mammalian cells. Free Radic Biol Med. 2013;65:882–94.

Wang X, Ji J, Zhang H, Fan Z, Zhang L, Shi L, Zhou F, Chen WR, Wang H, Wang X. Stimulation of dendritic cells by DAMPs in ALA-PDT treated SCC tumor cells. Oncotarget. 2015;6(42):44688.

Wang D, Wang T, Liu J, Yu H, Jiao S, Feng B, Zhou F, Fu Y, Yin Q, Zhang P. Acid-activatable versatile micelleplexes for PD-L1 blockade-enhanced cancer photodynamic immunotherapy. Nano Lett. 2016;16(9):5503–13.

Warloe T, Peng Q, Heyerdahl H, Moan J, Steen HB, Giercksky K-E. Photodynamic therapy with 5-aminoolevulinic acid-induced porphyrins and DMSO/EDTA for basal cell carcinoma. In: 5th international photodynamic association biennial meeting, international society for optics and photonics; 1995. p. 226–235.

Wen AM, Lee KL, Cao P, Pangilinan K, Carpenter BL, Lam P, Veliz FA, Ghiladi RA, Advincula RC, Steinmetz NF. Utilizing viral nanoparticle/dendron hybrid conjugates in photodynamic therapy for dual delivery to macrophages and cancer cells. Bioconjug Chem. 2016;27(5):1227–35.

Wieman T, Panella T, Lustig R. Photodynamic therapy (PDT) of locally recurrent breast cancer (LRBC) with lutetium texaphyrin (Lutrin): a phase IB/IIA trial. Proc ASCO. 1999:111a.

Williams DC, El Bejjani R, Ramirez PM, Coakley S, Kim SA, Lee H, Wen Q, Samuel A, Lu H, Hilliard MA. Rapid and permanent neuronal inactivation in vivo via subcellular generation of reactive oxygen with the use of KillerRed. Cell Rep. 2013;5(2):553–63.

Wöhrle D, Hirth A, Bogdahn-Rai T, Schnurpfeil G, Shopova M. Photodynamic therapy of cancer: second and third generations of photosensitizers. Russ Chem Bull. 1998;47(5):807–16.

Wojtovich AP, Foster TH. Optogenetic control of ROS production. Redox Biol. 2014;2:368–76.

Xu S, Chisholm AD. Highly efficient optogenetic cell ablation in C. elegans using membrane-targeted miniSOG. Sci Rep. 2016;6:21271.

Yan L, Kanada M, Zhang J, Okazaki S, Terakawa S. Photodynamic treatment of tumor with Bacteria expressing KillerRed. PLoS One. 2015,10(7).

Yang Y, Hu Y, Wang H. Targeting antitumor immune response for enhancing the efficacy of photodynamic therapy of cancer: recent advances and future perspectives. Oxid Med Cell Longev. 2016.

Yang G, Xu L, Xu J, Zhang R, Song G, Chao Y, Feng L, Han F, Dong Z, Li B. Smart nanoreactors for pH-responsive tumor homing, mitochondria-targeting, and enhanced photodynamic-immunotherapy of cancer. Nano Lett. 2018;18(4):2475–84.

Zheng Y, Yin G, Le V, Zhang A, Chen S, Liang X, Liu J. Photodynamic-therapy activates immune response by disrupting immunity homeostasis of tumor cells, which generates vaccine for cancer therapy. Int J Biol Sci. 2016;12(1):120.

Zheng Y, Li Z, Chen H, Gao Y. Nanoparticle-based drug delivery systems for controllable photodynamic cancer therapy. Eur J Pharm Sci. 2020;105213.

Effects of the Transformation of Metallic Nanoparticles in the Environment and Its Toxicity on Aquatic and Terrestrial Life Forms

Suman Das, Debayan Ghosh, Kunal Kerkar, Manisha Tiwari, and Dhermendra K. Tiwari

Contents

Abstract Nanotechnology opens avenue in various research fields including physics, chemistry, material science, environmental sciences, and biological sciences. Various households and industrial items/equipments utilized nanomaterial-based system to improve performance and efficiency. Biological sciences also utilized nanomaterial in research and medical applications for exploration of biological system and treatment of diseases. These developments definitely improve human life and make daily life cost-effective, easier, and smooth. On the other hand, too much exposure of human and other living system increases risk and side effects of nanomaterial-based system and creates awareness among scientific community to explore long-lasting side effect. In this chapter, we discuss the sources and related side effects of nanomaterial exposure on biological system.

Authors Suman Das, Debayan Ghosh and Kunal Kerkar have equally contributed to this chapter.

S. Das · K. Kerkar · M. Tiwari · D. K. Tiwari (✉)
Department of Biotechnology, Goa University, Taleigao, Goa, India
e-mail: dktiwari@unigoa.ac.in

D. Ghosh
Centre for Nanobiotechnology, Vellore Institute of Technology, Vellore, Tamil Nadu, India

© The Author(s), under exclusive license to Springer Nature Switzerland AG 2021
K. K. Kesari, N. K. Jha (eds.), *Free Radical Biology and Environmental Toxicity*,
Molecular and Integrative Toxicology, https://doi.org/10.1007/978-3-030-83446-3_3

Keywords Nanoparticles · Metal toxicity · Nanomaterial exposures

1 Introduction

In the present era, nanotechnology is an expeditiously flourishing area and is considered one of the highest priority research areas because of its tremendous potentiality and economic impact. Nanomaterial is a substance that possesses at least one structural dimension in the nano range scale (10^{-9} of a meter). Nanoparticles can be categorized into several categories such as metals (Cu, Zn, Ag, Au, and Fe), metal oxides (SiO_2, TiO_2, Fe_3O_4, CeO_2, and Al_2O_3), non-metals (silica and quantum dots), polymeric (alginate, chitosan, and PLGA [Poly (d,l-lactic-co-glycolic acid)]), and carbons (CNT and fullerenes). A recent study has shown that there are more than 1600 nanoproducts in the market around the world. Due to their unique size-dependent, tempting physicochemical properties they are employed in a wide range of applications such as engineering, medicines, communications, textiles, agriculture, food industry, environmental applications, and several other sectors. It has attracted a greater interest and attention of the scientific community and has received tremendous appreciation because of its wide range of applications. Today, nanoparticles are found in almost every product such as cosmetics, paints, personal care products, electronic equipment, fabrication units, environmental solutions, food products, and various other purification techniques. According to a US estimate, it is estimated that by the end of 2020, the global production and use of the nanoparticles is expected to reach 58,000 tons per year (Turan et al. 2019).

In nature, everything has its advantages and disadvantages. Despite having such supreme characteristics and beneficial applications, nanomaterials end up entering into the environmental media causing environmental pollution and other detrimental effects. Due to their unique physical and chemical properties, in the environment, the behavior of nanoparticles, distribution, transport, fate, and toxicity to living organisms are quite distinct as compared to conventional xenobiotics. In recent years, the scientific community is substantially concerned about the adverse impacts of nanoparticle's transformation and toxicity to both living forms and environment (Bundschuh et al. 2018).

The chances of NP getting released into the various environmental matrices are exceedingly high with their increased production and application. Multiple sources are present in the environment from which NPs are released. There are mainly two primary sources, namely (a) point sources and (b) nonpoint sources. All the intentional or voluntary discharges of NP such as engineering, automobiles and industrial exhaust, mining, drug manufacturing, sewage treatment, discharge of industrial effluents in natural water bodies, and several other activities are categorized under point sources nanoparticles in the environment. Nonpoint sources consist of all the unintentional or incidental activities such as storms, dust, forest fire, and volcanic eruptions. Many nanoparticles enter into the surface water from various activities such as wastewater discharges, soil leaching, and atmospheric deposits which en routes NP into different water bodies. Surface water can also be invaded by groundwater carrying NPs and various other pollutants with it. Therefore, both the aquatic and terrestrial environments are susceptible to NP contamination as their input sources are diverse (Abbas et al. 2020).

The transportation, fate, and toxicity of the nanoparticles bank on the physical–chemical and biological properties of both the NP and the surrounding medium. NPs transform once they are released into the soil and aqueous media. Several transformation processes such as agglomeration/aggregation (homo or hetero), sedimentation and deposition, redox reactions, dissolution, sulfidation, adsorption, and photochemical and biological mediated reactions change the characteristics of the NPs, thereby altering their mobility and bioavailability. To study the negative aspects of NPs and how they behave in the environment, it is crucial to understand the dynamic transformation process of several NPs to determine its fate and bioavailability, which will help to study its various adverse effects toward flora and fauna in the environment (Auvinen et al. 2017).

Desalination, disinfection, and reuse of wastewater are emerging market segments with a considerable turnover. Wastewater treatment plants (WWTP) act as the major intermediaries for transporting nanoparticles into the aquatic ecosystem. High concentration of most widely employed metal oxide nanoparticles (MeO) such as TiO_2, ZnO, AgCl, CNT, and fullerenes are present at a concentration of 250–950, 31–200, 0.02–0.16, 0.21–0.40, and 0.09–0.34 mg/kg in a sewage treatment plant which ultimately seeps into various water bodies and pollutes the aquatic environment resulting in increased toxicity and death of marine organisms. Al_2O_3, CeO_2, CuO, ZnO, and TiO_2 are the primarily studied and most widely used metal MeO present in a WWTP and act as a threat to the environment. It is emblematic that various aquatic life forms uptake different NPs which are internalized by the process of endocytosis and phagocytosis. Cellular protein upon interaction with the NPs causes its structural conformation to change, which favors the uptake of NPs through specialized receptors. Accumulation of these NPs inside the cellular organelles (mitochondria, vesicles, Golgi bodies, etc.) exhibits noxious and harmful response, mostly lethal. NPs generate reactive oxygen species (ROS) as they possess the large surface area to volume ratio, one of the most leading causes of NP toxicity. NPs show different types of toxicity at different organismal levels. At the cellular level, phenomena such as cytotoxicity, oxidative stress, and antioxidant activity are seen. Fibrosis, inflammation, and necrosis are observed at the organ level. Bacteria and other marine microorganisms frequently uptake several NPs like SiO_2, CeO_2, TiO_2, and ZnO which cause damage to their cells. A fascinating insight is that the accumulation of NPs doesn't occur in microorganisms and doesn't show any toxic response. Thus, it is evident that, in the aquatic environment, the presence of NPs poses an intense threat to the biotic community and escalates the overall toxicity within the aquatic organisms (Zhou et al. 2015; Yang et al. 2013; Tso et al. 2010).

2   Sources of Nanoparticle in the Environment

2.1   *Natural and Anthropogenic Sources*

Nanoparticles are not any new particles but are existing from the time of formation of the earth. There are various sources from which multiple nanoparticles are released into the environment. A recent study has shown that approximately 8100 metric tons of nanoparticles are released into the atmosphere every year, globally (Keller and Lazareva 2014). They are released mainly from two types of sources, point and nonpoint sources (Lin et al. 2010). Water treatment plant, industries, automobiles and transport, landfill site, garbage incineration, forest fire, etc. are placed under point sources (Gottschalk and Nowack 2011). Nonpoint sources of nanoparticles include painting, textile, dyeing, cleaning agents, personal hygiene products, weathering, and abrasion (Peng et al. 2017). Release of nanoparticles into the environment can be further categorized into natural and anthropogenic/intentional sources. Phenomena such as dust storm, forest fire, volcanic reduction, and soil erosion are naturally occurring that release nanoparticles in the environment. Contrastingly, activity such as drug manufacturing, energy production, environmental bioremediation, waste management, biomedical imaging, mining, automobile exhaust, burning of fossil fuels, construction, and demolition are the anthropogenic or intentional sources of various nanoparticles in the environment (Singh 2015). Examples of nanoparticle applications, which results in its international entry into the environment, are mentioned below.

2.1.1   Wastewater

Previous studies have shown that new word metallic nanoparticles and their oxides have been found in the urban and industrial sewage water. Engineered nanoparticles (ENPs) such as AgCl and Ag Np have been detected in the wastewater discharged from a nearby laundry station (Benn and Westerhoff 2008). It is reported that adsorption of the nanoparticle to other macro and micropollutants and other nanoparticles may result in the formation of agglomerates that exhibit increased or decreased reactivity (Kim et al. 2010).

2.1.2   Personal Care Products

ZnO, TiO_2, and Ag are the most frequently used nanoparticles in the personal care products such as mouthwash, cosmetics, and sunscreen. ZnO and TiO_2 in sunscreen act as filters that protect the skin from the sun's harmful ultraviolet rays. These particles are often coated with inert materials such as magnesium, silica, alumina, zinc, and zirconium, the lifetime of which is mostly unknown (Botta et al. 2011).

2.1.3 Anti-fouling Applications

Carbon nanotubes (CNT) recently have been widely used to prevent biofouling. However, in the aquatic ecosystem, the interaction of other toxic compounds with the nanoparticle intensifies its toxicity. The use of tributyltin (TBT), which is adsorbed on TiO_2 and used for 20 years in the anti-fouling paints, leads to the formation of toxic TiO_2–TBT complexes that have devastating effect toward the marine ecosystem (Upadhyayula and Gadhamshetty 2010).

3 Physicochemical Properties of Nanoparticles

In fact, physicochemical properties of nanoparticles are essential to understand the transformation, fate, and toxicity of nanoparticles in the aquatic and terrestrial environments. Each nanoparticle has different physicochemical properties that play a crucial part in controlling the nanoparticle's character. The bulk counterpart usually affects the nanoparticle's unique properties that either confer them with a beneficial character or affect the nanoparticle's toxicity. Properties such as particle size and shape and surface characteristics such as surface charge, area, and coating of various materials constitute the nanoparticles (Fig. 1) (Lead et al. 2018; Gatoo et al. 2014).

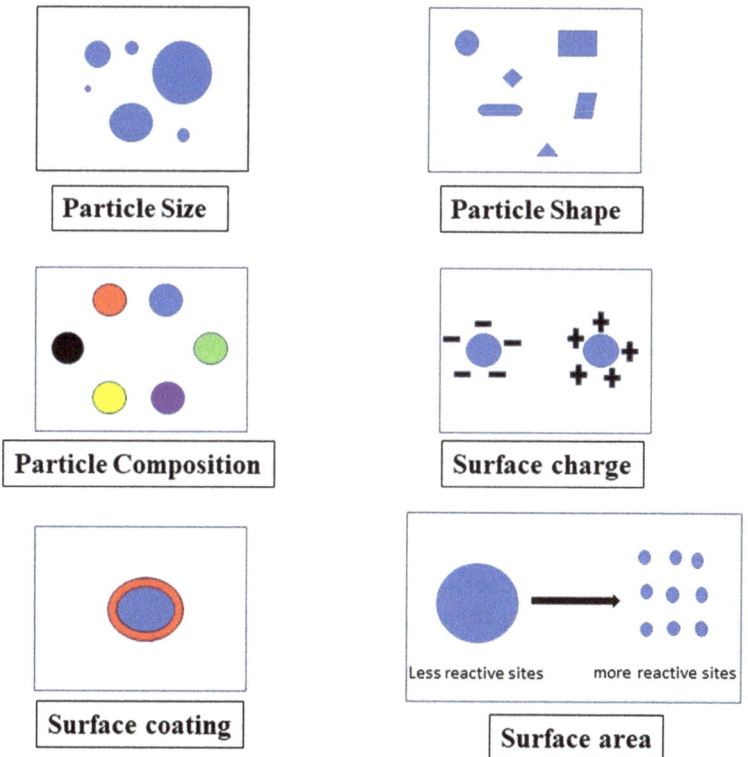

Fig. 1 Physicochemical properties of nanoparticles

3.1 Particle Size

Particle size is one of the critical characteristics that affect the nanoparticle's transformation and fate in the environment. Furthermore, it also involves their interaction with the biological system. As we know the size of a nanoparticle varies between 1 and 100 nanometer, smaller sized particles may remain suspended in the environmental matrices such as water or air as it prevents their gravitational settlement. Because of their small size, the force of gravity doesn't influence them to settle down. Therefore, they can settle down only when the size increases by the process aggregation and agglomeration with other particles. Additionally, particle size also affects various cellular mechanisms such as center uptake pathway and efficiency of the particle (Aillon et al. 2009).

3.2 Surface Area

As the particle size decreases, the surface area increases, which proves to be advantageous for the nanoparticle. A large surface area corresponds to a more extensive reactivity of the particle that affects the nanoparticle's transportation and fate in the environment (Powers et al. 2007). Previous studies have shown that damaging abilities and oxidation potential are higher in small-sized nanoparticles as they have higher surface area than the larger particles. It has been seen that particles with a size smaller than 50 nanometers are essentially toxic for the living organisms' tissue system. In contrast, reticuloendothelial system (RES) takes the larger particles and shifts their toxicity to liver and spleen to protect and safeguard other vital tissues and organs (De Jong et al. 2008). Nanoparticles with a size smaller than 10 nanometer deposits in the tracheobronchial region whereas particle having a diameter more than 100-millimetre deposits all over the issue system (Asgharian and Price 2007).

3.3 Surface Charge and Agglomeration

Zeta potential is the parameter by which a nanoparticle's surface charge is represented. The formation of charges occurs on the outermost part of the stern layer in a colloid (Sellers et al. 2008). Surface charge influences the nanoparticle's stability, which in turn governs the rate of aggregation and toxicity (Gatoo et al. 2014). It has been observed that positively charged nanoparticle exhibits better interaction with the biological systems, namely transmembrane permeability, selective adsorption, and integrity with the blood–brain barrier (Georgieva et al. 2011; Hoshino et al. 2004). Aggregation of nanoparticles is influenced by the surface charge alone but particle size and composition also play a significant role in it. Thus, it is inferred that as the concentration of nanoparticle increases the toxicity of the same decreases.

3.4   Particle Morphology

In nature, nanoparticle exists in different shapes such as spherical, triangles, cubes, rods, stars, and others. It is an essential aspect since it influences various cellular processes such as the membrane wrapping process during endocytosis or phagocytosis (Gatoo et al. 2014). The spherical nanoparticles are faster and easier to endocytose and are far less toxic than other nanoparticle forms (Lee et al. 2007).

3.5   Surface Coating

Surface coating plays a vital role in governing the nanoparticle's fate and transport in the environment as they considerably change the physicochemical properties of nanoparticle such as chemical reactivity, which in turn affects the cytotoxic properties. According to the type and nature of the surface coating used, it may induce toxicity or eliminate them based on the coating's chemical nature (Yin et al. 2005). To understand the toxicity induction by surface coating, let us take the example of silica nanoparticles known to generate reactive oxygen species (ROS) on their surface, which induces cytotoxic effects. Contrastingly if the nanoparticle is coated with hydrophilic polyethylene glycol (PEG) and surfactants, the nanoparticle's toxicity is mitigated, and the nanoparticle is stabilized in the biological system (Risom et al. 2005).

3.6   Composition of the Particle

The composition of the nanoparticle plays a role in inducing the toxicity. Griffit et al. studied the effect of nano-Ag, nano-Cu, and TiO_2 of same dimensions on aquatic organisms such as algal species and zebrafish where they found that TiO_2 didn't exhibit any toxicity. In contrast, nano-Ag and nano-Cu revealed toxicity to marine organisms (Griffitt et al. 2008).

4   Transformation of Nanoparticles in Various Environment

From the previous studies, it is well established that once the nanoparticle is released into the environment, it will inescapably undergo a series of transformation processes that comprise physical, chemical, and biological transformation in various environmental systems such as air, water, and soil (Lowry et al. 2012a; Nowack et al. 2012). The nanoparticle's transformation differs based on its nature and several other environmental factors such as pH, ionic strength, water chemistry, and presence of natural organic matter (NOM). NOM frequently creates a natural coating on the nanoparticle's surface, thereby altering their surface properties, which will affect their transformation and fate in the aqueous environment (Biswas and Sarkar 2019).

4.1 Physical Transformation

Physical transformation is an important process that can occur at all the nanoparticle's life cycle stages. It mainly occurs in three processes: (1) aggregation/agglomerations, (2) adsorption, and (3) deposition (Fig. 2) (Peijnenburg et al. 2015).

4.1.1 Aggregation/Agglomeration

When tiny nanoparticles come closer to each other in the aqueous environment, formation of a cluster occurs in which the particles are held together by Vander wall force of interaction, and the phenomenon of such cluster formation is called aggregation. The terms aggregation and agglomeration are interchangeably used in colloidal science. It isn't easy to distinguish between them as they hold the same meaning. Although, aggregation is a non-reversible process in which strong ionic forces and covalent bonds are responsible for the particles' adherence to one another, resulting in an increased cluster of particles. There are two major pathways by

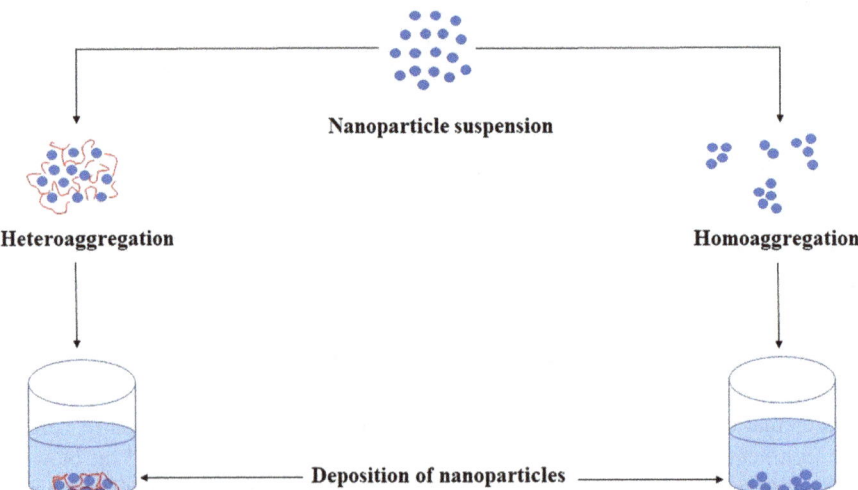

PHYSICAL TRANSFORMATION

Nanoparticle suspension

Heteroaggregation

Homoaggregation

Deposition of nanoparticles

Fig. 2 Physical transformation of MNPs

which aggregation mainly occurs in the environment, homo-aggregation (aggregation of similarly sized particles) and hetero-aggregation (aggregation of particles of dissimilar size) (Lei et al. 2018; Lowry et al. 2012a). In the environmental matrix, as the nanoparticles are present in trace amount, hetero-aggregation is predominant due to natural colloidal substances. Due to aggregation, the surface area to volume ratio of the nanoparticle reduces, which affects the reactivity. The increase in the nanoparticle aggregate size affects various phenomena such as reactivity, uptake by living organisms, transportation in the porous media, and sedimentation. According to the DLVO theory, aggregation occurs with the combined effect of Vander walls force of attraction and electrostatic repulsion due to the formation of the counter ions' electrical double layer. Several other interactions which are not DLVO theory interactions contribute significantly in the aggregation process such as magnetic interaction, hydrophobic interactions, and hydration force (Dwivedi et al. 2015; Adamczyk and Weroński 1999).

Aggregation leads to a decrease in the available surface area of the nanoparticle, thus lowering its reactivity. For example, TiO_2 (216 nm), Fe_2O_3 (40 nm), and Ag (60 nm) clustered together and form agglomerates of micrometer size, when they were released into different environmental waters, lowers their mobility and exposure to aquatic receptors (Chekli et al. 2015). Thus, the toxicity of nanoparticle decreases due to aggregation as it results from surface area mediated reaction such as dissolution or generation of reactive oxygen species (ROS). In a suspension, aggregation also decreases the concentration of nanoparticles and causes and increases the aggregate size through the time, which usually favors the process of deposition. Monex et al. showed that ZnO (21–34 nm) and CuO (1–100 nm) nanoparticle rapidly aggregated in the aquatic environment and led to the formation of larger sized aggregates of 250–800 nm and >810 nm respectively (Mouneyrac et al. 2014).

Aggregation significantly affects the bioavailability and persistence of nanoparticle in the environment. If aggregation decreases the rate of degradation or dissolution, the persistence of nanoparticles increases, although at a different location as compared to the dispersion nanoparticles. It affects the bioavailability such that when there is a formation of large size aggregates or hetero-aggregates, uptake or direct transport across the cell membrane and/or wall is hindered and thereby mechanisms such as phagocytosis may also be affected. On the other hand, when the hetero-aggregates are composed of soft biogenic particles, it increases the nanoparticles' bioavailability. Hu et al. studied that in the cells of *Isochrysis galbana*, a microalga, the agglomeration process directly reduced the accumulation and toxicity of TiO_2 nanoparticles by deposition of the particles. TiO_2 nanoparticles, in a recent study, agglomerated on the surface of *Microcystis aeruginosa,* a freshwater cyanobacteria causing a shading effect resulting in an inhibited algal growth when exposed to TiO_2 concentration of 50–100 mg/L (Wu et al. 2019; Hu et al. 2018).

Aggregation/agglomerations depend on the nature, type, and physicochemical behavior of the nanoparticles and several suspension medium conditions. Zhou and Keller investigated the aggregation kinetics of spherical and nano rod-shaped ZnO nanoparticles. They stated that aggregation kinetics depends on the size and shape and the solution chemistry (pH, ionic strength, and NOM). This may be due to the diffusion coefficient that is the effectiveness of the presence and attachment of different crystal phase with the shape (Zhou et al. 2012). In the suspension, the nanoparticle's initial concentration plays a pivotal role in the aggregation process because at higher initial concentration, the collision frequency is enhanced, leading to the rapid formation of agglomerates. By the formation of larger aggregates, the particles reduce their surface energy; thus, it is a thermodynamically driven process. It has been confirmed by several studies where faster aggregation rate was observed at the higher initial concentration for ZnO, TiO_2, and CeO_2. Therefore, as the concentration of nanoparticle increases, aggregates of larger size are expected. The surface coating of the nanoparticle and the nature of its coating material alter the aggregation behavior (Amde et al. 2017). Marie et al. found that CeO_2 (size: 3–4 nm; concentration: 10 mg/L) aggregated within few minutes unlike that of citrate coated CeO_2 nanoparticle which took much longer time of 3 days to obtain an aggregate of lesser size. Surface functionalization of metal oxide nanoparticle increases the significant effect on their behaviors of aggregation and deposition (Marie et al. 2014). Vikesland et al. reported that Fe_2O_3 (size: 10 nm) that has been functionalized by tetramethylammonium hydroxide aggregated more quickly than non-functionalized Fe_2O_3 particle. Similarly, when the surface of TiO_2 nanoparticle was coated with dodecenyl succinic anhydride and polyacrylic acid, lower aggregation of TiO_2 nanoparticle was observed when suspended in the aqueous medium (Elbasuney 2017; Vikesland et al. 2016).

In the aquatic environment, stability and aggregation of the nanoparticle depend on the ionic strength to a large extent. Consequently, there is a compression in the electric double layer and reduction in the electrostatic repulsion when there is high ionic strength in the surrounding medium, thereby increasing the aggregation process. Yung et al. reported that the aggregation of ZnO nanoparticle (20 nm) occurs by the influence of salinity. An increase in the aggregation size was observed (0.84–146 μm) as the ionic strength of the solution increases 12–22 PSU (Yung et al. 2015). A recent study showed that at high ionic strength, higher agglomeration of zinc oxide nanoparticle was observed (size: 130 nm) (Odzak et al. 2017). Higher aggregation of TiO_2 was observed in the presence of 45 ppm of Ca^{2+} and 5 ppm of Mg^{2+} ions. Quik et al. 2014 reported that high ionic strength exacerbated homo- and hetero-aggregation of CeO_2 nanoparticle (RD 175 nm) was observed in water (Quik et al. 2014).

pH of the solution also plays a vital role in determining the aggregation behavior, since pH affects the surface charge particle, which strongly influences their aggregation behavior. At pH point of zero charges (pH_{pzc}), when the pH of the colloidal solution becomes near the isoelectric point or point of zero charges, high homo- or hetero-aggregation is expected. As the colloidal system has minimum stability at this point as the net effect of the nanoparticles' surface charge and their mobility is zero (Zhou and Keller 2010). CuO nanoparticle having pH_{pzc} = 2.1 clustered rapidly into microscale agglomerates in the aqueous system. Higher aggregation of TiO_2 nanoparticle was observed at pH_{pzc} = 6.2 and the suspension is stable at pH <5 and >8. The agglomeration of GO NP at low pH with zeta potential of −4.25 mV where active agglomeration of graphene oxide nanoparticle was observed. Typically, nanoparticles, whose zeta potential values other than that of −30 mV to +30 mV, are considered stable in the colloidal system (Loosli et al. 2015). Another factor that plays a role in aggregation in the aqueous environment is natural organic matter (NOM), and other collide. NOM is defined as the large heterogeneous pool of various, naturally occurring organic compounds which includes macromolecules such as humic acid, fluvic acid, and EPS and lower molecular weight carbon-based compounds such as carboxylic acids, amines, and thiols that are frequently found in the aquatic and terrestrial environment which are generated from the dead and decay of microbes, plants, and animal residues in the environment. NOM usually gets adsorbed on nanoparticles' surface due to their physicochemical characteristics, which favors them to do so. Thus, it significantly influences the agglomeration, stability, and deagglomeration of nanoparticles. The surface chemistry in terms of charge, structure, and surface potential of the nanoparticle changes upon NOM coating which decreases the agglomerations and increases the stability of the particles in the suspension. Moreover, aggregations of nanoparticle are reduced and stability is enhanced in the presence of NOM (30 mg of C/L), which is an environmentally relevant concentration. High concentration of humic acid and alginate of Suwannee River induced deagglomerations of already agglomerated TiO_2, ZnO, and FeO nanoparticles in the natural environmental. The overall gross function of NOM in the process of agglomeration of nanoparticles depends on the inherent properties of the NOM such as age, origin, and natural environmental conditions (Amde et al. 2017).

4.1.2 Deposition

In the aqueous environment, when the aggregates from a solution settle down on the bottom surface, the phenomenon is called deposition. Aggregation/agglomeration favors the process of deposition in the aqueous environment. The physicochemical properties of nanoparticles and the environmental conditions, as discussed earlier, also influence the deposition process. It is also affected by the type of nanoparticles as the extent of aggregation varies with their nature. For example, in natural water CeO_2 undergoes greater than 95% of deposition within 7 days compared to ZnO (Van Koetsem et al. 2015). Aggregation increases the particle flock size in the aqueous environment, and therefore bigger particles settle down under the force of gravitation favoring the process of deposition. For estimation of the sedimentation rate, an equation has been derived as follows (Quik et al. 2014):

$$C_t = \left(C_0 - C_{ns}\right)e^{-(V_s/h+k_{dis})t} + C_{ns}$$

where,

C_{ns} = non-settling concentration [g/L],
V_s = sedimentation rate [m $^{d-1}$],
h = sedimentation length [m],
k_{dis} = dissolution rate
t = time.

Usually, it has been seen that the sedimentation process is generally faster in seawater rather than other types of natural water bodies such as freshwater, groundwater, and lake water. The sedimentation rate, along with aggregation, also varies with different kinds of natural water bodies. The similar sedimentation rate is exhibited by ZnO, NiO, and nZVI for freshwater, groundwater, and seawater. CeO_2, Fe_2O_3, and Fe_3O_4 in freshwater exhibit higher sedimentation rates than aggregation time. The phenomenon of deposition in the aqueous environment is an essential parameter as it affects the nanoparticles' overall transportation process by lowering it, which increases the chances of uptake by various benthic life forms. Thus, these different physical processes can influence the overall transformation, fate, persistence, and bioavailability of nanoparticles in the aqueous environment and porous media (Abbas et al. 2020).

4.2   Chemical Transformation

In the environment, nanoparticle undergoes various chemical transformation owing to their several active surface entities such as dissolution, redox reaction, complexation, adsorption, degradation, and photochemical reactions in soil and water and release of toxic ions (Fig. 3).

4.2.1   Dissolution and Release of Toxic Ions

This is a type of transformation process of the nanoparticle that affects several properties such as persistence, fate, reactivity, toxicity, and bioavailability. The dissolution of nanoparticle, like other processes, depends on the chemistry of the

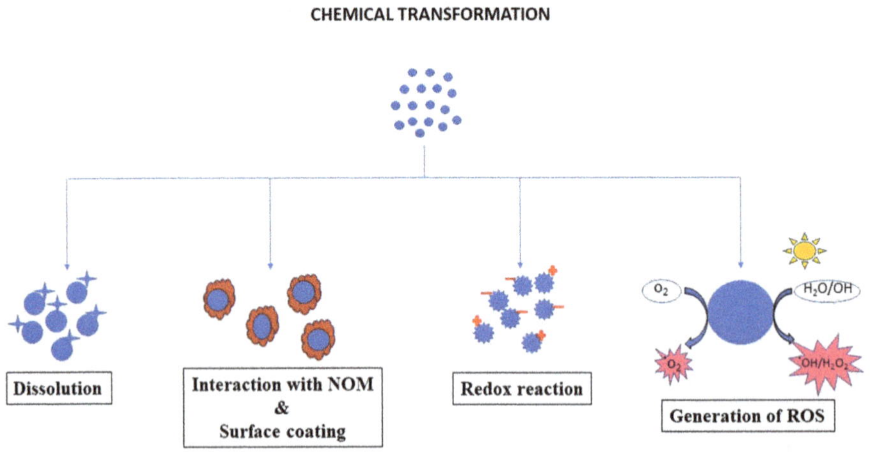

Fig. 3 Chemical transformation of MNPs

environment and the physicochemical properties. It is one of the vital chemical transformation processes that occur in the environment. It has been reported that dissolution occurs between a range of 1% and 80% in different environmental matrices (Tangaa et al. 2016; Lowry et al. 2012b).

The morphology and size of the particle's effects significantly in their dissolution property. Smaller particles possess higher surface area than the larger ones due to the difference within the surface area to volume ratio. For example, ZnO nanoparticles of size 4–130 nm at pH 7.5, the smaller sized particles show faster dissolution than the larger ones, which is good agreement with that of other reports (Mudunkotuwa et al. 2012). Another instance where Fe_2O_3 size 8 nm exhibited higher dissolution (3- to 10-folds) than that of the same larger 40 nm particles (Lanzl et al. 2012). The usage of stabilizing agents, such as citrate and other organic coatings, can drastically reduce dissolution. For example, uncoated ZnO exhibits maximum dissolution in the first hour, whereas zinc oxide nanoparticle when coated with organic substance exhibits slower dissolution rates where maximum dissolution is obtained in 7 days and stability is drastically modified (Gelabert et al. 2014). Based on the dissolution properties, nanoparticles can be categorized into highly soluble such as FeO, CuO, QDs, Ag, and ZnO; poorly soluble such as TiO_2 and CeO_2, and insoluble such as graphene, CNT, fullerenes, and carbon dots (Tangaa et al. 2016). The rate of dissolution and its extent depend upon nanoparticles' intrinsic properties such as size crystalline phase, chemical composition, shape and morphology, and surface coating. It is also dependent on the environmental media constituents such as temperature, pH, ionic strength, inorganic ligands, and natural organic matter (Amde et al. 2017). It has been seen that as the value of pH increased, dissolution of ZnO nanoparticle decreases as the value of the pH increase whereas the dissolution of the ZnO nanoparticle was observed 2 decrease. The higher dissolution rate was observed for ZnO nanoparticle of <130 nanometers at lower pH (<6.5) which is in good agreement with other studies (Odzak et al. 2017).

The ionic strength of the solution also affects the dissolution of nanoparticles. At low ionic strength, a rapid and high dissolution rate of the nanoparticle is expected. It has been seen that when the ionic strength of the solution increases by a range of 12–22 PSU, the dissolution of the ZnO nanoparticle and subsequent release of ions decreased. It is reported that the dissolution of the CeO_2 in seawater is insignificant (Buffet et al. 2011).

The presence of NOM in the environment affects the nanoparticle's dissolution as there is an occurrence of increase complexation free metal ions with NOM effects the process of dissolution. When the concentration of natural organic matter is between 0 and 40 ppm, an increased rate of dissolution is observed for zinc oxide nanoparticles; the aromatic carbon of the NOM plays a crucial role in determining the dissolution of the nanoparticles (Wang et al. 2016).

Other factors such as temperature and light, even though rarely studied, also affect the nanoparticles' dissolution property in the environment. ZnO NP shows temperature-dependent dissolution. At higher temperature (35 °C), a higher rate of aggregation and lower dissolution is observed than that of temperature range between 15 and 25 °C. Similarly, ZnO NP suspension at a concentration of 100 ppm exhibited higher dissolution at 20 °C than at 37 °C. The UV rays from the sun impact the dissolution behavior of ZnO NP. In the presence of UV (300–400 nm) and visible light (400–700 nm), the dissolution of ZnO NP is facilitated because these rays affect the nanoparticle's surface properties and coatings (Odzak et al. 2017; Ma et al. 2014).

The presence of ions and cheating agents in the aqueous environment affects the dissolution of metal oxide nanoparticles. The dissolution of ZnO NP of size 40 nm reduces drastically in the aqueous environment in low amounts of phosphate ions. Similarly, in the presence of 10 eq/L of SO_4^{2-} in the system, ZnO NP's high stability is observed because of inverted surface potential or surface neutralization. Extracellular polymeric substances (EPS) due to their excellent metal-binding capabilities and abundant functional groups, the presence of which in the aqueous environment can increase the dissolution of copper-based nanoparticles such as CuO (Peng et al. 2017).

Due to the process of dissolution, several toxic ions are released into the environment such as Ag NP undergoing oxidative dissolution in surface water and well-aerated soil due to which AG^0 converts into toxic AG^+ ions (Dwivedi et al. 2015). Similarly, Au nanoparticles are highly unstable and under ambient conditions, oxidative dissolution occurs to produce Au^{3+} ions which are highly toxic in nature (Unrine et al. 2010). Thus, dissolution of nanoparticles in various environmental matrices releases various types of toxic ions which have a very detrimental effect on various life forms.

4.2.2 Redox Reactions

This is a coupled process of oxidation–reduction reactions in the environment in which transfer of electron in the form of loss or gain occurs in between chemical moieties. The oxidizing agent, reducing agent, and pH are among the environmental conditions that govern redox reactions (Nowack et al. 2012). Metallic nanoparticles such as NZVI, Cu, Ag, and FeO are very much prone to redox reactions (Mitrano et al. 2016). As a major constituent of QDs, sulfur and selenium undergo oxidations and result in the release of harmful ions, which is cytotoxic in nature (Derfus et al. 2004). NZVI, which is highly prone to redox reactions, generates ROS, which is cytotoxic to various organisms in the environment. Over time, oxidation occurs on the surface of NZVI particle which transforms it into different iron oxides. As a result of redox activity, agglomeration, sedimentation, and biotoxicity are minimalized in the environment (Phenrat et al. 2009).

The coating substances play a significant role in the redox reactions of nanoparticles. In contrast to the uncoated counterparts, the silica-coated Fe_3O_4 NP exhibited strong stability toward the oxidation reactions. The redox reactions of the metal oxide nanoparticles are facilitated by various types of substances present in the environment, for instance, CeO_2 which possesses both Ce (III) and Ce (IV) and undergoes oxidation reaction upon interaction with macromolecules such as biological media and humic acid, which results in the altered ratio of Ce (III) and Ce (IV) ions on the surface of CeO_2 nanoparticles (Collin et al. 2014). In the aqueous environment, many cellular proteins are also involved in redox reactions. NiO NP was reduced to zero-valent Ni in the presence of soluble cellular proteins. In natural water bodies and well-aerated soils, oxidation acts as the dominant phenomenon whereas the process of reduction is dominant in oxygen-depleted zones such as groundwater and carbon-rich sediments. In some instances, an insoluble coating is formed on the surface of nanoparticle as a result of a redox reaction, which enhances its stability in the environment. On the other hand, the nanoparticle's stability may also decrease as the formation of a new layer on the surface of the particle may facilitate the interaction with other environmental components, thereby reducing the stability. Thus, redox reaction substantially affects the transformation, fate, and toxicity of metallic nanoparticles in the environment (Amde et al. 2017).

4.3   Biological Transformation

It is an inevitable transformation process as there is a constant interaction between nanoparticles with the living organisms. Biodegradation, formation of nanoparticles facilitated by macromolecules, and eco/bio corona and nano-bio interactions are the three main biologically mediated transformation processes. Different processes, such as interaction with macromolecules, changes related to surface, and core or redox reactions, facilitate the uptake of these nanoparticles by the living organisms, resulting in the biological transformation that will change the aggregation properties, surface charge, reactivity, and toxicity of the nanoparticles (Fig. 4).

4.3.1   Formation of Nanoparticles Facilitated by Macromolecules

In the environment, diverse group of flora and fauna produces and secrets macromolecules or biomolecules (Xu 2011). These biomolecules obtained from the tissue extracts contain various reductive enzymes and functional groups that react with the metal ions and produce engineered nanoparticles. For example, pomegranate peel extract has been used for the synthesis of bimetallic Fe–Ni nanoparticles. The phytochemicals present in the pomegranate peel extract act as a reducing agent and result in green synthesized bimetallic nanoparticles (Ravikumar et al. 2019). Ascorbic acid ($C_6H_8O_6$) possesses the ability to transform metal ions to synthesize corresponding nanoparticles (Adil et al. 2015). Biochemical extract of various microbes and plant species can synthesize nanoparticle from metal salts or ions. They possess multiple enzymes containing various phenolic or hydroxyl functionalities and other metabolites such as reducing sugars and terpenoids (Oza et al. 2020). These enzymes act as a reducing agent, singlet oxygen quencher ($^1O_2$), hydrogen donor, and metal chelator during the formation of nanoparticles in the environment (Iravani 2011). The silver nanoparticle has been synthesized from the crude methanolic extract of *D. pinnaculata* by reducing silver acetate (Rao et al. 2016). Au nanoparticles can be prepared from Ziziphus leaf extract. Green synthesis of Fe_2O_3 NP has been done using the bottle brush flower extract (Aljabali et al. 2018).

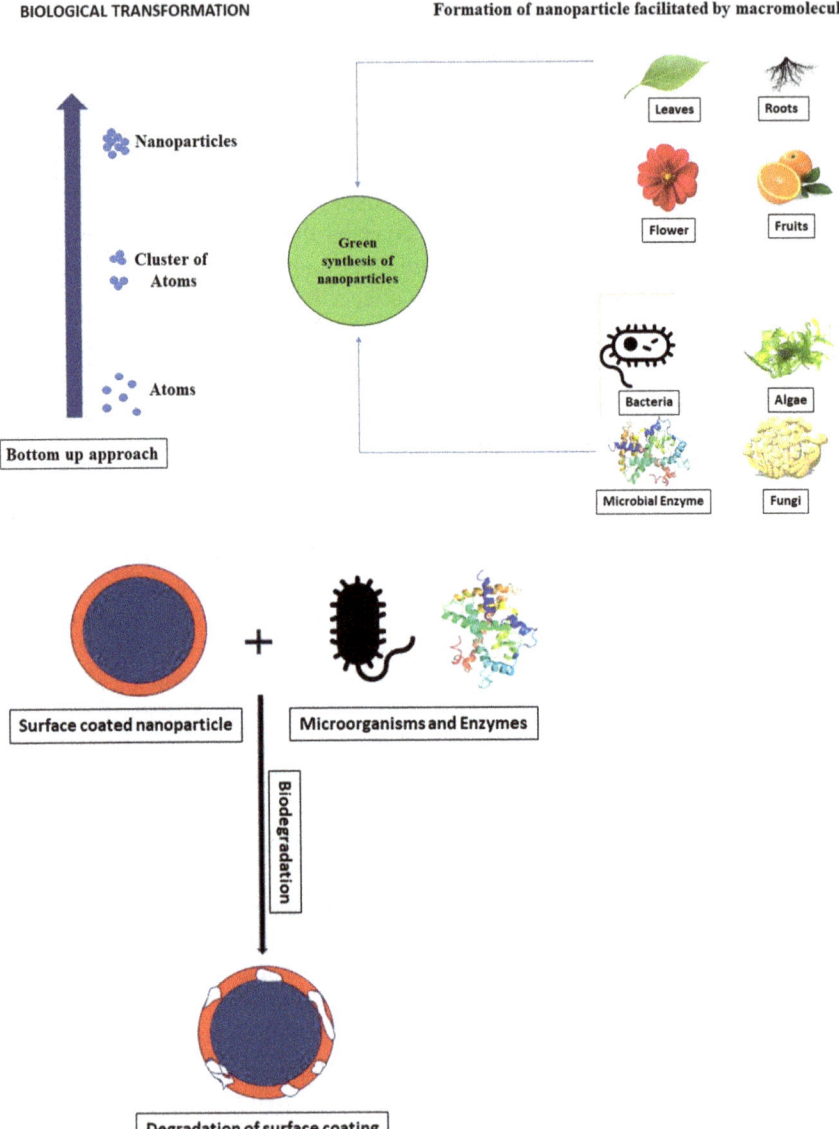

Fig. 4 Biological transformation of nanomaterials

Several microorganisms such as algae, bacteria, and fungi can transform metal ions into their corresponding nanoparticles by changing the oxidation state of the metallic element. *Geobacter sulfadurans* and *E. coli* can effectively reduce Pd (II) to Pd (0) and Ag, respectively (Kang et al. 2014; Yates et al. 2013). *Fusarium oxysporum* can synthesize CdS NP by converting Cd^{2+} and SO_4^{2-} ions with subsequent release of enzyme, sulfate reductase (Ahmad et al. 2002). *Fusarium solani* has been used to synthesize Au NP by the reduction of Au (III) to Au (0) (Gopinath and Arumugam 2014). *S. wegeti* and *S. mutans*, members of algae, are able to synthesize Au and ZnO NP (Azizi et al. 2014).

4.3.2   Biodegradation Process of Nanoparticles

The surface coating or the core of the nanoparticle can be biologically degraded. Carbon-based materials fullerenes and CNT are particularly relevant for this process. Hydroxylated fullerol can be enzymatically graded by the white-rot basidiomycetes fungi by the process of oxidation and is mineralized to CO_2 in 32 weeks (Schreiner et al. 2009). Another photo enzyme, horse radish peroxidase (HRP) present in the rhizosphere zone of horseradish, has been employed for the degradation of MWCNT (Zhao et al. 2011). The naturally occurring manganese peroxidase has biodegraded the close cage structure of SWCNT through the malonate mediated pathway or Mn^{3+} oxidation process (Zhang et al. 2014). Neutrophil nucleoperoxidase, an enzyme generated in human, can perform biological degradation of graphene oxide NP which results in the generation of graphene degraded product that is non-cytotoxic for human lungs (Mukherjee et al. 2018). As a matrix material, a large number of organic polymers are used for the surface coating of the nanoparticle to achieve the desired properties. The mechanism of microbial mediated degradation degrades these polymer coatings. Thus, after biodegradation of coating material, transformation and fate of the nanoparticle depend upon the physical and chemical properties of bare nanoparticle. However, very less studies have been done about the potential of microbes in the process of biodegradation of organic polymer coatings present on the surface of different engineered nanomaterials.

4.3.3   Eco/-Corona and Nano-Bio Interaction

The term corona refers to the layer formed on the nanoparticle's surface as a result of adsorption of biological and environmental constituents having a molecular weight between 10 to 2,000,000 Da (Nasser et al. 2020). From the simplest microbes to the complex plants and animals release various biomolecules in the environment which possess complex chemical structure and are composed of lipids, carbohydrates, proteins, and nucleic acid. Eco-corona is the interaction of the aforementioned biomolecules with that of the nanoparticles, which again depends on the physicochemical properties of the nanoparticles. Corona can be further categorized into two types, hard and soft corona. In hard corona, the nanoparticle interacts strongly with the particles' surface and loosely bound molecules of the outer layer. Whereas, weak corona are those in which surface of the particle is weakly associated with the biomolecules' rapidly exchanging layer. The aggregation of nanoparticles due to eco-corona formation occurs via different types of mechanisms, such as divalent cation bridging, inter-particle bridging, and localized charge (Pulido-Reyes et al. 2017; Lundqvist et al. 2011).

The eco-corona helps provide a biological identity that distinguishes the nanoparticle from the synthetic one as it modifies the size and the interfacial composition (Bergese and Colombo 2014). It is known that surface of the NP when coated with various natural biomolecules, the stability of the particle is altered which ultimately changes the level of toxicity to various receptors. Biomolecules such as EPS, proteins, and NOM alleviate NP's toxicity toward algae, zebrafish, and plants, thereby playing a protective role (Huang et al. 2020; Kansara et al. 2020; Zhou et al. 2016). EPS proteins of 20–80 kDa curtailed the toxicity of TiO_2 ENPs toward green alga *Dunaliella tertiolecta*. EPS derived from two algal species, *Chlamydomonas reinhardtii* and *Dunaliella tertiolecta*, created a surface coating that enhances the stability of nano-TiO_2, thereby increasing their persistence in the environment (Adeleye and Keller 2016). In some instances, eco-corona changes the aggregation and dissolution behavior that adds changes to the toxicity of the NP, thereby making them more alluring food source for various organisms. The accumulation and toxicity of polystyrene NPs are increased upon interaction with biomolecules secreted from *Daphnia magna* (Nasser et al. 2020).

In the biological system, the bio-corona is analogous to eco-corona in the ecological environment. Nanoparticle upon entering a complicated biological system interacts with all the available biological macromolecules and results in "bio-corona" formation (Neagu et al. 2017). The proteins present in the bio-corona govern all the interaction between the NP and internalization pathways, uptake by specific cellular receptors, and the immune response at the interface of the ENMs surface and the surrounding biological fluids (biofluids) or bio-structures. The interaction that occurs at the interface of the NP's surface and the surrounding biological fluids or bio-structures is called the nano-bio-interactions. Interactions such as adsorption of biomolecules, dissolution, surface restructuring to minimize free energy, and redox reactions are among the most critical and active interactions. Driving forces like hydrophobic, electrostatic, hydrogen bond, molecular recognition, metal-coordinate, and stereoselective interactions shape the boundary of different nano-bio-interfaces (Wang et al. 2019).

5 Toxic Effects of Nanomaterials on Aquatic and Terrestrial Life Forms

Metallic nanomaterials are most commonly and widely used engineered nanomaterial, whose effect in the environment is hardly studied upon. In order to understand the effect of nanomaterials, it is important to determine which species is highly susceptible to that particular nanomaterial in order to understand the overall effect of its toxicity in that particular environment. Nanomaterials have a wide number of applications in terms of available commercial products such as electronic gadgets, biomedicals, injectables, sports equipments, and kits but its toxic effects and its fate in a biological being have not been studied much. It has been noted that nanomaterials have a long-term consequence on human health and has been approved by many government agencies such as United States National Institute for Occupational Safety and Health and Japan's Ministry of Health. The potential exposure routes of nanomaterials in biological system are through ingestion, inhalation, and dermal penetration, but toxicity is mainly defined by the characteristics of nanomaterials such as its shape, size, surface area, and chemistry (Khan and Shanker 2015). Since nanomaterials have been used in daily life nowadays, it's important to protect the workers and the end product consumers from the toxic effect of nanomaterials (Sharifi et al. 2012). It has also been found out that engineered nanomaterials can lead to serious environment treats due to its overuse and increasing consumption (Oberdörster et al. 2005). Aquatic ecosystem serves as a sink for all the nanomaterials that have been introduced into the natural environment by various sources including human interference as well as natural sources (Moore 2006). Much of the current research on nanomaterials toxicity has been mainly focused on carbonaceous nanomaterials such as carbon nanotubes and fullerenes. Metallic nanomaterials also known as nanometals have found to adversely affect aquatic life forms tremendously and it being having massively diverse applications in industries and research institutes makes matters worse if it's released in natural environment. Although the nanomaterials are naturally available in the environment the problem arises due to the production of engineered nanomaterials which are water soluble as in the case of drug delivery system. Since the production of engineered nanomaterials results in the release in air, water, and soil, it is of utmost important to monitor its toxic effect; otherwise, an unclear toxicological data might results in adverse impact on aquatic, wildlife, and human health.

5.1 Mechanism of Toxicity

Toxicity of an engineered nanomaterial is due to single or combinations of many mechanisms. The most common model used to study the toxicity of the engineered metallic nanomaterials is the biotic ligand model (BLM), which describes metal toxicity as the concentration of free metal ion–biotic ligand complexes, where biotic ligand is a site for metal binding in the presence of competitive ions (Di Toro et al. 2001). Silver, copper, and zinc nanomaterials may exert toxic effect through dissolution and invariably releasing free metal ion in the organism. Whereas insoluble nanomaterials such as TiO_2 and gold are comparatively nontoxic as there is no free ion release, but this does not specify that non-soluble nanomaterials won't exert toxic effects as there can be chances of toxicity unrelated free metal ion release (Yin et al. 2011; Shoults-Wilson et al. 2011; Meyer et al. 2010).

In such a case, the toxicity can be due to the generation of reactive oxygen species (ROS) such as oxyradicals or superoxides, which are produced inside the organism and cause cellular damages (Yamakoshi et al. 2003). Toxicity can also be due to disruption of cell wall and cell membranes (Klaine et al. 2008).

5.2   Nanomaterials in Plants

Nanomaterials can be added to the soil directly through plants' growth enhancers such as fertilizers and plant protection products or indirectly through land treatment or wastewater treatment such as sludge. Most of the terrestrial life forms depend on plants for survival; hence, the degree of toxicity and bioavailability of nanomaterial to plants will have a major impact on terrestrial ecosystem. Nanomaterials are absorbed into the plants by different means as shown in Table 1.

Once the nanomaterial enters the plant cell it is either transported apoplastically or symplastically. The transportation across the cells is facilitated due to plasmodesmata. Nanoparticles in plants either get translocated or accumulated at specific sites. It has been found out that nanomaterials have a huge impact on plant growth, seed germination, elongation of root, and biomass production.

5.3   Nanomaterials in Aquatic Environment

Most of the toxic effect of the nanomaterial comes from the inhalation of the nanomaterial in air. The toxicity of the material increases as the size is reduced to nanometer scale even if it was inert at its bulk form. As the size is reduced the surface area increases which may result in more efficient binding with the biological system and cause greater toxicity.

The effect of nanomaterial on aquatic environment is of great concern because there is a runoff of waste water, industrial effluents, and surface runoff into the water body. Nanomaterials have also been used to treat wastewater (Vaseashta et al. 2007). When we talk about the environmental risks of nanomaterials it is of utmost

Table 1 Types of uptake of nanomaterial by plants

Types of uptake	Process of uptake	References
Through roots	For the plants to take up nanomaterial it should cross the cell wall or the plasma membrane of the endodermal or exodermal cells into the root stele. The cell walls of these cells contain layers of suberin and lignin also known as casparian bands, which inhibit the movement of water, ions, and nanomaterial into the root stele.	Enstone et al. (2002) and Hose et al. (2001)
Through cell wall pores	Uptake through cell wall pores is highly size selective and to pass the cell wall the nanomaterial has to cross it passively through the pore.	Adani et al. (2011)
Through endocytosis	Cellular studies indicate that endocytosis process can be a potential entrance pathway for nanomaterials because if the nanomaterial in question has an appropriate surface chemistry, it can activate the membrane receptors and hence the process of endocytosis.	Liu et al. (2009)

importance to have knowledge on its mobility, ecotoxicity, and its bioavailability, and all these factors are governed by the surface properties of the engineered nanomaterial (Navarro et al. 2008).

Once the nanomaterial enters the water body it becomes highly affected by the environmental surrounding and undergo transformation many a time; some of the examples include aggregation and agglomeration. The behavior of the nanomaterial in the aquatic environment depends not only on the surface properties of the nanomaterial but also on the physicochemical properties of the water body such as temperature, pH, salinity, and dissolved ions. Generally there is a robust amount of naturally occurring nanomaterial in the aquatic environment due to weathering of rocks, sediments, and presence of organic matter and minerals but the major concern is the interaction of manufactured nanomaterial with the components of the aquatic environment causing the modification of the surface of the nanomaterials causing adsorption of metals, colloids, and organic matter to the surface, hence changing its bioavailability in the environment (Glenn and Klaine 2013; Petersen et al. 2011). Hence altering the toxicity results obtained while researching the impacts of the manufactured nanomaterial in the environment.

The testing methods for toxicity of the nanomaterial is mainly focused on the water column but due to the setting behavior of the nanomaterials the sediments and benthic organisms are more highly prone for toxicity as compared to pelagic organisms. Also the published papers on toxicity are been biased toward pelagic organisms as there are very few studies on the toxicity effect of benthic organisms.

5.4   Nanomaterials in Terrestrial Animals and Human Beings

The main entry point for nanomaterials in the worms is through soil. Entry of engineered nanomaterial in the soil may affect earthworms which are critical component of soil and detrivores in ecology (Oades 1993). There have been several experiments conducted on the earthworms to check its effect. For example, labeled carbon nanotubes' presence in the soil made the PHAs less available for worms (Petersen et al. 2009). It has been well documented that the concentration of nanomaterial in soil is higher than in water, making it a major sink for nanomaterials (Dinesh et al. 2012; Gottschalk et al. 2009). The major advantage of nanomaterials is its small size, shape, surface properties, chemical composition, aggregation, and solubility which may have a negative impact once being exposed to the environment. Major toxic effect of nanomaterial was found to be that once being exposed to the environment it interacts with the biological matter and change its surface characteristics. The major route of exposure to nanomaterial for humans is through the food or indirectly through dissolution from food containers (Bergin and Witzmann 2013). Once it enters the blood stream it is circulated throughout the entire body. Since the size is small it has a large surface area which may lead to more chemical activity in the body. There are different sites in the body which are targeted by the nanomaterials, such as liver (hepatotoxicity), which is a hotspot since it is the area where detoxification of xenobiotics occurs. Liver acts as a filter for the blood and then is transported to the rest of the body. This may lead to accumulation of toxicants or nanomaterials in the organ. Usually the major site where the nanomaterials are entrapped is the reticuloendothelial system (Hillyer and Albrecht 2001). Apart from liver, kidney is another site for nanomaterial toxicity. It has been reported that kidney is the primary organ for the removal of nanomaterial from the body (Wang et al. 2009).

5.5 *Nanomaterials in Microbial Community*

Microbial communities are exposed to nanomaterials in soil, water, sewage, and sediments. Mechanism of action of nanomaterial on microbial communities can be a direct effect or indirect by interacting with other organic matter in the environment and enhancing its toxicity by changing its bioavailability (Simonet and Valcárcel 2009). Mode of effect on microbial communities may be cell damage, reactive oxygen species, and protein damage. The toxic effect of nanomaterial on microorganism at molecular level is not that much focused as compared to cellular level. For example, the cell wall composition of both Gram positive and Gram negative bacteria varies, so this could have a varied level of toxic effect based on the interaction on nanomaterial on the cell wall (Zou et al. 2016). The major concern is the effect of nanomaterial on beneficiary bacteria which helps in plant growth, nitrogen fixation, nutrient cycling, and conversion of organic matter in soil. Many nanomaterials have showed to have a negative impact on the soil microbial consortium, reducing its overall number in soil and beneficiary impact on the ecosystem.

6   Conclusion and Future Prospects

The rapid development and use of metal nanoparticles have given rise to a great concern over its adverse effect on environmental health and safety (EHS). In this chapter, we highlight the importance of toxicological effects related to transformation of MNPs with the aim to better understand the EHS of MNPs in the natural environments. The transformation of MNPs in various environments like terrestrial, air, and aquatic environments and the toxic properties of these transformed particles are unique from those of intact particles. Increased release of toxic metal ions, changes in physicochemical characteristics, synergistic effects of environmental pollutants and MNPs, alterations in bioavailability, and altered reactions with biological macromolecules like proteins are all potential mechanisms by which transformation might influence MNP toxicity.

However, many research studies have explored the process of transformation of MNPs in natural environments and their toxic impact, and there are still various challenges and knowledge gaps exist in evaluating the EHS of MNPs when exposed to environment. Researches should be focused on developing methods and instruments which can completely and partially remove transformed MNPs from biological and natural environments. Additional researches are also required to evaluate slow transformation of MNPs and their related toxicity at concentrations which are environmentally relevant despite the existing limitations in test methods and instrumental technology. Furthermore, the molecular mechanisms and properties of toxicity induced by transformed MNPs are still unclear, and therefore, further research is required to explicate the influence of transformation process on pure MNPs. Therefore, it is necessary to search for alternative and less toxic environment-friendly nanoparticles and thereby promote sustainable nanotechnology.

References

Abbas Q, Yousaf B, Amina AMU, Munir MAM, El-Naggar A, Rinklebe J, Naushad M. Transformation pathways and fate of engineered nanoparticles (ENPs) in distinct interactive environmental compartments: a review. Environ Int. 2020;138:105646. https://doi.org/10.1016/j.envint.2020.105646.

Adamczyk Z, Weroński P. Application of the DLVO theory for particle deposition problems. Adv Colloid Interf Sci. 1999;83(1):137–226. https://doi.org/10.1016/S0001-8686(99)00009-3.

Adani F, Papa G, Schievano A, Cardinale G, D'Imporzano G, Tambone F. Nanoscale structure of the cell wall protecting cellulose from enzyme attack. Environ Sci Technol. 2011;45(3):1107–13.

Adeleye AS, Keller AA. Interactions between algal extracellular polymeric substances and commercial TiO_2 nanoparticles in aqueous media. Environ Sci Technol. 2016;50(22):12258–65. https://doi.org/10.1021/acs.est.6b03684.

Adil SF, Assal ME, Khan M, Al-Warthan A, Siddiqui MRH, Liz-Marzán LM. Biogenic synthesis of metallic nanoparticles and prospects toward green chemistry. Dalton Trans. 2015;44(21):9709–17. https://doi.org/10.1039/C4DT03222E.

Ahmad A, Mukherjee P, Mandal D, Senapati S, Khan MI, Kumar R, Sastry M. Enzyme mediated extracellular synthesis of CdS nanoparticles by the fungus, Fusarium oxysporum. J Am Chem Soc. 2002;124(41):12108–9. https://doi.org/10.1021/ja027296o.

Aillon KL, Xie Y, El-Gendy N, Berkland CJ, Forrest ML. Effects of nanomaterial physico-chemical properties on in vivo toxicity. Adv Drug Deliv Rev. 2009;61(6):457–66. https://doi.org/10.1016/j.addr.2009.03.010.

Aljabali AAA, Akkam Y, Al Zoubi MS, Al-Batayneh KM, Al-Trad B, Abo Alrob O, Alkilany AM, Benamara M, Evans DJ. Synthesis of gold nanoparticles using leaf extract of Ziziphus zizyphus and their antimicrobial activity. Nanomaterials. 2018;8(3):174.

Amde M, Liu J-f, Tan Z-Q, Bekana D. Transformation and bioavailability of metal oxide nanoparticles in aquatic and terrestrial environments. A review. Environ Pollut. 2017;230:250–67. https://doi.org/10.1016/j.envpol.2017.06.064.

Asgharian B, Price OT. Deposition of ultrafine (NANO) Particles in the human lung. Inhal Toxicol. 2007;19(13):1045–54. https://doi.org/10.1080/08958370701626501.

Auvinen H, Gagnon V, Rousseau DPL, Du Laing G. Fate of metallic engineered nanomaterials in constructed wetlands: prospection and future research perspectives. Rev Environ Sci Biotechnol. 2017;16(2):207–22. https://doi.org/10.1007/s11157-017-9427-0.

Azizi S, Ahmad MB, Namvar F, Mohamad R. Green biosynthesis and characterization of zinc oxide nanoparticles using brown marine macroalga Sargassum muticum aqueous extract. Mater Lett. 2014;116:275–7. https://doi.org/10.1016/j.matlet.2013.11.038.

Benn TM, Westerhoff P. Nanoparticle silver released into water from commercially available sock fabrics. Environ Sci Technol. 2008;42(11):4133–9. https://doi.org/10.1021/es7032718.

Bergese P, Colombo I. Chapter 1 – Thermodynamics of (nano)interfaces. In: Berti D, Palazzo G, editors. Colloidal foundations of nanoscience. Amsterdam: Elsevier; 2014. p. 1–31. https://doi.org/10.1016/B978-0-444-59541-6.00001-1.

Bergin IL, Witzmann FA. Nanoparticle toxicity by the gastrointestinal route: evidence and knowledge gaps. Int J Biomed Nanosci Nanotechnol. 2013;3(1-2):163–210.

Biswas JK, Sarkar D. Nanopollution in the aquatic environment and ecotoxicity: no nano issue! Curr Pollut Rep. 2019;5(1):4–7. https://doi.org/10.1007/s40726-019-0104-5.

Botta C, Labille J, Auffan M, Borschneck D, Miche H, Cabié M, Masion A, Rose J, Bottero J-Y. TiO2-based nanoparticles released in water from commercialized sunscreens in a life-cycle perspective: structures and quantities. Environ Pollut. 2011;159(6):1543–50. https://doi.org/10.1016/j.envpol.2011.03.003.

Buffet P-E, Tankoua OF, Pan J-F, Berhanu D, Herrenknecht C, Poirier L, Amiard-Triquet C, Amiard J-C, Bérard J-B, Risso C, Guibbolini M, Roméo M, Reip P, Valsami-Jones E, Mouneyrac C. Behavioural and biochemical responses of two marine invertebrates Scrobicularia plana and Hediste diversicolor to copper oxide nanoparticles. Chemosphere. 2011;84(1):166–74. https://doi.org/10.1016/j.chemosphere.2011.02.003.

Bundschuh M, Filser J, Lüderwald S, McKee MS, Metreveli G, Schaumann GE, Schulz R, Wagner S. Nanoparticles in the environment: where do we come from, where do we go to? Environ Sci Eur. 2018;30(1):1–17.

Chekli L, Zhao YX, Tijing LD, Phuntsho S, Donner E, Lombi E, Gao BY, Shon HK. Aggregation behaviour of engineered nanoparticles in natural waters: characterising aggregate structure using on-line laser light scattering. J Hazard Mater. 2015;284:190–200. https://doi.org/10.1016/j.jhazmat.2014.11.003.

Collin B, Oostveen E, Tsyusko OV, Unrine JM. Influence of natural organic matter and surface charge on the toxicity and bioaccumulation of functionalized ceria nanoparticles in Caenorhabditis elegans. Environ Sci Technol. 2014;48(2):1280–9. https://doi.org/10.1021/es404503c.

De Jong WH, Hagens WI, Krystek P, Burger MC, Sips AJAM, Geertsma RE. Particle size-dependent organ distribution of gold nanoparticles after intravenous administration. Biomaterials. 2008;29(12):1912–9. https://doi.org/10.1016/j.biomaterials.2007.12.037.

Derfus AM, Chan WCW, Bhatia SN. Probing the Cytotoxicity of semiconductor quantum dots. Nano Lett. 2004;4(1):11–8. https://doi.org/10.1021/nl0347334.

Di Toro DM, Allen HE, Bergman HL, Meyer JS, Paquin PR, Santore RC. Biotic ligand model of the acute toxicity of metals. 1. Technical basis. Environ Toxicol Chem. 2001;20(10):2383–96.

Dinesh R, Anandaraj M, Srinivasan V, Hamza S. Engineered nanoparticles in the soil and their potential implications to microbial activity. Geoderma. 2012;173:19–27.

Dwivedi AD, Dubey SP, Sillanpää M, Kwon Y-N, Lee C, Varma RS. Fate of engineered nanoparticles: implications in the environment. Coord Chem Rev. 2015;287:64–78. https://doi.org/10.1016/j.ccr.2014.12.014.

Elbasuney S. Sustainable steric stabilization of colloidal titania nanoparticles. Appl Surf Sci. 2017;409:438–47. https://doi.org/10.1016/j.apsusc.2017.03.013.

Enstone DE, Peterson CA, Ma F. Root endodermis and exodermis: structure, function, and responses to the environment. J Plant Growth Regul. 2002;21(4):335–51.

Gatoo MA, Naseem S, Arfat MY, Mahmood Dar A, Qasim K, Zubair S. Physicochemical properties of nanomaterials: implication in associated toxic manifestations. Biomed Res Int 2014.

Gelabert A, Sivry Y, Ferrari R, Akrout A, Cordier L, Nowak S, Menguy N, Benedetti MF. Uncoated and coated ZnO nanoparticle life cycle in synthetic seawater. Environ Toxicol Chem. 2014;33(2):341–9. https://doi.org/10.1002/etc.2447.

Georgieva JV, Kalicharan D, Couraud P-O, Romero IA, Weksler B, Hoekstra D, Zuhorn IS. Surface characteristics of nanoparticles determine their intracellular fate in and processing by human blood–brain barrier endothelial cells in vitro. Mol Ther. 2011;19(2):318–25. https://doi.org/10.1038/mt.2010.236.

Glenn JB, Klaine SJ. Abiotic and biotic factors that influence the bioavailability of gold nanoparticles to aquatic macrophytes. Environ Sci Technol. 2013;47(18):10223–30.

Gopinath K, Arumugam A. Extracellular mycosynthesis of gold nanoparticles using Fusarium solani. Appl Nanosci. 2014;4(6):657–62. https://doi.org/10.1007/s13204-013-0247-4.

Gottschalk F, Nowack B. The release of engineered nanomaterials to the environment. J Environ Monit. 2011;13(5):1145–55. https://doi.org/10.1039/C0EM00547A.

Gottschalk F, Sonderer T, Scholz RW, Nowack B. Modeled environmental concentrations of engineered nanomaterials (TiO$_2$, ZnO, Ag, CNT, fullerenes) for different regions. Environ Sci Technol. 2009;43(24):9216–22. https://doi.org/10.1021/es9015553.

Griffitt RJ, Luo J, Gao J, Bonzongo J-C, Barber DS. Effects of particle composition and species on toxicity of metallic nanomaterials in aquatic organisms. Environ Toxicol Chem. 2008;27(9):1972–8. https://doi.org/10.1897/08-002.1.

Hillyer JF, Albrecht RM. Gastrointestinal persorption and tissue distribution of differently sized colloidal gold nanoparticles. J Pharm Sci. 2001;90(12):1927–36.

Hose E, Clarkson DT, Steudle E, Schreiber L, Hartung W. The exodermis: a variable apoplastic barrier. J Exp Bot. 2001;52(365):2245–64.

Hoshino A, Fujioka K, Oku T, Suga M, Sasaki YF, Ohta T, Yasuhara M, Suzuki K, Yamamoto K. Physicochemical Properties and cellular toxicity of nanocrystal quantum dots depend on their surface modification. Nano Lett. 2004;4(11):2163–9. https://doi.org/10.1021/nl048715d.

Hu J, Wang J, Liu S, Zhang Z, Zhang H, Cai X, Pan J, Liu J. Effect of TiO$_2$ nanoparticle aggregation on marine microalgae Isochrysis galbana. J Environ Sci. 2018;66:208–15. https://doi.org/10.1016/j.jes.2017.05.026.

Huang X, Li Y, Chen K, Chen H, Wang F, Han X, Zhou B, Chen H, Yuan R. NOM mitigates the phytotoxicity of AgNPs by regulating rice physiology, root cell wall components and root morphology. Environ Pollut. 2020;260:113942. https://doi.org/10.1016/j.envpol.2020.113942.

Iravani S. Green synthesis of metal nanoparticles using plants. Green Chem. 2011;13(10):2638–50. https://doi.org/10.1039/C1GC15386B.

Kang F, Alvarez PJ, Zhu D. Microbial extracellular polymeric substances reduce Ag+ to silver nanoparticles and antagonize bactericidal activity. Environ Sci Technol. 2014;48(1):316–22. https://doi.org/10.1021/es403796x.

Kansara K, Kumar A, Karakoti AS. Combination of humic acid and clay reduce the ecotoxic effect of TiO2 NPs: a combined physico-chemical and genetic study using zebrafish embryo. Sci Total Environ. 2020;698:134133. https://doi.org/10.1016/j.scitotenv.2019.134133.

Keller AA, Lazareva A. Predicted releases of engineered nanomaterials: from global to regional to local. Environ Sci Technol Lett. 2014;1(1):65–70. https://doi.org/10.1021/ez400106t.

Khan HA, Shanker R. Toxicity of nanomaterials. Biomed Res Int. 2015;2015:521014. https://doi.org/10.1155/2015/521014.

Kim B, Park C-S, Murayama M, Hochella MF. Discovery and characterization of silver sulfide nanoparticles in final sewage sludge products. Environ Sci Technol. 2010;44(19):7509–14. https://doi.org/10.1021/es101565j.

Klaine SJ, Alvarez PJ, Batley GE, Fernandes TF, Handy RD, Lyon DY, Mahendra S, McLaughlin MJ, Lead JR. Nanomaterials in the environment: behavior, fate, bioavailability, and effects. Environ Toxicol Chem. 2008;27(9):1825–51.

Lanzl CA, Baltrusaitis J, Cwiertny DM. Dissolution of hematite nanoparticle aggregates: influence of primary particle size, dissolution mechanism, and solution pH. Langmuir. 2012;28(45):15797–808. https://doi.org/10.1021/la3022497.

Lead JR, Batley GE, Alvarez PJJ, Croteau M-N, Handy RD, McLaughlin MJ, Judy JD, Schirmer K. Nanomaterials in the environment: behavior, fate, bioavailability, and effects – an updated review. Environ Toxicol Chem. 2018;37(8):2029–63. https://doi.org/10.1002/etc.4147.

Lee M-K, Lim S-J, Kim C-K. Preparation, characterization and in vitro cytotoxicity of paclitaxel-loaded sterically stabilized solid lipid nanoparticles. Biomaterials. 2007;28(12):2137–46. https://doi.org/10.1016/j.biomaterials.2007.01.014.

Lei C, Sun Y, Tsang DCW, Lin D. Environmental transformations and ecological effects of iron-based nanoparticles. Environ Pollut. 2018;232:10–30. https://doi.org/10.1016/j.envpol.2017.09.052.

Lin D, Tian X, Wu F, Xing B. Fate and transport of engineered nanomaterials in the environment. J Environ Qual. 2010;39(6):1896–908. https://doi.org/10.2134/jeq2009.0423.

Liu Q, Chen B, Wang Q, Shi X, Xiao Z, Lin J, Fang X. Carbon nanotubes as molecular transporters for walled plant cells. Nano Lett. 2009;9(3):1007–10.

Loosli F, Le Coustumer P, Stoll S. Impact of alginate concentration on the stability of agglomerates made of TiO_2 engineered nanoparticles: water hardness and pH effects. J Nanopart Res. 2015;17(1):1–9.

Lowry GV, Gregory KB, Apte SC, Lead JR. Transformations of nanomaterials in the environment. ACS Publications; 2012a.

Lowry GV, Gregory KB, Apte SC, Lead JR. Transformations of nanomaterials in the environment. Environ Sci Technol. 2012b;46(13):6893–9. https://doi.org/10.1021/es300839e.

Lundqvist M, Stigler J, Cedervall T, Berggård T, Flanagan MB, Lynch I, Elia G, Dawson K. The evolution of the protein corona around nanoparticles: a test study. ACS Nano. 2011;5(9):7503–9. https://doi.org/10.1021/nn202458g.

Ma H, Wallis LK, Diamond S, Li S, Canas-Carrell J, Parra A. Impact of solar UV radiation on toxicity of ZnO nanoparticles through photocatalytic reactive oxygen species (ROS) generation and photo-induced dissolution. Environ Pollut. 2014;193:165–72. https://doi.org/10.1016/j.envpol.2014.06.027.

Marie T, Mélanie A, Lenka B, Julien I, Isabelle K, Christine P, Elise M, Catherine S, Bernard A, Ester A, Jérôme R, Alain T, Jean-Yves B. Transfer, transformation, and impacts of Ceria nanomaterials in aquatic mesocosms simulating a pond ecosystem. Environ Sci Technol. 2014;48(16):9004–13. https://doi.org/10.1021/es501641b.

Meyer JN, Lord CA, Yang XY, Turner EA, Badireddy AR, Marinakos SM, Chilkoti A, Wiesner MR, Auffan M. Intracellular uptake and associated toxicity of silver nanoparticles in Caenorhabditis elegans. Aquat Toxicol. 2010;100(2):140–50.

Mitrano DM, Limpiteeprakan P, Babel S, Nowack B. Durability of nano-enhanced textiles through the life cycle: releases from landfilling after washing. Environ Sci Nano. 2016;3(2):375–87. https://doi.org/10.1039/C6EN00023A.

Moore M. Do nanoparticles present ecotoxicological risks for the health of the aquatic environment? Environ Int. 2006;32(8):967–76.

Mouneyrac C, Buffet P-E, Poirier L, Zalouk-Vergnoux A, Guibbolini M, Faverney CR, Gilliland D, Berhanu D, Dybowska A, Châtel A, Perrein-Ettajni H, Pan J-F, Thomas-Guyon H, Reip P, Valsami-Jones E. Fate and effects of metal-based nanoparticles in two marine invertebrates, the bivalve mollusc Scrobicularia plana and the annelid polychaete Hediste diversicolor. Environ Sci Pollut Res. 2014;21(13):7899–912. https://doi.org/10.1007/s11356-014-2745-7.

Mudunkotuwa IA, Rupasinghe T, Wu C-M, Grassian VH. Dissolution of ZnO nanoparticles at circumneutral pH: a study of size effects in the presence and absence of citric acid. Langmuir. 2012;28(1):396–403. https://doi.org/10.1021/la203542x.

Mukherjee SP, Gliga AR, Lazzaretto B, Brandner B, Fielden M, Vogt C, Newman L, Rodrigues AF, Shao W, Fournier PM, Toprak MS, Star A, Kostarelos K, Bhattacharya K, Fadeel B. Graphene oxide is degraded by neutrophils and the degradation products are non-genotoxic. Nanoscale. 2018;10(3):1180–8. https://doi.org/10.1039/C7NR03552G.

Nasser F, Constantinou J, Lynch I. Nanomaterials in the environment acquire an "eco-corona" impacting their toxicity to Daphnia Magna – a call for updating toxicity testing policies. Proteomics. 2020;20(9):1800412. https://doi.org/10.1002/pmic.201800412.

Navarro E, Baun A, Behra R, Hartmann NB, Filser J, Miao A-J, Quigg A, Santschi PH, Sigg L. Environmental behavior and ecotoxicity of engineered nanoparticles to algae, plants, and fungi. Ecotoxicology. 2008;17(5):372–86.

Neagu M, Piperigkou Z, Karamanou K, Engin AB, Docea AO, Constantin C, Negrei C, Nikitovic D, Tsatsakis A. Protein bio-corona: critical issue in immune nanotoxicology. Arch Toxicol. 2017;91(3):1031–48. https://doi.org/10.1007/s00204-016-1797-5.

Nowack B, Ranville JF, Diamond S, Gallego-Urrea JA, Metcalfe C, Rose J, Horne N, Koelmans AA, Klaine SJ. Potential scenarios for nanomaterial release and subsequent alteration in the environment. Environ Toxicol Chem. 2012;31(1):50–9. https://doi.org/10.1002/etc.726.

Oades J. The role of biology in the formation, stabilization and degradation of soil structure. In: Soil structure/soil biota interrelationships. Elsevier; 1993. p. 377–400.

Oberdörster G, Oberdörster E, Oberdörster J. Nanotoxicology: an emerging discipline evolving from studies of ultrafine particles. Environ Health Perspect. 2005;113:823–39.

Odzak N, Kistler D, Sigg L. Influence of daylight on the fate of silver and zinc oxide nanoparticles in natural aquatic environments. Environ Pollut. 2017;226:1–11. https://doi.org/10.1016/j.envpol.2017.04.006.

Oza G, Reyes-Calderón A, Mewada A, Arriaga LG, Cabrera GB, Luna DE, Iqbal HMN, Sharon M, Sharma A. Plant-based metal and metal alloy nanoparticle synthesis: a comprehensive mechanistic approach. J Mater Sci. 2020;55(4):1309–30. https://doi.org/10.1007/s10853-019-04121-3.

Peijnenburg WJGM, Baalousha M, Chen J, Chaudry Q, Von der Kammer F, Kuhlbusch TAJ, Lead J, Nickel C, Quik JTK, Renker M, Wang Z, Koelmans AA. A review of the properties and processes determining the fate of engineered nanomaterials in the aquatic environment. Crit Rev Environ Sci Technol. 2015;45(19):2084–134. https://doi.org/10.1080/10643389.2015.1010430.

Peng C, Zhang W, Gao H, Li Y, Tong X, Li K, Zhu X, Wang Y, Chen Y. Behavior and potential impacts of metal-based engineered nanoparticles in aquatic environments. Nanomaterials. 2017;7(1):21.

Petersen EJ, Pinto RA, Landrum PF, Weber J, Walter J. Influence of carbon nanotubes on pyrene bioaccumulation from contaminated soils by earthworms. Environ Sci Technol. 2009;43(11):4181–7.

Petersen EJ, Pinto RA, Mai DJ, Landrum PF, Weber WJ Jr. Influence of polyethyleneimine graftings of multi-walled carbon nanotubes on their accumulation and elimination by and toxicity to Daphnia magna. Environ Sci Technol. 2011;45(3):1133–8.

Phenrat T, Long TC, Lowry GV, Veronesi B. Partial oxidation ("aging") and surface modification decrease the toxicity of nanosized zerovalent iron. Environ Sci Technol. 2009;43(1):195–200. https://doi.org/10.1021/es801955n.

Powers KW, Palazuelos M, Moudgil BM, Roberts SM. Characterization of the size, shape, and state of dispersion of nanoparticles for toxicological studies. Nanotoxicology. 2007;1(1):42–51. https://doi.org/10.1080/17435390701314902.

Pulido-Reyes G, Leganes F, Fernández-Piñas F, Rosal R. Bio-nano interface and environment: a critical review. Environ Toxicol Chem. 2017;36(12):3181–93. https://doi.org/10.1002/etc.3924.

Quik JTK, Velzeboer I, Wouterse M, Koelmans AA, van de Meent D. Heteroaggregation and sedimentation rates for nanomaterials in natural waters. Water Res. 2014;48:269–79. https://doi.org/10.1016/j.watres.2013.09.036.

Rao NH, Lakshmidevi N, Pammi SVN, Kollu P, Ganapaty S, Lakshmi P. Green synthesis of silver nanoparticles using methanolic root extracts of Diospyros paniculata and their antimicrobial activities. Mater Sci Eng C. 2016;62:553–7. https://doi.org/10.1016/j.msec.2016.01.072.

Ravikumar KVG, Debayan G, Mrudula P, Chandrasekaran N, Amitava M. In situ formation of bimetallic FeNi nanoparticles on sand through green technology: application for tetracycline removal. Front Environ Sci Eng. 2019;14(1):16. https://doi.org/10.1007/s11783-019-1195-3.

Risom L, Møller P, Loft S. Oxidative stress-induced DNA damage by particulate air pollution. Mutat Res/Fundament Mol Mech Mutag. 2005;592(1):119–37. https://doi.org/10.1016/j.mrfmmm.2005.06.012.

Schreiner KM, Filley TR, Blanchette RA, Bowen BB, Bolskar RD, Hockaday WC, Masiello CA, Raebiger JW. White-rot basidiomycete-mediated decomposition of C60 fullerol. Environ Sci Technol. 2009;43(9):3162–8. https://doi.org/10.1021/es801873q.

Sellers K, Mackay C, Bergeson LL, Clough SR, Hoyt M, Chen J, Henry K, Hamblen J. Nanotechnology and the environment. CRC Press; 2008.

Sharifi S, Behzadi S, Laurent S, Forrest ML, Stroeve P, Mahmoudi M. Toxicity of nanomaterials. Chem Soc Rev. 2012;41(6):2323–43.

Shoults-Wilson W, Zhurbich OI, McNear DH, Tsyusko OV, Bertsch PM, Unrine JM. Evidence for avoidance of Ag nanoparticles by earthworms (Eisenia fetida). Ecotoxicology. 2011;20(2):385–96.

Simonet BM, Valcárcel M. Monitoring nanoparticles in the environment. Anal Bioanal Chem. 2009;393(1):17.

Singh AK. Engineered nanoparticles: structure, properties and mechanisms of toxicity. Academic; 2015.

Tangaa SR, Selck H, Winther-Nielsen M, Khan FR. Trophic transfer of metal-based nanoparticles in aquatic environments: a review and recommendations for future research focus. Environ Sci Nano. 2016;3(5):966–81. https://doi.org/10.1039/C5EN00280J.

Tso C-p, Zhung C-m, Shih Y-h, Tseng Y-M, Wu S-c, Doong R-a. Stability of metal oxide nanoparticles in aqueous solutions. Water Sci Technol. 2010;61(1):127–33.

Turan NB, Erkan HS, Engin GO, Bilgili MS. Nanoparticles in the aquatic environment: usage, properties, transformation and toxicity – a review. Process Saf Environ Prot. 2019;130:238–49. https://doi.org/10.1016/j.psep.2019.08.014.

Unrine JM, Hunyadi SE, Tsyusko OV, Rao W, Shoults-Wilson WA, Bertsch PM. Evidence for bioavailability of au nanoparticles from soil and biodistribution within earthworms (Eisenia fetida). Environ Sci Technol. 2010;44(21):8308–13. https://doi.org/10.1021/es101885w.

Upadhyayula VKK, Gadhamshetty V. Appreciating the role of carbon nanotube composites in preventing biofouling and promoting biofilms on material surfaces in environmental engineering: a review. Biotechnol Adv. 2010;28(6):802–16. https://doi.org/10.1016/j.biotechadv.2010.06.006.

Van Koetsem F, Verstraete S, Van der Meeren P, Du Laing G. Stability of engineered nanomaterials in complex aqueous matrices: settling behaviour of CeO_2 nanoparticles in natural surface waters. Environ Res. 2015;142:207–14. https://doi.org/10.1016/j.envres.2015.06.028.

Vaseashta A, Vaclavikova M, Vaseashta S, Gallios G, Roy P, Pummakarnchana O. Nanostructures in environmental pollution detection, monitoring, and remediation. Sci Technol Adv Mater. 2007;8(1-2):47.

Vikesland PJ, Rebodos R, Bottero J, Rose J, Masion A. Aggregation and sedimentation of magnetite nanoparticle clusters. Environ Sci Nano. 2016;3(3):567–77.

Wang F, Gao F, Lan M, Yuan H, Huang Y, Liu J. Oxidative stress contributes to silica nanoparticle-induced cytotoxicity in human embryonic kidney cells. Toxicol in Vitro. 2009;23(5):808–15.

Wang Z, Zhang L, Zhao J, Xing B. Environmental processes and toxicity of metallic nanoparticles in aquatic systems as affected by natural organic matter. Environ Sci Nano. 2016;3(2):240–55. https://doi.org/10.1039/C5EN00230C.

Wang Y, Cai R, Chen C. The nano–bio interactions of nanomedicines: understanding the biochemical driving forces and redox reactions. Acc Chem Res. 2019;52(6):1507–18. https://doi.org/10.1021/acs.accounts.9b00126.

Wu D, Yang S, Du W, Yin Y, Zhang J, Guo H. Effects of titanium dioxide nanoparticles on Microcystis aeruginosa and microcystins production and release. J Hazard Mater. 2019;377:1–7. https://doi.org/10.1016/j.jhazmat.2019.05.013.

Xu J. Biomolecules produced by mangrove-associated microbes. Curr Med Chem. 2011;18(34):5224–66. https://doi.org/10.2174/092986711798184307.

Yamakoshi Y, Umezawa N, Ryu A, Arakane K, Miyata N, Goda Y, Masumizu T, Nagano T. Active oxygen species generated from photoexcited fullerene (C60) as potential medicines: O_2-• versus $^1O_2$. J Am Chem Soc. 2003;125(42):12803–9.

Yang Y, Zhang C, Hu Z. Impact of metallic and metal oxide nanoparticles on wastewater treatment and anaerobic digestion. Environ Sci Process Impacts. 2013;15(1):39–48.

Yates MD, Cusick RD, Logan BE. Extracellular palladium nanoparticle production using Geobacter sulfurreducens. ACS Sustain Chem Eng. 2013;1(9):1165–71. https://doi.org/10.1021/sc4000785.

Yin H, Too HP, Chow GM. The effects of particle size and surface coating on the cytotoxicity of nickel ferrite. Biomaterials. 2005;26(29):5818–26. https://doi.org/10.1016/j.biomaterials.2005.02.036.

Yin L, Cheng Y, Espinasse B, Colman BP, Auffan M, Wiesner M, Rose J, Liu J, Bernhardt ES. More than the ions: the effects of silver nanoparticles on Lolium multiflorum. Environ Sci Technol. 2011;45(6):2360–7.

Yung MMN, Wong SWY, Kwok KWH, Liu FZ, Leung YH, Chan WT, Li XY, Djurišić AB, Leung KMY. Salinity-dependent toxicities of zinc oxide nanoparticles to the marine diatom Thalassiosira pseudonana. Aquat Toxicol. 2015;165:31–40. https://doi.org/10.1016/j.aquatox.2015.05.015.

Zhang C, Chen W, Alvarez PJJ. Manganese peroxidase degrades pristine but not surface-oxidized (carboxylated) single-walled carbon nanotubes. Environ Sci Technol. 2014;48(14):7918–23. https://doi.org/10.1021/es5011175.

Zhao Y, Allen BL, Star A. Enzymatic degradation of multiwalled carbon nanotubes. J Phys Chem A. 2011;115(34):9536–44. https://doi.org/10.1021/jp112324d.

Zhou D, Keller AA. Role of morphology in the aggregation kinetics of ZnO nanoparticles. Water Res. 2010;44(9):2948–56. https://doi.org/10.1016/j.watres.2010.02.025.

Zhou D, Abdel-Fattah AI, Keller AA. Clay particles destabilize engineered nanoparticles in aqueous environments. Environ Sci Technol. 2012;46(14):7520–6. https://doi.org/10.1021/es3004427.

Zhou X-h, Huang B-c, Zhou T, Liu Y-c, Shi H-c. Aggregation behavior of engineered nanoparticles and their impact on activated sludge in wastewater treatment. Chemosphere. 2015;119:568–76.

Zhou K, Hu Y, Zhang L, Yang K, Lin D. The role of exopolymeric substances in the bioaccumulation and toxicity of Ag nanoparticles to algae. Sci Rep. 2016;6(1):32998. https://doi.org/10.1038/srep32998.

Zou X, Zhang L, Wang Z, Luo Y. Mechanisms of the antimicrobial activities of graphene materials. J Am Chem Soc. 2016;138(7):2064–77.

Drug-Induced Oxidative Stress and Cellular Toxicity

Shalini Mani, Sakshi Tyagi, Km Vaishali Pal, Himanshi Jaiswal, Anvi Jain, Aaru Gulati, and Manisha Singh

Contents

Abstract Reactive oxygen species (ROS) are a derivative of usual metabolic activities and have proven to play a crucial role in different cell signaling pathways, crucial for cell survival and homeostasis. There are defined cellular mechanisms that help in regulating the ROS levels, and different antioxidant enzymes such as catalase (CAT) and superoxide dismutase (SOD) help in neutralizing these ROS and thus help in the regulation of their levels. However, under certain conditions such as excessive ROS production, which is beyond the neutralizing capacity of the cells, or deficient antioxidant enzymes, the accumulation of ROS takes place which consequently leads to oxidative stress (OS). There are various external factors known to trigger excessive ROS production and thus OS. Surprisingly, a large number of drugs which on the one hand help in curbing the disease pathogenesis are known to be responsible for OS and associated cytotoxicity in various diseases and thus exhibit their side effects. In this chapter, we review the various types of cellular

S. Mani (✉) · S. Tyagi · K. V. Pal · H. Jaiswal · A. Jain · A. Gulati · M. Singh
Centre for Emerging Diseases, Department of Biotechnology, Jaypee Institute of Information Technology, Noida, Uttar Pradesh, India
e-mail: shalini.mani@jiit.ac.in

© The Author(s), under exclusive license to Springer Nature Switzerland AG 2021
K. K. Kesari, N. K. Jha (eds.), *Free Radical Biology and Environmental Toxicity*,
Molecular and Integrative Toxicology, https://doi.org/10.1007/978-3-030-83446-3_4

stress generated due to different reactive ions such as ROS as well as RNS (reactive nitrogen species) and their effect on various cellular components. The chapter further includes the different drugs and various studies supporting their ROS/RNS generating effects. In the end, the significance of co-supplementations of various antioxidants along with these drugs has also been discussed in brief.

Keywords Reactive oxygen species · Reactive nitrogen species · Antioxidants · Oxidative stress · Drugs

1 Introduction

A drug is a chemical constituent that has a known physiological and psychological effect on humans. It is generally used for the cure, prevention, treatment, or diagnosis of disease. Depending on the disorder, the usage of drugs may vary. Based on its interaction in a biological system, it produced desired results. If the drug's effect on the body is positive, the drug is termed as medicine, whereas, if it leads to detrimental effects, the drug is categorized as poison. Though drugs can be used for the treatment of diverse types of diseases, certain medicinal drugs may lead to adverse effects as reported in several studies (Deavall et al. 2012). As far as the underlying mechanism is concerned, oxidative stress could also be one of the reasons. The toxicity induced in numerous tissues and organ systems including liver, kidney, ear, cardiovascular and nervous systems on exposure to certain drugs after a certain period is often termed as drug-induced oxidative stress. The drugs that lead to oxidative stress (OS) by excessive generation of reactive oxygen species (ROS), reactive nitrogen species (RNS), and/ or by disruption of the endogenous antioxidant system are collectively known as "oxidative drugs" (Moore 2002; Pereira et al. 2012). Well-characterized oxidative drugs include (1) non-steroidal anti-inflammatory drugs (NSAID) such as aspirin, phenacetin, and acetaminophen (Tylenol); (2) antiretroviral agents such as azidothymidine; (3) Antimalarial drugs, such as chloroquine and primaquine; (4) antipsychotic agents such as chlorpromazine; and (5) Anti-neoplastic drugs such as cisplatin (Daugaard and Abidgaard 1989). However, mechanisms of drug-induced OS may vary. ROS can be directly generated via drug metabolism. An example of drug-induced OS is chlorpromazine, and an adverse event associated with this is cutaneous phototoxicity caused by photoactivation in the skin (Moore 2002). There is a production of both singlet oxygen and superoxide species on photochlorination of chlorpromazine. These species may then target DNA and macromolecules and may result in detrimental effects on the skin. Under normal conditions, damage due to ROS is protected with the help of an antioxidant defense system. However, in the case of drug-induced OS, cellular antioxidant enzymes may be not able to prevent the damaging effects of OS on cellular processes.

ROS-induced damage to DNA and other macromolecules is related to the onset and progression of many disorders such as cardiovascular disease, Alzheimer's diseases, ischemic stroke, as well as normal aging processes (Maynard et al. 2009). Mutations due to ROS/RNS can also lead to the transformation of a normal cell to a malignant one and thus resulting in cancer development (Shacter 2000). Hence, in our current chapter, we are detailing different types of drug-induced toxicity and associated disorders/diseases. We have also briefly emphasized the use of antioxidant therapy as a supplement along with such OS inducing drugs.

2 Reactive Oxygen Species (ROS) and Reactive Nitrogen Species (RNS): Double-Edged Sword

Both ROS and RNS are important free radicals and are generated as a by-product of various biological reactions (Fig. 1). Both ROS and RNS are important for different cellular signaling pathways and regulate the viability and proliferation of these cells. However, the same free radicals become damaging, once their level goes high in the cellular system.

2.1 Production of ROS

Oxidative stress is a phenomenon that arises from an excess of the ROS formation and hence the antioxidant defense system responses. ROS is formed in a biological context as a by-product of the natural metabolism of oxygen through various

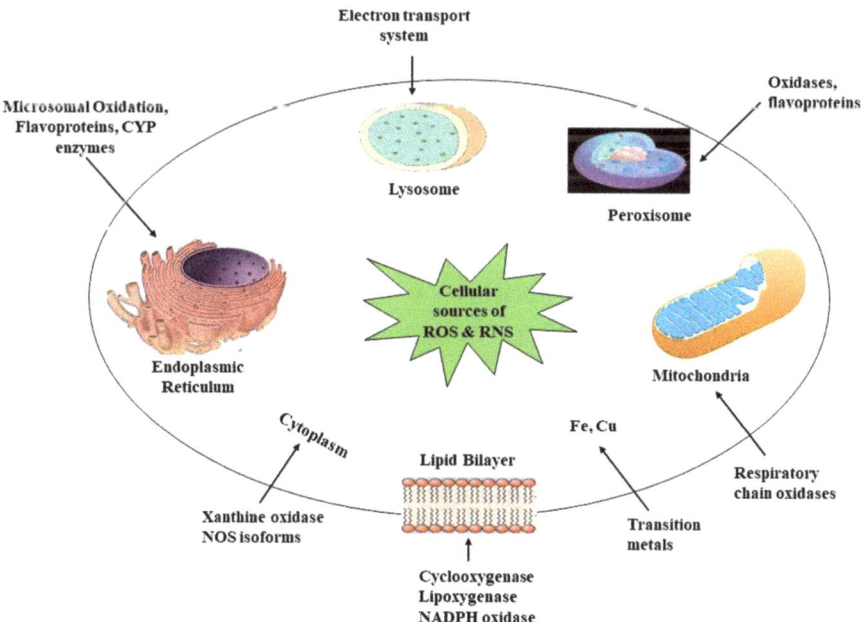

Fig. 1 Schematic representation of various cellular sources playing role in ROS and RNS generation

enzymatic reactions. The impact of ROS is strongly reliant on their tissue enormous amount, and ROS is being seen either as a healthy cell signal in addition to environmental obstacles or even as a product and source of oxidative damage when present at low and medium levels. ROS induces a cytotoxic response at low concentrations, whereas cell damage activities are induced at high concentrations. ROS is produced in living organisms in many cellular systems located mostly on the plasma membrane, in the peroxisomes, in the cytosol, and even on mitochondrial and endoplasmic reticulum (ER) membranes (Bergamini et al. 2004).

2.1.1 ROS Generation in Mitochondria

Mitochondria are important in aerobic organisms for various important components, which include respiration and production of oxidative energy, intracellular calcium concentration regulation, and β-oxidation control of fatty acids. And because of their role in energy production, mitochondria have been taken into account. The key source of ROS during normal metabolism is the mitochondrial electron-transport chain. Bypassing the electrons to electron acceptors such as FAD, FMN, and NAD+, they use almost 95% of the oxygen required by oxidative animals to generate nutrients by oxidizing compounds found in many foods.

Several redox centers are involved in the process, primarily grouped into four protein complexes implanted inside the inner membrane of the mitochondria. Electrons are passed to the lipid-soluble transporter ubiquinone by Complexes I and II. From this, the electrons are passed to oxygen by Complex III, cytochrome c (the other dynamic carrier), and Complex IV. A mechanism by which such a force allows protons via the mitochondrial ATP synthase to move to ATP synthesis back into the matrix is called oxidative phosphorylation (Chen et al. 2003).

The major ROS produced inside of mitochondria by uniform autooxidation of electron acceptors is O_2- that is incorporated by mitochondrial superoxide dismutase (SOD) into H_2O_2 that can also be transformed into OH- radicals through the Fenton reaction:

$$H_2O_2 + Fe^{2+} \rightarrow \bullet OH + OH - + Fe^{3+}$$

Complexes I and III are located at the key sites involved in mitochondrial ROS formation. For the determination of ROS results, the localization of the generators is critical as it evaluates whether O_2- is generated in the matrix of mitochondria or the space between intermembranes. Therefore, also Complex I and Complex III generators allow leaving O_2- inside the matrix in which mitochondrial DNA can be destroyed, while the Complex III generator as well releases O_2- into its space of intermembrane in which the cytosol is easier to enter (Di Meo et al. 2016).

2.1.2 ROS Production by Cytosol

So many soluble components of the cell, along with thiols, flavins, hydroquinones, and catecholamines, may lead to the formation of ROS in intracellular space as they can serve redox reactions. Also, during their catalytic action, some cytosolic enzymes generate ROS. Xanthine oxidase (XO) is a potential ROS-producing enzyme. In normal tissues, xanthine dehydrogenase (XDH), which uses NADP+ like an electron acceptor, is the enzyme catalyzing the oxidation of hypoxanthine to xanthine and xanthine to uric acid. Likewise, the enzyme is transformed through the dehydrogenase state into oxidase from inside weakened tissues, either through the reversible oxidation of the residues of cysteine or through permanent proteolysis induced by Ca^{2+}, which passes electrons during xanthine as well as hypoxanthine oxidation to molecular oxygen (O_2) producing superoxide radical.

2.1.3 ROS Production in ER

ER has participated in numerous functions, including the transport, synthesis, and folding of Golgi, lysosomal, secretory, and cell surface proteins, the accumulation of calcium, the digestion of lipids, and the detoxification of drugs in certain cell types. The smooth ER introduces an electron transport chain consisting of two structures dedicated to the metabolism of xenobiotics and the incorporation of fatty acid double bonds that are also capable of creating ROS. The metabolism of xenobiotics normally happens in two steps. Phase I reactions add a polar group using O_2- in a lipophilic substrate material (AH) and reducing agent (RH2):

$$AH + O_2 + RH2 \rightarrow AOH + R + H_2O$$

The reaction is established as a monooxygenase reaction and is associated with a flavoprotein (NADPH cytochrome P450 reductase) mechanism and cytochromes known collectively as P450 cytochromes (CYPs). The microsomal H_2O_2 generator may be the NADPH-cytochrome c reductase. Interestingly, it was because the decay of two catalytic cycle intermediates could form O_2- and H_2O_2. One of the major producers of ROS in liver cells is the microsomal cytochrome P450 relying on the monooxygenase system (Di Meo et al. 2016).

2.1.4 ROS Generation by Plasma Membrane

In many biological processes, including cell adhesion, cell signaling, and ion conductivity, the plasma membrane is involved. As it is typically subjected to an oxidizing environment, it is a major site of free radical reactions as well. ROS, which can be formed from dysfunctional cells in tissues, causes oxidative damage to the components of the membrane unless successful antioxidant mechanisms are in operation. Increased membrane permeability, affected by lipid or protein oxidation, can contribute to reduced ion gradients of transmembrane, loss of secretory function, and inactivation of metabolic cellular processes.

The primary cause of ROS is O_2- production by NADPH oxidases, a membrane-bound enzyme. A well-known mechanism is O_2- production by the phagocyte enzyme, which helps to destroy bacterial invaders. When activated, the phagocyte NADPH oxidase produces the total amount of O_2- and H_2O_2 that account for a large fraction (10–90%) of the overall oxygen intake of microglia, macrophages, and neutrophils, although it is less clear whether this and other NADPH oxidases contribute to the total amount of cellular ROS generation at rest or at the time of activation (Di Meo et al. 2016).

2.1.5 ROS Generation by Lysosomes

Increased ROS concentrations, which involve free radicals including O_2- and •OH and non-radical species such as H_2O_2, result in OS. The integrity of the lysosomes is compromised by ROS, and this lysosomal membrane permeabilization then causes caspase-mediated apoptosis or necrotic cellular damage mediated by cathepsin. Also, ROS causes cell death involved with autophagosome aggregation and, in response to OS, this type of autophagic cell death is triggered in the nervous system (Kubota et al. 2010). To preserve an optimum pH for acidic hydrolases, the function of the redox chain is to promote the accumulation of protons inside of lysosomes. •OH radical, which involved the transition of three electrons to molecular oxygen, tended to give rise to the electron transport chain, while O_2- was not observed. The intermediate presence of O_2- is not omitted because the acid pH-milieu within lysosomes favors the dismutation of O_2- to H_2O_2, which is broken into •OH by intralysosomal Fe^{2+} (Di Meo et al. 2016).

2.2 Effect of ROS on the Dysfunction of the Cellular System

ROS has cell-stimulatory and mitogenic functions, as described above. However, their action is mainly attributed to the fostering of cell death at higher concentrations, generally with the morphologic characteristics of apoptosis. At the molecular scale, it could be emphasized that ROS exerts its damaging effects, eventually creating •OH, with disruption to many cellular components (Fig. 2).

2.2.1 Lipid Damage

Due to the existence of double bonds, which certainly constitute inherent regions of increased reactivity, unsaturated fatty acids are especially prone to alteration by oxidants, such as ROS. OS, resulting in lipid peroxidation, which transforms unsaturated lipids into polar lipid hydroperoxides, may cause interaction in complex damage to cellular biomembranes. The lipid peroxidation is problematic because the mediated damage is readily exacerbated by the excessive secretion of fatty aldehydes or reactive radicals (Finaud et al. 2006). They can change DNA, protein, and other macromolecules. Malondialdehyde (MDA), 4-hydroxy-2-nonenaldehyde (HNE), 2-propenaldehyde (acrolein), and isoprostanes are examples that can be measured as an indirect OS index (Deavall et al. 2012).

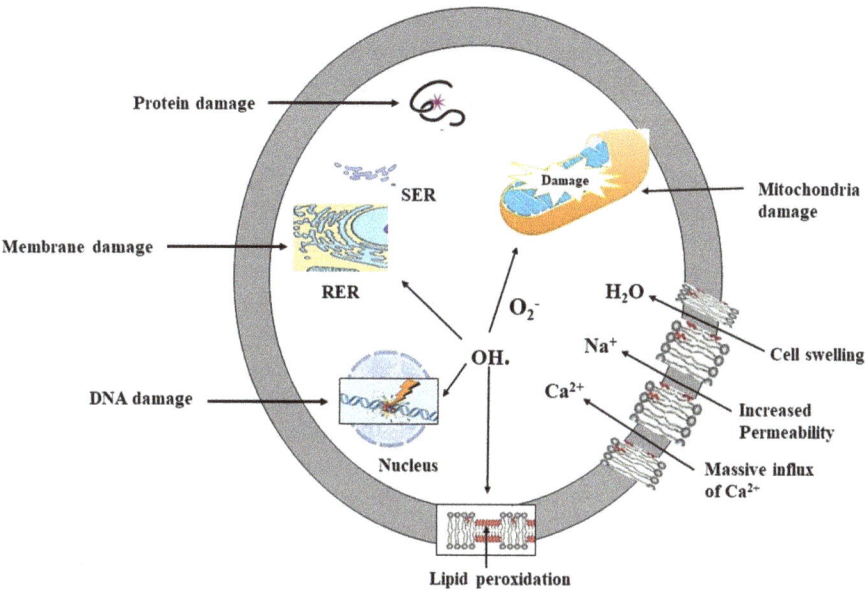

Fig. 2 Effect of oxidative stress on the cellular system

There is a simple straight peroxidative effect on polyunsaturated fatty acids (PUFA), close to those on most other cellular molecules. Classically, hydroxyl radical OH is the reactive species, which separates a hydrogen atom close to a double bond from the methylene moiety, providing relative stability of the associated radical by delocalization over the double bond itself. Molecular oxygen with the formation of an unsaturated conjugate peroxide undergoing spontaneous breakdown is further invaded by this intermediate, releasing aldehydes (primarily MDA and HNE) and hydrocarbons (pentane and hexane). Free fatty acids and even some esterified fatty acids, especially those found in cell membrane phospholipids, can be involved in such processes. The entire membrane system is disrupted in such cases and, due to the local aggregation of radicals the compromised cells can quickly undergo lysis. Furthermore, it must be understood that the breakdown of PUFA in phospholipids at position 2 creates two moieties, the external one that can diffuse away and be metabolized pretty rapidly, whereas the intrinsic one preserves its ester bonding to glycerol existing in the structure of the membrane until the entire phospholipid is degraded by itself or renovated by phospholipase interference (Bergamini et al. 2004).

2.2.2 Protein Damage

Phosphatases, kinases, transcription factors, and metabolic enzymes are oxidation-sensitive proteins, so protein oxidation can have a significant effect on cellular homeostasis by directly influencing cell signaling, cell structure, and metabolism-like enzymatic processes. Such proteins are more vulnerable than others to oxidation. Undesirable chemical alterations of the protein backbone, chain degradation,and oxidation of the side chains of the protein are the main alterations in cellular proteins caused by OS. The association of hydroxyl radicals with amino acids, such as lysine, proline, and arginine side-chains lead to the regular development of carbonyl moieties (aldehydes and ketones), almost all forms of amino acids, specifically in the context of lysine, proline, and arginine side-chains. Concerning this adaptive system of radical damage, oxidative damage could also occur from metal-catalytic changes at the metal-binding site and also from the production of adducts such as 4-hydroxy-nonenylaldehyde among proteins and reactive lipid peroxidation degradation products. Reaction of hydroxyl radical with the carbon of amino acids in the peptide is the fundamental mechanism causing damage to the protein backbone by oxidative radical-based damage. Peroxy compounds are also successfully synthesized by diamide and alpha-amidation pathways, which may turn into peptidyl cross-linkage products or result in protein degradation.

2.2.3 Nucleic Acids Damage

ROS-sensitive substances, and mostly hydroxyl radicals, react rapidly with nucleic acids, especially DNA, causing a wide range of structural reforms, including changes to the purine bases and the portion of sugar, with resulting glycoside bond modification, cross-linkage of the strand, and breakage of the strand. Almost all effects are highly responsible for mutation, specifically in the context of DNA replication in cells that are proliferating. •OH is the ROS species that is especially susceptible to nucleic acids, owing to the low oxygen tension in the nuclear region and the widespread nucleus compartmentation itself.

Alteration of the deoxyribose structure includes the immediate addition of hydroxyl radical to hydrogen, the discharge of water, and the production of a carbon-centered radical, which would be afterward either cycled to position 8 in purine nuclei or modified by the b-glycosidic bond or eventually by cleavage of carbon–carbon bonds in the sugar semi-acetal conformation. Strand-breakage with base release and development of two broken filaments on the same strand (ending with phosphate moieties of 3′ and 5′) with propenal release, quickly transformed into malondialdehyde, is the result of this final transition.

Purine or pyrimidine base changes are also based on hydroxyl radical conjugates on the electron-rich p orbitals of the side chain. The most vulnerable in the case of purine bases are broad guanine, which is modified into its 8-OH byproduct, which is extremely capable of mutations since its shape is comparatively stable with thymine–base pair. The most common alteration in pyrimidine bases is the adduct to glycol derivatives at positions 5 and 6. Malondialdehyde can alter bases quickly with C-O and C-N incorporations, repeatedly with a guanine side chain at positions 2 and 6, introducing an alternative cyclic system (Bergamini et al. 2004).

2.3 Generation of RNS and Its Damaging Effect on the Cellular System

RNS has a duplex role, like ROS, as it may either be disruptive to living systems or beneficial. RNS is obtained from the interactions of biologically produced free radicals to form more permanent species, resulting in several biological effects, such as ROS. The "RNS" contains such a wide variety of compounds with conflicting and distinct properties under strict chemical criteria that their only unifying feature can be drawn from NO•. NO• is derived from the amino acid, L-arginine metabolism. L-arginine is converted into L-citrulline and NO• by 5-electron oxidation of the guanidine nitrogen of L-arginine by the enzyme that catalyzes this oxidation, familiar as nitric oxide synthases (NOS). Two NOS isoenzymes usually known as low-output, calcium-dependent are neuronal NOS (nNOS or NOS1) and endothelial NOS (eNOS or NOS3). The high-output, calcium-independent, cytokine-inducible NOS (iNOS or NOS2) isoform is the third member of the NOS family.

Various roles are performed by NO• produced by NOS isoforms revealed in diverse cell types. NO• produced by nNOS in neurons serves in contact between nerve cells, whereas the free radical formed by iNOS in macrophages and smooth muscle cells contributes to one's destroying role, while NO• produced by eNOS in endothelial cells relaxes the cardiovascular system and maintains normal blood pressure in blood vessels. NO•, earlier detected in blood vessel regulation as a signaling molecule but now recognized as a transformer of essential physiological processes, can trigger harmful metabolic enzymes from cellular toxicity and produce peroxynitrite by superoxide reaction. NO• s rapid reactions to free radicals have arisen as one of the leading pathways to RNS formation and superoxide (O_2-) reactions to peroxynitrite formation (ONOO-) (Patel et al. 1999). Peroxynitrite is a potent oxidant; however, most of the biological molecules, such as DNA, proteins, and lipids, react at a comparatively slow rate. Under physiological conditions, peroxynitrite is chemically reactive, resulting in nitrate formation through isomerization. ONOO- is reactive to all large biomolecular groups and hence tends to mediate cytotoxicity independent of NO• or O_2 (Ryter et al. 2007). Nitrating proteins and lipids is a principal process by which peroxynitrite facilitates cellular impairment. Concerning protein modifications, peroxynitrite can induce the nitrosation of amines and thiols and the nitration of tryptophan and tyrosine. Protein modifications caused by peroxynitrite are often permanent and are usually associated with cellular injuries or death.

Some NO• functions are carried out by cGMP self-reliant pathways, including those related to mitochondria, in the signaling and modulation of cell activity. Most mitochondrial effects at physiological concentrations are attributed to NO• and are added to the respiratory chain (Benhar 2018). Firstly, NO• interacts with O_2- at the binding site just at cytochrome c oxidoreductase binuclear core, contributing to reverse the inhibition of the activity of cytochrome c oxidase. Secondly, NO•, interacting with respiratory Complex III, prevents the movement of electrons and increases the production of O_2-. NO• is also caused by protein nitrosation, preferentially interfering with nucleophilic sites in protein thiol residues, and mitochondria, assessed with NO-donors, show S-nitrosation and inactivation of Complex I (Di Meo et al. 2016).

3   Drugs-Induced OS and Associated Cytotoxicity

OS is known to exhibit a significant role in drug metabolism. The majority of the drugs are lipophilic in nature; hence, for proper targeting of these drugs to the specific site, they must be converted into hydrophilic molecules. Now the hydrophilic nature of these converted drugs will facilitate the proper absorption, distribution, and excretion too. The entire process which helps in converting these non-polar compounds into polar compounds is known as drug metabolism. The process of drug metabolism is divided into three different phases (Deavall et al. 2012). During phase I, cytochrome P450 enzymes oxidize these drugs and thus prepare these lipophilic drugs for their conversion into a hydrophilic drug. In phase II, these modified compounds are further conjugated to polar compounds in the presence of enzymes. Lastly, in phase III, the conjugated compounds may be further processed, before they are recognized by efflux transporters and secreted out of cells (Gandhi et al. 2012). CYP is a part of the microsomal monooxygenase system and is found in the endoplasmic reticulum (Zangar et al. 2004). Mainly, the poor coupling of the P450 catalytic cycle results in the continuous production of ROS which increases OS. The binding of substrate to CYP450 happens by the addition of one molecule of oxygen to the enzyme and thus forming its oxy complex. This oxy complex further gets reduced to a peroxy complex and accepts two protons and produces water through intermediate reaction. ROS is generated during the intermediate stages of CYP-mediated transformation of these drugs and hence the constant production results in the consumption of NADPH. The generated ROS continues to oxidize NADPH and further produce ROS even in the absence of any substrate (Barouki and Morel 2001; Bondy and Naderi 1994).

There are diverse types of drugs that are used for the treatment of different diseases/disorders. Medicinal drugs that are generally used for treating different diseases are categorized as antineoplastic, antipsychotic, antiretroviral, antipyretic, etc. However, in the long run, they are reported to exert severe side effects. There are also studies suggesting that during drug metabolism, these drugs lead to the generation of reactive intermediates which in turn may react with molecular oxygen/nitrogen thus generating ROS/RNS. This excessively generated ROS/RNS may impair the functioning of enzymes involved in drug metabolism and hence lead to toxicity. Therefore, some of the medicinal drugs that are involved in inducing OS and the associated toxicity which have been studied are summarized in Table 1.

3.1   Antipsychotic Drugs

These drugs are used for the treatment of psychiatric disorders such as schizophrenia, Parkinson's disease (PD), schizoaffective disorder, and delusional disorders. These are also involved in treating acute psychotic, manic, and psychotic-depressive disorders. dopamine (DA), precisely dopamine receptor 2 (D2), are the potential

Table 1 Examples of drugs involved in OS generation and associated toxicities

Class of drug	Name of drug	Structure of drug	Evidence for OS generation	Toxicity induced due to drug-induced OS	References
Antipsychotic	clozapine		Production of ROS and RNS	Neutrophil toxicity, cardiotoxicity	Deavall et al. (2012) and Wiciński and Węclewicz (2018)
	chlorpromazine		Production of singlet oxygen and superoxide in response to UV A/B irradiation	Hepatotoxicity, agranulocytosis, cutaneous phototoxicity	Otreba et al. (2015)
	olanzapine		Elevated ROS, mitochondrial dysfunction	Hepatotoxicity	Eftekhari et al. (2016)
	haloperidol		Downregulated levels of reduced glutathione, CAT and SOD, and elevated OS	Tardive dyskinesia, cardiotoxicity	Cho and Lee (2013)
	L-dihydroxyphe-nylalanine		Increased ROS production and downregulated antioxidant level	Neurotoxicity	Sabens et al. (2010)

(continued)

Table 1 (continued)

Class of drug	Name of drug	Structure of drug	Evidence for OS generation	Toxicity induced due to drug-induced OS	References
Antineoplastic	doxorubicin		The rise in levels of ROS and RNS leading to DNA damage, mitochondrial dysfunction	Cardiotoxicity	Sawyer et al. (2010) and Renu et al. (2018)
	bleomycin		Generate ROS resulting in DNA breakage, lipid peroxidation, GSH depletion, and oxidative damages to the lung	Pulmonary toxicity	Moeller et al. (2008)
	cisplatin		Increase in hydrogen peroxide, superoxide anion, and hydroxyl radical. Reduction of antioxidants, GSH-peroxidase, and GSH-reductase. Mitochondrial dysfunction	Ototoxicity, neurotoxicity (peripheral neuropathy), and renal toxicity (nephrotoxicity)	Miller et al. (2010) and Deavall et al. (2012)
	methotrexate		Elevated ROS due to downregulated levels of NADPH and GSH	Pulmonary toxicity, hepatotoxicity, and renal toxicity	Vardi et al. (2010) and Dalaklioglu et al. (2013)
	paclitaxel		Increased production of ROS, especially superoxide	Peripheral sensory neuropathy, cutaneous toxicity, and GI toxicity	Xiao et al. (2011)

Class of drug	Name of drug	Structure of drug	Evidence for OS generation	Toxicity induced due to drug-induced OS	References
Antimalarial	chloroquine		Elevated lipid peroxidation, mitochondrial dysfunction	Retinopathy, neuropathy, cardiomyopathy	Chaanine et al. (2015) and Redmann et al. (2017)
	primaquine		Production of secondary free radicals	Hematotoxicity	Bowman et al. (2005)
Antiretroviral	azidothymidine		Elevation of ROS/ RNS	Cardiotoxicity, neurotoxicity	Kline et al. (2009) and Akay et al. (2014)
	didanosine		Inhibition of mitochondrial DNA polymerase affecting mitochondrial function	Peripheral neuropathy, hepatic steatosis, pancreatitis	Kakuda (2000)
Analgesic and antipyretic	acetaminophen		CYP450 enzymes-mediated metabolism may induce OS leading to depleted glutathione, activation of proapoptotic proteins. Mitochondrial dysfunction, inflammation	Hepatic/ renal toxicity	Deavall et al. (2012)
	aspirin		Elevated ROS, reduced GSH pool resulted in the elevated OS, mitochondrial dysfunction	GI toxicity, hepatic toxicity, cerebral bleeding	Raja et al. (2011)

(continued)

Table 1 (continued)

Class of drug	Name of drug	Structure of drug	Evidence for OS generation	Toxicity induced due to drug-induced OS	References
Drugs of abuse	cocaine		Increased levels of H_2O_2, lipid peroxidation	Neurotoxicity	Cunha-Oliveira et al. (2013)
	amphetamine		Elevated production of ROS due to mitochondrial mutation	Neurotoxicity	Frey et al. (2006)
Antibiotics	levofloxacin		Induction of ROS due to mitochondrial dysfunction	Hepatotoxicity	Kohanski et al. (2017)
	ciprofloxacin		Increased ROS production	Liver and kidney damage	Afolabli and Oyewo (2014)
Anti-tuberculosis	isoniazid		Production of free radicals via CYP2E1-induced OS, mitochondrial dysfunction	Hepatotoxicity, neurotoxicity, nephrotoxicity	Yue et al. (2009) and Ahadpour et al. (2016)
	rifampin		Elevated levels of CYPs and GSTs	Hepatotoxicity	Ramappa and Aithal (2013)

binding sites for the antipsychotic drugs (Maurya et al. 2017). Drugs that are frequently being used as antipsychotic drugs are clozapine, chlorpromazine, L-dihydroxyphenylalanine (L-DOPA), olanzapine (OLZ), haloperidol (HP), etc. However, in the long run, they lead to adverse effects such as cardiotoxicity and neurotoxicity. Thus, the evidence for the same has been presented below with the help of some examples.

3.1.1 Clozapine

It is a commonly prescribed antipsychotic drug used for treating schizophrenia. This drug mainly binds to central dopamine (D1, D2, and D4) receptors and exerts its therapeutic action. It also has a high affinity to α-adrenergic (Kalkman et al. 1998), muscarinic (Richelson and Souder 2000), and histamine receptors (Bymaster et al. 1996) in addition to dopamine and serotonin receptors (Monsma et al. 1993). However, the usage of clozapine has been restricted as it has been reported to cause severe side effects such as neutropenia and agranulocytosis (Fehsel et al. 2005; Li and Cameron 2012). It has been suggested in studies that OS could be one of the reasons for the associated toxicity. During the metabolism of clozapine, reactive intermediates are generated resulting in excess ROS formation (Dragovic et al. 2013). CYP450 enzyme mediates oxidative bioactivation of clozapine and leads to the production of reactive nitrite ions and clozapine-N-oxide. There are also peroxidases-mediated bioactivation of clozapine that led to the generation of reactive nitrenium ion. This nitrenium ion then forms protein adducts and aids in neutrophil toxicity (Li and Cameron 2012; Wiciński and Węclewicz 2018). It has also been observed in several studies that prolong treatment of clozapine is associated with cardiotoxicity; however, the exact mechanism has not been deciphered yet (Ronaldson et al. 2010; Curto et al. 2016; Yuen et al. 2018).

3.1.2 Chlorpromazine

It belongs to the largest class of antipsychotic drugs and is used for treating psychotic disorders such as schizophrenia and manic depression in adults. It mediates its antipsychotic action via blocking postsynaptic dopamine receptors present in cortical and limbic areas of the brain, hence preventing the surplus of dopamine in the brain (Mailman and Murthy 2010; Ginovart and Kapur 2012). However, according to studies, chlorpromazine therapy led to side effects such as hepatotoxicity, agranulocytosis, hyperprolactinemia, and jaundice (Antherieu et al. 2013; Drucker and Rosen 2011; Mitkov et al. 2014), and it was suggested that OS imparts a crucial role in associated toxicity. There is a conversion of chlorpromazine via CYPD1 and CYP1A2 into quinone imine metabolite which is toxic in nature. This molecule also leads to the generation of toxic radicals via peroxidase catalyzed oxidation, thus accounting for hepatic pathophysiology (Macallister et al. 2013). Another well-known side effect of this drug is cutaneous phototoxicity (Moore 2002). On photochlorination, chlorpromazine is converted to an excited state, which on reaction with molecular oxygen as a subsequent energy transfer leads to the production of both excited singlet oxygen and superoxide species. These reactive species in turn can interact with DNA and other biomolecules and elicit toxic reactions in these cells. Sebastein et al. suggested that chlorpromazine-induced intrahepatic cholestasis in human hepatocytes and OS plays a pivotal role in this toxicity as both primary and aggravating factor (Antherieu et al. 2013). In a study conducted by Otreba et al. the effect of chlorpromazine on the level of ROS concentration in HEMn-DP melanocytes was estimated using electron paramagnetic resonance spectroscopy. It was found that chlorpromazine triggered the elevation of OS, which could also be the reason for its associated undesirable side effects (Otreba et al. 2015).

3.1.3   Olanzapine (OLZ)

This antipsychotic drug aims to treat early onset schizophrenia and other depressive disorders with or without symptoms (Han et al. 2013; Wang and Si 2013). It acts as a serotonin-dopamine receptor antagonist (Geddes et al. 2000). Besides its superior efficacy, long-term administration of olanzapine results in adverse toxic effects such as dyslipidemia, metabolic syndrome, and cognitive dysfunction (Muench and Hamer 2010; Zhang et al. 2017). The liver is the major organ for the detoxification of drugs and is also the prone site for damage due to drug-induced OS. One of the severe adverse effects associated with olanzapine treatment is hepatotoxicity (Dumortier et al. 2002; Raz et al. 2001). Numerous studies have documented OLZ-induced hepatotoxicity (Hung et al. 2009; Jadallah et al. 2003; Manceaux et al. 2011; Tchernichovsky and Sirota 2004); however, the exact mechanism regarding the hepatotoxicity has not been clear yet. In a study of human neuronal cell line and mouse brain, it was reported that OLZ via increased OS caused neurotoxicity, mitochondrial membrane depolarization, and damage (Vucicevic et al. 2014). It was also observed that OLZ administration at high doses may increase the ROS formation and impair mitochondrial functioning as well resulting in several toxicities associated with drug-induced OS (Eftekhari et al. 2016).

3.1.4   Haloperidol (HP)

It is a typical antipsychotic drug. It provides cure for tics in the case of Tourette syndrome, mania in bipolar disorder (BD), vomiting, nausea, acute psychosis, and schizophrenia (Rifkin et al. 1991; Dold et al. 2015). This drug acts as a high-affinity antagonist to the D2 receptor. It exerts its antipsychotic action by blockage of D2 receptors, particularly in the mesolimbic and mesocortical regions. Like other antipsychotic drugs, HP has also been reported to show toxic effects after treatment.

The most common adverse effect associated with HP treatment is tardive dyskinesia (impairment of extrapyramidal nerve tract) (Shivakumar and Ravindranath 1992). In addition to Tardive Dyskinesia, HP treatment is often associated with cardiotoxicity. Cardiotoxicity represents the most severe adverse effect of this drug as it prolongs the QT interval and thereby elevates the risk of arrhythmias of the Torsade de Pointes (TdP) type that may lead to sudden cardiac death (Fayer 1986; Hunt and Stern 1995). It has been suggested from studies that OS has also been involved in the pathogenesis of Tardive Dyskinesia (Cho and Lee 2013). Thakur et al. 2015 reported that prolong administration of HP also resulted in elevated OS, and the parameters used were elevated levels of lipid peroxidation and nitrite concentration, and downregulated the reduced glutathione, catalase (CAT), and SOD in comparison to the vehicle-treated animals in the cortex as well as in the striatal region. Hence, due to the elevated OS in these regions after HP treatment, there can be an association between OS and a decrease in the antioxidant brain defense system. It was inferred that elevated concentration of dopamine metabolism may result in the production of free radicals and induce OS (Goff and Coyle 2001; Lieberman et al. 2008; Cho and Lee 2013).

3.1.5   L-Dihydroxyphenylalanine (L-DOPA)

Most of the people suffering from Parkinson's disease (PD) are administered this drug. L-DOPA lowers down the levels of DA in the striatum and is also involved in the inhibition of DA transporter (DAT), present on the nerve terminals in the brain. It is also indirectly involved in the activation of the receptors of D1(D1, D5 receptors) and D2 (D2, D3, D4 receptors) families. However, usage of this drug has been limited due to its associated toxicity such as neurotoxicity. The association of neurotoxicity with PD is involved in dopamine oxidation leading to free radicals and quinone species generation (Graham 1978). The dopamine is changed to dopamine quinone which further undergoes oxidation and forms dopamine o-semiquinone. There is the contribution of this dopamine-derived quinone toward the basic phenomenon of OS-associated mitochondrial defects in the pathogenesis of PD (Gluck and Zeevalk 2004; Berman and Hastings 1999). It has been suggested that excess generation of ROS, as well as downregulated levels of antioxidant enzymes, may be due to elevated levels of oxidative processes resulting in the alteration of nucleic acids during PD treatment via L-DOPA (Dorszewska et al. 2005; Dorszewska et al. 2009). Administration of L-DOPA is also suggested to decrease the cell viability and length of a neurite, in a dose-dependent manner. In these studies, it was also observed to exhibit the OS and damage to DNA as well (Sabens et al. 2010; Stansley and Yamamoto 2013).

3.2   Antineoplastic Drugs

It refers to an array of medicinal drugs that are clinically used for decreasing the growth of an abnormal mass in specific tissues. The abnormal mass is clinically termed as "tumor" or "neoplasm" which may be benign or malignant in nature.

These drugs exert their action via several mechanisms; however, a common mechanism of action is preventing cell replication or growth. There are numerous types of antineoplastic drugs based on their different mechanisms of action such as (1) alkylating drugs (e.g., cisplatin, chlorambucil, procarbazine, carmustine, etc.); (2) antimicrotubule drugs (e.g., vinblastine, paclitaxel, etc.); (3) antimetabolites (e.g., methotrexate, cytarabine, etc.); (4) topoisomerase inhibitors (etoposide, doxorubicin, etc.); and (5) cytotoxic agents (bleomycin, mitomycin, etc.). Some of the antineoplastic drugs with their pharmacological effect and side effects have been discussed below:

3.2.1   Doxorubicin (DOX)

It is an anthracycline antineoplastic drug. It is used in many chemotherapy treatments for hematological and solid tumors (Damiani et al. 2016). This drug exerts its mechanism of action by inserting into the DNA strand of cancer cell and inhibiting the topoisomerase II, thus preventing DNA replication. It may also lead to the production of free radical formation resulting in the generation of OS in the cancer cell. These two actions of DOX ultimately lead to the death of the cancerous cell (Thorn et al. 2011). However, it is quite effective as an antineoplastic drug but often side effects of DOX treatment are observed such as dose-dependent cardiotoxicity. It has been reported in many studies that the reason behind DOX cardiotoxicity could be primarily because of extreme ROS production from the enzymatic redox cycle. There can be different mechanisms by which DOX can generate ROS (Renu et al. 2018). One of the main mechanisms is the one-electron reduction mechanism by the Complex I of the mitochondrial ETS system in which the quinone moiety of doxorubicin molecule is converted into semiquinone moiety (Marcillat et al. 1989; Octavia et al. 2012). There is superoxide radical formation when this semiquinone moiety reacts with molecular O_2, and the original quinone form is retained by the doxorubicin molecule. The cycling process of quinone to semiquinone conversion results in the production of a large amount of superoxide radical (O_2-), which in turn leads to the generation of active ROS/RNS species (such as H_2O_2, • OH, and ONOO-) (Chen et al. 2007).

DOX is also known to generate ROS by interfering with iron metabolism. The semiquinone moiety of DOX, O_2-, and their byproduct H_2O_2 is known to trigger the release of Iron from ferritin (an iron storage protein) and cytoplasmic aconitase (which contains 4Fe-4S cluster) (Minotti et al. 2004). Due to the loss of the (4Fe-4S) cluster, the cytoplasmic aconitase gets converted into iron regulatory protein (IRP)-1. This IRP-1 protein is known to exhibit a very strong binding toward a highly conserved iron-responsive element, present in the untranslated regions of transferrin receptor (TfR) (Chen et al. 2007). Consequently, iron uptake exceeds iron sequestration and leads to increased levels of free iron at the cellular level. This in turn leads to the more generation of •OH via Fenton chemistry reaction, resulting in OS and cytotoxicity (Renu et al. 2018).

3.2.2 Bleomycin (BLM)

This drug is mainly used in treating lymphomas, squamous cell carcinomas, testicular tumors, and malignant pleural effusions. It mediates its cytotoxic effect by the formation of a complex with oxygen and other metal ions, and this complex then binds to DNA and cleaves to the phosphodiester–deoxyribose backbone. Though it has wide applications as a chemotherapeutic agent, clinically it has been limited due to its severe adverse effects such as pulmonary fibrosis (PF) (Froudarakis et al. 2013; Yu et al. 2016). BLM is known to produce ROS such as superoxide and hydroxyl radicals. DNA damage, lipid peroxidation, epithelial cell injury, and excessive deposition of extracellular matrix and lung collagen synthesis contribute to the production of ROS in the lung tissue. BLM administration induces inflammatory and fibrotic reactions thereby promoting collagen production via fibroblasts (Moeller et al. 2008; Chaudhary et al. 2006). Due to free radicals and cytokines, initially, BLM pulmonary toxicity appears as pneumonia starting with damage to vascular tissues and then eventually progressing toward fibrosis (Sleijfer 2001). Thus, it can be suggested that although bleomycin is known to produce ROS, when this creates an imbalance between intracellular ROS and the antioxidant system then it progresses toward lung injury often termed as pulmonary fibrosis.

3.2.3 Cisplatin

It is used for treating testicular, lung, bladder, ovarian, and gastrointestinal cancers. At the molecular level, cisplatin is involved in various processes such as the generation of OS via ROS production and lipid peroxidation, activation of p53 signaling and cell cycle arrest, decreased activity of protooncogenes, and anti-apoptotic proteins, and induction of apoptotic pathways (Dasari and Tchounwou 2014). Although its antineoplastic role is well known, however, it has been reported to exert multiorgan toxicity. The mechanism underlying this multiorgan toxicity is observed to be redox imbalance. Based on clinical studies, ototoxicity, neurotoxicity (peripheral neuropathy), and renal toxicity (nephrotoxicity) have been observed as adverse effects of this therapy, and severity may increase with a span of 5–20 years after therapy (Sprauten et al. 2012; Dasari and Tchounwou 2014).

In the case of nephrotoxicity, this drug mainly affects the S3 segment of the proximal tubule (PT) (Cristofori et al. 2007; Miller et al. 2010). It has been evident from studies that Organic cation transporter (OCT) 2 is the entry route for the cisplatin into the cells (Burger et al. 2010) as OCT2 is known to be present at the basolateral surface of PT cells in the human kidney. Thus, it can be proposed that the reason behind the cisplatin accumulation in PT could be the transport via OCT2. Furthermore, cisplatin-induced OS via generating superoxide anion, H_2O_2, and hydroxyl radical has been observed both in vitro and in vivo (Masuda et al. 1994; Tsutsumishita et al. 1998). It has also been suggested that elevated ROS generation is involved in ototoxicity in response to cisplatin administration. Increased levels of nicotinamide adenine dinucleotide phosphate oxidases (NOX-1 and NOX4) are implicated for the cisplatin-induced ototoxicity (Kim et al. 2010a, b; Sheth et al. 2017). Hepatotoxicity is also observed as one of the side effects of a high dosage of cisplatin administration. Cisplatin is observed to decrease the level of reduced glutathione GSH and thus increase the level of OS (Yilmaz et al. 2005; Miller et al. 2010). Various in-vivo studies have also reported that there was a significant increase in levels of hepatic malonaldehyde (MDA) and downregulation of antioxidant enzymes when treated with cisplatin (Yilmaz, et al. 2005; Mansour et al. 2006).

3.2.4 Methotrexate (MTX)

MTX has been extensively used as an antimetabolite as well as an antineoplastic agent for treating various types of cancer of the head, neck, skin, and breast. Currently, it has also been used as an alternative therapy in the case of psoriasis and rheumatoid arthritis (Bergner et al. 2012). It exerts its cytotoxic effect by activating apoptosis in cells with a high proliferation rate and also hinders DNA synthesis and replication by blocking the action of dihydrofolate reductase at the S-phase of the cell cycle (Wiczer et al. 2016). Long-term administration of MTX treatment is often associated with severe side effects including cytopenias, pulmonary toxicity, hepatotoxicity, and renal toxicity (Gaies et al. 2012). Hepatic injury is often observed as a fatal and most common side effect; however, the exact mechanism has not been clearly understood in the literature (Hemeida and Mohafez 2008; West 1997; Conway and Carey 2017). It has been demonstrated in some studies that OS could contribute to the mechanism of MTX-induced hepatotoxicity (Gaies et al. 2012; Vardi et al. 2010). It was observed that excess production of ROS induced hepatotoxicity and is responsible for the depletion of both enzymatic and non-enzymatic mitochondria antioxidant defense systems (Gouveia et al. 2016; Moghadam et al. 2015). It has been evident from several studies that MTX administration downregulates the levels of NADPH, GSH, and increases ROS and associated hepatic toxicity with decreased cellular antioxidant machinery (Uraj et al. 2008; Hemeida and Mohafez 2008; Vardi et al. 2010; Dalaklioglu et al. 2013). It has also been evident from past studies that OS is also involved in MTX-induced testicular toxicity (Gokce et al. 2011; Vardi et al. 2009; Armagan et al. 2008). In a study by Yulug et al. it was demonstrated that excess ROS production results in sperm abnormalities and infertility (Yulug et al. 2013). An evident association between sperm damage induced due to MTX and OS was reported (Vardi et al. 2009). MDA is a biomarker for lipid peroxidation, and in this study it has been presented that MTX induces OS in tissues by elevated MDA levels (Miyazono et al. 2004; Sener et al. 2006; Armagan et al. 2008; Kose et al. 2012).

3.2.5 Paclitaxel

It is a member of the taxane family of antineoplastic drugs and is generally used as a first-line treatment for various cancers of breast, lung, and ovary (Ferlini et al. 2003). It exerts its antineoplastic action by interference in the mitotic spindle and microtubule dynamics leading to apoptosis. Despite its wide application, it is found to have various side effects including peripheral sensory neuropathy, cutaneous toxicity, and gastrointestinal toxicity (Rowinsky and Donehower 1995; Hanna et al. 2002; He et al. 2016). Peripheral sensory neuropathy is most commonly observed as a side effect of paclitaxel treatment which becomes more severe with cumulative dosing (Postma et al. 1995). It has been demonstrated in past studies that there is an association between paclitaxel-induced neuropathy and elevated ROS concentration (Flatters and Bennett 2006; Fidanboylu et al. 2011; Duggett et al. 2016). From the recent studies, the involvement of ROS and RNS has been observed in paclitaxel-induced neuropathy via behavioral studies with pharmacological scavenging agents (Doyle et al. 2012; Kim et al. 2010a, b). Paclitaxel is also known to target and affect mitochondria functioning. Mitochondrial dysfunction will result in an imbalance of intracellular redox potential and increased production of ROS, especially superoxide (Xiao et al. 2011). These superoxides can cause many modifications in peripheral neurons such as redundant mitochondrial injury-inducing apoptosis, inflammation, and lastly neurodegeneration. Hence, it has been inferred that various metabolic, energetic, and functional deficiencies in neurons cause peripheral neuropathic damage.

3.3 Antimalarial Drugs

It belongs to the category of drugs that are used for the treatment of malaria. Malaria is a fatal disease that is transmitted by the female Anopheles mosquito. This fatal disease is caused by the plasmodium, a genus of pathogenic protozoans. The life cycle of plasmodium is very complex; hence, various drugs are required at different stages to kill or resist this parasite. Two crucial antimalarial drugs, chloroquine and primaquine, have been discussed below with their mechanism of action as well as their toxic effects.

3.3.1 Chloroquine (CQ)

Chloroquine has been widely used for the treatment of malaria for many years. Several studies have documented the protective effects of chloroquine in case of cancer, systemic lupus erythematosus (SLE), rheumatoid arthritis as well as in viral infections (Thome et al. 2013; Al-Bari 2017; Zhang et al. 2015; Aguiar et al. 2018). CQ exerts its antimalarial effect during blood stages or liver stages of the parasite life cycle (Milner 2018). However, its clear mechanism of antimalarial action has not been elucidated. In the same regard, to date, the most accepted theory is that CQ interferes with the hemoglobin digestion by the parasite thus preventing further growth of plasmodium (Skrzypek and Callaghan 2017). Despite its beneficial effects, severe toxic effects have been reported such as retinopathy, neuropathy, cardiomyopathy, hypoglycemia, dermatological reactions, and bone marrow suppression (Ruiz-Irastorza et al. 2010; Blignaut et al. 2019). The effect of chloroquine treatment at the cellular level is the elevation in mitochondrial fragmentation and cristate destruction (Chaanine et al. 2015). In a study of cortical neurons, similar results were observed where after chloroquine therapy there was a significant increase in mitochondrial DNA damage, suppression of key mitochondrial functions, and as a consequence, it leads to significant changes in cellular metabolism (Redmann et al. 2017). In another study, it was shown that CQ treatment leads to elevated lipid peroxidation and downregulated activity of antioxidant enzymes in rat's retina (Bhattacharyya et al. 1983). Various studies have also reported the adverse effect of CQ treatment on lysosomal function (Mackenzie 1983; Slater 1993).

3.3.2 Primaquine (PQ)

It is an antimalarial drug that is used for treating malaria. Despite its high toxicity, it is extensively used as an antimalarial drug due to its mechanism of action of blocking the transmission of plasmodium and thus preventing further transmission of the disease. However, beyond its usefulness and adaptability, it has been always been associated with harmful side effects. Hematotoxicity is one of the most severe adverse effects associated with primaquine therapy. Hematotoxicity is the condition of hemolytic anemia that can be induced especially in people with genetic deficiency of certain enzymes such as glucose-6-phosphate dehydrogenase deficiency (G-6-PDD) or NADH methemoglobin reductase deficiency. Though the mechanism underlying PQ-mediated toxicity has not been resolved yet, however, the hematotoxicity is observed to be linked with its active metabolite, 5- hydroxy primaquine. This active metabolite leads to the generation of secondary free radicals after undergoing the intracellular redox cycle, thus inducing OS (Bowman et al. 2005).

3.4 Antiretroviral Drugs

Antiretroviral drugs are the drugs that are used primarily for the treatment of retroviral infections such as the Human Immunodeficiency Virus (HIV). Due to the high complexity of the HIV infection mechanism, there are more chances of mutation during the transcription process which may lead to drug resistance. Thus, an HIV therapy regimen is used in combination with different drugs targeting different steps within the HIV life cycle thus aiming at synergistic antiviral effect. Presently, there are six main classes of antiretroviral agents that act on HIV infection and are clinically relevant. These include nucleoside reverse transcriptase inhibitors (NRTIs), non-nucleoside reverse transcriptase inhibitors (NNRTIs), fusion inhibitors, integrase inhibitors, protease inhibitors, and chemokine receptor antagonists (Pau and George 2014). Commonly used antiretroviral drugs are abacavir (ABC), didanosine, delavirdine, azidothymidine (AZT), maraviroc, etc. Many of these are observed to be involved in the elevation of ROS/RNS and induced toxicity after chronic administration (Akay et al. 2014). For instance, two of them have been explained below with their evidence on ROS/RNS and related side effects.

3.4.1 Azidothymidine (AZT)

AZT is used for the treatment of HIV. It is a member of class NRTIs, and it was the first approved antiretroviral drug for treating HIV. It acts on the HIV as a strong nucleoside reverse transcriptase inhibitor. AZT helps in preventing viral replication by attaching to DNA polymerase of reverse transcriptase, thus leading to termination of the chain (Pau and George 2014). It is used chronically in addition to other antiretroviral agents in "Highly Active Antiretroviral Therapy" treatments (Broder 2010). However, its long-term administration is accompanied by multiple side effects such as neuropathy, skeletal myopathy, and cardiac dysfunction (Dalakas et al. 1990). It is evident from many studies that chronic administration of AZT elevates mitochondrial ROS resulting in mitochondrial dysfunction. This mitochondrial dysfunction induced due to OS is often responsible for associated toxicities such as cardiotoxicity and neurotoxicity (Hantson 2019; Koczor and Lewis 2010; Gardner et al. 2014; Akay et al. 2014). As per the study conducted by Gardner et al. (2014), AZT is reported to reduce cell proliferation, mitochondrial metabolism, and cytochrome C oxidase activity but increased the lactate production in human muscle cell lines. In-vivo studies conducted in different mice models also indicated that AZT administration may lead to overproduction of hydrogen peroxide and thus AZT-induced toxicity too (Kline et al. 2009; Pereira et al. 2012). Another study conducted in 2010 by Amatore et al. also indicated that AZT may trigger the production of ROS and RNS in mouse macrophage model systems. These cells incubated with AZT were observed to release reactive species including peroxide and peroxinitrate. It is further suggested that probably thymidine alone did not escalate the release of ROS/RNS, thus indicating the significance of azido moiety AZT for generating OS (Amatore et al. 2010).

3.4.2 Didanosine

This drug is a nucleoside reverse transcriptase inhibitor that was generally used in combination with other antiretroviral drugs in the treatment of HIV. It is a deoxyadenosine analog that exerts its therapeutic action by competitive inhibition of HIV via chain termination. However, due to its adverse toxic effects, it is rarely used. It has been proposed that didanosine induces mitochondrial toxicity via inhibition of mitochondrial DNA polymerase leading to peripheral neuropathy, hepatic steatosis, pancreatitis, and lactic acidosis in the long run (Kakuda 2000).

3.5 Analgesic and Antipyretic Drugs

These are named analgesic and antipyretic drugs as they work in combination to relieve the body pain (analgesic action) and lower the body temperature in fever (pyrexia). In general, these agents are used for the treatment of fever, headaches, muscle aches, toothaches, common cold, etc. Examples of commonly used analgesic and antipyretic drugs are non-steroidal anti-inflammatory drugs (NSAIDs) like aspirin, Ibuprofen, etc., and acetaminophen (Tylenol). However, acetaminophen is not considered as an NSAID due to its less inflammatory effect but has been represented as NSAIDs due to a similar mechanism of action. NSAIDs exert their action by blocking the prostaglandin production via suppression of cyclooxygenase enzymes. For instance, two drugs have been discussed below with their desired as well as adverse effects.

3.5.1   Acetaminophen (Tylenol)

This drug is the most commonly used analgesic and antipyretic drug. Its mechanism of action is through suppression of cyclooxygenase 3 in the central nervous system and prostaglandin E2. Its overdose is associated with adverse effects such as acute hepatic toxicity and renal toxicity causing hepatic/renal failure, and in extreme conditions it may also lead to a patient's death. Most of it is metabolized (app. 63%) in the liver via glucuronidation and the remaining (~34%) by sulfation. It has been reported in adults, a very sparse amount of therapeutic dosing (~1%) is excreted via urination (Hendrickson and Bizovi 2006). At the therapeutic doses, there is oxidation of only less than 5% APAP to the reactive intermediate, electrophilic N-Acetyl-P-Benzoquinone Imine (NAPQI) mediated by the microsomal P-450 enzyme system. This reactive intermediate further gets reduced via glutathione and afterward gets excreted from the body in the form of a benign compound, mercapturic acid. However, at higher doses, there is depletion of glutathione and sulfate stores, which leads to the movement of more APAP to the CYP-450 mixed-function oxidase system, and as a result, there is the formation of more reactive intermediates (NAPQI). These free electrophilic intermediates participate in adduct formation with sulfhydryl and glutathione moieties on cellular protein resulting in cellular homeostasis disruption and ultimately triggering cell death (Mazer and Perrone 2008).

3.5.2   Aspirin

Aspirin (acetyl salicylic acid, ASA) is a member of non-steroidal anti-inflammatory drugs (NSAIDs). It has been extensively used as analgesic, antipyretic, and antiinflammatory all over the world (Fuster and Sweeny 2011). Aspirin mediates its pharmacological action by inhibition of COX-1 and COX-2, key regulators in the prostaglandin (PG) biosynthesis (Vane and Botting 1997; Vane and Botting 2003). Besides its analgesic and antipyretic action, this drug is also known to prevent cardiovascular diseases (CVD) at low doses (Patrono et al. 2005; Cryer and Mahaffey 2014). However, it has been reported from clinical studies that even at low doses in case of prevention from CVD, it has been associated with side effects such as gastric ulcer and bleeding (Sorensen et al. 2000; Garcia Rodriguez et al. 2001). Adverse effects such as gastritis (Garcia Rodriguez et al. 2001; Lavie et al. 2017), hepatotoxicity (Agúndez et al. 2011), cerebral bleeding (Ge et al. 2011) are often observed upon chronic exposure to aspirin. Gastritis or gastrointestinal (GI) toxicity is the most common and severe side effect associated with long-term usage of low-dose aspirin (Valkhoff et al. 2012; Cryer and Mahaffey 2014). It has been documented in several studies that there can be an association between ASA-induced toxicity and the elevated OS, mitochondrial dysfunction, and apoptosis activation (Goel et al. 2003; Zimmermann et al. 2000; Dikshit et al. 2006). Raja et al. using HepG2 cells in culture, have reported that elevated ROS, the decline in levels of GSH pool, and elevated OS followed by mitochondrial dysfunction are implicated with ASA-induced toxicity (Raja et al. 2011). ASA is converted to salicylic acid in the serum, liver, and GI tract via hydrolysis process by esterases. In a study by Doi et al. using an animal model, the effect of salicylic acid was investigated, and the parameters were ATP content, oxygen consumption, and lipid peroxidation. It was found that there was a decrease in ATP levels and an increase in lipid peroxidation (Doi and Horie 2010). Hence, it can be suggested that mitochondrial dysfunction and lipid peroxidation could be the triggering factor for ASA-induced hepatic toxicity.

3.6 Drugs of Abuse

In the present scenario, drug addiction is a serious issue all over the world (Cunha-Oliveira et al. 2008). Cannabis, amphetamines, cocaine, and opioids are often abused as illicit drugs. Among these, cannabis is the most abused drug worldwide. Depending on the path of administration chosen by the user, this drug abuse leads to adverse effects in different organs (Carvalho et al. 2012). It has been evident and supported from many studies that OS plays an important role in the toxicity induced due to abuse of these drugs. The major organs affected due to these drugs are the brain, heart, kidney, or liver (Cunha-Oliveira et al. 2013; Riezzo et al. 2012). Cocaine and amphetamine, for instance, have been explained below with their targets as well as adverse effects.

3.6.1 Cocaine

Cocaine is an alkaloid obtained from the leaves of South America's native Erythroxylon coca plant. By soaking the alkaloid in hydrochloric acid, cocaine hydrochloride is formed which produces a water-soluble salt that is useful for medical purposes. It is sold in crystals, granules, or a white substance that is mildly bitter and deadens the tongue and lips. Cocaine hydrochloride, which is 89% cocaine by weight, decomposes at 195 °C on heating and melts. A colorless, odorless, clear crystalline form that is nearly insoluble in water, but the cocaine alkaloid is soluble in alcohol, acetone, oils, and ether (Cregler and Mark 1986). By preventing the reuptake of neurotransmitters (norepinephrine, dopamine, and serotonin) at synaptic junctions, cocaine works, leading to increased concentration levels of neurotransmitters. Since norepinephrine is the main sympathetic nervous system neurotransmitter, it leads to sympathetic activation and contributes to vascular constriction, tachycardia, mydriasis, and hyperthermia (Warner 1993). Cocaine's internal stimulus induces an acute euphoria that is sometimes compared with orgasm. The mechanism responsible for this phenomenon is unclear and may be linked to the capacity of cocaine to inhibit the central nervous system's dopamine and serotonin reuptake. Local anesthetic activity, which arises from its capacity to obstruct the sodium channel in neuronal cells, is another potential pharmacological impact of cocaine.

Exposure to cocaine raises hydrogen peroxide (H_2O_2) in the prefrontal cortex and striatum of rats and also results in brain oxidative injury, as shown by elevated levels of lipid peroxidation in the hippocampus of rats exposed to cocaine in utero and by protein oxidation in human neuronal progenitor cells exposed to cocaine (Cunha-Oliveira et al. 2013). Cyclin A downregulation triggers the effect of cocaine on neuronal progenitor cell proliferation. Fortunately, a microarray review reported that among all the known modulators/key regulators of the G1-to-S transition, only cyclin A was observed to be downregulated by cocaine. Further studies also indicate ROS-induced ER stress facilitated the transcriptional downregulation of cyclin A. The downregulation of cyclin A represents at least one molecular pathway by which cocaine induces the dysfunction of neural progenitor cells (Lee et al. 2008).

3.6.2   Amphetamine

Amphetamine (1-methyl-2-phenethylamine, AMPH) is a well-known psychostimulant, usually recommended for the treatment of various disorders, such as attention deficit, narcolepsy, as well as obesity too. In vivo studies conducted by Frey et al. (2006) revealed that continuous exposure to d-amphetamine raised the level of superoxide and TBARS (product of lipid peroxidation) in prefrontal and hippocampal submitochondrial particles. The amphetamine-induced superoxide and TBARS may be generated from different sources. One of the important sources is the inhibition of mitochondrial OXPHOS (Burrows et al. 2000). Also, amphetamine is known to raise the release of DA from cytoplasmic vesicles. This increase in DA level leads to its increased metabolism by monoamine oxidase (located in the outer mitochondrial membrane), and as a byproduct of this metabolic activity hydrogen peroxide and dihydroxyphenylacetic acid (Berman and Hastings 1999) are generated. DA is also known to undergo spontaneous auto-oxidation and thus produce highly reactive DA quinones. Other studies have also supported these observations and indicated that a single injection of higher doses of AMPH (5–7.5 mg/kg) increased the level of lipid peroxidation in the rat cortex and striatum (Bashkatova et al. 2006; Wan et al. 2000). In humans, the use of 3,4-methylenedioxymethamphetamine has also been reported to be associated with elevated lipid peroxidation and compromised SOD as well as CAT activity in their erythrocytes (Zhou et al. 2003). Interestingly, studies further indicate that the use of antioxidants α-phenyl-N-tert-butyl nitrone and N-acetylcysteine may partially inhibit the AMPH-induced oxidative stress and neurotoxicity in rat striatum (Wan et al. 2006).

3.7 Antibiotics

Antibiotics are drugs that help stop bacteria-caused infections. Substances generated by different species of microorganisms (bacteria, fungi) are antimicrobial agents that inhibit the growth of other microorganisms and it may end up destroying them. Popularly used, however, the term "antibiotics" includes non-microbial synthetic or semi-synthetic antibacterial agents, including some sulfonamides and metronidazole (Soares et al. 2012). Antibiotics have made many contemporary medical interventions accessible, including cancer therapy, organ transplants, and open-heart surgery, concerning the treatment of infectious diseases. To understand the implications of resistance, it is worthwhile to acknowledge that, depending on their mode of action, antibiotics are categorized as shown below:

1. Agents that block bacterial cell wall synthesis.
2. Agents, which impair permeability, interact with the cell membrane of the microorganism.
3. Agents which, by influencing the role of 30S or 50S ribosomal subunits, obstruct protein synthesis.
4. Agents that obstruct microorganism's major metabolic steps.

It is possible to classify antibiotics into two broad categories: bacteriostatic drugs that prevent bacterial growth and bactericidal drugs that destroy bacteria. Most prominently, bacteriostatic antibiotics work through the inactivation of pathways of bacterial protein synthesis. Bacteriostatic drugs block ribosome activity primarily, attacking both the ribosome subunits 30S (tetracycline family and aminocyclitol family) and 50S (macrolide family and chloramphenicol). Bactericidal antibiotics attack DNA replication and repair by the ability to bind to DNA-complexed DNA gyrase, which promotes the formation of double-strand DNA breaks and cell death (Kohanski ct al. 2007).

3.7.1 Levofloxacin

Levofloxacin is an antibiotic of fluoroquinolone which is the S-(−) optical isomer of the racemic compound ofloxacin. It has a wide variety of action against Gram-positive and Gram-negative bacteria in vitro, and even certain other pathogens, including *Mycoplasma*, *Chlamydia*, *Legionella*, and *Mycobacteria* species (Fish and Chow 1997). Quinolones are highly active antibiotics attacking DNA gyrase bacteria as part of their main bacteria-killing mechanism. Bactericidal antibiotics also facilitate altered bacterial metabolic processes, respiration, and iron homeostasis in addition to the traditional modes of action, resulting in the development of ROS that leads to a portion of the cell death. Bactericidal antibiotics are also capable of inducing ROS formation in mammalian cells via mitochondrial dysfunction, building on this widespread mechanism of cell death. This contributes to the buildup of weakened DNA, proteins, and lipids, which may have long-term influences on patient systems (Kohanski et al. 2017). Drug-induced hepatotoxicity in patients referred to emergency care with acute liver failure is a serious cause of hepatocellular damage. The most common side effects are 0.8% nausea, 0.5% acne, 0.4% stomach pain, 0.3% diarrhea, dizziness, and vomiting. Less common are phototoxicity, cardiotoxicity, side effects of the central nervous system, and hepatotoxicity (Gulen et al. 2015). Studies have suggested diagnosis with levofloxacin enhances the formation of ROS and the activity of caspase-3. In human Sino-nasal epithelial cells (SNECs), levofloxacin induces a strong proapoptotic response (Kohanski et al. 2017).

3.7.2 Ciprofloxacin

Ciprofloxacin is the carboxylic acid of 1-cyclopropyl-6-fluoro-1,4-dihydro 4-oxo-7-(1-piperazinyl)-3 quinoline. In conjunction with the 1-piperazinyl group substitution at position C-7, the fluorine substituent at position C-6 greatly improves the potency and expands the scope of antimicrobial action (Vance-Bryan et al. 1990). The DNA gyrase enzyme is the primary target of quinolone activity. Ciprofloxacin, including some Gram-positive and most Gram-negative microbes, is much more effective against a wide range of pathogenic bacteria. Studies have also established that the generation of ROS by fluoroquinolones has contributed to liver and kidney cell damage. Ciprofloxacin enhanced the concentration of MDA in the liver of rats at systemic doses used over a therapeutic duration, suggesting increased lipid peroxidation in these species. Ciprofloxacin has been documented to cause an increase in advanced oxidative protein products (AOPP). Intracellular oxidized protein generation is related to increased ROS production, resulting from a disturbance in the equilibrium between antioxidants and pro-oxidants (Afolabli and Oyewo 2014). The production of these oxidized proteins could contribute to the development of aggregates of cytotoxic proteins that are essential pathological considerations in cell long-term damage.

3.8 Antituberculosis Drugs

Antituberculosis or antimycobacterial agents comprise a diverse group of compounds that are used for treating tuberculosis and leprosy. Tuberculosis (TB) is a fatal disease caused by *Mycobacterium tuberculosis*. Due to the ability of Mycobacterium to establish persistent infection, the need to prolong antibiotic treatment is a prerequisite to control and cure TB. Thus, anti-tuberculosis therapy mainly aims at the rapid destruction of actively multiplying bacilli, prevention of acquired drug resistance, and disinfection of host tissues to avoid clinical relapse. For achieving the same, the combination of anti-tuberculosis drugs is used in the therapy regimen. Common examples of these drugs are isoniazid, streptomycin rifampin, ethambutol, and kanamycin. Out of these drugs, Isoniazid and Rifampin with their therapeutic as well as associated toxic effects have been presented below.

3.8.1 Isoniazid (INH)

Isoniazid is a widely used drug for the treatment of tuberculosis (TB). Tuberculosis is categorized as a communicable disease spread by infection of *Mycobacterium tuberculosis* or other strains of bacteria. INH exerts its therapeutic action by entering the mycobacterial cell as a non-toxic "precursor" through the process of passive diffusion (Timmins and Deretic 2006). Afterward, there is oxidative activation of INH by KatG. KatG is a multifunctional enzyme and primarily behaves as catalase and peroxidase. However, it may also behave as peroxynitritase and NADH oxidase (Ng et al. 2004). During "normal" conditions, KatG breaks the peroxide generated by NAD(P)H oxidase of phagocytes of infected patients. Interestingly, after its entrapment by INH, the role of KatG gets reversed and now instead of neutralizing peroxides, it starts behaving as an intracellular "manufacturer" of toxic INH metabolites.

Although INH is quite effective as an antituberculosis drug, however, severe side effects such as hepatotoxicity, neurotoxicity, nephrotoxicity have been reported with its long-term usage. The most commonly reported toxicities with the chronic administration of INH are hepatotoxicity and nephrotoxicity. It has been reported in humans, and in vitro models of the human hepatocellular carcinoma cell line, that there has been an association of INH and its metabolites in inducing hepatic toxicity (Huang et al. 2003). Based on past studies, it has been suggested that cytochrome P450-2E1 (CYP2E1) via the production of free radicals plays an important role in hepatic toxicity caused due to INH. Hydrazine and acetyl hydrazine are the chief toxic metabolites of INH produced via CYP2E1, and these two metabolites are documented in several studies to induce OS and associated hepatic toxicity in humans, animals, microsomes, and HepG2 cells (Shen et al. 2006; Chowdhury et al. 2006; Zhai et al. 2008). It has also been proposed in numerous studies that OS can be the underlying mechanism for INH-induced hepatotoxicity (Shen et al. 2007; Hasiloglu et al. 2012; Rao et al. 2012). However, various mechanisms may be involved in the toxicity induced due to INH but it has been suggested that it is the imbalance between production and elimination of free radicals after prolong INH administration that leads to the pathogenesis of hepatic toxicity (Yue et al. 2009). It has been reported that prolong INH administration causes OS, mitochondria dysfunction, and apoptosis in HepG2 cells hence resulting in associated toxicity (Schwab and Tuschl 2003; Bhadauria et al. 2007, 2010; Ahadpour et al. 2016).

3.8.2 Rifampin

It is a semisynthetic derivative of rifamycin with wide antimycobacterial activity. It is one of the most potent anti-tuberculosis drugs. It exerts its mechanism of action by inhibition of DNA-dependent polymerase which further blocks the synthesis and transcription of bacterial protein (Shi et al. 2007). This drug is normally well tolerated but has been reported to be implicated with dose-dependent and dose-independent toxic effects. GI symptoms such as nausea, anorexia, and diarrhea are commonly observed as dose-dependent side effects. Dose-independent side effects include hepatic toxicity, hypersensitive reactions such as urticaria, thrombocytopenia, hemolysis, and renal toxicity. However, hepatotoxicity is the most commonly observed side effect of rifampin when used in combination with INH. rifampin is involved in the activation of pregnane X receptor (PXR), a member of the nuclear receptor superfamily of ligand-dependent transcription factors followed by elevated levels of Phase I and Phase II drug-metabolizing enzymes, that includes CYPS and GSTs, as well as drug/substrate transporters, such as ATP binding cassette transporters (ABCB) (Ramappa and Aithal 2013). It has been suggested that the activation of hydrolases and other enzymes, via rifampin, is involved in elevated production of reactive metabolites from isoniazid that is toxic to hepatic cells, hence contributing to the implicated toxicity of isoniazid and rifampin (Askgaard et al. 1995).

4   Potential of Antioxidant Therapy for OS Generating Drugs

Generally, it is hypothesized that antioxidants may assist in lowering the OS-linked adversarial effects of various therapeutic drugs and prevent the cellular system. The antioxidant system is composed of enzymatic factors (CAT, SOD, and Gpx), non-enzymatic low molecular weight compounds (glutathione, n-acetylcysteine, vitamins E, A and C, coenzyme Q10, carnitine, myo-inositol, lycopene, etc.), and micro-nutrients (selenium, zinc, and copper). SOD is considered to be the first line of defense of the body against ROS. It has been found that up to some extent antioxidants exert a positive effect. The biological antioxidants studied so far are classified into two main categories: low molecular weight and high molecular weight compounds. Low molecular weight antioxidant compounds are known to reduce the free radicals by oxidizing themselves. For instance, ascorbic acid, tocopherol, reduced glutathione, etc. are known to get oxidized and thus help in protecting the cellular system from the damaging effects of free radicals. The relatively high molecular weight compounds/complexes including polyunsaturated macrocyclic aromatics are known to form a coordination bond with transition metal ions. These aromatics mimic the structure as well as the function of endogenous antioxidant enzymes (CAT, SOD, Gpx, GR, GST) and thus also known as "catalytic antioxidant mimetics." Other compounds are not chemically reductive agents but may exert intracellular "antioxidant" effects. Although endogeneous antioxidant defense system imparts a crucial role in evading the OS-induced damage, excessive ROS accumulation can overcome this defense mechanism too. Thus, there is a need for the administration of antioxidant molecules that may exert beneficial effects and may help in reducing the damage induced due to excessive OS generation. A lot of studies have been conducted on numerous plant extracts for estimating their protective effect against oxidative stress-mediated damages. These plant extracts are a mixture of polyphenols, flavonoids, as well as thiol and act as a natural antioxidant solution. Resveratrol is a natural antioxidant phytoestrogen that is reported to mediate its shielding effect against estrogen-induced breast cancer by triggering the NRF2.

Activation of NRF2 leads to the increased expression of antioxidant molecules such as NAD(P)H Quinone oxidoreductase (NQO1) and superoxide dismutase 3 (SOD3) (Singh et al. 2014).

Mangiferin (2-C-β-D-glucopyranosyl-1,3,6,7-tetrahydroxyxanthone) is another naturally occurring polyphenol present in different plants belonging to the Anacardiaceae and Gentianaceae families (e.g., mango). It is known to decrease galactosamine-induced hepatotoxicity by reducing both ROS and NO formation (Yoshimi et al. 2001; Das et al. 2012). Another compound, taurine (2-aminoethanesulfonic acid), is also observed to curb the alloxan-induced diabetic nephropathy. Taurine is also proposed to inhibit p47phox/CYP2E1 pathway and lower the renal OS as well as the activity of xanthine oxidase (Das et al. 2012). This compound is also reported to prevent testis from its oxidative damage, induced by doxorubicin. The functions of CAT, SOD, glutathione reductase, and glutathione S-transferase have been restored by taurine, while also reducing OS. In addition, ER stress-induced testicular cell death is also observed to improve. This molecule also defends against liver and heart damage induced by arsenic, as well as toxicity generated by sodium fluoride in murine hepatocytes (Das et al. 2008; Ghosh et al. 2009). Carotenoids are other natural antioxidants that always defend the cell membrane's structure, control epithelial cell proliferation, and promote healthy spermatogenesis too (Barati et al. 2020). It has been studied that there are different carotenoids (lycopene, β carotene, lutein, etc.) that are involved in reducing drug-induced OS and mediated cytotoxicity (Rao and Rao 2007). It was demonstrated in animal models of Alzheimer's disease (AD) that on lycopene administration there was a reduction in mitochondrial oxidative damage (Prakash and Kumar 2014). The downregulated expression of NF-κB and proinflammatory cytokines in the brain (Sachdeva and Chopra 2015) were also observed altogether contributing to the inhibition of Aβ formation and improvement of memory retention (Prakash and Kumar 2014; Sachdeva and Chopra 2015). It has also been reported that lycopene acts as a potent antioxidant by reducing smoke-generated ROS and modulating the redox-sensitive cell targets such as protein kinases, protein tyrosine phosphatases (PTP), MAP kinase (MAPKs), and transcription factors (Kaulmann and Bohn 2014). It has been shown in studies that β-carotene exerts dose-dependent pro-oxidant effect by interfering with the gene expression of heme oxygenase 1 in human dermal fibroblasts (FEK4) (El-Agamey et al. 2004; Rao and Rao 2007). Lutein behaves as a strong antioxidant; it mitigates OS induced by benzo(a)pyrene (Vijayapadma et al. 2014) and H_2O_2 (Gao et al. 2011). Curcumin acts as an antioxidant by scavenging ROS such as H_2O_2 and O_2- (Yadav et al. 2005). In a study by Al-Karawi et al. antidepressant effects of curcumin were observed in patients suffering from major depressive disorders. On curcumin administration, there was an improvement in symptoms of depression. However, its administration and effect on patients are dose-dependent as for any significant effect it needs to be administered for a longer period (Al-Karawi et al. 2016). Arjunolic acid (AA: 2,3,23-trihydroxyolean-12-en-28-oic acid) is a naturally occurring chiral triterpenoid saponin. It is isolated from the bark of *Terminalia arjuna* and well known for multi-functional therapeutic benefits (Ghosh and Sil 2013). It has been reported that AA mitigates renal and

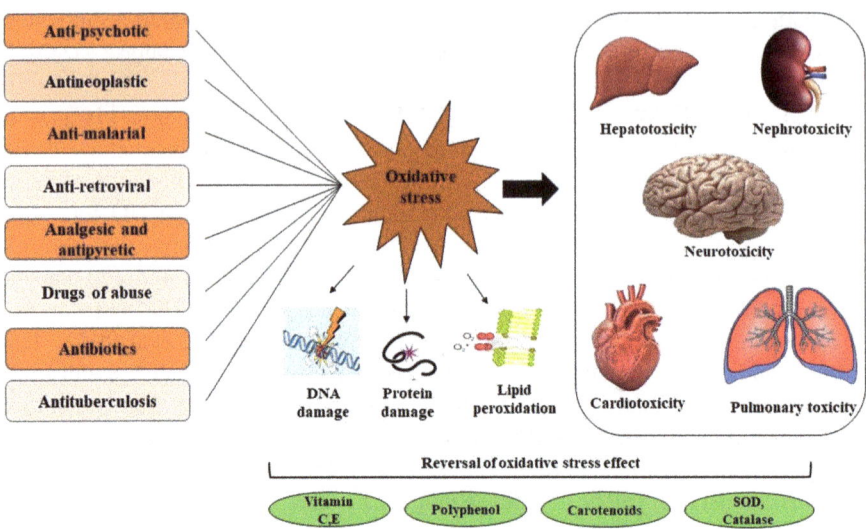

Fig. 3 Schematic representation of different classes of drugs, their toxicities on different organs, and suggestions to reverse the effect of drug-induced OS by using various antioxidants

hepatic OS induced due to sodium fluoride (Sinha et al. 2007). It also reduced the sodium nitrite-induced cardiac pathophysiology by interfering with the expression of pro-inflammatory cytokines and intrinsic apoptotic pathways (Al-Gayyar et al. 2014). This molecule alleviates cardiotoxicity induced due to DOX (Ghosh et al. 2011). Thus, based on studies mentioned so far, it can be suggested that antioxidant therapy can be effective in alleviating OS-induced cytotoxicity. In a similar context, Fig. 3 summarizes the different classes of drugs, their toxicities on different organs, and the role of antioxidants in reversing the effect of drug-induced OS.

5 Conclusion

In conclusion, the pharmaceutical companies are trying maximally to provide the most effective and safest drugs to the patients. However, a great number of drugs are being failed as such drugs, during their metabolisms, are reported to induce cytotoxicity and tissue damage. As per literature, even a lot of well-characterized drugs are also known to cause drug-induced cytotoxicity. For instance, doxorubicin, cisplatin, and AZT are well-characterized drugs and their clinical side effects are also well recognized. However, the precise mechanisms by which these drugs may persuade their lethal effects are not completely known. Various studies indicate that these drugs increase the level of either ROS and/or RNS and damage various cells and organs by different mechanisms. As per the current scenario, a lot of studies are further required to provide insight into drug-induced cytotoxicity and its relationship with ROS and ROS generation. Also, to curb the effect of OS generated by such drugs, the supplementation of various natural and/or synthetic antioxidants may also be recommended. However, the studies related to the co-supplementation of suitable antioxidants and drugs require a lot of attention.

Acknowledgments The authors acknowledge Jaypee Institute of Information Technology, Noida, for providing the entire infrastructure to complete this chapter.

Conflict of Interest The authors declare no conflict of interest.

References

Afolabli OK, Oyewo EB. Effects of ciprofloxacin and levofloxacin administration on some oxidative stress markers in the rat. Int J Biol Vet Agric Food Eng. 2014;8:31–9.

Aguiar ACC, Murce E, Cortopassi WA, et al. Chloroquine analogs as antimalarial candidates with potent in vitro and in vivo activity. Int J Parasitol Drugs Drug Resist. 2018;8:459–64.

Agúndez JA, Lucena MI, Martínez C, et al. Assessment of nonsteroidal anti-inflammatory drug-induced hepatotoxicity. Expert Opin Drug Metab Toxicol. 2011;7:817–28.

Ahadpour M, Eskandari MR, Mashayekhi V, et al. Mitochondrial oxidative stress and dysfunction induced by isoniazid: study on isolated rat liver and brain mitochondria. Drug Chem Toxicol. 2016;39:224–3.

Akay C, Cooper M, Odeleye A, et al. Antiretroviral drugs induce oxidative stress and neuronal damage in the central nervous system. J Neuro-Oncol. 2014;20:39–53.

Al-Bari MAA. Targeting endosomal acidification by chloroquine analogs as a promising strategy for the treatment of emerging viral diseases. Pharmacol Res Perspect. 2017;5(1):e00293.

Al-Gayyar MM, Al Youssef A, Sherif IO, et al. Protective effects of arjunolic acid against cardiac toxicity induced by oral sodium nitrite: effects on cytokine balance and apoptosis. Life Sci. 2014;111:18–26.

Al-Karawi D, Al Mamoori DA, Tayyar Y. The role of curcumin administration in patients with major depressive disorder: mini meta-analysis of clinical trials. Phytother Res. 2016;30:175–83.

Amatore C, Arbault S, Jaouen G, et al. Pro-oxidant properties of AZT and other thymidine analogues in macrophages: implication of the azido moiety in oxidative stress. Chem Med Chem. 2010;5:296–301.

Antherieu S, Azzi P, Dumont J, et al. Oxidative stress plays a major role in chlorpromazine-induced cholestasis in human HepaRG cells. Hepatology. 2013;57:1518–29.

Armagan A, Uzar E, Uz E, et al. Caffeic acid phenethyl ester modulates methotrexate-induced oxidative stress in testes of rat. Hum Exp Toxicol. 2008;27:547–52.

Askgaard DS, Wilcke T, Døssing M. Hepatotoxicity caused by the combined action of isoniazid and rifampicin. Thorax. 1995;50:213–4.

Barati E, Nikzad H, Karimian M. Oxidative stress and male infertility: current knowledge of pathophysiology and role of antioxidant therapy in disease management. Cell Mol Life Sci. 2020;77:93–113.

Barouki R, Morel Y. Repression of cytochrome P4501A1 gene expression by oxidative stress: mechanisms and biological implications. Biochem Pharmacol. 2001;61:511–6.

Bashkatova V, Mathieu A, Durand C, et al. Neurochemical changes and neurotoxic effects of an acute treatment with sydnocarb, a novel psychostimulant. Ann N Y Acad Sci. 2006;965(1):180–92.

Benhar M. Roles of mammalian glutathione peroxidase and thioredoxin reductase enzymes in the cellular response to nitrosative stress. Free Radic Biol Med. 2018;127:160–4.

Bergamini CM, Gambetti S, Dondi A, et al. Oxygen, reactive oxygen species and tissue damage. Curr Pharm Des. 2004;10:1611–26.

Bergner N, Monsef I, Illerhaus G, et al. Role of chemotherapy additional to high-dose methotrexate for primary central nervous system lymphoma (PCNSL). Cochrane Database Syst Rev. 2012;11:CD009355.

Berman SB, Hastings TG. Dopamine oxidation alters mitochondrial respiration and induces permeability transition in brain mitochondria: implications for Parkinson's disease. J Neurochem. 1999;73:1127–37.

Bhadauria S, Singh G, Sinha N, et al. Isoniazid induces oxidative stress, mitochondrial dysfunction and apoptosis in Hep G2 cell. Cell Mol Biol. 2007;53:102–14.

Bhadauria S, Mishra R, Kanchan R, et al. Isoniazid-induced apoptosis in HepG2 cells: generation of oxidative stress and Bcl-2 down-regulation. Toxicol Mech Methods. 2010;20:242–51.

Bhattacharyya B, Chatterjee TK, Ghosh JJ. Effects of chloroquine on lysosomal enzymes, NADPH induced lipid peroxidation, and antioxidant enzymes of rat retina. Biochem Pharmacol. 1983;32:2965–8.

Blignaut M, Espach Y, Vuuren MV, et al. Revisiting the cardiotoxic effect of chloroquine. Cardiovasc Drugs Ther. 2019;33:1–11.

Bondy SC, Naderi S. Contribution of hepatic cytochrome P450 systems to the generation of reactive oxygen species. Biochem Pharmacol. 1994;48:155–9.

Bowman ZS, Morrow JD, Jollow DJ, et al. Primaquine-induced hemolytic anemia: role of membrane lipid peroxidation and cytoskeletal protein alterations in the hemotoxicity of 5-hydroxyprimaquine. J Pharmacol Exp Ther. 2005;314:838–45.

Broder S. The development of antiretroviral therapy and its impact on the HIV-1/AIDS pandemic. Antivir Res. 2010;85:1–18.

Burger H, Zoumaro-Djayoon A, Boersma AWM, et al. Differential transport of platinum compounds by the human organic cation transporter hOCT2 (hSLC22A2). Br J Pharmacol. 2010;159:898–908.

Burrows K, Gudelsky G, Yamamoto B. Rapid and transient inhibition of mitochondrial function following methamphetamine or 3,4-methylenedioxymethamphetamine administration. Eur J Pharmacol. 2000;398(1):11–8.

Bymaster FP, Calligaro DO, Falcone JF, et al. Radioreceptor binding profile of the atypical antipsychotic olanzapine. Neuropsychopharmacology. 1996;14:87–96.

Carvalho M, Carmo H, Costa VM, et al. Toxicity of amphetamines: an update. Arch Toxicol. 2012;86:1167–231.

Chaanine AH, Gordon RE, Nonnenmacher M, et al. High-dose chloroquine is metabolically cardiotoxic by inducing lysosomes and mitochondria dysfunction in a rat model of pressure overload hypertrophy. Phys Rep. 2015;3:1–18.

Chaudhary NI, Schnapp A, Park JE. Pharmacologic differentiation of inflammation and fibrosis in the rat bleomycin model. Am J Respir Crit Care Med. 2006;173:769–76.

Chen Q, Vazquez EJ, Moghaddas S, et al. Production of reactive oxygen species by mitochondria: central role of complex III. J Biol Chem. 2003;278:36027–31.

Chen Y, Jungsuwadee P, Vore M, et al. Collateral damage in cancer chemotherapy: oxidative stress in nontargeted tissues. MolInterv. 2007;7:147–15.

Cho CH, Lee HJ. Oxidative stress and tardive dyskinesia: pharmacogenetic evidence. Prog Neuro-Psychopharmacol Biol Psychiatry. 2013;46:207–13.

Chowdhury A, Santra A, Bhattacharjee K, et al. Mitochondrial oxidative stress and permeability transition in isoniazid and rifampicin induced liver injury in mice. J Hepatol. 2006;45:117–26.

Conway R, Carey JJ. Risk of liver disease in methotrexate treated patients. World J Hepatol. 2017;9:1092–100.

Cregler LL, Mark H. Medical complications of cocaine abuse. N Engl J Med. 1986;315:1495–500.

Cristofori P, Zanetti E, Fregona D, et al. Renal proximal tubule segment-specific nephrotoxicity: an overview on biomarkers and histopathology. Toxicology. 2007;35:270–5.

Cryer B, Mahaffey KW. Gastrointestinal ulcers, role of aspirin, and clinical outcomes: pathobiology, diagnosis, and treatment. J Multidiscip Healthc. 2014;7:137–46.

Cunha-Oliveira T, Rego AC, Oliveira CR. Cellular and molecular mechanisms involved in the neurotoxicity of opioid and psychostimulant drugs. Brain Res Rev. 2008;58:192–208.

Cunha-Oliveira T, Rego AC, Carvalho F, et al. Chapter 17: medical toxicology of drugs of abuse. In: Miller P, editor. Principles of addiction- comprehensive addictive behaviors and disorders, vol. 1. Academic; 2013. p. 159–75.

Curto M, Girardi N, Lionetto L, et al. Systematic review of clozapine cardiotoxicity. Curr Psychiatry Rep. 2016;18:68.

Dalakas MC, Illa I, Pezeshkpour GH, et al. Mitochondrial myopathy caused by long-term zidovudine therapy. N Engl J Med. 1990;322:1098–105.

Dalaklioglu S, Genc GE, Aksoy NH, et al. Resveratrol ameliorates methotrexate-induced hepatotoxicity in rats via inhibition of lipid peroxidation. Hum Exp Toxicol. 2013;32:662–71.

Damiani RM, Moura DJ, Viau CM, et al. Pathways of cardiac toxicity: comparison between chemotherapeutic drugs doxorubicin and mitoxantrone. Arch Toxicol. 2016;90:2063–76.

Das J, Sil PC. Taurine ameliorates alloxan-induced diabetic renal injury, oxidative stress-related signaling pathways and apoptosis in rats. Amino Acids. 2012;43:1509–23.

Das J, Ghosh J, Manna P, et al. Taurine provides antioxidant defense against NaF-induced cytotoxicity in murine hepatocytes. Pathophysiology. 2008;15:181–90.

Das J, Ghosh J, Roy A, et al. Mangiferin exerts hepatoprotective activity against D-galactosamine induced acute toxicity and oxidative/nitrosative stress via Nrf2-NFκB pathways. Toxicol Appl Pharmacol. 2012;260:35–47.

Dasari S, Tchounwou PB. Cisplatin in cancer therapy: molecular mechanisms of action. Eur J Pharmacol. 2014;740:364–78.

Daugaard G, Abildgaard U. Cisplatin nephrotoxicity. Cancer Chemother Pharmacol. 1989;25(1):1–9.

Deavall DG, Martin EA, Horner JM, et al. Drug-induced oxidative stress and toxicity. J Toxicol. 2012;2012:1–13.

Di Meo S, Reed TT, Venditti P, et al. Role of ROS and RNS sources in physiological and pathological conditions. Oxidative Med Cell Longev. 2016;2016:1–44.

Dikshit P, Chatterjee M, Goswami A, et al. Aspirin induces apoptosis through the inhibition of proteasome function. J Biol Chem. 2006;281:29228–35.

Doi H, Horie T. Salicylic acid-induced hepatotoxicity triggered by oxidative stress. Chem Biol Interact. 2010;183:363–8.

Dold M, Samara MT, Li C, et al. Haloperidol versus first-generation antipsychotics for the treatment of schizophrenia and other psychotic disorder. Cochrane Database Syst Rev. 2015;1.

Dorszewska J, Florczak J, Rozycka A, et al. Polymorphisms of the CHRNA4 gene encoding the alpha4 subunit of nicotinic acetylcholine receptor as related to the oxidative DNA damage and the level of apoptotic proteins in lymphocytes of the patients with Alzheimer's disease. DNA Cell Biol. 2005;24:786–94.

Dorszewska J, Kempisty B, Jaroszewska-Kolecka J, et al. Expression and polymorphisms of gene 8-oxoguanine glycosylase 1 and the level of oxidative DNA damage in peripheral blood lymphocytes of patients with Alzheimer's disease. DNA Cell Biol. 2009;28:579–88.

Doyle T, Chen Z, Muscoli C, et al. Targeting the overproduction of peroxynitrite for the prevention and reversal of paclitaxel-induced neuropathic pain. J Neurosci. 2012;32:6149–60.

Dragovic S, Gunness P, Ingelman-Sundberg M, et al. Characterization of human cytochrome P450s involved in the bioactivation of clozapine. Drug Metab Dispos. 2013;41:651–8.

Drucker AM, Rosen CF. Drug-induced photosensitivity: culprit drugs, management and prevention. Drug Saf. 2011;34:821–37.

Duggett NA, Griffiths LA, McKenna OE, et al. Oxidative stress in the development, maintenance and resolution of paclitaxel-induced painful neuropathy. Neuroscience. 2016;333:13–26.

Dumortier G, Cabaret W, Stamatiadis L, et al. Hepatic tolerance of atypical antipsychotic drugs. L'Encéphale. 2002;28:542–51.

Eftekhari A, Azarmi Y, Parvizpur A, et al. Involvement of oxidative stress and mitochondrial/lysosomal cross-talk in olanzapine cytotoxicity in freshly isolated rat hepatocytes. Xenobiotica. 2016;46:369–78.

El-Agamey A, Lowe GM, McGarvey DJ, et al. Carotenoid radical chemistry and antioxidant/pro-oxidant properties. Arch Biochem Biophys. 2004;430:37–48.

Fayer SA. Torsades de pointes ventricular tachyarrhythmia associated with haloperidol. J Clin Psychopharmacol. 1986;6:375.

Fehsel K, Loeffler S, Krieger K, et al. Clozapine induces oxidative stress and proapoptotic gene expression in neutrophils of schizophrenic patients. J Clin Psychopharmacol. 2005;25:419–26.

Ferlini C, Raspaglio G, Mozzetti S, et al. Bcl-2 down-regulation is a novel mechanism of paclitaxel resistance. Mol Pharmacol. 2003;64:51–8.

Fidanboylu M, Griffiths LA, Flatters SJL. Global inhibition of reactive oxygen species (ROS) inhibits paclitaxel-induced painful peripheral neuropathy. PLoS One. 2011;6:25212.

Finaud J, Lac G, Filaire E. Oxidative stress. Sports Med. 2006;36:327–58.

Fish DN, Chow AT. The clinical pharmacokinetics of levofloxacin. Clin Pharmacokinet. 1997;32(2):101–19.

Flatters SJL, Bennett GJ. Studies of peripheral sensory nerves in paclitaxel-induced painful peripheral neuropathy: evidence for mitochondrial dysfunction. Pain. 2006;122:245–57.

Frey BN, Valvassori SS, Gomes KM, et al. Increased oxidative stress in submitochondrial particles after chronic amphetamine exposure. Brain Res. 2006;1097:224–9.

Froudarakis M, Hatzimichael E, Kyriazopoulou L, et al. Revisiting bleomycin from pathophysiology to safe clinical use. Crit Rev Oncol Hematol. 2013;87:90–100.

Fuster V, Sweeny JM. Aspirin: a historical and contemporary therapeutic overview. Circulation. 2011;123:768–78.

Gaies E, Jebabli N, Trabelsi S, et al. Methotrexate side effects: review article. Drug Metab Toxicol. 2012;3:4–8.

Gandhi A, Moorthy B, Ghose R. Drug disposition in pathophysiological conditions. Curr Drug Metab. 2012;13:1327–44.

Gao S, Qin T, Liuetal Z. Lutein and zeaxanthin supplementation reduces H2O2-induced oxidative damage in human lens epithelial cells. Mol Vis. 2011;17:3180–90.

Garcia Rodriguez LA, Hernandez-Diaz S, De Abajo FJ. Association between aspirin and upper gastrointestinal complications: systematic review of epidemiologic studies. Br J Clin Pharmacol. 2001;52:563–71.

Gardner K, Hall PA, Chinnery PF, et al. HIV treatment and associated mitochondrial pathology: review of 25 years of in vitro, animal, and human studies. Toxicol Pathol. 2014;42:811–22.

Ge L, Niu G, Han X, et al. Aspirin treatment increases the risk of cerebral microbleeds. Can J Neurol Sci. 2011;38:863–8.

Geddes J, Freemantle N, Harrison P, et al. Atypical antipsychotics in the treatment of schizophrenia: systematic overview and meta-regression analysis. BMJ. 2000;321:1371–6.

Ghosh J, Sil PC. Arjunolic acid: a new multifunctional therapeutic promise of alternative medicine. Biochimie. 2013;95(6):1098–109.

Ghosh J, Das J, Manna P, et al. Taurine prevents arsenic-induced cardiac oxidative stress and apoptotic damage: role of NF-kappa B, p38 and JNK MAPK pathway. Toxicol Appl Pharmacol. 2009;240:73–87.

Ghosh J, Das J, Manna P, Sil PC. The protective role of arjunolic acid against doxorubicin induced intracellular ROS dependent JNK-p38 and p53- mediated cardiac apoptosis. Biomaterials. 2011;32:4857–66.

Ginovart N, Kapur S. Role of dopamine D 2 receptors for antipsychotic activity. Curr Antipsychotics. 2012;212:27–52.

Gluck MR, Zeevalk GD. Inhibition of brain mitochondrial respiration by dopamine and its metabolites: implications for Parkinson's disease and catecholamine-associated diseases. J Neurochem. 2004;91:788–95.

Goel A, Chang DK, Ricciardiello L, et al. A novel mechanism for aspirin-mediated growth inhibition of human colon cancer cells. Clin Cancer Res. 2003;9:383–90.

Goff DC, Coyle JT. The emerging role of glutamate in the pathophysiology and treatment of schizophrenia. Am J Psychiatry. 2001;158:1367–77.

Gokce A, Oktar S, Koc A, et al. Protective effects of thymoquinone against methotrexate-induced testicular injury. Hum Exp Toxicol. 2011;30:897–903.

Gouveia LDAV, Cardoso CA, Rosa G, et al. Effects of the intake of sesame seeds (Sesamum indicum L.) and derivatives on oxidative stress: a systematic review. J Med Food. 2016;19:337–45.

Graham DG. Oxidative pathways for catecholamines in the genesis of neuromelanin and cytotoxic quinones. Mol Pharmacol. 1978;14:633–43.

Gulen M, Ay MO, Avci A, et al. Levofloxacin-induced hepatotoxicity and death. Am J Ther. 2015;22:93–6.

Han C, Wang SM, Kato M. Second-generation antipsychotics in the treatment of major depressive disorder: current evidence. Expert Rev Neurother. 2013;13(7):851–70.

Hanna YM, Baglan KL, Stromberg JS, et al. Acute and subacute toxicity associated with concurrent adjuvant radiation therapy and paclitaxel in primary breast cancer therapy. Breast J. 2002;8:149–53.

Hantson P. Mechanisms of toxic cardiomyopathy. Clin Toxicol. 2019;57:1–9.

Hasiloglu ZI, Albayram S, Asik M, et al. MRI findings of isoniazid-induced central nervous system toxicity in a child. Clin Radiol. 2012;67:932–5.

He YJ, Winham SJ, Hoskins JM, et al. Carboplatin/taxane-induced gastrointestinal toxicity: a pharmacogenomics study on the SCOTROC1 trial. Pharm J. 2016;16:243–8.

Hemeida RAM, Mohafez OM. Curcumin attenuates methotraxate-induced hepatic oxidative damage in rats. J Egyptian Natl Canc Inst. 2008;20:141–8.

Hendrickson RG, Bizovi KE. Acetaminophen Goldfrank's toxicologic emergencies. 8th ed. New York: McGraw-Hill; 2006. p. 333–43.

Huang YS, Chern HD, Su WJ, et al. Cytochrome P450 2E1 genotype and the susceptibility to antituberculosis drug-induced hepatitis. Hepatology. 2003;37:924–30.

Hung CC, Wei IH, Huang CC. Late-onset cholestatic hepatitis induced by olanzapine in a patient with schizophrenia. Prog Neuro-Psychopharmacol Biol Psychiatry. 2009;33:1574–5.

Hunt N, Stern TA. The association between intravenous haloperidol and torsades-de-pointes - 3 cases and a literature review. J Consult Liaison Psychiatry. 1995;36:541–9.

Jadallah KA, Limauro DL, Colatrella AM. Acute hepatocellular cholestatic liver injury after olanzapine therapy. Ann Intern Med. 2003;138:357–8.

Kakuda TN. Pharmacology of nucleoside and nucleotide reverse transcriptase inhibitor-induced mitochondrial toxicity. Clin Ther. 2000;22(6):685–708.

Kalkman HO, Neumann V, Hoyer D, et al. The role of alpha2-adrenoceptor antagonism in the anticataleptic properties of the atypical neuroleptic agent, clozapine, in the rat. Br J Pharmacol. 1998;124:1550–6.

Kaulmann A, Bohn T. Carotenoids, inflammation, and oxidative stress-implications of cellular signaling pathways and relation to chronic disease prevention. Nutr Res. 2014;34:907–29.

Kim HJ, Lee JH, Kim SJ, et al. Roles of NADPH oxidases in cisplatin-induced reactive oxygen species generation and ototoxicity. J Neurosci. 2010a;30(11):3933–46.

Kim HK, Zhang YP, Gwak YS, et al. Phenyl N-tert-butylnitrone, a free radical scavenger, reduces mechanical allodynia in chemotherapy-induced neuropathic pain in rats. Anesthesiology. 2010b;112:432–9.

Kline ER, Bassit L, Hernandez-Santiago BI, et al. Longterm exposure to AZT, but not d4T, increases endothelial cell oxidative stress and mitochondrial dysfunction. Cardiovasc Toxicol. 2009;9:1–12.

Koczor CA, Lewis W. Nucleoside reverse transcriptase inhibitor toxicity and mitochondrial DNA. Expert Opin Drug Metab Toxicol. 2010;6:1493–504.

Kohanski MA, Dwyer DJ, Hayete B, et al. A common mechanism of cellular death induced by bactericidal antibiotics. Cell. 2007;130:797–810.

Kohanski MA, Tharakan A, London NR, et al. Bactericidal antibiotics promote oxidative damage and programmed cell death in sinonasal epithelial cells. Int Forum Allergy Rhinol. 2017;7:359–64.

Kose E, Sapmaz HI, Sarihan E, et al. Beneficial effects of montelukast against methotrexate-induced liver toxicity: a biochemical and histological study. Sci World J. 2012;2012:1–6.

Kubota C, Torii S, Hou N, et al. Constitutive reactive oxygen species generation from autophago-
some/lysosome in neuronal oxidative toxicity. J Biol Chem. 2010;285:667–74.

Lavie CJ, Howden CW, Scheiman J, et al. Upper gastrointestinal toxicity associated with long-
term aspirin therapy: consequences and prevention. Curr Probl Cardiol. 2017;42:146–64.

Lee CT, Chen J, Hayashi T, et al. A mechanism for the inhibition of neural progenitor cell proliferation by cocaine. PLoS Med. 2008;5(6):e117.

Li X, Cameron MD. Potential role of a quetiapine metabolite in quetiapine induced neutropenia
and agranulocytosis. Chem Res Toxicol. 2012;25:1004–11.

Lieberman JA, Bymaster FP, Meltzer HY, et al. Antipsychotic drugs: comparison in animal models
of efficacy, neurotransmitter regulation, and neuroprotection. Pharmacol Rev. 2008;60:358–403.

MacAllister SL, Young C, Guzdek A, et al. Molecular cytotoxic mechanisms of chlorpromazine in
isolated rat hepatocytes. Can J Physiol Pharmacol. 2013;91:56–63.

Mackenzie AH. Pharmacological actions of 4-aminoquinoline compounds. Am J Med.
1983;75:5–10.

Mailman RB, Murthy V. Third generation antipsychotic drugs: partial agonism or receptor functional selectivity? Curr Pharm Des. 2010;16:488–501.

Manceaux P, Constant E, Zdanowicz N, et al. Management of marked liver enzyme increase
during olanzapine treatment: a case report and review of the literature. Psychiatr Danub.
2011;23:S15–7.

Mansour HH, Hafez HF, Fahmy NM, et al. Silymarin modulates Cisplatin induced oxidative stress
and hepatotoxicity in rats. J Biochem Mol Biol. 2006;39:656–61.

Marcillat O, Zhang Y, Davies KJ. Oxidative and non-oxidative mechanisms in the inactivation
of cardiac mitochondrial electron transport chain components by doxorubicin. Biochem
J. 1989;259:181–9.

Masuda H, Tanaka T, Takahama U. Cisplatin generates superoxide anion by interaction with DNA
in a cell-free system. Biochem Biophys Res Commun. 1994;203:1175–80.

Maurya PK, Rizzo LB, Xavier G, et al. Shorter leukocyte telomere length in patients at ultra-high
risk for psychosis. Eur Neuropsychopharmacol. 2017;27(5):538–42.

Maynard S, Schurman SH, Harboe C, et al. Base excision repair of oxidative DNA damage and
association with cancer and aging. Carcinogenesis. 2009;30(1):2–10.

Mazer M, Perrone J. Acetaminophen-induced nephrotoxicity: pathophysiology, clinical manifestations, and management. J Med Toxicol. 2008;4:2–6.

Miller RP, Tadagavadi RK, Ramesh G, et al. Mechanisms of cisplatin nephrotoxicity. Toxins.
2010;2:2490–518.

Milner DA. Malaria pathogenesis. Cold Spring Harb Perspect Med. 2018;8(1):a025569.

Minotti G, Menna P, Salvatorelli E, et al. Anthracyclines: molecular advances and pharmacologic
developments in antitumor activity and cardiotoxicity. Pharmacol Rev. 2004;56:185–229.

Mitkov MV, Trowbridge RM, Lockshin BN, et al. Dermatologic side effects of psychotropic medications. Psychosomatics. 2014;55:1–20.

Miyazono Y, Gao F, Horie T. Oxidative stress contributes to methotrexate-induced small intestinal
toxicity in rats. Scand J Gastroenterol. 2004;39:1119–27.

Moeller A, Ask K, Warburton D, et al. The bleomycin animal model: a useful tool to investigate
treatment options for idiopathic pulmonary fibrosis? Int J Biochem Cell Biol. 2008;40:362–82.

Moghadam AR, Tutunchi S, Namvaran-Abbas-Abad A, et al. Preadministration of turmeric prevents methotrexate-induced liver toxicity and oxidative stress. BMC Complement Altern Med.
2015;15:246–58.

Monsma FJ, Shen Y, Ward RP, et al. Cloning and expression of a novel serotonin receptor with high
affinity for tricyclic psychotropic drugs. Mol Pharmacol. 1993;43:320–7.

Moore DE. Drug-induced cutaneous photosensitivity: incidence, mechanism, prevention and management. Drug Saf. 2002;25:345–72.

Muench J, Hamer AM. Adverse effects of antipsychotic medications. Am Fam Physician.
2010;81:617–22.

Ng VH, Cox JS, Sousa AO, et al. Role of KatG catalase-peroxidase in mycobacterial pathogenesis:
countering the phagocyte oxidative burst. Mol Microbiol. 2004;52(5):1291–302.

Octavia Y, Tocchetti CG, Gabrielson KL, et al. Doxorubicin-induced cardiomyopathy: from molecular mechanisms to therapeutic strategies. J Mol Cell Cardiol. 2012;52:1213–25.

Otreba M, Zdybel M, Pilawa B, et al. EPR spectroscopy of chlorpromazine-induced free radical formation in normal human melanocytes. Eur Biophys J. 2015;44:359–65.

Patel RP, McAndrew J, Sellak H, et al. Biological aspects of reactive nitrogen species. BBA-Bioenergetics. 1999;1411(2–3):385–400.

Patrono C, Garcia Rodriguez LA, Landolfi R, et al. Low-dose aspirin for the prevention of athero-thrombosis. N Engl J Med. 2005;353:2373–83.

Pau AK, George JM. Antiretroviral therapy: current drugs. Infect Dis Clin. 2014;28:371–402.

Pereira CV, Nadanaciva S, Oliveira PJ, et al. The contribution of oxidative stress to drug-induced organ toxicity and its detection in vitro and in vivo. Expert Opin Drug Metab Toxicol. 2012;8:219–37.

Postma TJ, Vermorken JB, Liefting AJ, et al. Paclitaxel-induced neuropathy. Ann Oncol. 1995;6(5):489–94.

Prakash A, Kumar A. Implicating the role of lycopene in restoration of mitochondrial enzymes and BDNF levels in β-amyloid induced Alzheimer' s disease. Eur J Pharmacol. 2014;741:104–11.

Ramappa V, Aithal GP. Hepatotoxicity related to anti-tuberculosis drugs: mechanisms and management. J Clin Exp Hepatol. 2013;3:37–49.

Rao AV, Rao LG. Carotenoids and human health. Pharmacol Res. 2007;55:207–16.

Rao CV, Rawat AKS, Singh AP, et al. Hepatoprotective potential of ethanolic extract of Ziziphus oenoplia (L.) Mill roots against antitubercular drugs induced hepatotoxicity in experimental models. Asian Pac J Trop Med. 2012;5:283–8.

Raz A, Bergman R, Eilam O, et al. A case report of olanzapine-induced hypersensitivity syndrome. Am J Med Sci. 2001;321:156–8.

Redmann M, Benavides GA, Berryhill TF, et al. Inhibition of autophagy with bafilomycin and chloroquine decreases mitochondrial quality and bioenergetic function in primary neurons. Redox Biol. 2017;11:73–81.

Renu K, Abilash VG, Pichiah PBT, et al. Molecular mechanism of doxorubicin-induced cardiomyopathy–an update. Eur J Pharmacol. 2018;818:241–53.

Richelson E, Souder T. Binding of antipsychotic drugs to human brain receptors focus on newer generation compounds. Life Sci. 2000;68:29–39.

Riezzo I, Fiore C, De CD, et al. Side effects of cocaine abuse: multiorgan toxicity and pathological consequences. Curr Med Chem. 2012;19:5624–46.

Rifkin A, Doddi S, Karajgi B, et al. Dosage of haloperidol for schizophrenia. Arch Gen Psychiatry. 1991;48:166–70.

Ronaldson KJ, Taylor AJ, Fitzgerald PB, et al. Diagnostic characteristics of clozapine-induced myocarditis identified by an analysis of 38 cases and 47 controls. J Clin Psychiatry. 2010;71:976–81.

Rowinsky EK, Donehower RC. Paclitaxel (taxol). N Engl J Med. 1995;332:1004–14.

Ruiz-Irastorza G, Ramos-Casals M, Brito-Zeron P, et al. Clinical efficacy and side effects of anti-malarials in systemic lupus erythematosus: a systematic review. Ann Rheum Dis. 2010;69:20–8.

Ryter SW, Kim HP, Hoetzel A, et al. Mechanisms of cell death in oxidative stress. Antioxid Redox Signal. 2007;9:49–89.

Sabens EA, Distler AM, Mieyal JJ. Levodopa deactivates enzymes that regulate thiol– disulfide homeostasis and promotes neuronal cell death: implications for therapy of Parkinson's disease. Biochemist. 2010;49(12):2715–24.

Sachdeva AK, Chopra K. Lycopene abrogates Aβ (1–42)-mediated neuroinflammatory cascade in an experimental model of Alzheimer's disease. J Nutr Biochem. 2015;26(7):736–44.

Sawyer DB, Peng X, Chen B, et al. Mechanisms of anthracycline cardiac injury: can we identify strategies for cardioprotection? Prog Cardiovasc Dis. 2010;53:105–13.

Schwab CE, Tuschl H. In vitro studies on the toxicity of isoniazid in different cell lines. Hum Exp Toxicol. 2003;22:607–15.

Sener G, Ekşioğlu-Demiralp E, Cetiner M et al β-glucan ameliorates methotrexate-induced oxidative organ injury via its antioxidant and immunomodulatory effects. Eur J Pharmacol. 2006;542:170–8.

Shacter E. Protein oxidative damage. Methods Enzymol. 2000;319:428–36.

Shen C, Zhang H, Zhang G, et al. Isoniazid-induced hepatotoxicity in rat hepatocytes of gel entrapment culture. Toxicol Lett. 2006;167:66–74.

Shen C, Zhang G, Meng Q. An in vitro model for long-term hepatotoxicity testing utilizing rat hepatocytes entrapped in micro-hollow fiber reactor. Biochem Eng J. 2007;34:267–72.

Sheth S, Mukherjea D, Rybak LP, et al. Mechanisms of cisplatin-induced ototoxicity and otoprotection. Front Cell Neurosci. 2017;11:338.

Shi R, Itagaki N, Sugawara I. Overview of anti-tuberculosis (TB) drugs and their resistance mechanisms. Mini-Rev Med Chem. 2007;7:1177–85.

Shivakumar BR, Ravindranath V. Oxidative stress-induced by administration of the neuroleptic drug haloperidol is attenuated by higher doses of haloperidol. Brain Res. 1992;595:256–62.

Singh B, Shoulson R, Chatterjee A, et al. Resveratrol inhibits estrogen-induced breast carcinogenesis through induction of NRF2- mediated protective pathways. Carcinogenesis. 2014;35:1872–80.

Sinha M, Manna P, Sil PC. Aqueous extract of the bark of Terminalia arjuna plays a protective role against sodium-fluoride-induced hepatic and renal oxidative stress. J Nat Med. 2007;61:251–60.

Skrzypek R, Callaghan R. The 'pushmi-pullyu' of resistance to chloroquine in malaria. Essays Biochem. 2017;61:167–75.

Slater AFG. Chloroquine: mechanism of drug action and resistance in Plasmodium falciparum. Pharmacol Ther. 1993;57:203–35.

Sleijfer S. Bleomycin-induced pneumonitis. Chest. 2001;120(2):617–24.

Soares GMS, Figueiredo LC, Faveri M, et al. Mechanisms of action of systemic antibiotics used in periodontal treatment and mechanisms of bacterial resistance to these drugs. J Appl Oral Sci. 2012;20:295–309.

Sorensen HT, Mellemkjaer L, Blot WJ, et al. Risk of upper gastrointestinal bleeding associated with use of low-dose aspirin. Am J Gastroenterol. 2000;95:2218–24.

Sprauten M, Darrah TH, Peterson DR, et al. Impact of long-term serum platinum concentrations on neuro- and ototoxicity in cisplatin-treated survivors of testicular cancer. J Clin Oncol. 2012;30:300–7.

Stansley BJ, Yamamoto BK. L-dopa-induced dopamine synthesis and oxidative stress in serotonergic cells. Neuropharmacology. 2013;67:243–51.

Tchernichovsky E, Sirota P. Hepatotoxicity, leucopenia and neutropenia associated with olanzapine therapy. Int J Psychiatry Clin Pract. 2004;8:173–7.

Thakur KS, Prakash A, Bisht R, et al. Beneficial effect of candesartan and lisinopril against haloperidol-induced tardive dyskinesia in rat. J Renin-Angiotensin-Aldosterone Syst. 2015;16:917–29.

Thome R, Lopes SCP, Costa FTM, et al. Chloroquine: modes of action of an undervalued drug. Immunol Lett. 2013;153:50–7.

Thorn CF, Oshiro C, Marsh S, et al. Doxorubicin pathways: pharmacodynamics and adverse effects. Pharmacogenet Genomics. 2011;21:440–6.

Timmins GS, Deretic V. Mechanisms of action of isoniazid. Mol Microbiol. 2006;62:1220–7.

Tsutsumishita Y, Onda T, Okada K, et al. Involvement of H2O2 production in cisplatin-induced nephrotoxicity. Biochem Biophys Res Commun. 1998;242:310–2.

Uraz S, Tahan V, Aygun C, et al. Role of ursodeoxycholic acid in prevention of methotrexate-induced liver toxicity. Dig Dis Sci. 2008;53:1071–7.

Valkhoff VE, Sturkenboom MC, Kuipers EJ. Risk factors for gastrointestinal bleeding associated with low-dose aspirin. Best Pract Res Clin Gastroenterol. 2012;26:125–40.

Vance-Bryan K, Guay DRP, Rotschafer JC. Clinical pharmacokinetics of ciprofloxacin. Clin Pharmacokinet. 1990;19:434–61.

Vane JR, Botting RM. Mechanism of action of aspirin-like drugs. Semin Arthritis Rheum. 1997;26:2–10.

Vane JR, Botting RM. The mechanism of action of aspirin. Thromb Res. 2003;110:255–8.

Vardi N, Parlakpinar H, Ates B, et al. Antiapoptotic and antioxidant effects of β-carotene against methotrexate-induced testicular injury. Fertil Steril. 2009;92:2028–33.

Vardi N, Parlakpinar H, Cetin A, et al. Protective effect of beta-carotene on methotrexate–induced oxidative liver damage. Toxicol Pathol. 2010;38:592–7.

Vijayapadma V, Ramyaa P, Pavithra D, et al. Protective effect of lutein against benzo(a)pyrene-induced oxidative stress in human erythrocytes. Toxicol Ind Health. 2014;30:284–93.

Vucicevic L, Misirkic-Marjanovic M, Paunovic V, et al. Autophagy inhibition uncovers the neurotoxic action of the antipsychotic drug olanzapine. Autophagy. 2014;10:2362–78.

Wan F, Lin H, Huang K, et al. Systemic administration of d-amphetamine induces long-lasting oxidative stress in the rat striatum. Life Sci. 2000;66(15):205–12.

Wan F, Tung C, Shiah I, et al. Effects of α-phenyl-N-tert-butyl nitrone and N-acetylcysteine on hydroxyl radical formation and dopamine depletion in the rat striatum produced by d-amphetamine. Eur Neuropsychopharmacol. 2006;16(2):147–53.

Wang P, Si T. Use of antipsychotics in the treatment of depressive disorders. Shanghai Arch Psychiatry. 2013;25(3):134.

Warner EA. Cocaine abuse. Ann Intern Med. 1993;119:226–35.

West SG. Methotrexate hepatoxicity. Rheum Dis Clin N Am. 1997;24:883–915.

Wiciński M, Węclewicz MM. Clozapine-induced agranulocytosis/granulocytopenia: mechanisms and monitoring. Curr Opin Hematol. 2018;25:22–8.

Wiczer T, Dotson E, Tuiten A, et al. Evaluation of incidence and risk factors for high-dose methotrexate-induced nephrotoxicity. J Oncol Pharm Pract. 2016;22:430–6.

Xiao WH, Zheng H, Zheng FY, et al. Mitochondrial abnormality in sensory, but not motor, axons in paclitaxel-evoked painful peripheral neuropathy in the rat. Neuroscience. 2011;199:461–9.

Yadav VS, Mishra KP, Singh DP, et al. Immunomodulatory effects of curcumin. Immunopharmacol Immunotoxicol. 2005;27:485–97.

Yilmaz HR, Sogut S, Ozyurt B, et al. The activities of liver adenosine deaminase, xanthine oxidase, catalase, superoxide dismutase enzymes and the levels of malondialdehyde and nitric oxide after cisplatin toxicity in rats: protective effect of caffeic acid phenethyl ester. Toxicol Ind Health. 2005;21(1–2):67–73.

Yoshimi N, Matsunaga K, Katayama M, et al. The inhibitory effects of mangiferin, a naturally occurring glucosylxanthone, in bowel carcinogenesis of male F344 rats. Cancer Lett. 2001;163:163–70.

Yu Z, Yan B, Gao L, et al. Targeted delivery of bleomycin: a comprehensive anticancer review. Curr Cancer Drug. 2016;16:509–21.

Yue J, Dong G, He C, et al. Protective effects of thiopronin against isoniazid-induced hepatotoxicity in rats. Toxicology. 2009;264:185–91.

Yuen JWY, Kim DD, Procyshyn RM, et al. Clozapine-induced cardiovascular side effects and autonomic dysfunction: a systematic review. Front Neurosci. 2018;12:203.

Yulug E, Turedi S, Alver A, et al. Effects of resveratrol on methotrexate-induced testicular damage in rats. Sci World J. 2013;2013:1–6.

Zangar RC, Davydov DR, Verma S. Mechanisms that regulate production of reactive oxygen species by cytochrome P450. Toxicol Appl Pharmacol. 2004;199:316–31.

Zhai Q, Lu SR, Lin Y, et al. Oxidative stress potentiated by diallylsulfide, a selective CYP2E1 inhibitor, in isoniazid toxic effect on rat primary hepatocytes. Toxicol Lett. 2008;183:95–8.

Zhang Y, Liao Z, Zhang LJ, et al. The utility of chloroquine in cancer therapy. Curr Med Res Opin. 2015;31(5):1009–13.

Zhang C, Fang X, Yao P, et al. Metabolic adverse effects of olanzapine on cognitive dysfunction: a possible relationship between BDNF and TNF-alpha. Psychoneuroendocrinology. 2017;81:138–43.

Zhou J, Chen P, Zhou Y, et al. 3,4-Methylenedioxymethamphetamine (MDMA) abuse may cause oxidative stress and potential free radical damage. Free Radic Res. 2003;37(5):491–7.

Zimmermann KC, Waterhouse NJ, Goldstein JC, et al. Aspirin induces apoptosis through release of cytochrome c from mitochondria. Neoplasia. 2000;2:505–13.

Adversities of Nanoparticles in Elderly Populations

Arti Devi, Gaurav Mudgal, and Zaved Ahmed Khan

Contents

Abstract In literature, several studies on nanotoxicity have been published which proved that nanoparticle entry inside the cell leads to many biochemical events, ultimately causing free radical generation, mitochondrial damage, cell injury, and cell death. Many in vivo and in vitro studies have proved that the use of nanoparticles in various fields requires prior safety and toxicology evaluation. In this study, the possible toxic effects of nanoparticles on different diseases are explored and various side effects of different types of nanoparticles have been reviewed. After doing a literature study, it has been found that nanoparticles affect elderly and susceptible population more as compared to young and adults.

A. Devi · G. Mudgal
University Institute of Biotechnology, Chandigarh University, Mohali, Punjab, India

Z. A. Khan (✉)
Faculty of Science, Baba Farid Group of Institutions, Bathinda, Punjab, India

© The Author(s), under exclusive license to Springer Nature Switzerland AG 2021
K. K. Kesari, N. K. Jha (eds.), *Free Radical Biology and Environmental Toxicity*,
Molecular and Integrative Toxicology, https://doi.org/10.1007/978-3-030-83446-3_5

115

Keywords Nanoparticles · Elderly · Toxicity · Free radical · Diseases ·
Respiratory · Cardiovascular · Infectious

1 Introduction

*Nano is a term used for all the processes like scientific and technical, which are
done at the nano level (*Bawa et al. 2016*).* The International Organization for
Standardization defines a nanoparticle as a nano-object with all three external
dimensions in the nanoscale, which is approximately 1–100 nanometers. Clay, mineral, and products of bacteria exist in nature all over the olden times of the earth and
used intentionally like finely ground metal colorants, but engineered nanoparticles
have given the impression only in the last few decades. Their novel physicochemical, thermal, and electrical properties facilitate their application in clothing,medicine,
and cosmetics thereby increasing the probability for human and environmental contact with these nanomaterials. Due to these unique properties, they can also interact
with living cells, and due to their small size, they can penetrate freely inside the
cellular system, which can ultimately increase their toxic effects. There are several
nanoparticles-based products available in the market, but only a few nano-based
products are used in large quantities. These nanoparticles enter our lifestyle as the
use of nanoparticle-based products in our daily life has become very common, but
we remain unaware of this (Buzea et al. 2007). There are several new properties in
nanoparticles which make them suitable to use in daily life in cosmetics like chemical reactivity, solubility, transparency, and color (Raj et al. 2012). Easy absorption
of cosmetics is provided by nanoparticles by the process of diffusion through the
skin layer (Schäfer-Korting et al. 2007). Nevertheless, nanoparticles of silver, gold,
and titanium dioxide pass directly inside a living organism and can be the reason for
toxicity. Several sectors in the industry have increased the use of nanoparticles. The
industry has been using nanoparticles in food, environment, cosmetics, diagnosis,
treatment, therapeutic approaches, and several other sectors also. In the medical
field, nanoparticles have provided extraordinary applications because they can be
used to diagnose the disease with the help of biosensors, the drug can be released
sustainably and can also be used to deliver the drug. Due to several benefits of
nanoparticles, they can be used for targeted drug delivery and can be used for early
detection of the disease. There are several benefits of nanoparticles, but excessive
exposure to nanoparticles proved to be dangerous. Nanoparticles pose a danger to
living organisms by breaking DNA strands and altering gene expression, and excessive formation of free radicals can cause mitochondrial dysfunction also at the cellular level (Bawa et al. 2016).

At the organism level, nanoparticles can stimulate inflammatory cytokines, suppress or induce a defense system, and lead to inflammation (Raj et al. 2012; Sengul
and Asmatulu 2020). Nevertheless, there is a need for better knowledge of nanotoxicity. Many toxicological studies mainly revolve around nanoparticle toxicity in
young animals, but only little is known for the susceptible population. This is very

important to analyze the adverse effects of nanoparticles in a vulnerable population like cardiovascular patients, patients with lung problems, patients with liver disease, patients suffering from infectious diseases, and some other diseases. People with pre-exciting ailment have been reported with structural and functional abnormalities; therefore, nanoparticles can accumulate and distribute into the body, which can cause severe consequences. The people with the pre-existing ailment are devoid of the good immune system; therefore, they cannot clear nanoparticles from their system. The bad repair mechanism of the susceptible population cannot repair the damage caused by free radicles; hereon conceived oxidative stress due to nanoparticle accumulation (post several internalizations) may cause inflammation corroborating tissue injury (Li et al. 2014). Hasty advancement in the nanoparticle industry leads to excessive exposure of workers to nanoparticles which can cause serious health issues (Iavicoli et al. 2014). So, there is a need to outline potential hazards and precautionary measures for unprotected workers and susceptible populations (Schulte and Trout 2011; Trout and Schulte 2010). In such a situation, preventive hazard management can be boosted by defining vulnerable people who show severe side effects from nanoparticle exposure due to weak immune system and not able to respond to nanoparticles (Manno et al. 2014). Furthermore, according to recent studies, nanoparticles can cause respiratory problems and cardiovascular problems in elderly residents (Peters et al. 1997; Von Klot et al. 2002; Oberdörster et al. 2005). So, from the above discussion, it is clear that the elderly population is more susceptible to nanoparticles due to weaker antioxidative mechanisms and immune system. In this study, we have summarized how nanoparticles affect elderly people and focused on various comebacks of nanoparticle contact with a vulnerable population and elderly people.

2   Mechanism of Nanoparticle Toxicity

Nanoparticle interaction with cell and cellular components like mitochondria, endoplasmic reticulum, and other organelles leads to the formation of reactive oxygen species which include superoxide anions, hydroxyl anions. These ultimately activate the oxidative enzyme cascade causing oxidative stress as shown in Fig. 1 (Cho et al. 2012).

Many cell signaling pathways are activated when cell interacts with nanoparticles, which can further cause adverse effects on the cell and cell damage takes place. A cell can uptake these nanoparticles through the cell membrane, which start accumulating in the cytoplasm, Golgi bodies, endoplasmic reticulum, and nucleus, and persistence of these nanoparticles in the cellular organelles causes the generation of free radicals and oxidative stress. As oxidative stress is responsible for the activation of oxidative enzyme cascade, it leads to cell damage through inflammation and apoptosis (Marano and Guadagnini 2012). It has been reported that reactive oxygen species generation and cell damage are initial phenomena in nanoparticles induced cell injury. Carbon nanotubes and nanoparticles based on metals can cause

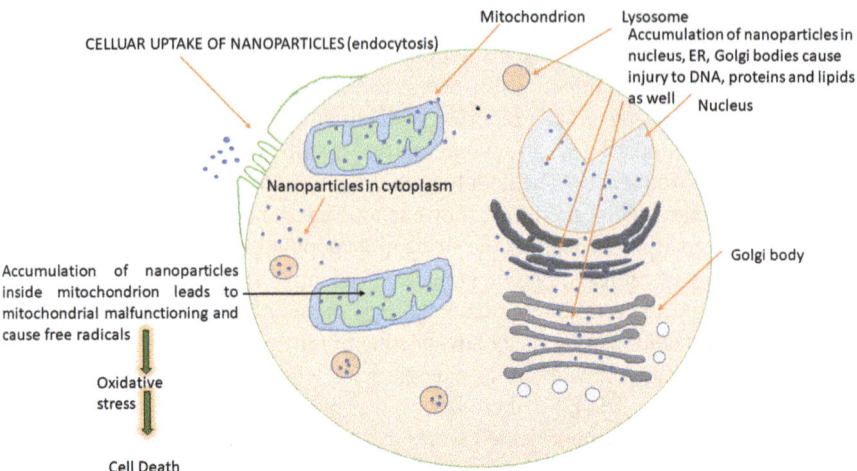

Fig. 1 Mechanism of nanoparticle toxicity in cell. Nanoparticles accumulate inside organelles like mitochondria, endoplasmic reticulum, nucleus, and Golgi bodies and lead to oxidative stress and cell death

oxidative stress by reacting with the cell. Excessive generation of free radicals can render the cells deficient for antioxidants and exposure of mitochondrial machinery to nanoparticles, in turn causing apoptosis led by upregulation of NADPH oxidase system, following which dysregulated calcium regulatory processes corroborate cause cell death. Genotoxicity, inflammation, cancer, and fibrosis are the results of excessive release of cytokines and chemokines induced by free radical generation. Therefore, the main reason for cell death when cells exposed to nanoparticle is contributed to free excessive radical generation. At this point, this is very important to assess the toxic effects of nanoparticles before using these nanoparticles for treatment. Nanoparticles assessment can be done with the help of several in vitro and in vivo models, and this mechanism of nanoparticle toxicity should be considered by doing safety evaluation of nanoparticles (Khlebtsov and Dykman 2011). Figure 1 shows the nanoparticle effect on cell and cellular organelles.

3   Nanoparticle Toxicity in Susceptible People

The intrinsic ability of the nanoparticles to interact with the biomolecules and the type of feedback of biomolecules when interacting with nanoparticles collectively decide biocompatibility for nanoparticles. The interaction between nanoparticles and biomolecules includes availability, degradation, and excretion events further in the human body. Another system which protects the nerve cells from toxicity is the blood–brain barrier. But the liver is exposed to nanoparticles which can take up most nanoparticles in circulation and in healthy individuals these nanoparticles are excreted successfully (Longmire et al. 2008). In most of the healthy individuals, nanoparticle toxicity is minimum, but in diseased people and aged people due to abnormal functioning of the body organs nanoparticles can cause damage. This section emphasizes CVD (cardiovascular diseases), lung diseases, and liver diseases. In a study, gadolinium orthovanadate nanoparticles have been shown to cause acute toxicity in aging rats (Karpenko et al. 2013).

4 Nanoparticle Toxicity in the Elderly Population

Aging is the time-related deterioration of the physiological functions necessary for survival and fertility. The characteristics of aging as distinguished from diseases of aging (such as cancer and heart disease) affect all the individuals of a species and the changes that occur during an organisms' life span, though the rate at which these take place varies widely (Lemoine 2020). Due to several environmental factors and other reasons, there is a continuous increase in the aging population. Due to this, the human is continuously entering the era of the so-called global aging (Mathers and Loncar 2006). With aging, the capacity of the cell to repair and do all the metabolic functions decreases and becomes more vulnerable to neurological diseases and cardiovascular diseases, and the immune system becomes weak. Therefore, elderly people are more prone to side effects of nanoparticle exposure (Neupane et al. 2010). According to a recent study, nanoparticles show more severe side effects in the elderly population as compared to the young population. In a study, 24.1 mg/m^3 of SiO_2 nanoparticles have been given to a 20-month-old rat by nasal administration for at least 4 weeks; after this, cardiovascular dysfunction was induced which is characterized by an atrioventricular blockage, increased concentration of fibrinogens, myocardial ischemic damage, and increased blood viscosity (Chen et al. 2008). The severe symptoms are mostly absent in young organisms. Furthermore, if bronchoalveolar-lavage parameters and serum histamine levels are considered, it would interpret the respiratory system of the elderly more prone to nanoparticle led maladies (Chen et al. 2008).

5 Nanoparticle Toxicity in Cardiovascular Diseases

In cardiovascular diseases, blood vessels and the heart are affected, which can lead to death also. A WHO survey inferred that cardiovascular diseases stood as the critical factor for 20 million cases in 2015 (Mathers 2008). According to the literature, it has been found that suspended particulate matter having nano dimensions is the main activator of heart diseases (Nogueira 2009; Sun et al. 2010). After this outcome, the main concern is nanoparticles, whether inhaled or taken through other routes, can also increase the risk of cardiovascular diseases. It has been reported in an in vivo study that the coronary arteries become dysfunctional due to nanoparticle accumulation (Nurkiewicz et al. 2008; Minarchick et al. 2013). Nanoparticle accumulation (iron oxide) in mitochondrial cells has been found to cause swelling in mitochondria in cell culture studies. Cytoplastic vacuolation and cell death have also been found in human aortic cells. After inhalation of nanoparticles starts the production of nitric oxide (NO) with the attachment of monocytes, and these two things are the early stages of heart problem (Zhu et al. 2011). It is clear from the above studies that inhalation of nanoparticles can lead to the development of cardiovascular disease, and the persons already suffering from this disease are at high risk of mortality. Atherosclerosis (clotting in blood vessels) is very common in heart diseases, and several animal studies are available in which scientists have used the atherosclerosis animal model to study the effect of nanoparticles in cardiovascular diseases. Animal studies are very useful to find out various consequences of inhaling nanoparticles (Agmon et al. 2002; Klaus 2000; Taute et al. 2002). Transgenic animal models of atherosclerosis have been produced by deleting apolipoprotein E as this gene is responsible for clearing chylomicrons and VLDL from blood provides a useful tool to study various cardiovascular diseases (Meir and Leitersdorf 2004). This model has been used to analyze the effect of nanoparticle inhalation in the population suffering from cardiovascular disease. It was found that apoE-/- mice exposure to nickel nanoparticles causes the progression of clotting in blood vessels (Kang et al. 2011). Single wall carbon nanotubes have been found to cause mitochondrial DNA damage, depletion of glutathione, increase in protein carbonyls, and increase in plaque deposition in blood vessels in apolipoprotein E deficient mice (Li et al. 2007). The effect of nanoparticle inhalation has also been studied in the atherosclerosis model, which was established by feeding Sprague-Dawley rats vitamin D3 and high-lipid chow (Pang et al. 2010). Intravenous injection of multiwall carbon nanotubes (MWCNTs) in these animals led to aggravated atherosclerosis and increased calcification. Exposure of tight junctions present in endothelial cells to multiwall carbon nanotubes also stimulates atherosclerosis as it disrupts the function of tight junctions (Xu et al. 2012). Another model used to investigate the effect of nanoparticles on cardiovascular diseases is the pulmonary inflammation animal model. Titanium dioxide, carbon nanotubes, and carbon black have also been found to cause coagulatory dysfunction which is mainly due to intensification of pulmonary vessel permeability (Inoue et al. 2006, 2007, 2008). Mitochondrial DNA damage (Ballinger et al. 2000; Ballinger et al. 2002), inflammation (Libby et al. 2002), and oxidative stress (Harrison et al. 2003) are the main factors that can cause damage to blood vessels and cause atherosclerosis (Choksi et al. 2004). Nanoparticles are responsible for all these factors (Zhu et al. 2011; Su et al. 2012) which increase the risk of platelet accumulation in blood vessels and cause atherosclerosis (Libby et al. 2011).

6 Nanoparticle Toxicity in Chronic Respiratory Disease

Nanoparticles are negatively affecting several people from kids to the elderly (Sha et al. 2013). Most of the elderly people, people who are addicted to tobacco, are suffering from COPD (chronic obstructive pulmonary disease) like asthma and bronchitis (Mathers and Loncar 2006). Approximately 300 million people were affected by asthma (Masoli et al. 2004) and COPD in 2004, which becomes the cause of death of more than 3 million people in 2011 (http://www.who.int/media-centre/factsheets/fs310/en/). The airways of patients suffering from chronic respiratory diseases are sensitive to irritants such as airborne particles. Elderly people suffering from COPD are extremely sensitive to airborne allergens. Mast cells and basophils are stimulated by these allergens, leading to inflammation in airways (Agrawal and Shao 2010; Jin et al. 2018; Feeney et al. 2004). On one side, nanoparticles play an important role in imaging, respiratory disease treatment, and diagnosis (Card et al. 2008), but on the other side, some people are extremely sensitive to nanoparticles. Ultrafine nanoparticles affect more asthmatics and elderly populations as compared to healthy and young patients (Kim and Kang 1997). According to the literature, it has been found that asthmatic patients and the elderly population are more prone to nanoparticle deposition in their lungs as compared to healthy persons. This is indicated by a study where 74% deposition was found during the resting stage and approximately 40% during physical activity (Chalupa et al. 2004). The consequence of this increased nanoparticle deposition in the lungs can cause increased lung residual volume. This increased accumulation of nanoparticles in asthmatic and elderly continuously increases swelling in the lungs, which leads to severe consequences (Kamata et al. 2011). According to the literature, it has been found that inhalation of nanoparticles like ultrafine particles that come from diesel exhaust can cause increased symptoms of asthma in in vivo models (Inoue et al. 2007; Takano et al. 2002). As soon as the subject inhales nanoparticles the process of inflammation starts in the lungs (Santhanam et al. 2008; Sung et al. 2008; Grassian et al. 2007; Ambalavanan et al. 2013) which cause severe destruction of epithelial cells of the lungs airway (Khatri et al. 2013). Multiwall carbon nanotubes can cause increased and abnormal cell growth of goblet cells, which leads to increased secretion of mucus in the lung airways (Inoue et al. 2009). Once nanoparticles are inhaled, they led to a series of chemical events like phagocytosis by macrophages; chemotaxis leads to activation of complement system cycle and ultimately clearance of this inflammatory response takes time (Oberdörster et al. 2005). A lipopolysaccharide treated model has been generated to study the effect of nanoparticles in asthmatic subjects (Nikula and Green 2000).

7 Nanoparticle Toxicity in Liver Diseases

As a major detoxification organ, the liver is susceptible to injury from multiple risk factors (Long 2020; Roberts et al. 2007). In the previous sections of this review, the side effects of copper nanoparticles, silicon, and titanium dioxide which cause damage to the heart and lungs have been discussed, but the most affected organ is the liver as nanoparticles are translocated from all these organs to the liver (Gustafson et al. 2015). Nanoparticles tend to accumulate more in the liver as compared to other organs (Almeida et al. 2011). Magnetic nanoparticle accumulation in the liver cells causes interaction of nanoparticles with liver cells and leads to morphological as well as physiological changes. A number of different types of cells are present in the liver like hepatocytes, liver endothelial cells, and Kupffer cells, and effect of nanoparticles has been studied on a number of in vivo and in vitro models (Chinde and Grover 2017). It has been found in literature that gold nanorods induce severe liver damage in chronic liver injury animal models (C57BL/6 mice). Gold nanorods were found to induce liver damage by initializing hepatic macrophages which ultimately cause the death of hepatocytes, higher levels of ALT (alanine aminotransferase), and mononuclear cell infiltration. This liver injury animal model was produced by treating the mice with carbon tetrachloride and concanavalin A (Bartneck et al. 2012). Concanavalin A and carbon tetrachloride induced liver damage, which was further intensified by gold nanorods. Another model used to analyze the effect of nanoparticles was the nonalcoholic steatohepatitis model which was produced by treating mice with methionine and choline-deficient diet for approximately 4 weeks. Treatment of such mice with nanoparticles has been shown to induce inflammatory response and apoptosis in the liver. It is clear from the above discussion that patients suffering from liver problems should be well aware of nanoparticle exposure (Hwang et al. 2012).

8 Nanoparticle Toxicity in Infectious Diseases

Approximately 20% of death is due to infectious diseases, and mainly viral diseases cause approximately one-third of these deaths. SARS (severe acute respiratory syndrome), MERS (Middle East respiratory syndrome), and COVID-19 (coronavirus disease) are highly infectious types of viral diseases (Gurunathan et al. 2020). Coronavirus term came into existence in 1968, and the term is given mainly due to its morphology under a microscope (Almeida et al. 1968). Coronaviruses are sphere-shaped and can occur in various distinct shapes 80–120 nm diameter. When observed under the electron microscope, the exterior of coronavirus is decked with spikes made up of glycol-proteins (Masters 2006). Some short projections of coronavirus are composed of HE (hemagglutinin-esterase), which is dimeric and can be found in beta-coronaviruses, for example, in human coronaviruses (De Groot 2006). The genome of these viruses is composed of a number of open reading frames and single-stranded genome RNA, sizes ranging from 27 to 32 kb (Masters 2006). A new risk for human beings has been established with the emergence of a new strain of coronavirus, i.e., COVID-19 (Singhal 2020). In the twentieth century, no coronavirus COVID-19 became a major risk factor (Prajapat et al. 2020). It was suspected that in December 2019, in China (Wuhan, Hubei province), the transmission of this virus in humans is through unidentified animal sources, probably bats (Singhal 2020). By taking this in view, new effective and bioavailable medicine should be synthesized to prevent this disease. Therefore, scientists are working on synthesizing nanomedicines. In a recent study, it has been discussed that zinc nanoparticles can reduce inflammation and decrease mucus in the lungs. But further clinical and experimental data is required to analyze the safety evaluation of zinc nanoparticles (Skalny et al. 2020). In a recent study, it has been published that particulate matter in the air increases the spread of coronavirus in the air. There is a correlation between increased morbidity and particulate matter. Particulate matter is also shown to increase lung inflammation and mucus and leads to lung cell death during viral infections (Qing et al. 2019). In another study, again particulate matter installation in ACE2 knockout mice has been reported to induce lung injury (Lin et al. 2018). A wide range of nanomaterials is available, but due to their nanotoxicity, safety evaluation of nanomaterials is important before their execution for the treatment of the disease (Xu et al. 2020).

9 Effect of Nanoparticles on Some Other Diseases

In addition to causing severity of CVD, pulmonary infection, and liver function alteration, nanoparticles cause a number of other diseases. Ultrafine particles of silica are known to cause multiple organ injuries (Niu et al. 2016). Silica nanoparticles have been found to cause an increase in creatinine and urea levels after investigation of kidney functions. In this study, the nanotoxicity of silicon nanoparticles has been assayed with the help of adult male rats for 28 days. By analyzing the results of this study, silica nanoparticles were found to cause malfunctioning of kidneys after histopathological studies (Azouz and Korany 2020). In a study, gold nanoparticles have been found to increase the release of inflammatory cytokines in plasma, leading to inflammation (Feng et al. 2020). In another study, chromium and copper nanoparticles have been shown to cause DNA damage (Bhabra et al. 2009). Silica and titanium dioxide particles have been shown to cause complications in pregnancy (Yamashita et al. 2011). In another study, carbon nanotubes were found to accumulate in testes of mice, increase oxidative stress, and decrease the thickness of seminiferous tubes (Bai et al. 2010). Titanium dioxide nanoparticles are widely used in creams due to their whitish color and opacity, and are also widely used in the food industry. High doses of titanium dioxide nanoparticles were given intragastrically to male Wistar rats to analyze the effect on the brain of rats. After analyzing the results, it was found that titanium dioxide nanoparticles are causing downregulation of acetylcholinesterase in the brain, with an increase in interleukin-6 ultimately causes inflammation in the brain. The above study concludes that titanium dioxide nanoparticles are neurotoxic (Heidari et al. 2019). In another study mice brains were exposed to titanium dioxide nanoparticles with a dose level of 2.5, 5, and 10 mg/kg for 90 days continuously. After analyzing the results, it was found that titanium nanoparticles stimulate TLRs/TNF-α/NF-κB pathway and cause neuroinflammation (Ze et al. 2014). Due to special properties of iron nanoparticles are used in cell biology as well as experimental biology. These are also used as drug delivery systems in anticancer therapy (Nikolovski et al. 2018). No doubt they are one of the good choices in the drug delivery system, but in literature, they are also proved to cause neurotoxicity. The effect of iron oxide nanoparticles has been studied and they are found to cause striatum and hippocampus toxicity in the brain (Wu et al. 2013). From the above literature review, it is clear that the use of nanoparticles requires intensive nanotoxicology evaluation.

Table 1 Evidence for the side effects of nanoparticles in elderly and susceptible population

Disease under study	Model used	Outcome of the study	Authors
Nanoparticle toxicity concerning elderly	Male Fisher F344 rats, 12 weeks old weighing 300–320 g (young adults) and 19 months old weighing approximately 400–425 g, aged group	The concentration of titanium dioxide nanoparticles was found to be higher in aged rats as compared to young rats.	Gaté et al. (2017)
	Male, aged Wister rats 300–350 mg	Gadolinium orthovanadate nanoparticles have been shown to cause acute toxicity.	Karpenko et al. (2013)
	Human subjects	Particulate matter causes more severity of disease in elderly people.	Wang et al. (2015)
Nanoparticle toxicity concerning cardiovascular diseases	Wistar rats weighing 250–300 g	Administration of silicon dioxide nanoparticles was found to enter in rat heart mitochondria, causing damage to cardiomyocytes.	Lozano et al. (2020)
	Mouse model	Titanium dioxide nanoparticles disturbed the antioxidant system in the mouse heart.	Sheng et al. (2013)
	C57BL/6 mice	Titanium dioxide nanoparticles–treated mouse heart shows activation of complement system cascade.	Husain et al. (2015)
	Rabbit model	Titanium dioxide nanoparticles cause myocarditis in mice.	Yu et al. (2016)
	Male Sprague-Dawley rats (3 weeks, 8 weeks, 20 weeks)	An increase in fibrinogen and blood viscosity was found to be high in old rats as compared to young and adult.	Chen et al. (2008)
Nanoparticle toxicity concerning pulmonary disease	Human fibroblast cells (HFL-1)	Exposure of HFL-1 cells to copper nanoparticles leads to activation of interleukins which cause mitochondrial damage.	Lerner et al. (2015)
	Male Wistar rats	Titanium dioxide nanoparticles cause changes in surfactant protein levels.	Okada et al (2019)
	Lung cancer cells A549	DNA damage, cytotoxicity, and cell death induced by titanium dioxide nanoparticles.	Wang et al. (2015)
	Cellular model	The cytotoxic and inflammatory response in the lungs when exposed to biodegradable nanoparticles.	Fattal et al. (2014)
	Male Sprague-Dawley rats (3 weeks, 8 weeks, 20 weeks)	Silicon dioxide nanoparticles were found to cause pulmonary inflammation in lungs of old rats as compared to young and adult.	Chen et al. (2008)
Nanoparticle toxicity concerning liver diseases	Human liver carcinoma cells	Needle-shaped PLGA-PEG nanoparticles cause cytotoxicity	Zhang et al. (2017)
	Human bronchiolar carcinoma cells	Silica nanoparticles cause an increase in reactive oxygen species, increased lactate dehydrogenase, and malondialdehyde.	Chen et al. (2013)
	Male rats	Decreased cellular viability in liver.	Arefian et al. (2015)
	HepG2 cell line	Decreased cellular viability.	Faedmaleki et al. (2014)
	Rat liver	Administration of titanium dioxide leads to hepatotoxicity.	Chen et al. (2019)
Nanoparticles concerning kidney	Male Sprague-Dawley rats	Elevation of creatinine and urea level in kidneys.	Azouz and Korany (2020)

10   Conclusions

Nanoparticle advancement has offered several opportunities in treatment, diagnosis, and early detection. Nanoparticles are also used in food products to increase color and bioavailability and also to preserve the food. There is no doubt that nanoparticles' advancement has revolutionized the medical field and other sectors like food and environment, but the side effects of these nanoparticles cannot be ignored as it has been reported from the literature that nanoparticles can intensify symptoms of cardiovascular diseases, pulmonary issues, liver diseases, and all other diseases. Therefore, this is our responsibility; while synthesizing nanoparticles, safety and toxicity of nanoparticles should be evaluated by using different in vitro and in vivo approaches for maximum benefit from these small-sized particles.

Acknowledgments   The authors thank the fraternity of University Institute of Biotechnology and also the University Center for Research and Development (UCRD) at Chandigarh University (CU) for support. All the authors have equally contributed to the manuscript and revised the manuscript. The institute to which the authors are currently affiliated, however, has no role in shaping the manuscript. All the authors declare no conflict of interest.

References

Agmon Y, Khandheria BK, Meissner I, Schwartz GL, Petterson TM, O'Fallon WM, et al. Relation of coronary artery disease and cerebrovascular disease with atherosclerosis of the thoracic aorta in the general population. Am J Cardiol. 2002;89(3):262–7.

Agrawal DK, Shao Z. Pathogenesis of allergic airway inflammation. Curr Allergy Asthma Rep. 2010;10(1):39–48.

Almeida J, Berry D, Cunningham C, Hamre D, Hofstad M, Mallucci L, et al. Coronaviruses. Nature. 1968;220(650):2.

Almeida JPM, Chen AL, Foster A, Drezek R. In vivo biodistribution of nanoparticles. Nanomedicine. 2011;6(5):815–35.

Ambalavanan N, Stanishevsky A, Bulger A, Halloran B, Steele C, Vohra Y, et al. Titanium oxide nanoparticle instillation induces inflammation and inhibits lung development in mice. Am J Phys Lung Cell Mol Phys. 2013;304(3):L152–L61.

Arefian Z, Pishbin F, Negahdary M, Ajdary M. Potential toxic effects of Zirconia Oxide nanoparticles on liver and kidney factors. 2015.

Azouz RA, Korany RM. Toxic impacts of amorphous silica nanoparticles on liver and kidney of male adult rats: an in vivo study. Biol Trace Elem Res. 2020:1–10.

Bai Y, Zhang Y, Zhang J, Mu Q, Zhang W, Butch ER, et al. Repeated administrations of carbon nanotubes in male mice cause reversible testis damage without affecting fertility. Nat Nanotechnol. 2010;5(9):683–9.

Ballinger SW, Patterson C, Yan C-N, Doan R, Burow DL, Young CG, et al. Hydrogen peroxide– and peroxynitrite-induced mitochondrial DNA damage and dysfunction in vascular endothelial and smooth muscle cells. Circ Res. 2000;86(9):960–6.

Ballinger SW, Patterson C, Knight-Lozano CA, Burow DL, Conklin CA, Hu Z, et al. Mitochondrial integrity and function in atherogenesis. Circulation. 2002;106(5):544–9.

Bartneck M, Ritz T, Keul HA, Wambach M, Bornemann J, Gbureck U, et al. Peptide-functionalized gold nanorods increase liver injury in hepatitis. ACS Nano. 2012;6(10):8767–77.

Bawa R, Audette GF, Rubinstein I. Handbook of clinical nanomedicine: nanoparticles, imaging, therapy, and clinical applications. CRC Press; 2016.

Bhabra G, Sood A, Fisher B, Cartwright L, Saunders M, Evans WH, et al. Nanoparticles can cause DNA damage across a cellular barrier. Nat Nanotechnol. 2009;4(12):876–83.

Buzea C, Pacheco II, Robbie K. Nanomaterials and nanoparticles: sources and toxicity. Biointerphases. 2007;2(4):MR17–71.

Card JW, Zeldin DC, Bonner JC, Nestmann ER. Pulmonary applications and toxicity of engineered nanoparticles. Am J Phys Lung Cell Mol Phys. 2008;295(3):L400–L11.

Chalupa DC, Morrow PE, Oberdörster G, Utell MJ, Frampton MW. Ultrafine particle deposition in subjects with asthma. Environ Health Perspect. 2004;112(8):879–82.

Chen Z, Meng H, Xing G, Yuan H, Zhao F, Liu R, et al. Age-related differences in pulmonary and cardiovascular responses to SiO2 nanoparticle inhalation: nanotoxicity has susceptible population. Environ Sci Technol. 2008;42(23):8985–92.

Chen Q, Xue Y, Sun J. Kupffer cell-mediated hepatic injury induced by silica nanoparticles in vitro and in vivo. Int J Nanomedicine. 2013;8:1129.

Chen Z, Zhou D, Han S, Zhou S, Jia G. Hepatotoxicity and the role of the gut-liver axis in rats after oral administration of titanium dioxide nanoparticles. Part Fibre Toxicol. 2019;16(1):1–17.

Chinde S, Grover P. Toxicological assessment of nano and micron-sized tungsten oxide after 28 days repeated oral administration to Wistar rats. Mutat Res Genet Toxicol Environ Mutagen. 2017;819:1–13.

Cho W-S, Duffin R, Thielbeer F, Bradley M, Megson IL, MacNee W, et al. Zeta potential and solubility to toxic ions as mechanisms of lung inflammation caused by metal/metal oxide nanoparticles. Toxicol Sci. 2012;126(2):469–77.

Choksi K, Boylston W, Rabek J, Widger W, Papaconstantinou J. Oxidatively damaged proteins of heart mitochondrial electron transport complexes. Biochimica et Biophysica Acta (BBA)-Molecular Basis of Disease. 2004, 1688;(2):95–101.

De Groot RJ. Structure, function and evolution of the hemagglutinin-esterase proteins of corona- and toroviruses. Glycoconj J. 2006;23(1-2):59–72.

Faedmaleki F, Shirazi FH, Salarian A-A, Ashtiani HA, Rastegar H. Toxicity effect of silver nanoparticles on mice liver primary cell culture and HepG2 cell line. Iran J Pharm Res. 2014;13(1):235.

Fattal E, Grabowski N, Mura S, Vergnaud J, Tsapis N, Hillaireau H. Lung toxicity of biodegradable nanoparticles. J Biomed Nanotechnol. 2014;10(10):2852–64.

Feeney AS, Fendrick AM, Quintiliani R. Acute exacerbation of chronic bronchitis: a primary care consensus guideline. Am J Manag Care. 2004;10:689–96.

Feng L, Ning R, Liu J, Liang S, Xu Q, Liu Y, et al. Silica nanoparticles induce JNK-mediated inflammation and myocardial contractile dysfunction. J Hazard Mater. 2020;391:122206.

Gaté L, Disdier C, Cosnier F, Gagnaire F, Devoy J, Saba W, et al. Biopersistence and translocation to extrapulmonary organs of titanium dioxide nanoparticles after subacute inhalation exposure to aerosol in adult and elderly rats. Toxicol Lett. 2017;265:61–9.

Grassian VH, O'Shaughnessy PT, Adamcakova-Dodd A, Pettibone JM, Thorne PS. Inhalation exposure study of titanium dioxide nanoparticles with a primary particle size of 2 to 5 nm. Environ Health Perspect. 2007;115(3):397–402.

Gurunathan S, Qasim M, Choi Y, Do JT, Park C, Hong K, et al. Antiviral potential of nanoparticles—Can nanoparticles fight against coronaviruses? Nanomaterials. 2020;10(9):1645.

Gustafson HH, Holt-Casper D, Grainger DW, Ghandehari H. Nanoparticle uptake: the phagocyte problem. Nano Today. 2015;10(4):487–510.

Harrison D, Griendling KK, Landmesser U, Hornig B, Drexler H. Role of oxidative stress in atherosclerosis. Am J Cardiol. 2003;91(3):7–11.

Heidari Z, Mohammadipour A, Haeri P, Ebrahimzadeh-bideskan A. The effect of titanium dioxide nanoparticles on mice midbrain substantia nigra. Iran J Basic Med Sci. 2019;22(7):745.

Husain M, Wu D, Saber AT, Decan N, Jacobsen NR, Williams A, et al. Intratracheally instilled titanium dioxide nanoparticles translocate to heart and liver and activate complement cascade in the heart of C57BL/6 mice. Nanotoxicology. 2015;9(8):1013–22.

Hwang JH, Kim SJ, Kim Y-H, Noh J-R, Gang G-T, Chung BH, et al. Susceptibility to gold nanoparticle-induced hepatotoxicity is enhanced in a mouse model of nonalcoholic steatohepatitis. Toxicology. 2012;294(1):27–35.

Iavicoli I, Leso V, Manno M, Schulte PA. Biomarkers of nanomaterial exposure and effect: current status. J Nanopart Res. 2014;16(3):2302.

Inoue K-i, Takano H, Yanagisawa R, Hirano S, Sakurai M, Shimada A, et al. Effects of airway exposure to nanoparticles on lung inflammation induced by bacterial endotoxin in mice. Environ Health Perspect. 2006;114(9):1325–30.

Inoue K-i, Takano H, Yanagisawa R, Hirano S, Kobayashi T, Fujitani Y, et al. Effects of inhaled nanoparticles on acute lung injury induced by lipopolysaccharide in mice. Toxicology. 2007;238(2-3):99–110.

Inoue K, Takano H, Ohnuki M, Yanagisawa R, Sakurai M, Shimada A, et al. Size effects of nanomaterials on lung inflammation and coagulatory disturbance. Int J Immunopathol Pharmacol. 2008;21(1):197–206.

Inoue K-i, Koike E, Yanagisawa R, Hirano S, Nishikawa M, Takano H. Effects of multi-walled carbon nanotubes on a murine allergic airway inflammation model. Toxicol Appl Pharmacol. 2009;237(3):306–16.

Jin C, Shelburne CP, Li G, Riebe KJ, Sempowski GD, Foster WM, et al. Particulate allergens potentiate allergic asthma in mice through sustained IgE-mediated mast cell activation. J Clin Invest. 2018;121(3):941–55.

Kamata H, Tasaka S, Inoue K-i, Miyamoto K, Nakano Y, Shinoda H, et al. Carbon black nanoparticles enhance bleomycin-induced lung inflammatory and fibrotic changes in mice. Exp Biol Med. 2011;236(3):315–24.

Kang GS, Gillespie PA, Gunnison A, Moreira AL, Tchou-Wong K-M, Chen L-C. Long-term inhalation exposure to nickel nanoparticles exacerbated atherosclerosis in a susceptible mouse model. Environ Health Perspect. 2011;119(2):176–81.

Karpenko N, Malukin YV, Koreneva E, Klochkov V, Kavok N, Smolenko N et al., editors. The effects of chronic intake of nanoparticles of cerium dioxide or gadolinium ortovanadate into aging male rats. Proceedings of the International Conference Nanomaterials: Applications and Properties; 2013: Sumy State University Publishing.

Khatri M, Bello D, Pal AK, Cohen JM, Woskie S, Gassert T, et al. Evaluation of cytotoxic, genotoxic and inflammatory responses of nanoparticles from photocopiers in three human cell lines. Part Fibre Toxicol. 2013;10(1):42.

Khlebtsov N, Dykman L. Biodistribution and toxicity of engineered gold nanoparticles: a review of in vitro and in vivo studies. Chem Soc Rev. 2011;40(3):1647–71.

Kim CS, Kang TC. Comparative measurement of lung deposition of inhaled fine particles in normal subjects and patients with obstructive airway disease. Am J Respir Crit Care Med. 1997;155(3):899–905.

Klaus D. Atherosclerosis and arteriosclerosis in hypertension. Nieren Hochdruckkrankh. 2000;29:1–16.

Lemoine M. Defining aging. Biol Philos. 2020;35(5):1–30.

Lerner CA, Sundar IK, Watson RM, Elder A, Jones R, Done D, et al. Environmental health hazards of e-cigarettes and their components: oxidants and copper in e-cigarette aerosols. Environ Pollut. 2015;198:100–7.

Li Z, Hulderman T, Salmen R, Chapman R, Leonard SS, Young S-H, et al. Cardiovascular effects of pulmonary exposure to single-wall carbon nanotubes. Environ Health Perspect. 2007;115(3):377–82.

Li Y, Zhang Y, Yan B. Nanotoxicity overview: nano-threat to susceptible populations. Int J Mol Sci. 2014;15(3):3671–97.

Libby P, Ridker PM, Maseri A. Inflammation and atherosclerosis. Circulation. 2002;105(9):1135–43.

Libby P, Ridker PM, Hansson GK. Progress and challenges in translating the biology of atherosclerosis. Nature. 2011;473(7347):317–25.

Lin C-I, Tsai C-H, Sun Y-L, Hsieh W-Y, Lin Y-C, Chen C-Y, et al. Instillation of particulate matter 2.5 induced acute lung injury and attenuated the injury recovery in ACE2 knockout mice. Int J Biol Sci. 2018;14(3):253.

Longmire M, Choyke PL, Kobayashi H. Clearance properties of nano-sized particles and molecules as imaging agents: considerations and caveats. 2008.

Lozano O, Silva-Platas C, Chapoy-Villanueva H, Pérez BE, Lees JG, Ramachandra CJ, et al. Amorphous SiO2 nanoparticles promote cardiac dysfunction via the opening of the mitochondrial permeability transition pore in rat heart and human cardiomyocytes. Part Fibre Toxicol. 2020;17:1–16.

Manno M, Sito F, Licciardi L. Ethics in biomonitoring for occupational health. Toxicol Lett. 2014;231(2):111–21.

Marano F, Guadagnini R. Cellular Mechanisms of nanoparticle's toxicity. Encyclopedia of nanotechnology. 2012.

Masoli M, Fabian D, Holt S, Beasley R, Program GIFA. The global burden of asthma: executive summary of the GINA Dissemination Committee report. Allergy. 2004;59(5):469–78.

Masters PS. The molecular biology of coronaviruses. Adv Virus Res. 2006;66:193–292.

Mathers C. The global burden of disease: 2004 update. World Health Organization; 2008.

Mathers CD, Loncar D. Projections of global mortality and burden of disease from 2002 to 2030. PLoS Med. 2006;3(11):e442.

Meir KS, Leitersdorf E. Atherosclerosis in the apolipoprotein E–deficient mouse: a decade of progress. Arterioscler Thromb Vasc Biol. 2004;24(6):1006–14.

Minarchick VC, Stapleton PA, Porter DW, Wolfarth MG, Çiftyürek E, Barger M, et al. Pulmonary cerium dioxide nanoparticle exposure differentially impairs coronary and mesenteric arteriolar reactivity. Cardiovasc Toxicol. 2013;13(4):323–37.

Neupane B, Jerrett M, Burnett RT, Marrie T, Arain A, Loeb M. Long-term exposure to ambient air pollution and risk of hospitalization with community-acquired pneumonia in older adults. Am J Respir Crit Care Med. 2010;181(1):47–53.

Nikolovski D, Jeremic M, Paunovic J, Vucevic D, Radosavljevic T, Radojević-Škodrić S, et al. Application of iron oxide nanoparticles in contemporary experimental physiology and cell biology research. Rev Adv Mater Sci. 2018;53(1):74–8.

Nikula KJ, Green FH. Animal models of chronic bronchitis and their relevance to studies of particle-induced disease. Inhal Toxicol. 2000;12(sup 4):123–53.

Niu Y-M, Zhu X-L, Chang B, Tong Z-H, Cao W, Qiao P-H, et al. Nanosilica and polyacrylate/ nanosilica: a comparative study of acute toxicity. Bio Med Res Int. 2016;2016

Nogueira JB. Air pollution and cardiovascular disease. Revista portuguesa de cardiologia: orgao oficial da Sociedade Portuguesa de Cardiologia= Portuguese journal of cardiology: an official journal of the Portuguese Society of Cardiology. 2009; 28(6):715.

Nurkiewicz TR, Porter DW, Hubbs AF, Cumpston JL, Chen BT, Frazer DG, et al. Nanoparticle inhalation augments particle-dependent systemic microvascular dysfunction. Part Fibre Toxicol. 2008;5(1):1.

Oberdörster G, Oberdörster E, Oberdörster J. Nanotoxicology: an emerging discipline evolving from studies of ultrafine particles. Environ Health Perspect. 2005;113(7):823–39.

Okada T, Lee BW, Ogami A, Oyabu T, Myojo T. Inhalation of titanium dioxide (P25) nanoparticles to rats and changes in surfactant protein (SP-D) levels in bronchoalveolar lavage fluid and serum. Nanotoxicology. 2019;13(10):1396–408.

Pang J, Xu Q, Xu X, Yin H, Xu R, Guo S, et al. Hexarelin suppresses high lipid diet and vitamin D3-induced atherosclerosis in the rat. Peptides. 2010;31(4):630–8.

Peters A, Wichmann HE, Tuch T, Heinrich J, Heyder J. Respiratory effects are associated with the number of ultrafine particles. Am J Respir Crit Care Med. 1997;155(4):1376–83.

Prajapat M, Sarma P, Shekhar N, Avti P, Sinha S, Kaur H, et al. Drug targets for corona virus: a systematic review. Indian J Pharm. 2020;52(1):56.

Qing H, Wang X, Zhang N, Zheng K, Du K, Zheng M, et al. The effect of fine particulate matter on the inflammatory responses in human upper airway mucosa. Am J Respir Crit Care Med. 2019;200(10):1315–8.

Raj S, Jose S, Sumod U, Sabitha M. Nanotechnology in cosmetics: opportunities and challenges. J Pharm Bioall Sci. 2012;4(3):186.

Roberts RA, Ganey PE, Ju C, Kamendulis LM, Rusyn I, Klaunig JE. Role of the Kupffer cell in mediating hepatic toxicity and carcinogenesis. Toxicol Sci. 2007;96(1):2–15.

Santhanam P, Wagner JG, Elder A, Gelein R, Carter J, Driscoll K, et al. Effects of subchronic inhalation exposure to carbon black nanoparticles in the nasal airways of laboratory rats. Int J Nanotechnol. 2008;5(1):30–54.

Schäfer-Korting M, Mehnert W, Korting H-C. Lipid nanoparticles for improved topical application of drugs for skin diseases. Adv Drug Deliv Rev. 2007;59(6):427–43.

Schulte PA, Trout DB. Nanomaterials and worker health: medical surveillance, exposure registries, and epidemiologic research. J Occup Environ Med. 2011;53:S3–7.

Sengul AB, Asmatulu E. Toxicity of metal and metal oxide nanoparticles: a review. Environ Chem Lett. 2020:1–25.

Sha B, Gao W, Wang S, Li W, Liang X, Xu F, et al. Nano-titanium dioxide induced cardiac injury in rat under oxidative stress. Food Chem Toxicol. 2013;58:280–8.

Sheng L, Wang X, Sang X, Ze Y, Zhao X, Liu D, et al. Cardiac oxidative damage in mice following exposure to nanoparticulate titanium dioxide. J Biomed Mater Res A. 2013;101(11):3238–46.

Singhal T. A review of coronavirus disease-2019 (COVID-19). Indian J Pediatr. 2020:1–6.

Skalny AV, Rink L, Ajsuvakova OP, Aschner M, Gritsenko VA, Alekseenko SI, et al. Zinc and respiratory tract infections: Perspectives for COVID-19. Int J Mol Med. 2020;46(1):17–26.

Su L, Han L, Ge F, Zhang SL, Zhang Y, Zhao BX, et al. The effect of novel magnetic nanoparticles on vascular endothelial cell function in vitro and in vivo. J Hazard Mater. 2012;235:316–25.

Sun Q, Hong X, Wold LE. Cardiovascular effects of ambient particulate air pollution exposure. Circulation. 2010;121(25):2755–65.

Sung JH, Ji JH, Yoon JU, Kim DS, Song MY, Jeong J, et al. Lung function changes in Sprague-Dawley rats after prolonged inhalation exposure to silver nanoparticles. Inhal Toxicol. 2008;20(6):567–74.

Takano H, Yanagisawa R, Ichinose T, Sadakane K, Yoshino S, Yoshikawa T, et al. Diesel exhaust particles enhance lung injury related to bacterial endotoxin through expression of proinflammatory cytokines, chemokines, and intercellular adhesion molecule-1. Am J Respir Crit Care Med. 2002;165(9):1329–35.

Taute B, Feller S, Hansgen K, Podhaisky H. Carotid atherosclerosis in patients with peripheral arterial disease. Perfusion. 2002;15(5):183–+.

Trout DB, Schulte PA. Medical surveillance, exposure registries, and epidemiologic research for workers exposed to nanomaterials. Toxicology. 2010;269(2-3):128–35.

Von Klot S, Wölke G, Tuch T, Heinrich J, Dockery D, Schwartz J, et al. Increased asthma medication use in association with ambient fine and ultrafine particles. Eur Respir J. 2002;20(3):691–702.

Wang Y, Cui H, Zhou J, Li F, Wang J, Chen M, et al. Cytotoxicity, DNA damage, and apoptosis induced by titanium dioxide nanoparticles in human non-small cell lung cancer A549 cells. Environ Sci Pollut Res. 2015;22(7):5519–30.

Wu J, Ding T, Sun J. Neurotoxic potential of iron oxide nanoparticles in the rat brain striatum and hippocampus. Neurotoxicology. 2013;34:243–53.

Xu YY, Yang J, Shen T, Zhou F, Xia Y, Fu JY, et al. Intravenous administration of multi-walled carbon nanotubes affects the formation of atherosclerosis in sprague-dawley rats. J Occup Health. 2012;54(5):361–9.

Xu H, Zhong L, Deng J, Peng J, Dan H, Zeng X, et al. High expression of ACE2 receptor of 2019-nCoV on the epithelial cells of oral mucosa. Int J Oral Sci. 2020;12(1):1–5.

Yamashita K, Yoshioka Y, Higashisaka K, Mimura K, Morishita Y, Nozaki M, et al. Silica and titanium dioxide nanoparticles cause pregnancy complications in mice. Nat Nanotechnol. 2011;6(5):321–8.

Yao Y, Long M. The biological detoxification of deoxynivalenol: a review. Food Chem Toxicol. 2020;145:111649.

Yu X, Hong F, Zhang YQ. Bio-effect of nanoparticles in the cardiovascular system. J Biomed Mater Res A. 2016;104(11):2881–97.

Ze Y, Sheng L, Zhao X, Hong J, Ze X, Yu X, et al. TiO$_2$ nanoparticles induced hippocampal neuro-inflammation in mice. PLoS One. 2014;9(3):e92230.

Zhang B, Lung PS, Zhao S, Chu Z, Chrzanowski W, Li Q. Shape dependent cytotoxicity of PLGA-PEG nanoparticles on human cells. Sci Rep. 2017;7(1):1–8.

Zhu M-T, Wang B, Wang Y, Yuan L, Wang H-J, Wang M, et al. Endothelial dysfunction and inflammation induced by iron oxide nanoparticle exposure: Risk factors for early atherosclerosis. Toxicol Lett. 2011;203(2):162–71.

Toxicity of Titanium Dioxide Nanoparticles and Oxidative Stress

Mohammad Rafiq Wani and G. G. H. A. Shadab

Contents

Abstract Titanium dioxide nanoparticles (TiO$_2$ NPs) are being used in food and other commonly used products on a large scale. As such, exposure to humans and environment has raised concerns regarding their toxic potential. Studies have reported that upon exposure to humans, TiO$_2$ NPs accumulate and induce oxidative damage, genotoxicity, and apoptosis in lungs, alimentary tract, liver, heart, spleen, kidneys, and cardiac muscle and these tissues. In this chapter, we highlight various applications of TiO$_2$ NPs, their toxicity potential as reported in published literature, and discuss their mechanism of toxicity.

Keywords TiO$_2$ NPs · Oxidative stress · Genotoxicity · Systemic toxicity

M. R. Wani · G. G. H. A. Shadab (✉)
Cytogenetics and Molecular Toxicology Laboratory, Section of Genetics, Department of Zoology, Aligarh Muslim University, Aligarh, Uttar Pradesh, India
e-mail: ggha.shadab.zo@amu.ac.in

131

1 Introduction

The branch of science dealing with the production and use of nanomaterials is nano-technology. The extensive use of nanomaterials in various fields, such as biomedical, cosmetic, and agriculture, has made nanotechnology one of the fast advancing field of science and technology. Nanomaterials are substances whose dimensions are in nanoscale, i.e., 1–100 nm (ISO 2015). A nanometer is one-billionth of a meter. Nanoparticles, nanofibers, nanotubes, and nanocomposites are all nanomaterials. Nanoparticles are extremely small in size but possess large surface-to-volume ratio, size, and size-dependent optical properties, these unique properties have made the nanoparticles suitable for use in a wide variety of products (Sajid et al. 2015). Cobalt nanoparticles find use in drug delivery and other biomedical applications (Ansari et al. 2017). In electronic and various biotechnological products, silica nanoparticles are extensively used (Kaphle et al. 2018; Zhou et al. 2019). Due to such large uses, the inevitable human and environment exposure has raised concerns about their toxicity inducing potential. Nanotoxicology refers to the study of toxicity due to nanomaterials. Recent reports have provided evidence of toxicity-inducing ability of various types of nanoparticle exposures to human and the environment.

2 Types and Applications of Nanoparticles

Carbon-based nanoparticles such as carbon nanotubes or graphene nanoparticles have versatile shapes and excellent thermal conductivity and hence are used in nanoelectronics (Ema et al. 2016; Sardoiwala et al. 2018). Micelles and liposomes are derived from organic materials and these nanoparticles serve as biosensors/diagnostics and are also used in drug delivery and DNA transfecting agents (Farré and Barceló 2012; Sandeep 2013). Composite-based nanoparticles have biomedical applications such as drug delivery and are also widely used for packaging of materials. The most frequently used nanoparticles are metal-based which are used in paints, coatings, cosmetics, and environmental remediation (EPA 2017). Nanoparticles of gold, silver, titanium, zinc, etc. are examples of metal nanoparticles. Oxides of metals are also source of nanoparticles. TiO_2 NPs are manufactured from oxides of titanium metal.

3 Nanoparticle Characterization

Before testing for toxicity, NPs are characterized to understand their physico-chemical properties that can then be correlated with their biological response. Electron microscopy (both scanning and transmission) is used to determine the morphological features of nanoparticles. SEM together with EDX (energy-dispersive X-ray) is used to determine the elemental constituents of NPs whereas atomic force microscopy (AFM) is used to image NPs in 3D. When NPs are dispersed in the dispersion medium, such as DMSO or water or saline, the particle size is measured by dynamic light scattering (DLS), but in the dry state, the particle size is measured by X-ray diffraction (XRD) powder. The NPs are also characterized for surface area using Brunauer–Emmet–Teller (BET) and zeta potential.

4 Factors Affecting NP Toxicity Potential

4.1 Particle Size

Owing to nano size, NPs can easily penetrate cell membranes and interact with cellular components, thereby causing cell damage. As compared to the bulk form, nano forms of substances have been reported to induce increased toxicity. TiO_2 in nano form caused more tissue damage and toxicity than its bulk form (Li et al. 2009). Phototoxicity of TiO_2 NPs of particle size 25 nm was more than TiO_2 NPs of 142 nm particle size (Sanders et al. 2012).

4.2 Shape and Morphology

TiO_2 NPs exist in different shapes in the form of particles, spheres, rods, cubes, etc. TiO_2 NPs in the form of nanorods were more toxic than spherical NPs of same size and surface area (Hsiao and Huang 2011).

4.3 Aggregation and Dissolution

While in aquatic medium, most NPs do not dissolve but tend to agglomerate or disperse and this affects their toxicity potential. This largely depends upon NP size and surface properties as well as stability of dispersion medium. This stability is dependent upon pH, temperature, as well as ionic strength of the medium.

4.4 Surface Charge

NP dispersion in the aquatic medium is also dependent on its surface charge. Changing the surface charges of TiO_2 NPs influenced the photocatalytic degradation activity of these NPs (Azeez et al. 2018).

5   Titanium Dioxide Nanoparticles (TiO$_2$ NPs)

5.1   Occurrence

Titanium dioxide is an oxide of titanium. In nature, titanium dioxide (TiO$_2$) exists in three crystalline forms: rutile is the most common form, while the other two forms (anatase and brookite) are relatively less common. TiO$_2$ NPs can also be synthesized in industries and laboratories. Some of the widely used procedures include: (i) hydrolyzing acidic titanium salt solutions (Wang et al. 2010) and (ii) sol–gel method for shape- and size-controlled synthesis (Vijayalakshmi and Rajendran 2012).

5.2   Applications/Uses

Titanium dioxide nanoparticles (TiO$_2$ NPs) constitute one of the major nanoparticles (NPs) produced and used in the world (Emerich and Thanos 2003). About 4 million tons of TiO$_2$ NPs are consumed annually globally (Ortlieb 2010). TiO$_2$ NPs are used in cosmetic products including sunscreens and toothpastes, food items such as chewing gums, and also in ceramics, paints, and many other day-by-day and commercial products (Xie et al. 2011), besides having medical and antibacterial applications (Li et al. 2010). These NPs are used to whiten food items and various pharmaceutical drugs (Kaida et al. 2004; Wolf et al. 2003). Coatings and fillings of various food items consist of about 20 mg TiO$_2$/g food, which is more than the highest level found in chewing gums (16 mg TiO$_2$/g) (Ropers et al. 2017). TiO$_2$ particles have a half-life of 12.7 days (Elgrabli et al. 2015), continuous exposure/intake can lead to bioaccumulation (Jovanovic 2015). TiO$_2$ NPs under the name E171 is also used as food additive which has been approved by the United States Food and Drug Administration (US FDA) and the European Food Safety Authority (EFSA) (FDA (2012), "Re-evaluation of titanium dioxide (E 171) as a food additive" 2016). These NPs also have biomedical applications and photocatalytic applications (Gélis et al. 2003; Sun et al. 2004; Wang et al. 2008a, b). As a result of such extensive usage, human and environment exposure to these nanoparticles is inevitable (Nel et al. 2006).

5.3   Toxicity of Titanium Dioxide Nanoparticles

The wide use of nanoparticles such as TiO$_2$ NPs can be attributed to their unique properties such as high surface-area to mass ratio and a high redox potential. These same properties also confer these NPs the ability to be toxic to biological systems (Chan 2006; Dusinska et al. 2011). Despite the widespread applications, studies suggest that exposure to TiO$_2$ NPs induces toxicity. These NPs have been classified as potential carcinogen (IARC 2002; NIOSH 2009). Animal models have shown liver toxicity after treatment with TiO$_2$ NPs (Liu et al. 2009; Xu et al. 2013; Shakeel et al. 2016). However, liver is not the only target of these NPs. In animal models, administration of TiO$_2$ NPs has been shown to reduce immune response (Duan et al. 2010) and induce inflammation in spleen (Chen et al. 2009), besides causing renal damage (Liang et al. 2009; Zhao et al. 2010) and genotoxicity in bone marrow cells (Sycheva et al. 2011; Dobrzyńska et al. 2014). In humans, TiO$_2$ NPs can induce oxidative damage to cellular components (Gurr et al. 2005), DNA damage (Jugan et al. 2012; Petković et al. 2011; Shukla et al. 2013), membrane damage and apoptosis (Acar et al. 2015; Coccini et al. 2015; Valdiglesias et al. 2013), damage to small intestine (Faust et al. 2014), heart (Kan et al. 2014), and brain (Allen 2016; Feng et al. 2015; Wang et al. 2008a, b; Márquez-Ramírez et al. 2012). Human exposure to NPs occurs chiefly through inhalation, ingestion, and injection of nanomedicine (Oberdörster et al. 2005). TiO$_2$ NPs have been found to accumulate in various body organs, including lungs, liver, heart, spleen, kidneys, and even in heart tissue, and cause oxidative damage in these tissues (Baranowska-Wojcik et al. 2020).

5.4 Genotoxicity of Titanium Dioxide Nanoparticles

TiO$_2$ NPs have been reported to induce genotoxicity in various *in vivo* and *in vitro* models (reviewed by Wani and Shadab 2020).

5.4.1 The Comet Assay

The most widely used test to determine DNA damage breaks in eukaryotic cells is the comet assay (also known as single-cell gel electrophoresis assay) (Fairbairn et al. 1995; Singh et al. 1988; Tice et al. 2000). The test is based on the principle that when DNA is damaged and a part is broken, this small fragment is pulled out of nucleus due to applied electric current, leaving behind the larger undamaged DNA in the nucleus. Cells are treated with a certain chemical, single cell suspensions are put on microscopic slide layered with agarose, a detergent is added for cell lysis, and electrophoresis is carried out. When these slides are viewed under fluorescence microscope and if the DNA is broken, comet-like structures are seen, the tail of the comet is the broken DNA and the head is the undamaged DNA. The greater the DNA damage, the longer will be the tail. "Olive tail moment (OTM)" is a parameter to assess the DNA damage (Courbiere et al. 2013). It is calculated by multiplying the percentage of DNA in the comet tail by the tail length. Tice and colleagues modified the traditional comet assay so that it can be used to determine special cases of DNA damage, for instance, by employing endonuclease III enzyme, oxidized pyrimidines can be determined whereas damaged purines are determined by the use of formamidopyrimidine DNA glycosylase (Tice et al. 2000).

5.4.2   The Micronucleus Assay (MN Assay)

The MN assay allows measurement of chromosome breaks (Fenech 2000). On exposure to some chemical substances, chromosome parts break forming micronuclei. At the time of cell division, these broken parts or micronuclei remain out of daughter cell nuclei. A modified version of MN assay is cytokinesis-block micronucleus (CBMN) assay. In this test, when cells undergo nuclear division, addition of cytochalasin-B prevents cells from cytokinesis (Fenech 2000; Schmid 1975). Such cells are then identified by their binucleate appearance. After cytochalasin is added, the cells grow for some time so that to induce chromosomal damage by a test chemical. For micronuclei detection, the cells are then stained and visualized under microscope (Fenech 2000; Kirsch-Volders et al. 1997). For determination of micronuclei *in vivo*, the bone marrow of the tested animal is used. FBS is used to flush out the bone marrow, the suspension is centrifuged, and the pellet is dropped on slides which are air-dried and stained for visualizing micronuclei under a microscope (Zhurkov et al. 1996).

5.4.3   In Vivo Studies on TiO$_2$ NP Genotoxicity

Exposure of 200–500 mg/kg bw of TiO$_2$ NPs induced DNA damage in Swiss-Albino mice (Chakrabarti et al. 2019). A 28-day exposure to TiO$_2$-NP at low concentrations induced genotoxicity in liver, spleen, and thymus cells of mice (Manivannan et al. 2019). Ali et al. (2019) reported that exposure to TiO$_2$ NP induced chromosomal aberrations in mice. Exposure of *Drosophila melanogaster* for 20 generations induced increasing DNA damage in each subsequent generation (Jovanović et al. 2018). Exposure of golden mussel (*Limnoperna fortunei*) to TiO$_2$ NPs damaged DNA in hemocytes (Girardello et al. 2016). Intragastric administration of TiO$_2$ NPs damaged DNA and induced micronuclei formation in bone marrow cells of Wistar rats (Grissa et al. 2015). About 2–3 weeks post exposure to TiO$_2$ NPs, blood cells in zebrafish observed DNA damage. TiO$_2$ NPs damaged DNA in the blood cells of fish (*Trachinotus carolinus*) (Vignardi et al. 2015).

5.4.4   In Vitro Studies on TiO$_2$ NP Genotoxicity

TiO$_2$ NPs induced DNA damage and micronuclei formation in human umbilical vein endothelial cells (Liao et al. 2019). TiO$_2$ NPs exposure induced genotoxicity in human amniocytes (Mottola et al. 2019), DNA fragmentation in PNT1A, A375 human melanoma, and SH-SY5Y human cell lines (Skubalova et al. 2019). TiO$_2$ NP-induced DNA damage has also been reported in RAW 264.7 cells (Chakrabarti et al. 2019), human peripheral blood lymphocytes (Osman et al. 2018; Patel et al. 2017; Kurzawa-Zegota et al. 2017), intestinal and hepatic cells (Jalili et al. 2018), human nasal mucosa cells (Hackenberg et al. 2017), bronchial epithelial cells (Ghosh et al. 2017), A549 and BEAS-2B cell lines (Biola-Clier et al. 2017), Chinese hamster lung fibroblast (V-79) cells (Jain et al. 2017), human colon carcinoma cell line Caco-2 (Proquin et al. 2017), human lung alveolar carcinoma cell line (A549) (Armand et al. 2016; Kansara et al. 2015; Wang et al. 2015), mouse fibroblasts, human keratinocytes and human fibroblast cells (Tomankova et al. 2015), and BEAS-2B cells (Vales et al. 2015).

5.4.5 Mechanisms of TiO$_2$ Nanoparticle-Induced Toxicity: Generation of Oxidative Stress

One of the most widely accepted mechanism of nanoparticle toxicity is generation of oxidative stress. Oxidative stress occurs when pro-oxidant production and anti-oxidant production is imbalanced. When such imbalance is in favor of the prooxidants such as free radicals and reactive oxygen species (ROS), oxidative stress is said to occur (Sies 2000).

A free radical is a molecular species containing an unpaired electron, and it can donate this electron to other species or keep this electron and accept another electron from other electron-donating species; thus, it acts as an oxidant or a reductant (Lobo et al. 2010; Cheeseman and Slater 1993). Being highly unstable, free radicals accept electrons in order to attain stability but the electron donor itself is converted into a free radical. When O$_2$ donates its electron, the free radicals so formed such as H$_2$O$_2$ or superoxide radicals are termed as reactive oxygen species (ROS). When generated in excess, free radicals including ROS can attack many cellular components including biomolecules such as DNA (Young and Woodside 2001).

That NPs induce genotoxicity via oxidative stress has been confirmed through various experimental studies. Various in vitro studies have been done in which cells have been treated with TiO$_2$ NPs to investigate the genotoxic effect (Petković et al. 2011; Tomankova et al. 2005; Kansara et al. 2015; Jugan et al. 2012; Ghosh et al. 2013; Jaeger et al. 2012; Xue et al. 2010; Sanders et al. 2012; Yin et al. 2012; Liu et al. 2010; Meena et al. 2012; Winter et al. 2011). These studies used various cell types including A549 cells, human lymphocytes, HaCaT cells, PC12 cells, human embryonic kidney (HEK 293) cells, and dendritic cells. All these studies reported that TiO$_2$ NPs treatment to various cell types lead to increased production of oxidative stress which damaged the genetic material DNA. Even TiO$_2$ used as food additive E171 has been reported to damage DNA by generating oxidative stress (Dorier et al. 2017; Proquin et al. 2017).

5.4.6 Antioxidant Enzymes

To prevent ROS build-up and hence oxidative stress, organisms employ a variety of antioxidants. Enzymatic antioxidant SOD converts O$_2^-$ into H$_2$O$_2$ and O$_2$, H$_2$O$_2$ is further reduced by CAT to H$_2$O and O$_2$ (Beauchamp and Fridovich 1972; Buege and Aust 1978), thus keeping ROS level low and preventing oxidative damage to cells. Various studies have reported effect of TiO$_2$ NP treatment on antioxidant enzyme activity levels. TiO$_2$ NPs reduce the levels of CAT and SOD activities, thereby increasing the levels of ROS and free radicals high, generating oxidative stress.

5.4.7 Measurement of NP-Induced ROS and Oxidative Stress

The amount of ROS generated in biosystems, in vivo or in vitro, as a result of exposure to NPs can be measured by various methods. One very reliable method is electron spin resonance (ESR) spectroscopy, in which a certain molecule termed as spin trap is used that can chemically react with the free radical generating a product the amount of which can then be measured. Another method involves the use of fluorescent probes such as 2', 7'-dichlorodihydrofluorescein diacetate (H$_2$DCF-DA), which can measure the amount of NP-induced intracellular ROS as well as ROS in a cell-free system. H$_2$DCF–DA has been used to measure intracellular ROS in various studies that have reported an increase in intracellular ROS on exposure to nanoparticles (Roesslein et al. 2013; Setyawati et al. 2015; Wan et al. 2012; Yu et al. 2010; Capasso et al. 2014; Ahamed et al. 2015). Inside the cell, an esterase converts DCFH-DA into nonfluorescent DCFH. When ROS is present in the system, oxidation of DCFH occurs and dichlorofluorescein (DCF) is generated which emits fluorescence (LeBel et al. 1992).

Early signs of oxidative stress include ROS generation, depletion of endogenous antioxidants such as GSH, reduced CAT and SOD activity, and rise in malonyldialdehyde (MDA) levels due to increased lipid peroxidation. Previous studies have measured the levels of GSH and MDA as well as activity of CAT and SOD after exposure to TiO$_2$ NPs: GSH is measured using Ellman's reagent, MDA by thiobarbituric acid (TBA) assay, CAT activity by Aebi's method (1974), and SOD activity by Markland and Markland (1974) procedure.

5.5 Nanoparticle-Induced Inflammation

Inflammation in body is a normal response to injury. Presence of any pathogens in the body also leads to inflammation as a defensive response. However, excess inflammation activates inflammatory cells, which in turn lead to cytokine release and generation of ROS. Chronic inflammation is the cause of several inflammatory disorders, including those that affect the lung and heart (Manda-Handzlik and Demkow 2015; Konior et al. 2014); Thannickal et al. 2004). One of the causes of excess inflammatory response and neutrophils and macrophage activation is the phagocytosis of nanoparticles, as nanoparticle exposure induces ROS and NOS (Bakand et al. 2012; Madl et al. 2014; Sabella et al. 2014 ; Manke et al. 2013). Various studies have associated nanoparticle exposure with inflammation (Madl et al. 2014; Zhang et al. 1998; Zhang et al. 2000; Zhang et al. 2003) and reported that nanoparticles-induced oxidative stress leads to release of cytokines (Bakand et al. 2012; Madl et al. 2014; Sabella et al. 2014 ; Manke et al. 2013). Some studies have reported that nanoparticles exposure lead to mitochondrial damage and cell death without inducing inflammation (Xia et al. 2006).

Nanoparticles-induced oxidative stress can cause inflammation by regulating the expression of proinflammatory genes via intracellular signaling pathways. The proinflammatory genes are regulated by NF-κB and AP-1 transcription factors, which are redox-sensitive proteins and various studies have reported their activation in NP-exposed cell (Nel et al. 2006; Li et al. 2008).

In macrophages, NP exposure leads to inflammasome activation and IL-1β secretion (Reisetter et al. 2011; Yazdi et al. 2010). IL-1β is involved in acute and chronic inflammation (van de Veerdonk et al. 2011; Dinarello 1997, 2009; Kapoor et al. 2011). IL-1β results from pro-IL-1β in a cleavage process regulated by inflammasome complexes (Tschopp and Schroder 2010; Mariathasan et al. 2006; Martinon et al. 2009; Bauernfeind et al. 2011). The NLRP3, the most elusive and widely studied inflammasome, consists of various domains including caspase recruitment domain (ASC) and plays an essential role in the formation of mature IL-1β (Xia et al. 2006; Said-Sadier et al. 2010). NLRP3 is activated in response to a variety of signals including bacterial toxins and environmental toxins, and since it is involved in many inflammatory diseases, much of the research has focused on it. Various studies have reported that IL-1β is released in response to TiO_2 NP exposure causing pulmonary inflammation (Yazdi et al. 2010; Riteau et al. 2010). Gui et al. (2011) and Trouiller et al. (2009) have also reported increased production of TNF-α, INF-g, and IL- 8 in blood in response to TiO_2 NP exposure. Such inflammatory molecules are pivotal in nanoparticle-induced toxic effects.

5.6 Systemic Toxicity

Nanoparticles can translocate via blood and reach various internal organs, causing systemic toxicity to them, although the amount of NPs translocated and the extent of damage may vary. When administered intraperitoneally, TiO_2 NPs are absorbed in the vascular system in the gastrointestinal tract, and are then transported via blood to various organs including liver, heart, lungs, and brain (Afaq et al. 1998; Kandeil et al. 2019). While in blood, these NPs cause cell cytotoxicity and oxidatively damage blood components and cause RBS hemolysis (Shakeel et al. 2016). On intraperitoneal administration of TiO_2 NPs in rats for 14 days, Ti content was found to accumulate in liver, kidneys, spleen, lung, brain, and heart (Wani et al. 2021). TiO_2 NPs damaged liver, kidney, and spleen of rats and induced biochemical disturbance in serum (Wani et al. 2021). TiO_2 NPs treatment increased ALP, ALT, and AST enzyme activity in serum indicating liver malfunction and damage (Liu et al. 2009). Chen et al. 2009 also reported spleen damage due to TiO_2 NPs treatment. TiO_2 NPs treatment in rats leads to ROS generation and inflammation in spleen (Li et al. 2010).

Oral treatment of TiO_2 NPs in mice was found to induce spleen, liver, and kidney damage, besides causing blood clots in lungs Chen et al. (2015), while another study observed inflammation, apoptosis, and oxidative stress resulting in chronic gastritis in mice after treatment with TiO_2 NPs orally (Mohamed 2015). In an *in vitro* model of human intestine, TiO_2 NPs were reported to destroy the villi in the small intestine (Faust et al. 2014). Intragastric administration of TiO_2 NPs caused weight loss in mice which might be probably due to reduced number of villi in the mice intestine (Duan et al. 2010). Both heart arrhythmia and changes in blood pressure have also been observed in rats due to TiO_2 NPs (Chen et al. 2015). Long-term studies involving mice have reported heart muscle damage and pneumonia by TiO_2 NPs, which could be caused by dysregulation of cytokines in the heart Hong et al. (2015). In addition, TiO_2 NPs in high amounts have been reported to disturb energy and amino acid metabolism as well as disturbances in intestinal microflora (Bu et al. 2010).

On inhalation, TiO_2 NPs accumulate in the nose, trachea, and alveoli, from where they can be translocated via sensory nerves to the brain (Czajka et al. 2015; Simkó and Mattsson 2010), and owing to their nano size, they translocate through the blood–brain barrier into the brain areas (Allen 2016; Simkó and Mattsson 2010; Bramini et al. 2014) inducing oxidative stress and reducing antioxidative potential (Allen 2016; Wang et al. 2008a, b; Feng et al. 2015). TiO_2 NPs are apoptotic in in vitro cultures of murine microglia N9 cells (Czajka et al. 2015; Li et al. 2009) and can be toxic to rat and human glial cells (Huerta-García et al. 2014), which indicates that these NPs can cause brain damage. Cytotoxicity of TiO_2 NPs has been observed in in vitro model of neurons (PC 12 cells). Exposure to TiO_2 NPs leads to symptoms of Parkinson's disease, such as increased expression of genes associated with Lewy bodies formation and loss of dopaminergic neurons in zebrafish larvae brains (Hu et al. 2017). Besides apoptosis in the hippocampal region, mice exposed to TiO_2 NPs were found to have impaired spatial memory (Hu et al. 2011). TiO_2 NPs have been found to stimulate ROS production in the brain cells (Long et al. 2006), lower mitochondrial membrane potential and impair mitochondrial function (Freyre-Fonseca et al. 2011), and lower antioxidant enzyme activity including acetylcholine esterase in the brain (Jeon et al. 2011; Hu et al. 2010).

While in aquatic medium, TiO_2 NPs can interact with sediments and toxins and can induce damage to aquatic organisms (Ghosh et al. 2010). When adult zebra fish were exposed to TiO_2 NPs for 3 weeks, the number of viable embryos was reduced (Ramsden et al. 2013). TiO_2 NPs also inhibited the growth of goldfish (*Carassius auratus*) (Ates et al. 2013), induced oxidative damage to the liver of carp (*Cyprinus carpio*) (Hao et al. 2009), and reduced cell viability and chlorophyll content in freshwater microalgae, *Chlorella* sp. (Iswarya et al. 2015).

6 Conclusions

TiO_2 NPs are extensively used in a wide variety of day-to-day use products as well as in industries worldwide, reports of their toxic potential is of concern. In order to safely use these NPs with no or lower risk of hazardous effect on ecosystem, it is of utmost importance to understand their sources, how they interact with environment, and what possible risk they entail prior to use. Besides, before approving them for clinical use, their interactions with biomolecules and their possible toxic effects need to be investigated in full detail.

The toxic potential of NPs is affected by various factors such as their size, dispersion, shape, zeta potential, and dispersion in different media. These factors need to be considered for the safer use of NPs. In order to facilitate safe use of NPs in consumer products and pharmaceuticals, such particles must be assessed for health risks.

References

Acar MS, Bulut ZB, Ates A, Nami B, Koçak N, Yildiz B. Titanium dioxide nanoparticles induce cytotoxicity and reduce mi-totic index in human amniotic fluid-derived cells. Hum Exp Toxicol. 2015;34:174–82.

Aebi H. Catalases. In: Bergmeyer HU, editor. Methods of enzymatic analysis. New York: Chemic Academic Press Inc., Verlag Chemie International; 1974. p. 673–84.

Afaq F, Abidi P, Matin R, Rahman Q. Cytotoxicity, pro-oxidant effects and antioxidant depletion in rat lung alveolar macrophages exposed to ultrafine titanium dioxide. J Appl Toxicol. 1998;18:307–12.

Ahamed M, Akhtar MJ, Alhadlaq HA, Khan MA, Alrokayan SA. Comparative cyto- toxic response of nickel ferrite nanoparticles in human liver HepG2 and breast MFC-7 cancer cells. Chemosphere. 2015;135:278–88.

Ali SA, Rizk MZ, Hamed MA, et al. Assessment of titanium dioxide nanoparticles toxicity via oral exposure in mice: effect of dose and particle size. Biomarkers. 2019;24(5):492–8.

Allen R. The cytotoxic and genotoxic potential of titanium dioxide (TiO$_2$) nanoparticles on human SH-SY5Y neuronal cells in vitro. Plymouth Stud Sci. 2016;9:5–28.

Ansari SM, Bhor RD, Pai KR, Sen D, Mazumder S, Ghosh K, Kolekarv YD, Ramana CV. Cobalt nanoparticles for biomedical applications: facile synthesis, physiochemical characterization, cytotoxicity behavior and biocompatibility. Appl Surf Sci. 2017;414:171–87.

Armand L, Tarantini A, Beal D, et al. Long-term exposure of A549 cells to titanium dioxide nanoparticles induces DNA damage and sensitizes cells towards genotoxic agents. Nanotoxicology. 2016;10:913–23.

Ates M, Demir V, Adiguzel R, Arslan Z. Bioaccumulation, sub-acute toxicity, and tissue distribution of engineered titanium dioxide (TiO$_2$) nanoparticles in goldfish (Carassius auratus). J Nanomater. 2013.

Azeez F, Al-Hetlani E, Arafa M, Abdelmonem Y, Nazeer AA, Amin MO, Madkour M. The effect of surface charge on photocatalytic degradation of methylene blue dye using chargeable titania nanoparticles. Sci Rep. 2018;8(1):1–9.

Bakand S, Hayes A, Dechsakulthorn F. Nanoparticles: a review of particle toxicology following inhalation exposure. Inhal Toxicol. 2012;24:125–35.

Baranowska-Wojcik E, Szwajgier D, Oleszczuk P, et al. Effects of titanium dioxide nanoparticles exposure on human health -a review. Biol Trace Elem Res. 2020;193:118–29. https://doi.org/10.1007/s12011-019-01706-6.

Bauernfeind F, Bartok E, Rieger A, Franchi L, Nunez G, Hornung V. Cutting edge: reactive oxygen species inhibitors block priming, but not activation, of the NLRP3 inflammasome. J Immunol. 2011;187:613–7.

Beauchamp C, Fridovich I. Superoxide dismutase: improved assays and assay applicable to acrylamide gels. Anal Biochem. 1972;44:276–86. https://doi.org/10.1016/0003-2697(71)90370-8.

Biola-Clier M, Beal D, Caillat S, et al. Comparison of the DNA damage response in BEAS-2B and A549 cells exposed to titanium dioxide nanoparticles. Mutagenesis. 2017;32:161–72.

Bramini M, Ye D, Hallerbach A, Raghnaill MN, Salvati A, Aberg C, Dawson KA. Imaging approach to mechanistic study of nanoparticle interactions with the blood-brain barrier. ACS Nano. 2014;8:4304–12.

Bu Q, Yan G, Deng P, Peng F, Lin H, Xu Y, Cao Z, Zhou T, Xue A, Wang Y, Cen X, Zhao YL. NMR-based metabonomic study of the sub-acute toxicity of titanium dioxide nanoparticles in rats after oral administration. Nanotechnology. 2010;21:125105.

Buege JA, Aust SD. Microsomal lipid peroxidation. Methods Enzymol. 1978;52:302–10.

Capasso L, Camatini M, Gualtieri M. Nickel oxide nanoparticles induce inflammation and genotoxic effect in lung epithelial cells. Toxicol Lett. 2014;226:28–34.

Chakrabarti S, Goyary D, Karmakar S, et al. Exploration of cytotoxic and genotoxic endpoints following sub-chronic oral exposure to titanium dioxide nanoparticles. Toxicol Ind Health. 2019;35(9):577–92.

Chan VSW. Nanomedicine: an unresolved regulatory issue. Regul Toxicol Pharmacol. 2006;46:218–24.

Cheeseman KH, Slater TF. An introduction to free radical biochemistry. Br Med Bull. 1993;49:481–93.

Chen J, Dong X, Zhao J, Tang G. In vivo acute toxicity of titanium dioxide nanoparticles to mice after intraperitoneal injection. J Appl Toxicol. 2009;29:330–7.

Chen Z, Wang Y, Zhuo L, Chen S, Zhao L, Luan X, Wang H, Jia G. Effect of titanium dioxide nanoparticles on the cardiovascular system after oral administration. Toxicol Lett. 2015;239:123–30.

Coccini T, Grandi S, Lonati D, Locatelli C, De Simone U. Comparative cellular toxicity of titanium dioxide nanoparticles on human astrocyte and neuronal cells after acute and prolonged exposure. Neurobehav Toxicol. 2015;48:77–89.

Courbiere B, Auffan M, Rollais R, et al. Ultrastructural interactions and genotoxicity assay of cerium dioxide nanoparticles on mouse oocytes. Int J Mol Sci. 2013;14:21613–28.

Czajka M, Sawicki K, Sikorska K, Popek S, Kruszewski M, Kapka-Skrzypczak L. Toxicity of titanium dioxide nanoparticles in central nervous system. Toxicol in Vitro. 2015;29:1042–52.

Demir E, Burgucu D, Turna F, et al. Determination of TiO_2, ZrO_2, and Al_2O_3 nanoparticles on genotoxic responses in human peripheral blood lymphocytes and cultured embyronic kidney cells. J Toxic Environ Health A. 2013;76:990–1002.

Dinarello CA. Interleukin-1. Cytokine Growth Factor Rev. 1997;8:253–65.

Dinarello CA. Interleukin-1beta and the autoinflammatory diseases. N Engl J Med. 2009;360:2467–70.

Dobrzyńska MM, Gajowik A, Radzikowska J, et al. Genotoxicity of silver and titanium dioxide nanoparticles in bone marrow cells of rats in vivo. Toxicology. 2014;315:86–91.

Dorier M, Béal D, Marie-Desvergne C, et al. Continuous in vitro exposure of intestinal epithelial cells to E171 food additive causes oxidative stress, inducing oxidation of DNA bases but no endoplasmic reticulum stress. Nanotoxicology. 2017;11:751–61.

Duan Y, Liu J, Ma L, Li N, Liu H, Wang J, Zheng L, Liu C, Wang X, Zhao X, Yan J, Wang S, Wang H, Zhang X, Hong F. Toxicological characteristics of nanoparticulate anatase titanium dioxide in mice. Biomaterials. 2010;31:894–9.

Dusinska M, Fjellsbø LM, Magdolenova Z, et al. Safety of nanoparticles in medicine. In: Nanomedicine in health and disease. 2011.

Elgrabli D, Beaudouin R, Jbilou N, Floriani M, Pery A, Rogerieux F, Lacroix G. Biodistribution and clearance of TiO_2 nanoparticles in rats after intravenous injection. PLoS One. 2015;10(4)

Ema M, Hougaard KS, Kishimoto A, Honda K. Reproductive and developmental toxicity of carbon-based nanomaterials: a literature review. Nanotoxicology. 2016;10(4):391–412.

Emerich DF, Thanos CG. Nanotechnology and medicine. Expert Opin Biol Ther. 2003;3:655–63.

EPA U. Technical fact sheet: nanomaterials. 2017. Available at https://www.epa.gov/sites/production/files/2014-03/documents/ffrrofactsheet_emergingcontaminant_nanomaterials_jan2014_final.pdf. Accessed 1 Jan 2021.

Fairbairn DW, Olive PL, O'Neill KL. The comet assay: a comprehensive review. Mutat Res. 1995;339:37–59.

Farré M, Barceló D. Introduction to the analysis and risk of nanomaterials in environmental and food samples. In: Barcelo D, Farre M, editors. Comprehensive analytical chemistry, 1st edn, vol. 59. Amsterdam: Elsevier; 2012. p. 1–32.

Faust JJ, Doudrick K, Yang Y, Westerhoff P, Capco DG. Food grade titanium dioxide disrupts intestinal brush border micro- villi in vitro independent of sedimentation. Cell Biol Toxicol. 2014;30:169–88.

FDA. Listing of color additives exempt from certification. 2012. Available at: https://www.access-data.fda.gov/scripts/cdrh/cfdocs/cfcfr/CFRSearch.cfm?CFRPart=73. Accessed 3 Mar 2018.

Fenech M. The in vitro micronucleus technique. Mutat Res. 2000;455:81–95.

Feng X, Chen A, Zhang Y, Wang J, Shao L, Wei L. Central nervous system toxicity of metallic nanoparticles. Int J Nanomedicine. 2015;10:4321–40.

Freyre-Fonseca V, Delgado-Buenrostro NL, Gutiérrez-Cirlos EB, Calderón-Torres CM, Cabellos-Avelar T, Sánchez-Pérez Y, Pinzón E, Torres I, Molina-Jijón E, Zazueta C, Pedraza-Chaverri J, García-Cuéllar CM, Chirino YI. Titanium dioxide nanoparticles impair lung mitochondrial function. Toxicol Lett. 2011;202:111–9.

Gélis C, Girard S, Mavon A, et al. Assessment of the skin photoprotective capacities of an organo-mineral broad-spectrum sunblock on two ex vivo skin models. Photodermatol Photoimmunol Photomed. 2003;19:242–53.

Ghosh M, Bandyopadhyay M, Mukherjee A. Genotoxicity of titanium dioxide (TiO$_2$) nanoparticles at two trophic levels: plant and human lymphocytes. Chemosphere. 2010;81(10):1253–62.

Ghosh M, Chakraborty A, Mukherjee A. Cytotoxic, genotoxic and the hemolytic effect of titanium dioxide (TiO$_2$) nanoparticles on human erythrocyte and lymphocyte cells in vitro. J Appl Toxicol. 2013;33:1097–110.

Ghosh M, Öner D, Duca RC, et al. Cyto-genotoxic and DNA methylation changes induced by different crystal phases of TiO$_2$ -np in bronchial epithelial (16-HBE) cells. Mutat Res/Fund Mol Mech Mutagen. 2017;796:1–12.

Girardello F, Custódio CL, Vianna IV, et al. Titanium dioxide nanoparticles induce genotoxicity but not mutagenicity in golden mussel *Limnoperna fortunei*. Aquat Toxicol. 2016;170:223–8.

Grissa I, Elghoul J, Ezzi L, et al. Anemia and genotoxicity induced by sub-chronic intragastric treatment of rats with titanium dioxide nanoparticles. Mutat Res Genet Toxicol Environ Mutagen. 2015;794:25–31.

Gui S, Zhang Z, Zheng L, Cui Y, Liu X, Li N, Sang X, Sun Q, Gao G, Cheng Z, Cheng J, Wang L, Tang M, Hong F. Molecular mechanism of kidney injury of mice caused by exposure to titanium dioxide nanoparticles. J Hazard Mater. 2011;195:365–70.

Gurr JR, Wang AS, Chen CH, Jan KY. Ultrafine titanium dioxide particles in the absence of photoactivation can induce oxidative damage to human bronchial epithelial cells. Toxicology. 2005;213:66–73.

Hackenberg S, Scherzed A, Zapp A, et al. Genotoxic effects of zinc oxide nanoparticles in nasal mucosa cells are antagonized by titanium dioxide nanoparticles. Mutat Res Genet Toxicol Environ Mutagen. 2017;816–817:32–7.

Hao L, Wang Z, Xing B. Effect of sub-acute exposure to TiO$_2$ nanoparticles on oxidative stress and histopathological changes in Juvenile Carp (*Cyprinus carpio*). J Environ Sci (China). 2009;21(10):1459–66.

Hong FS, Wang L, Yu XH, Zhou YJ, Hong J, Sheng L. Toxicological effect of TiO$_2$ nanoparticle induced myocarditis in mice. Nanoscale Res Lett. 2015;10:326.

Hsiao IL, Huang YJ. Effects of various physicochemical characteristics on the toxicities of ZnO and TiO nanoparticles toward human lung epithelial cells. Sci Total Environ. 2011;409(7):1219–28.

Hu R, Gong X, Duan Y, Li N, Che Y, Cui Y, Zhou M, Liu C, Wang H, Hong F. Neurotoxicological effects and the impairment of spatial recognition memory in mice caused by exposure to TiO$_2$ nanoparticles. Biomaterials. 2010;31:8043–50.

Hu R, Zheng L, Zhang T, Gao G, Cui Y, Cheng Z, Cheng J, Hong M, Tang M, Hong F. Molecular mechanism of hippocampal apoptosis of mice following exposure to titanium dioxide nanoparticles. J Hazard Mater. 2011;191:32–40.

Hu Q, Guo F, Zhao F, Fu Z. Effects of titanium dioxide nanoparticles exposure on parkinsonism in zebrafish larvae and PC12. Chemosphere. 2017;173:373–9.

Huerta-García E, Pérez-Arizti JA, Márquez-Ramírez SG, Delgado-Buenrostro NL, Chirino YI, Iglesias GG, López-Marure R. Titanium dioxide nanoparticles induce strong oxidative stress and mitochondrial damage in glial cells. Free Radic Biol Med. 2014;73:84.

IARC (International Agency for Research on Cancer). IARC monographs on the evaluation of carcinogenic risks to humans. 2002;56. https://doi.org/10.1002/food.19940380335. Accessed 3 Mar 2020.

ISO. Vocabulary: part 1 – core terms. In: International Standardisation Organisation (ISO), Technical Specification ISO/TS. 2015. Available at: https://www.iso.org/standard/68058.html. Accessed 3 Dec 2020.

Iswarya V, Bhuvaneshwari M, Alex SA, Iyer S, Chaudhuri G, Chandrasekaran PT, Bhalerao GM, Chakravarty S, Raichur AM, Chandrasekaran N, Mukherjee A. Combined toxicity of two crystalline phases (anatase and rutile) of Titania nanoparticles towards freshwater microalgae: Chlorella sp. Aquat Toxicol. 2015;161:154–69.

Jaeger A, Weiss DG, Jonas L, et al. Oxidative stress-induced cytotoxic and genotoxic effects of nano-sized titanium dioxide particles in human HaCaT keratinocytes. Toxicology. 2012;296:27–36.

Jain AK, Senapati VA, Singh D, et al. Impact of anatase titanium dioxide nanoparticles on mutagenic and genotoxic response in Chinese hamster lung fibroblast cells (V-79): The role of cellular uptake. Food Chem Toxicol. 2017;105:127–39.

Jalili P, Gueniche N, Lanceleur R, et al. Investigation of the in vitro genotoxicity of two rutile TiO_2 nanomaterials in human intestinal and hepatic cells and evaluation of their interference with toxicity assays. NanoImpact. 2018;11:69–81.

Jeon YM, Park SK, Lee MY. Toxicoproteomic identification of TiO_2 nanoparticle-induced protein expression changes in mouse brain animal. Anim Cells Syst. 2011;15:107–14.

Jovanović B (2015) Critical review of public health regulations of titanium dioxide, a human food additive. Integr. Environ. Assess. Manag. 11(1):10–20.

Jovanović B, Jovanović N, Cvetković VJ, et al. The effects of a human food additive, titanium dioxide nanoparticles E171, on Drosophila melanogaster-a 20 generation dietary exposure experiment. Sci Rep. 2018;8(1):1–2.

Jugan ML, Barillet S, Simon-Deckers A, et al. Titanium dioxide nanoparticles exhibit genotoxicity and impair DNA repair activity in A549 cells. Nanotoxicology. 2012;6:501–13.

Kaida T, Kobayashi K, Adachi M, et al. Optical characteristics of titanium oxide interference film and the film laminated with oxides and their applications for cosmetics. J Cosmet Sci. 2004;55:219–20.

Kan H, Wu Z, Lin YC, Chen TH, Cumpston JL, Kashon ML, Leonard S, Munson AE, Castranova V. The role of nodose ganglia in the regulation of cardiovascular function following pulmonary exposure to ultrafine titanium dioxide. Nanotoxicology. 2014;8:447–54.

Kandeil MA, Mohammed ET, Hashem KS, Aleya L, Abdel-Daim MM. Correction to: moringa seed extract alleviates titanium oxide nanoparticles (TiO_2-NPs)-induced cerebral oxidative damage, and increases cerebral mitochondrial viability. Environ Sci Pollut Res. 2019;27:19185. https://doi.org/10.1007/s11356-019-06077-y.

Kansara K, Patel P, Shah D, et al. TiO_2 nanoparticles induce DNA double strand breaks and cell cycle arrest in human alveolar cells. Environ Mol Mutagen. 2015;56:204–17.

Kaphle A, Navya PN, Umapathi A, Daima HK. Nanomaterials for agriculture, food and environment: applications, toxicity and regulation. Environ Chem Lett. 2018;16(1):43–58.

Kapoor M, Martel-Pelletier J, Lajeunesse D, Pelletier JP, Fahmi H. Role of proinflammatory cytokines in the pathophysiology of osteoarthritis. Nat Rev Rheumatol. 2011;7:33–42.

Kirsch-Volders M, Elhajouji A, Cundari E, et al. The in vitro micronucleus test: a multi-endpoint assay to detect simultaneously mitotic delay, apoptosis, chromosome breakage, chromosome loss and non-disjunction. Mutat Res. 1997;392:19–30.

Konior A, Schramm A, Czesnikiewicz-Guzik M, Guzik TJ. NADPH oxidases in vascular pathology. Antioxid Redox Signal. 2014;20:2794–814.

Kurzawa-Zegota M, Sharma V, Najafzadeh M, et al. Titanium dioxide nanoparticles induce DNA damage in peripheral blood lymphocytes from polyposis coli, colon cancer patients and healthy individuals: an ex vivo/in vitro study. J Nanosci Nanotechnol. 2017;17:9274–85.

LeBel CP, Ischiropoulos H, Bondy SC. Evaluation of the probe 2′,7′-dichlorofluorescin as an indicator of reactive oxygen species formation and oxidative stress. Chem Res Toxicol. 1992;5:227.

Li N, Xia T, Nel AE. The role of oxidative stress in ambient particulate matter- induced lung diseases and its implications in the toxicity of engineered nanoparticles. Free Radic Biol Med. 2008;44:1689–99.

Li XB, Xu SQ, Zhang ZR, Schluesener HJ. Apoptosis induced by titanium dioxide nanoparticles in cultured murine microglia N9 cells. Chin Sci Bull. 2009;54:3830–6.

Li N, Duan Y, Hong M, Zheng L, Fei M, Zhao X, Wang J, Cui Y, Liu H, Cai J, Gong S. Spleen injury and apoptotic pathway in mice caused by titanium dioxide nanoparticles. Toxicol Lett. 2010;165:161–8.

Liao F, Chen L, Liu Y, et al. The size-dependent genotoxic potentials of titanium dioxide nanoparticles to endothelial cells. Environ Toxicol. 2019;34(11):1199–207.

Liang G, Pu Y, Yin L, Liu R, Ye B, Su Y, Li Y (2009) Influence of different sizes of titanium dioxide nanoparticles on hepatic and renal functions in rats with correlation to oxidative stress. J. Toxic. Environ. Health A. 72: 740–745. https://doi.org/10.1080/15287390902841516.

Liu H, Ma L, Zhao J, Liu J, Yan J, Ruan J, Hong F. Biochemical toxicity of nano-anatase TiO$_2$ particles in mice. Biol Trace Elem Res. 2009;29:170–80. https://doi.org/10.1007/s12011-008-8285-6.

Liu S, Xu L, Zhang T, et al. Oxidative stress and apoptosis induced by nanosized titanium dioxide in PC12 cells. Toxicology. 2010;267:172–7.

Lobo V, Patil A, Phatak A, Chandra N. Free radicals, antioxidants and functional foods: impact on human health. Pharmacogn Rev. 2010;4:118–26.

Long TC, Saleh N, Tilton RD, Lowry GV, Veronesi B. Titanium dioxide (P25) produces reactive oxygen species in immortalized brain microglia (BV2): implications for nanoparticle neurotoxicity. Environ Sci Technol. 2006;40:4346–52.

Madl AK, Plummer LE, Carosino C, Pinkerton KE. Nanoparticles, lung injury, and the role of oxidant stress. Annu Rev Physiol. 2014;76:447–65.

Manda-Handzlik A, Demkow U. Neutrophils: the role of oxidative and nitrosative stress in health and disease. Adv Exp Med Biol. 2015;857:51–60.

Manivannan J, Banerjee R, Mukherjee A. Genotoxicity analysis of rutile titan dioxide nanoparticles in mice after 28 days of repeated oral administration. Nucleus. 2019:1–8.

Manke A, Wang L, Rojanasakul Y. Mechanisms of nanoparticle-induced oxidative stress and toxicity. Biomed Res Int. 2013;942916

Mariathasan S, Weiss DS, Newton K, McBride J, O'Rourke K, Roose-Girma M, et al. Cryopyrin activates the inflammasome in response to toxins and ATP. Nature. 2006;440:228–32.

Marklund S, Marklund G. Involvement of the superoxide anion radical in the autoxidation of pyrogallol and a convenient assay for superoxide dismutase. Eur J Biochem. 1974;47:469–74.

Márquez-Ramírez SG, Delgado-Buenrostro NL, Chirino YI, Iglesias GG, López-Marure R. Titanium dioxide nanoparticles inhibit proliferation and induce morphological changes and apoptosis in glial cells. Toxicology. 2012;302:146–56.

Martinon F, Mayor A, Tschopp J. The inflammasomes: guardians of the body. Annu Rev Immunol. 2009;27:229–65.

Meena R, Rani M, Pal R, et al. Nano-TiO$_2$-induced apoptosis by oxidative stress-mediated DNA damage and activation of p53 in human embryonic kidney cells. Appl Biochem Biotechnol. 2012;167:791–808.

Mohamed HRH. Estimation of TiO$_2$ nanoparticle-induced genotoxicity persistence and possible chronic gastritis-induction in mice. Food Chem Toxicol. 2015;83:76–83.

Mottola F, Iovine C, Santonastaso M, et al. NPs-TiO$_2$ and lincomycin coexposure induces DNA damage in cultured human amniotic cells. Nanomaterials. 2019;9(11):1511.

Nel A, Xia T, Mädler L, et al. Toxic potential of materials at the nano level. Science. 2006;311:622–7.

NIOSH. Approaches to safe nanotechnology: managing the health and safety concerns associated with engineered nanomaterials (2009-125). 2009. Available at: https://www.cdc.gov/niosh/docs/2009-125/default.html. Accessed 3 Dec 2020.

Oberdörster G, Oberdörster E, Oberdörster J. Nanotoxicology: an emerging discipline evolving from studies of ultrafine particles. Environ Health Perspect. 2005;113:823–39.

Ortlieb M. White Giant or White Dwarf? Particle size distribution measurements of TiO_2. GIT Lab J Eur. 2010;14:42–3.

Osman IF, Najafzadeh M, Sharma V, et al. TiO_2 NPs induce DNA damage in lymphocytes from healthy individuals and patients with respiratory diseases – an ex vivo/ in vitro study. J Nanosci Nanotechnol. 2018;18:544–55.

Patel S, Patel P, Bakshi SR. Titanium dioxide nanoparticles: an in vitro study of DNA binding, chromosome aberration assay, and comet assay. Cytotechnology. 2017;69:245–63.

Petković J, Zegura B, Stevanović M, et al. DNA damage and alterations in expression of DNA damage responsive genes induced by TiO(2) nanoparticles in human hepatoma HepG2 cells. Nanotoxicology. 2011;5:341–53.

Proquin H, Rodríguez-Ibarra C, Moonen CGJ, et al. Titanium dioxide food additive (E171) induces ROS formation and genotoxicity: contribution of micro and nano-sized fractions. Mutagenesis. 2017;32:139–49.

Ramsden CS, Henry TB, Handy RD. Sub-lethal effects of titanium dioxide nanoparticles on the physiology and reproduction of zebrafish. Aquat Toxicol. 2013;126:404–13.

Re-evaluation of titanium dioxide (E 171) as a food additive. EFSA (Eur Food Saf Auth) J. 2016;14. https://doi.org/10.2903/j.efsa.2016.4545.

Reisetter AC, Stebounova LV, Baltrusaitis J, Powers L, Gupta A, Grassian VH, et al. Induction of inflammasome-dependent pyroptosis by carbon black nanoparticles. J Biol Chem. 2011;286:21844–52.

Riteau N, Gasse P, Fauconnier L, Gombault A, Couegnat M, Fick L, et al. Extracellular ATP is a danger signal activating P2X7 receptor in lung inflammation and fibrosis. Am J Respir Crit Care Med. 2010;182:774–83.

Roesslein M, Hirsch C, Kaiser JP, Krug HF, Wick P. Comparability of in vitro tests for bioactive nanoparticles: a common assay to detect reactive oxygen species as an example. Int J Mol Sci. 2013;14:24320–37.

Ropers MH, Terrisse H, Mercier-Bonin M, Humbert B (2017) Titanium dioxide as food additive, in: Janus M (Ed), Application of titanium dioxide. IntechOpen. https://doi.org/10.5772/intechopen.68883

Sabella S, Carney RP, Brunetti V, Malvindi MA, Al-Juffali N, Vecchio G, et al. A general mechanism for intracellular toxicity of metal-containing nanoparticles. Nanoscale. 2014;6:7052–61.

Said-Sadier N, Padilla E, Langsley G, Ojcius DM. Aspergillus fumigatus stimulates the NLRP3 inflammasome through a pathway requiring ROS production and the Syk tyrosine kinase. PLoS One. 2010;5:e10008.

Sajid M, Ilyas M, Basheer C, Tariq M, Daud M, Baig N, Shehzad F. Impact of nanoparticles on human and environment: review of toxicity factors, exposures, control strategies, and future prospects. Environ Sci Pollut Res Int. 2015;22(6):4122–43.

Sandeep KV. Nanomaterials-based health care and bioanalytical applications: trend and prospects. J Nanomater Mol Nanotechnol. 2013;2(2):2–6.

Sanders K, Degn LL, Mundy WR, et al. In vitro phototoxicity and hazard identification of nano-scale titanium dioxide. Toxicol Appl Pharmacol. 2012;258:226–36.

Sardoiwala MN, Kaundal B, Roy Choudhury S. Development of engineered nanoparticles expediting diagnostic and therapeutic applications across blood–brain barrier. In: Hussain CM, editor. Handbook of nanomaterials for industrial applications. 1st ed. Amsterdam: Elsevier; 2018. p. 696–709.

Schmid W. The micronucleus test. Mutat Res. 1975;31:9–15.

Setyawati MI, Tay CY, Leong DT. Mechanistic investigation of the biological effects of SiO_2, TiO_2, and ZnO nanoparticles on intestinal cells. Small. 2015;11:3458–68.

Shakeel M, Jabeen F, Qureshi NA, Fakhr-e-Alam M. Toxic effects of titanium dioxide nanoparticles and titanium dioxide bulk salt in the liver and blood of male Sprague-Dawley rats assessed by different assays. Biol Trace Elem Res. 2016;173:405–26.

Shukla RK, Kumar A, Gurbani D, et al. TiO$_2$ nanoparticles induce oxidative DNA damage and apoptosis in human liver cells. Nanotoxicology. 2013;7:48–60.

Sies H. What is oxidative stress? In: Keaney Jr JF, editor. Oxidative stress and vascular disease. Boston: Kluwer Academic Publishers; 2000. p. 1–8.

Simkó M, Mattsson MO. Risks from accidental exposures to engineered nanoparticles and neurological health effects: a critical review. Part Fibre Toxicol. 2010;7:42.

Singh NP, McCoy MT, Tice RR, et al. A simple technique for quantitation of low levels of DNA damage in individual cells. Exp Cell Res. 1988;175:184–91.

Skubalova Z, Michalkova H, Michalek P, et al. Prevalent anatase crystalline phase increases the cytotoxicity of biphasic titanium dioxide nanoparticles in mammalian cells. Colloids Surf B: Biointerfaces. 2019;182:110391.

Sun D, Meng TT, Loong TH, et al. Removal of natural organic matter from water using a nano-structured photocatalyst coupled with filtration membrane. Water Sci Technol. 2004:103–10.

Sycheva LP, Zhurkov VS, Iurchenko VV, et al. Investigation of genotoxic and cytotoxic effects of micro- and nanosized titanium dioxide in six organs of mice in vivo. Mutat Res Genet Toxicol Environ Mutagen. 2011;726:8–14.

Thannickal VJ, Toews GB, White ES, Lynch JP III, Martinez FJ. Mechanisms of pulmonary fibrosis. Annu Rev Med. 2004;55:395–417.

Tice RR, Agurell E, Anderson D, et al. Single cell gel/comet assay: guidelines for in vitro and in vivo genetic toxicology testing. Environ Mol Mutagen. 2000;35:206–21.

Tomankova K, Horakova J, Harvanova M, et al. (2015) Cytotoxicity, cell uptake and microscopic analysis of titanium dioxide and silver nanoparticles in vitro. Food and Chemical Toxicology 82: 106–115.

Trouiller B, Reliene R, Westbrook A, Solaimani P, Schiestl RH. Titanium dioxide nanoparticles induce DNA damage and genetic instability in vivo in mice. Cancer Res. 2009;69:8784–878969.

Tschopp J, Schroder K. NLRP3 inflammasome activation: the convergence of multiple signalling pathways on ROS production? Nat Rev Immunol. 2010;10:210–5.

Valdiglesias V, Costa C, Sharma V, Kilic G, Pásaro E, Teixeira JP, Dhawan A, Laffon B. Comparative study on effects of two different types of titanium dioxide nanoparticles on human neuronal cells. Food Chem Toxicol. 2013;57:352–61.

Vales G, Rubio L, Marcos R. Long-term exposures to low doses of titanium dioxide nanoparticles induce cell transformation, but not genotoxic damage in BEAS-2B cells. Nanotoxicology. 2015;9:568–78.

van de Veerdonk FL, Netea MG, Dinarello CA, Joosten LA. Inflammasome activation and IL-1beta and IL-18 processing during infection. Trends Immunol. 2011;32:110–6.

Vignardi CP, Hasue FM, Sartório PV, et al. Genotoxicity, potential cytotoxicity and cell uptake of titanium dioxide nanoparticles in the marine fish *Trachinotus carolinus* (Linnaeus, 1766). Aquat Toxicol. 2015;158:218–29.

Vijayalakshmi R, Rajendran V. Synthesis and characterization of nano-TiO$_2$ via different methods. Scholars Res Libr. 2012;4:1183–90.

Wan R, Mo Y, Feng L, Chien S, Tollerud DJ, Zhang Q. DNA damage caused by metal nanoparticles: involvement of oxidative stress and activation of ATM. Chem Res Toxicol. 2012;25:1402–11.

Wang J, Chen C, Liu Y, et al. Potential neurological lesion after nasal instillation of TiO$_2$ nanoparticles in the anatase and rutile crystal phases. Toxicol Lett. 2008a;183:72–80.

Wang J, Liu Y, Jiao F, Lao F, Li W, Gu Y, Li Y, Ge C, Zhou G, Li B, Zhao Y, Chai Z, Chen C. Time-dependant translocation and potential impairment on central nervous system by intranasally instilled TiO$_2$ nanoparticles. Toxicology. 2008b;254:82–90.

Wang TH, Navarrete-López AM, Li S, et al. Hydrolysis of TiCl$_4$: initial steps in the production of TiO$_2$. J Phys Chem A. 2010;114:7561–70.

Wang Y, Cui H, Zhou J, et al. Cytotoxicity, DNA damage, and apoptosis induced by titanium dioxide nanoparticles in human non-small cell lung cancer A549 cells. Environ Sci Pollut Res. 2015;22:5519–30.

Wani MR, Shadab GG. Titanium dioxide nanoparticle genotoxicity: a review of recent in vivo and in vitro studies. Toxicol Ind Health. 2020;36:514–30.

Wani MR, Maheshwari N, Shadab G. Eugenol attenuates TiO_2 nanoparticles-induced oxidative damage, biochemical toxicity and DNA damage in Wistar rats: an in vivo study. Environ Sci Pollut Res. 2021; https://doi.org/10.1007/s11356-020-12139-3.

Winter M, Beer HD, Hornung V, et al. Activation of the inflammasome by amorphous silica and TiO_2 nanoparticles in murine dendritic cells. Nanotoxicology. 2011;5:326–40.

Wolf R, Matz H, Orion E, et al. Sunscreens – the ultimate cosmetic. Acta Dermatovenerol Croat. 2003;11:158–62.

Xia T, Kovochich M, Brant J, Hotze M, Sempf J, Oberley T, et al. Comparison of the abilities of ambient and manufactured nanoparticles to induce cellular toxicity according to an oxidative stress paradigm. Nano Lett. 2006;6:1794–807.

Xie G, Wang C, Sun J, Zhong G. Tissue distribution and excretion of intravenously administered titanium dioxide nanoparticles. Toxicol Lett. 2011;205:55–61. https://doi.org/10.1016/j.toxlet.2011.04.034.

Xu J, Shi H, Ruth M, et al. Acute toxicity of intravenously administered titanium dioxide nanoparticles in mice. PLoS One. 2013;8:e70618.

Xue C, Wu J, Lan F, et al. Nano titanium dioxide induces the generation of ROS and potential damage in HaCaT cells under UVA irradiation. J Nanosci Nanotechnol. 2010;10:8500–7.

Yazdi AS, Guarda G, Riteau N, Drexler SK, Tardivel A, Couillin I, et al. Nanoparticles activate the NLR pyrin domain containing 3 (Nlrp3) inflammasome and cause pulmonary inflammation through release of IL-1alpha and IL-1beta. Proc Natl Acad Sci U S A. 2010;107:19449–54.

Yin JJ, Liu J, Ehrenshaft M, et al. Phototoxicity of nano titanium dioxides in HaCaT keratinocytes – generation of reactive oxygen species and cell damage. Toxicol Appl Pharmacol. 2012;263:81–8.

Young IS, Woodside JV. Antioxidants in health and disease. J Clin Pathol. 2001;54:176–86.

Yu M, Mo Y, Wan R, Chien S, Zhang X, Zhang Q. Regulation of plasminogen activator inhibitor-1 expression in endothelial cells with exposure to metal nanoparticles. Toxicol Lett. 2010;195:82–9.

Zhang Q, Kusaka Y, Sato K, Nakakuki K, Kohyama N, Donaldson K. Differences in the extent of inflammation caused by intratracheal exposure to three ultrafine metals: role of free radicals. J Toxicol Environ Health A. 1998;53:423–38.

Zhang Q, Kusaka Y, Donaldson K. Comparative pulmonary responses caused by exposure to standard cobalt and ultrafine cobalt. J Occup Health. 2000;42:179–84.

Zhang Q, Kusaka Y, Zhu X, Sato K, Mo Y, Kluz T, et al. Comparative toxicity of standard nickel and ultrafine nickel in lung after intratracheal instillation. J Occup Health. 2003;45:23–30.

Zhou F, Liao F, Chen L, Liu Y, Wang W, Feng S. The size dependent genotoxicity and oxidative stress of silica nanoparticles on endothelial cells. Environ Sci Pollut Res Int. 2019;26(2):1911–20.

Zhao J, Li N, Wang S, Zhao X, Wang J et al (2010) The mechanism of oxidative damage in the nephrotoxicity of mice caused by nano-anatase TiO2. J. Exp. Nanosci. 5:447–462. https://doi.org/10.1080/1745808100362893.

Zhurkov VS, Sycheva LP, Salamatova O, et al. Selective induction of micronuclei in the rat/mouse colon and liver by 1,2-dimethylhydrazine: a seven-tissue comparative study. Mutat Res/Genetic Toxicol. 1996;368:115–20.

Role of Arsenic in Carcinogenesis

Stephen James, Saniya Arfin, Manish K. Mishra, Arun Kumar,
Niraj Kumar Jha, Saurabh Kumar Jha, Kavindra Kumar Kesari,
Prabhanshu Kumar, Ashutosh Srivastava, and Dhruv Kumar

Contents

S. James
Department of Computer Applications, Mar Athanasios College for Advanced Studies
Tiruvalla (MACFAST), Tiruvalla, Kerala, India

Amity Institute of Molecular Medicine and Stem Cell Research (AIMMSCR), Amity
University, Noida, Uttar Pradesh, India

S. Arfin · D. Kumar (✉)
Amity Institute of Molecular Medicine and Stem Cell Research (AIMMSCR), Amity
University, Noida, Uttar Pradesh, India
e-mail: dkumar13@amity.edu

M. K. Mishra
Environmental Radioactivity Measurement Section, Environmental Monitoring & Assessment
Division, Bhabha Atomic Research Centre, Mumbai, India

A. Kumar
Mahavir Cancer Institute & Research Centre, Patna, Bihar, India

N. K. Jha · S. K. Jha
Department of Biotechnology, School of Engineering & Technology, Sharda University,
Greater Noida, Uttar Pradesh, India

K. K. Kesari
Department of Applied Physics & Department of Bioproducts and Biosystems, School of
Science, Aalto University, Espoo, Finland

P. Kumar
Amity Institute of Biotechnology, Amity University, Uttar Pradesh, Noida, India

A. Srivastava
Amity Institute of Marine Science and Technology, Amity University, Uttar Pradesh, Noida, India

© The Author(s), under exclusive license to Springer Nature Switzerland AG 2021   149
K. K. Kesari, N. K. Jha (eds.), *Free Radical Biology and Environmental Toxicity*,
Molecular and Integrative Toxicology, https://doi.org/10.1007/978-3-030-83446-3_7

Abstract Arsenic is one of the most abundant metals in the Earth's crust. It exists in both organic and inorganic forms. Contamination of drinking water globally by significant environmental exposure of Arsenic is one of the major public health problems. It is found that more than 100 million people are exposed to Arsenic worldwide. Drinking water Arsenic concentration has been reported very high in many countries around the world, leading to Arsenic poisoning in the human population. International Agency of Research on Cancer (IARC) has classified Arsenic as a class I human carcinogen. There are several types of cancers associated with Arsenic toxicity, such as skin, bladder, kidney, prostate, liver and lungs cancer. Among them, skin cancer is the most common neoplasm associated with Arsenic, and lung cancer is the most deadly. Arsenic concentration in high doses has been found out in various places around the world. Though Arsenic-induced cancer is well studied, several mechanisms of the same are unknown. Here, we have investigated the mechanism of carcinogenesis induced by inorganic Arsenic and its role in oxidative stress, apoptosis, Arsenic-induced DNA damage, role in affecting signal transduction pathways, epigenetic modifications due to Arsenic, genotoxicity caused by Arsenic, and also the role of Arsenic affecting Micro RNA. We also discussed in detail in this chapter the molecular mechanism involved in Arsenic-induced cancers such as skin, lungs, liver, prostrate, kidney and bladder. A better understanding of molecular mechanisms of Arsenic induced cancer will help us to develop effective therapeutic approach for Arsenic induced cancer.

Keywords Arsenic toxicity · Cancer · Oxidative stress · DNA damage · Apoptosis

1   Introduction

One of the most abundant elements in the Earth's crust is Arsenic. It is a metalloid exhibiting organic and inorganic forms. Common oxidation states of Arsenic in the environment are +3, As^{III}, also known as arsenite and +5 or As^V or arsenate (Vahter 2002). Its organic form is created when linked with carbon and hydrogen, whereas oxygen, chlorine and sulphur, among other elements combine with Arsenic to create inorganic form. Inorganic (iA) occurs in the natural soil especially in rocks. Arsenic and Arsenic-containing compounds are cancer-causing agents in humans. Several industries, including mining, pesticide, pharmaceutical, glass and microelectronics, are causes of Arsenic exposure. It is also found in natural sources such as ground-water. The principal route of Arsenic exposure in the occupational environment is inhalation. Contaminated drinking water is the predominant source of exposure to Arsenic. Contamination of drinking water globally by significant environmental exposure of Arsenic is one of the major public health problems (Page 2013). It is found that more than 100 million people are exposed to Arsenic worldwide (Shahid et al. 2018). Drinking water Arsenic concentration has been reported in many countries of the world, leading to Arsenic poisoning in the population. Arsenic exposure may cause diabetes, hearing loss, portal fibrosis and several cardiovascular and peripheral vascular diseases. Several developmental anomalies and multiple cancers are also reported due to Arsenic exposure. It may also result in neurologic and neurobehavioral disorders, hematologic disorders such as anaemia, leukopenia and eosinophilia. Significant mortality rates more than usual standardized and cumulative mortality rates exist for certain cancers such as skin, lung, liver, urinary bladder, kidney, and colon in those who are exposed. Though Arsenic-induced cancer is well studied, several mechanisms of the same are unknown (Tchounwou et al. 2003). The pathobiology of Arsenic-induced diseases and toxicological pathology of Arsenic in various organ systems have majority of studies and still underexplored as the information is critical for understanding the scale of health effects connected with Arsenic contact throughout the world. International Agency of Research on Cancer (IARC) has classified Arsenic as a class I human carcinogen (Rousseau et al. 2005). There are several types of cancers associated with Arsenic toxicity, such as skin, bladder, kidney, prostate, liver and lungs. Among them, skin cancer is the most common neoplasm associated with Arsenic and lung cancer is the deadliest. It is estimated that a daily consumption of 1.6 litre of water with inorganic Arsenic concentrations of 50 Microgram/Litre may cause cancer-related deaths in 21/1000 (Martinez et al. 2011c). World Health Organization and the U.S. Environmental Protection Agency had given a threshold of 10 Microgram/Litre (Yamamura 2001). India, Bangladesh, China and Taiwan are some of the major regions across the world whose significant populations are exposed to inorganic Arsenic (Heikens 2006). More than 21 million people in India and Bangladesh are exposed to more than 50 Microgram/Litre of Inorganic Arsenic which is a major cause of concern (Martinez et al. 2011c). A highest ever concentration of Arsenic in the drinking water has been reported in a recent study in Simri Village of Buxar District, Bihar, in Eastern India. Arsenic concentration in the groundwater sample found during the study was 1929 µg/L and in blood sample it was 664.7 µg/L (Rahman et al. 2019) (Fig. 1).

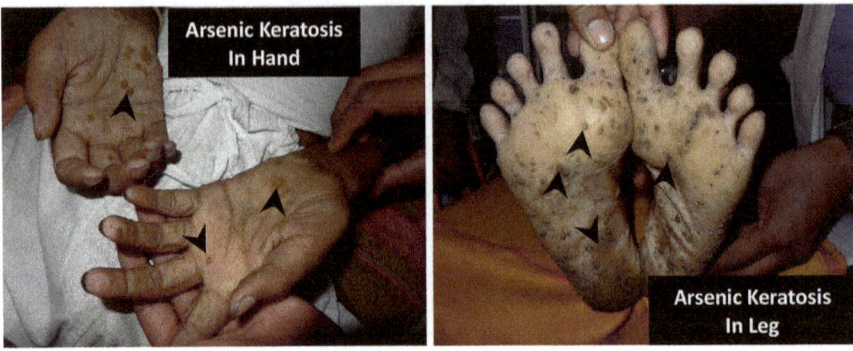

Fig. 1 Arsenic toxicity in the people of Devkuli Village of Brahmpur Block of Buxar district of Bihar

2 Arsenic and Skin Cancer

Much research has been going on the topic Arsenic-induced skin cancers and is well documented. The first observations are made as an increased frequency of skin cancer cases following treatment with Flower solution (1% potassium arsenate) (Fowler et al. 2015). It was widely used for skin and haematological disorders. Intraepithelial carcinoma or carcinoma in situ or Bowen`s disease basal cell carcinoma (BCC) and squamous cell carcinoma (SqCC) are commonly found in patients with long-term Arsenic exposure (Yu et al. 2006). Squamous cell carcinoma (SqCC) due to Arsenic can develop either de novo or progress from Bowen's disease (Yeh et al. 1968). Basal cell carcinoma due to Arsenic exposure occurs in multiple foci. Latency period of intraepithelial carcinoma is usually 10 years whereas that of other types of skin cancers may vary from 20 to 30 years. Pre-malignant lesions appear in the skin due to Arsenic exposure and is considered as precursor to BCC and SqCC (Martinez et al. 2011a). Accumulation of Arsenic in the skin may tend to skin hyperpigmentation and hyperkeratosis (Rahman et al. 2006; Yu et al. 2006). Arsenic-induced Bowen's disease may lead to SqCC and BCC. Arsenic-induced Bowen's cancer can be distinguished from other types by region of occurrence. Mostly it occurs in the sun-protected areas of the body, whereas other non-AS-BD originates in exposed regions of the skin. Most non-AS-BD completely recovers after surgery whereas many AS-BD reoccurs after a few years. Many AS-BD lesions may lead to more invasive BCC and SCC and to other types of cancers. There is also documented evidence that AS-BD may lead to pulmonary cancers even after 30 years. It has been identified that accumulation of AS-BD is associated with mutant form of P53. Most of the mutations occur in the exon 5 and exon 8 of the P53 protein (Dela Cruz et al. 2011). It has also been identified that Arsenic exposure was associated with G2/M cell cycle arrest and DNA aneuploidy in both in vitro and As-BD lesion studies

(Liao et al. 2017). Clinically, As-BD-induced skin lesion usually occurs in sun-protected regions. It has been identified that UVB has an effect on Arsenic-induced carcinogenesis. It plays a modulatory role in the cancer in Arsenic-related skin carcinogenesis. UVB irradiation has an inhibitory effect on cell proliferation as it reduces mutant p53 and ki-67 expressions, as well as decreases the number of apoptotic cells in As-BD lesions. Arsenic is not identified as a mutagenic in bacterial and mammalian cells, but it has been identified that it reinforces mutations caused by other mutagens including UVB. It has been identified that immunological dysfunction occurs in Arsenic-induced skin cancers. This may be due to decreased number in CD4+ cells (T helper). It occurs both in adults and in children. The association of impaired cellular immunity may also be attributed to the effects of Arsenic on human lymphocytes (Huang et al. 2019).

3   Arsenic and Lung Cancer

Arsenic concentration in water and incidence of lung cancer and other malignancies for both men and women are reported. It mainly has a relation between dose and response. There are several case-control and cohort-type studies which demonstrate an increased lung cancer risk following exposure to Arsenic in drinking water. Arsenic-induced lung cancer is basically sub type specific. High incidence of SCC occurs in non-smokers who have been exposed to cancer. The large mass of population who are exposed to Arsenic in the groundwater as in Bangladesh where in the groundwater used for drinking contains inorganic Arsenic. In southwestern Taiwan, severe systemic arteriosclerosis and spontaneous gangrene resulting in amputation, known as Blackfoot disease, are common in individuals who are exposed to Arsenic. Other places include Andean zone in South America, central regions of Argentina, and Northern Chile, where severe cases of Arsenic exposure are reported resulting in lung cancer. The relationship between lung cancer and Arsenic is basically dose–response relation where Arsenic concentrations in water supplies were >100 μg/L, even reaching as high as 2000 μg/L. There are several suggestions on the mechanism associated with progression to lung cancer after being exposed to Arsenic. But the exact genetic pathway associated with Arsenic exposure is not fully understood. Methylarsonic acid (MMA) and dimethylarsinic acid (DMA) metabolites are produced from Arsenic after being ingested into the body (Vahter and Concha 2001). Even though most are excreted through urine, some are deposited in tissues of the lung, liver, kidney, nails and hair. MMA and DMA are cytotoxic and genotoxic in cell lines. Highly biologically active MMA and DMA metabolites may possibly cause chromosomal abnormalities, oxidative stress, altered DNA repair, altered DNA methylation patterns, altered growth factors, enhanced cell proliferation, promotion/ progression, gene amplification and suppression of p53 (Khairul et al. 2017). Occupational exposure to inorganic Arsenic is associated with an increased risk of developing lung cancer especially in miners. In contrast to the drinking water studies which suggest a linear dose–response relationship between Arsenic exposure and lung cancer mortality, the occupational studies indicate a supralinear dose–response relationship. This is because of the fact that Arsenic is low soluble in water which results in more rapid metallization and rapidly removed in the case ingested Arsenic, dust particles which may contain Arsenic less rapidly eliminated from the body.

4 Arsenic and Kidney Cancer

Several cases of Renal Cell Carcinoma (RCC) have been reported in different studies from 1980s. Especially from the regions of Taiwan, Chile, Australia, South Eastern Michigan, USA, Utah, USA, Copenhagen, Denmark, among the other regions. All these regions have Arsenic presence in high doses more or less up to 100 ug/L or more, especially in water bodies. In a study conducted in Taiwan, it was found that that higher urinary Arsenic level (> 15.95 µg/g creatinine) is a strong predictor of RCC, and low estimated glomerular filtration rate (eGFR) (< 0.001) (Chen et al. 2011). There are several chromosomal aberrations in the form of chromosomal loss especially in the region spanning 3p14 to 3p26. It shows a direct relation between *VHL* tumour suppressor genes which are located in the said region. Abnormalities are found in the *VHL* gene in 50 to 60 percentage of RCC patients. Gene product from the *VHL* gene, pVHL, functions as a tumour suppressor protein and targets several proteins for degradation by proteasomes (Tanriover 2012).

5 Arsenic and Bladder Cancer

Incidence of Bladder cancer is reported in several places around the world, especially in and across the regions of Taiwan. Other regions include Argentina, Chile, Australia and the USA. A study in Chile shows a dose–response relationship in well water Arsenic concentration and rates of mortality from bladder cancer. There are also studies that show high exposure to Arsenic in high population regions of Taiwan, Argentina and Chile. People who are exposed to Arsenic in their drinking water are subjected to higher chromosomal aberrations, compared to those who are less exposed. Chromosomal alterations are seen in 3q and 11q gain and 8p, 17p and 9q loss. 3q alteration may depend on the tumour stage. Arsenic exposure with gains on 11q may be the main cause of amplification of Cyclin D1 and *INT2* genes as seen in bladder cancer patients. Loss on the chromosome 17p plays a major role in Arsenic-induced bladder tumorigenesis. It is also not related to tumour grade and not with stage or smoking history. The most likely effect on the deletion on 17p may be *p53* gene. In one study, it has been found that codon 175 of *p53* is a mutational hotspot (Moore et al. 2002).

6 Arsenic and Prostate Cancer

Early evidence of association of Arsenic and prostate cancer is found in Taiwan in 1980s. It is understood that there is a dose–response relationship between Arsenic and cancer. It is seen that in a group who are exposed to Arsenic at the highest levels in drinking water, there is nearly six-fold increase of prostate cancer incidence. Prostate cancer is an old age disease that has a relatively low case-fatality rate. This may result in overshadowing by other cancers over prostate cancer. The role of drinking water Arsenic in prostate cancer is studied in the USA also. Among the residents of Millard County, Utah, USA, drinking water Arsenic exposure time with respect to residence time is studied. Based on the records of Church of Jesus Christ of Latter-day Saints, prostate cancer mortality was expressively elevated. A region in Australia where Arsenic > 100 mg/kg and/or drinking-water concentrations > 0.01 mg/L is identified for the study. In the region of Victoria, Australia, 22 areas of elevated Arsenic exposure are analysed again with all cancer cases in the place. In all areas with elevated Arsenic in soil or water or both, a significant increase is shown in prostate cancer. Prostate-specific antigen (PSA), which is a well-known marker of prostate cancer, positive correlation occurred between urinary Arsenic concentration and serum prostate-specific antigen in a population of male workers related to copper foundry. There is an overexpression of PSA in prostate epithelial cells in in vitro cultures of prostate tumours transformed by Arsenic. Molecular events underlying prostate cancer involve DNA methylation. SAM gene which is used by Arsenic for biomethylation may result in hypo methylation as SAM will act as the methyl donor (Ajees and Rosen 2015; Reichard and Puga 2010). A decrease in DNA methyltransferases activity is an early event occurring before malignant transformation, which may account for the subsequent genomic DNA methylation. Another molecular event is the overexpression of unmutated K-ras which is seen in CAsE-PE cells. It is understood that overexpression of K-ras is one of the key events in the molecular change of the Arsenic-induced CAsE-PE cells (Benbrahim-Tallaa et al. 2007). Chronic Arsenic exposure induced malignant transformation, hyper proliferation and overexpression of K-ras, which may result in several pathway Raf/ MEK/ERK pathway (Benbrahim-Tallaa and Waalkes 2008).

7 Liver and Arsenic Cancer

Arsenic exposure to liver and carcinogenesis of liver cells are considered controversial. Liver cancer, primarily HCC, and its relation to Arsenic are first reported in Taiwan's southwest region. Other studies also found out Arsenic as a liver carcinogen in different areas of the world, especially in the region where Arsenic is reported in drinking water. Such regions include China, Chile, Argentina and Bangladesh. The effects of Arsenic include oxidative DNA damage, impaired DNA repair, apoptotic tolerance and hyper proliferation, DNA methylation and genomic instability, aberrant estrogen signalling (Liu and Waalkes 2008) (Fig. 2).

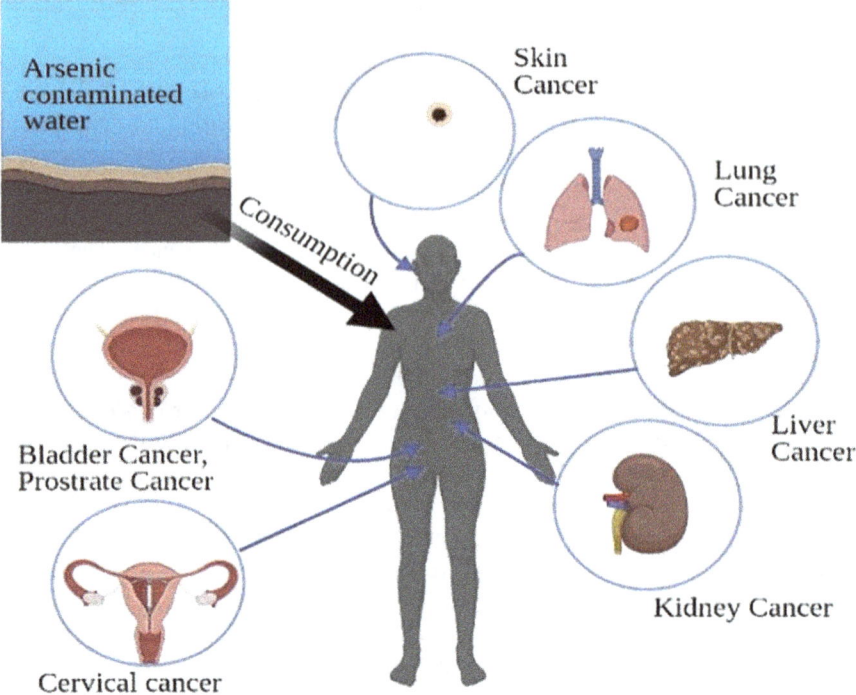

Fig. 2 Role of Arsenic in carcinogenesis in humans (skin cancer, lung cancer, liver cancer, kidney cancer, cervical cancer, bladder and prostate cancer)

8　Role of Arsenic in Oxidative Stress

The reactive oxygen species that are generated during normal signalling processes participate in important physiological functions such as regulation of redox reactions, cell-signalling in the cell cycle process, cellular proliferation, differentiation and apoptosis (Martindale and Holbrook 2002). The ROS generated within the cell are maintained at optimal levels by the cell's defence system called antioxidants which exist in a balance within the cells. This balance is disrupted during oxidative stress causing oxidative damage of proteins, lipids, DNA, etc.

A lot of studies have been carried out to support that Arsenic generates ROS in several cell types through a variety of below-mentioned mechanisms:

(I) Mitochondria: $O_2^{-\bullet}$ is generated by the electron transport chain in mitochondrial complexes I and III. The ATP generated by the electron transport chain provides energy to the cell. NADH (or nicotinamide adenine dinucleotide) helps to "jump-start" the electron transport chain by acting as the electron donor (i.e., the H in the NADH). NADH donates an electron to the electron transport chain after moving into the mitochondria from the cytosol, where the electrons pass "down the line" forming ATP with the help of some proteins and lipids. Some electrons after escaping the electron transport chain combine with oxygen forming a very unstable superoxide radical ($O_2^{-\bullet}$), a reactive oxygen species. Superoxide dismutase (SOD) can scavenge this radical by reducing superoxide to form hydrogen peroxide (H_2O_2). Catalase then detoxifies this hydrogen peroxide to water and O_2. Arsenic inhibits succinic dehydrogenase activity and brings about uncoupling oxidative phosphorylation releasing O_2 and giving rise to other forms of ROS leading to mitochondrial toxicity (Ellinsworth 2015).

(II) Nicotinamide adenine dinucleotide phosphate oxidase (NADPH oxidase or Nox). Nox, a membrane-associated enzyme, has been found to be involved in ROS generation in response to Arsenic. Nox isoforms synthesize ROS [$O_2^{-\bullet}$ or hydrogen peroxide (H_2O_2)] as their sole product (Ellinsworth 2015). Superoxide ($O_2^{-\bullet}$) forms toxic reactive nitrogen species peroxynitrite ($ONOO^-$) on rapidly reacting with endothelium-derived nicotic oxide, thereby reducing the bioavailability of nitric oxide. This peroxynitrite impairs nitric oxide by oxidizing the essential endothelial NOS (eNOS) cofactor (6R)-5, 6, 7, 8-tetrahydrobiopterin, leading to the uncoupling of eNOS and an increase in $O_2^{-\bullet}$ generation by the oxygenase component of the enzyme, thereby aggravating oxidative stress (Förstermann and Münzel 2006).

(III) Generation of ROS during the formation of intermediate arsine species (Yamanaka et al. 1997). For example, dimethyl Arsenic peroxyl radical is formed during the metabolic processing of DMAA. This dimethyl Arsenic peroxyl radical [$(CH_3)_2AsOO^{\bullet}$] and active oxygens produced in the further metabolic processing of the DMAA (dimethyl Arsenic acid) lead to DNA damage (Kato et al. 1994; Yamanaka and Okada 1994).

(IV) Methylated Arsenic species release redox-active iron from ferritin. Exogenous methylated Arsenic species cause the release of iron from ferritin, an iron-dependent formation of reactive oxygen species causing DNA damage. This reactive oxygen species pathway has been suggested to be the mechanism of action of Arsenic carcinogenesis in humans ("Arsenic species cause release of iron from ferritin generating activated oxygen," 1999).

(V) Under physiological conditions, Arsenic produces ROS by the oxidation of arsenite to arsenate (Del Razo et al. 2001). Inorganic Arsenic in the aqueous environment appears as arsenous acid (As (III)), Arsenic acid (As (V)), and their salts and tends to be more toxic than organic Arsenic. It was observed that As (III) is oxidized in parallel to the oxidation of Fe (II) by O_2 and by H_2O_2 generating •OH radicals at low pH. The Fenton reaction then produces a highly toxic hydroxyl free radical from hydrogen peroxide, a product of mitochondrial oxidative respiration (Hug and Leupin 2003).

(VI) Endoplasmic reticulum (ER): ROS is generated in the ER by the DMA III. The trivalent dimethylarsinous acid (DMAIII) exposure induces protein kinase-like endoplasmic reticulum kinase (PERK) phosphorylation in cells leading to the generation of ROS. The induction of ER stress occurred through the inhibition of the protein folding process. All three Arsenic species induced the activation of transcription factor 4 (ATF4) and C/EBP

homologous protein (CHOP) mRNA all leading to DMA III-induced cytotoxicity (Naranmandura et al. 2012).

(VII) Interference with cellular antioxidants like superoxide dismutase (SOD),
catalase (CAT) (Nordenson and Beckman 1991), glutathione (GSH) and
GSH-related enzymes (Chouchane and Snow 2001), which indirectly results
in increased ROS levels. The ability of arsenite (As (III)) to bind protein thiols such as glutathione (GSH) may also contribute to one of the mechanisms
of Arsenic toxicity. GSH-related enzymes such as glutathione reductase
(GR), glutathione peroxidase (GPx) and glutathione S-transferase (GST) are
important antioxidants playing a significant role in the detoxification of
Arsenic along with other carcinogens. GPx shows highest sensitivity to
Arsenic treatment while GST the least. It was also found that Arsenicals or
arsenothiols had the capability to inhibit GSH reductase. Also, biomethylation of As could yield species that inhibit reduction of GSSG and alter the
cell's redox status as methylated trivalent Arsenicals and arsenothiols have
greater potency than inorganic trivalent.

The Arsenic-induced ROS produced by these mechanisms can alter DNA structure by insertion, deletion and by several mutations leading to the manifestation of
cancer (Martinez et al. 2011b). Arsenic reduces the activity of antioxidative enzymes
and increases the number of peroxide products leading to oxidative injuries. Arsenic
exposure depletes glutathione (GSH) levels, increases oxidized glutathione (GSSG)
and malondialdehyde (MDA) which contribute to Arsenic-induced oxidative stress
(Flora 1999). One of the important mechanisms of Arsenic poisoning caused by
oxidative stress is lipid peroxidation (Das et al. 2018). Recent studies have demonstrated that Arsenic (III) administration creates redox imbalance by upregulating the
levels of malondialdehyde (MDA) and H2O2, thus downregulating the activity of
the antioxidant enzyme, including glutathione peroxidase (GSH-Px) and catalase
(CAT) (Masuda et al. 2018) (Fig. 3).

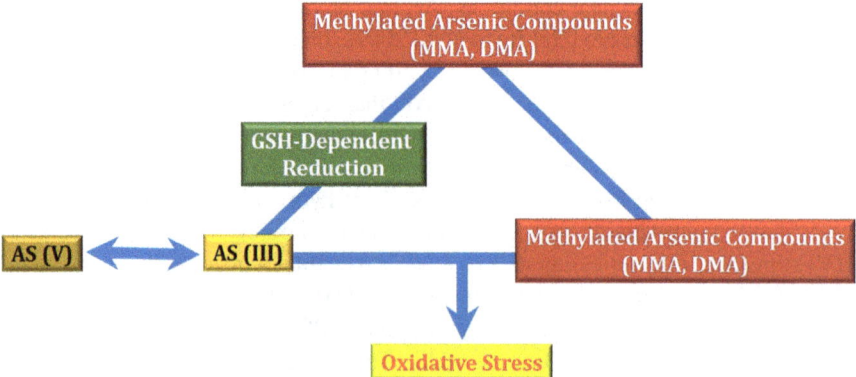

Fig. 3 Role of Arsenic (V), Arsenic (III) and methylated Arsenic compounds (MMA, DMA) in
oxidative stress generation

9   Arsenic-Induced ROS and Apoptosis

Increased oxidative stress has been found to be directly related to Arsenic-induced apoptosis. Excess ROS increases the mitochondrial membrane permeability and causes damage to the respiratory chain, further increasing ROS production (Chen et al. 2001). This mitochondrial membrane disruption leads to cytochrome c release from the mitochondria, initiating events leading to apoptosis. ROS have also been found to play an important role in tumor necrosis factor receptor (TNFR) and Fas receptor-mediated apoptosis (Krammer 1998). Arsenic toxicity includes the interaction of the trivalent Arsenic with the sulfhydryl (SH) group of biomolecules making trivalent Arsenic a known carcinogen and inducer of apoptosis. Due to this property of Arsenic, it is currently a standard anti-cancer treatment in certain tumour cell lines at low concentrations (1–3 μM). A study found an association of the induction of ROS upon Arsenic stimulation with the inhibition of Akt-mediated cell survival by caspase 3 activation ("Arsenic trioxide-induced apoptosis in U937 cells involve generation of reactive oxygen species and inhibition of Akt – PubMed," n.d.). Therefore, the activation of p38, JNK and caspase 3 by the Arsenic-induced ROS is concomitant with the occurrence of apoptosis independent of MAPK pathways. The accumulation of intracellular ROS resulting from Arsenic binding to the SH-group of glutathione (GSH), an important cellular antioxidant, leads to activation of caspases. Also in cancer cells, the level of their intracellular glutathione (GSH) inversely determines their sensitivity to Arsenic-induced apoptosis (Oketani et al. 2002).

10   Arsenic-Induced ROS and DNA Damage

Arsenic has been reported to be mutagenic to endogenous genes in mammalian cells inducing large multilocus deletions mediated through ROS. Other DNA damages related to Arsenic toxicity include gene amplification, breakage of DNA strand, DNA base modifications, DNA repair inhibition, mitotic cell arrest, induction of c-fos gene expression, and heme oxygenase an oxidative stress protein in mammalian cells (Lee et al. 1988). Hydroxyl radicals have been implicated to be mediators of arsenite mutagenicity (Hei et al. 1998). To back the role of ROS, studies were carried out in cultured mammalian cells in which it was found that superoxide dismutase, glutathione peroxidase and catalase reduced the incidence of arsenite-induced sister chromatid exchanges and chromosomal aberrations (Nordenson and Beckman 1991). Arsenic-induced ROS activates p53, which downregulates ROS by the activation of a series of antioxidant genes like *GPx-1*, sestrin 1 and sestrin 2 (Sablina et al. 2005), and also activates apoptosis or growth arrest in the Arsenic-exposed cells (Green and Kroemer 2009; Riley et al. 2008; Fig. 4). Chronic exposure to low levels of Arsenic in rats has shown to result in depletion of glutathione and increase in oxidized glutathione and lipid peroxidation. This depletion of

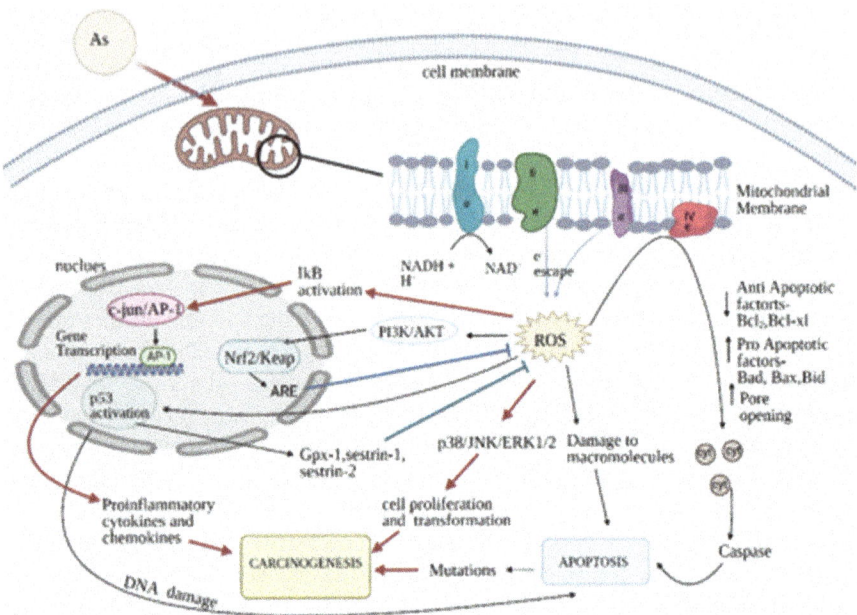

Fig. 4 Arsenic induces mitochondria to generate ROS after e⁻ escape from the electron transport chain (ETC). These free radicals increase the expression of proapoptotic factors. ROS also cause calcium-dependent decrease in the mitochondrial membrane potential and opening of the membrane pore allowing the release of cytochrome *c* from the mitochondria triggering caspase-dependent apoptosis. ROS causes apoptosis by directly activating p53 which also activates antioxidant genes like *Gpx-1*, sestrin-1 and -2 thereby downregulating ROS. Additionally, ROS activates IκB complex, jun, p38, and ERK which activate pro- or anti-inflammatory cytokines as well as promote cell proliferation and transformation and hence promote carcinogenesis

intracellular nonprotein sulfhydryls (mainly glutathione) increases Arsenic-induced mutations up to five-fold (Liu et al. 2001). Therefore, Arsenic might induce DNA damage through the following cascade:

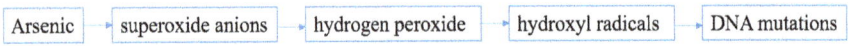

Studies have reported that Arsenic exposure along with zinc deficiency augments markers of oxidative stress leading to enhanced immune activation and proinflammatory cytokine production. In human studies, oxidative stress as a consequence of exposure to Arsenic leads to immune cell dysregulation and promotes an inflammatory response. One of the oxidative stress associated genes *HMOX1* was found to be strongly upregulated in a dose-dependent manner upon Arsenic treatment (Menzel et al. 1998; Cooper et al. 2007).

11   Arsenic-Induced ROS and Signal Transduction Pathways

Arsenite may interfere with signal transduction pathways either by direct kinase/ phosphatase-enzyme inhibition or by redox control of the regulatory molecules. Studies found that exposure of the cells to Arsenic increased total cellular tyrosine phosphorylation in a dose-dependent manner which is directly related to aberrant cell signalling (Fig. 4), uncontrolled cell growth in culture, and in vivo cancer development (Gozin et al. 1998). The induction of tyrosine phosphorylation is mediated by the activation of protein tyrosine kinases and/or the inhibition of protein tyrosine phosphatases. Arsenic activates epidermal growth factor receptor (EGFR), one of the receptor tyrosine kinases (RTKs) and nonreceptor tyrosine kinases (NTKs), which include Src family members, through ROS generation activating Ras and the mitogen-activated protein kinases (MAPKs) cell signalling (Chen et al. 1998; Wu et al. 2002). An increasing amount of evidence suggests that Arsenic can activate all three classes of MAPKs, including extracellular signal-regulated kinases (ERKs), c-Jun N-terminal kinases (JNK) and p38, in a variety of human cell lines in a time-dependent manner (Bode and Dong 2002). Arsenic also activates ERKs and p38 by the Ras/Raf/Mek pathway, while the activation of JNKs has proven to involve the Ras, Rho and MEKK3-4 (Wu et al. 2002; Ludwig et al. 1998; Porter et al. 1999). The activation of ERKs by Arsenic can induce cell proliferation and transformation leading to carcinogenesis, whereas JNKs and p38 activation can induce cell growth arrest and apoptosis leading to the inhibition of carcinogenesis (Bode and Dong 2002). Arsenic-mediated ROS also activate NF-κB which mediates amplification of pathogenic signals, and initiation or acceleration of carcinogenesis (Chen and Shi 2002). This NF-κB activation is time-, dose- and cell type–dependent. NF-κB is activated by low and noncytotoxic concentrations of Arsenic (1–10 μM); however, it is inhibited by high concentrations of Arsenic (>10 μM) (Chen and Shi 2002).

12   Arsenic-Induced Genotoxicity

Several DNA-related aberrations are caused by Arsenic, which include chromosomal aberrations, deletions, mutations, sister chromatid exchange, micronuclei formation, aneuploidy and DNA-protein cross-linking. Arsenic by itself is a poor mutagen, but it affects mutagenicity of other carcinogens. During biotransformation, ROS is generated which is primarily linked to Arsenic genotoxicity. ROS causes DNA strand breaks, crosslinks and chromosomal aberrations. Oxidative stress is the primary mechanism of genetic damage induced by Arsenic oxidative damage to DNA causes the DNA base modification (Cooke et al. 2003). Generation of 8-oxoguanine (8-OHdG), one of the most commonly formed DNA nucleobase modifications, is an indication for oxidative stress. G: C to T: A transversion mutations are instigated by 8-Oxoguanine. Oxidative DNA adducts 8-OHdG upon Arsenic exposure in several tissues which are well documented (Yamauchi et al. 2004;

Hinhumpatch et al. 2013). Single-strand breaks induced by Arsenic is one of the important DNA damaging effects caused by ROS on the DNA bases. Moreover, single-strand breaks may also cause indirectly by base excision repair (BER) mechanism (Hegde et al. 2008). Both clastogenic and aneuploidogenic effects, chromosomal and chromatid aberrations are induced by Arsenic. An increase of incidence of chromosomal aberrations in individuals having continuous exposure to Arsenic is found in several studies. In a study by Chile Martinez et al., an increased incidence of micronuclei formation is seen in peripheral blood lymphocytes due to Arsenic exposure. Another main Arsenic-induced genotoxicity is inhibition of DNA repair processes. Nucleotide Excision Repair (NER) and Base Excision Repair (BER) are the processes convoluted in the repair of DNA base damage brought by ROS. Arsenic may repress the BER mechanism. BER and NER genes expression levels changes are seen in human exposed populations (Muenyi et al. 2015). Decreased expression of Excision repair cross-complementing rodent repair deficiency, complementation group 1 (ERCC1). Several other genes are involved in Arsenic mechanism and detoxification. Precise genetic polymorphisms in genes such as Arsenic (III) methyltransferases (ASIIIMT), glutathione S-transferases (GST) and methylenetetrahydrofolate reductase (MTHFR) enzymes which are involved in the same are studied well. SNPs in DNA repair pathway genes such as *hOGG1, APE1, XRCC1, XRCC3* genes are also studied; all of them have shown a reduced capacity to repair the oxidative damages (Frankenberg 2012). Arsenic (III) methyltransferases (ASIIIMT) are associated with G7395A, G12390C, T35587C, G12390C, C14215T, A35991G are seen associated (Zeng et al. 2017) SNPs. The M287T polymorphism in ASIIIMT is involved in inter-individual difference in the Arsenic detoxification. This may lead to increase in MMA percentage and a greater risk to develop Arsenic-related cancers. Phase II detoxifications reaction enzyme GST group of enzymes are also involved. Four classes of GSTs – GST-P1, GST-O1, GST-M1 and GST-T1 – are convoluted in the biotransformation of Arsenic. GST-O2 Asn142Asp, GST-P1 Ile105Val variant, A222V of *MTHFR* gene also affect the biotransformation (Faita et al. 2013).

13 Arsenic-Induced Carcinogenesis: Role of miRNA

13.1 Micro RNA: miRNA

Micro RNA is a small RNA, 21–22 nucleotide long, which is involved in posttranscriptional gene silencing. It guides Argonaute (AGO) protein to mRNA to silence the expression of the gene. MiRNA biogenesis occurs in the cytoplasm and the nucleus. It involves two steps. The primary transcript (pri-miRNA) is cleaved in the nucleus by the microprocessor complex into 85-nucleotide pre-miRNA hairpin (pre-miRNA) (Graves and Zeng 2012). Trimmed pre-miRNA is transported from the nucleus to the cytoplasm where it is processed to a 22-nucleotide miRNA, which is of a single strand. RNA-induced silencing complex (RISC) is guided to the target mRNA after incorporating mature miRNA into the RISC complex. MiRNA may act as transducer/mediator among oncogene and tumour suppressor genes leading to

Fig. 5 Mechanism (s) of miRNA in gene silencing

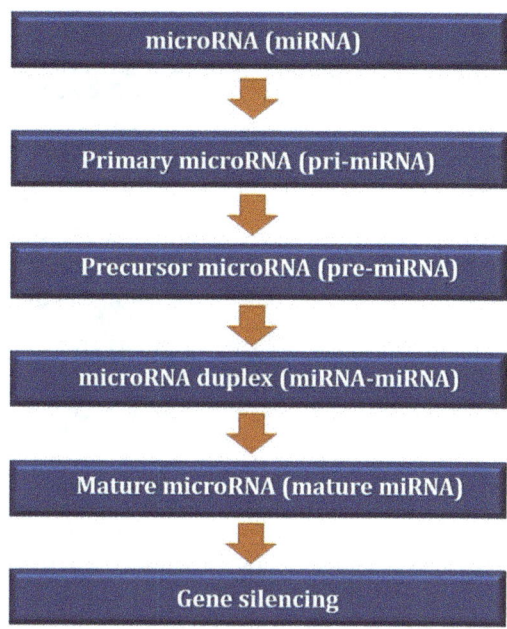

tumour formation. MiRNA binds to the 30-untranslated regions (UTRs) called the "seed region" leading to decreased translation, deadenylation, or degradation of the mRNA. The pairing of the miRNA to the target binds to the mRNA leading to the degradation of mRNA (Fig. 5). The role of miRNA lies in developmental processes, apoptosis, cell proliferation and regulating translation of most protein-coding genes (Ledgerwood et al. 2016; O'Brien et al. 2018).

13.2 Role of Arsenic in miRNA Lead Carcinogenesis

From 2006 onwards, there are a great number of studies specifying the role of Arsenic in cancer progression affecting miRNA. miRNAs mediating Arsenic-induced carcinogenesis, include miR-200, miR-190, miR-21 and miR-122 (Fig. 6). Majority of these effects are studied in different cell lines; but the the functional consequences the functional consequences are not identified yet. In a particular in HaCaT cells (immortalized human keratinocytes), are overexpressed. MiR-21 and miR-141 are also associated with human tumours. MiR-200 family has shown to have a role in epithelial–mesenchymal transition (EMT), and progression to cancer. Epithelial–mesenchymal transition (EMT) is identified as a manifestation on cancer progression. Several signalling pathways such as mitogen-activated protein kinase (MAPK), Jak-STAT and Wnt pathways are affected by Arsenic. Cyclin-dependent kinase 6 (CKD6), which has an important function in cell cycle progression which is one of the genes controlled by simultaneous regulation of miR-21, miR-200a and miR-141, may have some role to play in the cancer progression (Bustos et al. 2017).

Fig. 6 Human
Argonaute2-miR-122
bound to a seed and
supplementary paired
target

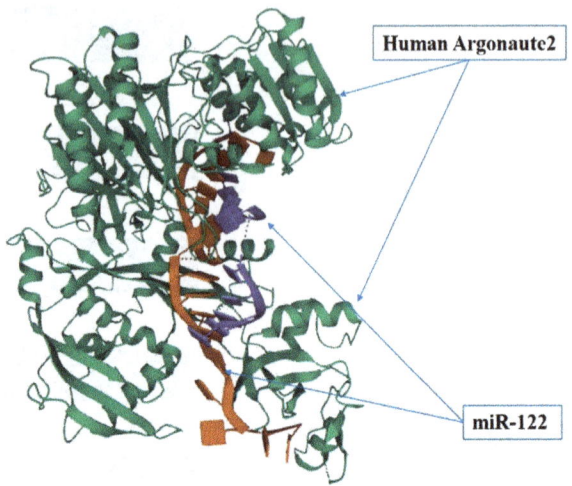

CKD6 also regulates tumour suppressor (RB) gene (Vélez-Cruz and Johnson 2017). In several studies, mir-21 is shown to be unregulated in different cases for those, who are exposed to Arsenic. Phosphatase and Tensin homolog (PTEN) and programmed cell death 4 (PDCD4) which are the downstream targets of mir-21 and survival proteins phosphorylated protein kinase B (pAKT) and phosphatidylinositol 3 kinase (PI3K) are studied. The relation of mir-21 with PTEN and PDCD4 is inverse, whereas pAKT and PI3K proteins have found to be increased. Let-7 family, tumour suppressor miRNA have been identified in induced neoplastic transformation in HaCaT cells. Let-7a, let-7b and let-7c have shown to have decreased level in these cell lines (Chirshev et al. 2019). In another study on HBE (Human Bronchial Epithelial) Cells Arsenic-induced malignant transformation shows an increase of mir-155. Nuclear factor (erythroid-derived 2)-like 2 (NRF2) is shown to have decreased level in HBE-transformed cancer cells. It is also identified that levels of PDCD4, a tumour suppressor as well as a target of mir21, have decreased in HBE cells (Gonzalez et al. 2015). In BEAS-2B cells after exposure to Arsenic, increased levels of interleukin 6 (IL-6), transcriptional activation of signal transducer and activator of transcription3 (STAT3), expression increase and STAT3 phosphorylation were reported. This shows that ROS-STAT3-miR-21-PDCD4 signalling pathway is affected by Arsenic-induced BEAS-2B malignant transformation. Another miRNA involved is miR-199a which is downregulated in BEAS-2B malignant transformation after been exposed to Arsenic for 26 weeks. Human urothelial cell transformation which is exposed to Arsenic may result in malignant cell transformation and EMT acquisition. In HUC1 cells, miR-200a, miR-200b and miR-200c were downregulated after been exposed to Arsenic. Micro RNAs miR-200a, miR-200b and miR-200c are also found to decrease in the urine of the individuals who are exposed to Arsenic (Humphries and Yang 2015), suggesting, miR-200 family members have a role to play. In the case of study on Liver cells, human hepatic epithelial (L-02) cells, upregulation of mir21 miRNA, resulting in the less expression of PTEN,

PDCD4 and sprouty RTK signalling antagonist 1 (SPRY1) which are all found to be the target protein of mir-21. (Buscaglia and Li 2011; Zhang et al. 2020) in transformed L-02 cells, miR-191, an oncogenic miRNA has found to be in the increased state. Hypoxia-inducible factor (HIF)-2α (HIF-2α) upregulates miR-191 levels to exert a function in the arsenite-induced malignant transformation of HBE cells (Chen et al. 2018). In a study on Sprague Dawley (SD) rats exposed for 60 days to sodium arsenite, after analysing miRNA expression profiles using RNA-Seq and qPCR profile data of liver samples, miR-151 and miR-183 were upregulated and miR-423, miR-26a and miR-148b were downregulated by Arsenic exposure (Ren et al. 2016). All these findings show that different deregulated miRNAs have a significant role in cancer development and need to have further study on the topic (Cardoso et al. 2018).

14 Conclusion

From various studies it is known that Arsenic exposure may result in cancer. Arsenic poisoning from Arsenic exposure especially from the groundwater is a major cause of concern across the world population. Arsenic exposure may cause several cardiovascular and peripheral vascular diseases, diabetes, hearing loss and portal fibrosis. Several developmental anomalies and multiple cancers are also reported due to Arsenic exposure. It may also result in neurologic and neurobehavioral disorders, hematologic disorders such as anaemia, leukopenia and eosinophilia. Inorganic Arsenic is a major carcinogen which may cause several cancers, including skin, bladder, kidney, prostate, liver and lungs among the other areas. Even though several studies have been done on Arsenic to understand its role as a major carcinogen, its role in the genotoxicity, oxidative stress, induction of DNA damage, miRNA inhibition are poorly understood.

Conflicts of interest The authors have nothing to declare.

References

Ajees AA, Rosen BP. As(III) S-adenosylmethionine methyltransferases and other Arsenic binding proteins. Geomicrobiol J. 2015;32:570–6. https://doi.org/10.1080/01490451.2014.908983.

Arsenic species cause release of iron from ferritin generating activated oxygen. Free Radic Biol Med. 1999; https://doi.org/10.1016/s0891-5849(99)90549-x.

Arsenic trioxide-induced apoptosis in U937 cells involve generation of reactive oxygen species and inhibition of Akt – PubMed [WWW Document]; n.d.

Benbrahim-Tallaa L, Waalkes MP. Inorganic Arsenic and human prostate cancer. Environ Health Perspect. 2008;116:158–64. https://doi.org/10.1289/ehp.10423.

Benbrahim-Tallaa L, Webber MM, Waalkes MP. Mechanisms of acquired androgen independence during Arsenic-induced malignant transformation of human prostate epithelial cells. Environ Health Perspect. 2007;115:243–7. https://doi.org/10.1289/ehp.9630.

Bode AM, Dong Z. The paradox of Arsenic: molecular mechanisms of cell transformation and chemotherapeutic effects. Crit Rev Oncol Hematol. 2002; https://doi.org/10.1016/S1040-8428(01)00215-3.

Buscaglia LEB, Li Y. Apoptosis and the target genes of microRNA-21. Chin J Cancer. 2011;30:371–80. https://doi.org/10.5732/cjc.30.0371.

Bustos MA, Ono S, Marzese DM, Oyama T, Iida Y, Cheung G, Nelson N, Hsu SC, Yu Q, Hoon DSB. MiR-200a regulates CDK4/6 inhibitor effect by targeting CDK6 in metastatic melanoma. J Invest Dermatol. 2017;137:1955–64. https://doi.org/10.1016/j.jid.2017.03.039.

Cardoso APF, Al-Eryani L, Christopher States J. Arsenic-induced carcinogenesis: the impact of miRNA dysregulation. Toxicol Sci. 2018;165:284–90. https://doi.org/10.1093/toxsci/kfy128.

Chen F, Shi X. Signaling from toxic metals to NF-κB and beyond: not just a matter of reactive oxygen species. Environ Health Perspect. 2002; https://doi.org/10.1289/ehp.02110s5807.

Chen W, Martindale JL, Holbrook NJ, Liu Y. Tumor promoter Arsenite activates extracellular signal-regulated kinase through a signaling pathway mediated by epidermal growth factor receptor and Shc. Mol Cell Biol. 1998; https://doi.org/10.1128/mcb.18.9.5178.

Chen F, Vallyathan V, Castranova V, Shi X. Cell apoptosis induced by carcinogenic metals. Mol Cell Biochem. 2001;222:183–8. https://doi.org/10.1023/A:1017970330982.

Chen J-W, Chen H-Y, Li W-F, Liou S-H, Chen C-J, Wu J-H, Wang S-L. The association between total urinary Arsenic concentration and renal dysfunction in a community-based population from central Taiwan. Chemosphere. 2011;84:17–24. https://doi.org/10.1016/j.chemosphere.2011.02.091.

Chen C, Yang Q, Wang D, Luo F, Liu X, Xue J, Yang P, Xu H, Lu J, Zhang A, Liu Q. MicroRNA-191, regulated by HIF-2α, is involved in EMT and acquisition of a stem cell-like phenotype in arsenite-transformed human liver epithelial cells. Toxicol in Vitro. 2018;48:128–36. https://doi.org/10.1016/j.tiv.2017.12.016.

Chirshev E, Oberg KC, Ioffe YJ, Unternaehrer JJ. Let – 7 as biomarker, prognostic indicator, and therapy for precision medicine in cancer. Clin Transl Med. 2019;8 https://doi.org/10.1186/s40169-019-0240-y.

Chouchane S, Snow ET. In vitro effect of Arsenical compounds on glutathione-related enzymes. Chem Res Toxicol. 2001;14:517–22. https://doi.org/10.1021/tx000123x.

Cooke MS, Evans MD, Dizdaroglu M, Lunec J. Oxidative DNA damage: mechanisms, mutation, and disease. FASEB J. 2003;17:1195–214. https://doi.org/10.1096/fj.02-0752rev.

Cooper KL, Liu KJ, Hudson LG. Contributions of reactive oxygen species and mitogen-activated protein kinase signaling in arsenite-stimulated hemeoxygenase-1 production. Toxicol Appl Pharmacol. 2007; https://doi.org/10.1016/j.taap.2006.09.020.

Das S, Joardar S, Manna P, Dua TK, Bhattacharjee N, Khanra R, Bhowmick S, Kalita J, Saha A, Ray S, De Feo V, Dewanjee S. Carnosic acid, a natural diterpene, attenuates Arsenic-induced hepatotoxicity via reducing oxidative stress, MAPK activation, and apoptotic cell death pathway. Oxidative Med Cell Longev. 2018; https://doi.org/10.1155/2018/1421438.

Del Razo LM, Quintanilla-Vega B, Brambila-Colombres E, Calderón-Aranda ES, Manno M, Albores A. Stress proteins induced by Arsenic. Toxicol Appl Pharmacol. 2001;177:132–48. https://doi.org/10.1006/taap.2001.9291.

Dela Cruz CS, Tanoue LT, Matthay RA. Lung cancer: epidemiology, etiology, and prevention. Clin Chest Med. 2011;32:605–44. https://doi.org/10.1016/j.ccm.2011.09.001.

Ellinsworth DC. Arsenic, reactive oxygen, and endothelial dysfunction. J Pharmacol Exp Ther. 2015; https://doi.org/10.1124/jpet.115.223289.

Faita F, Cori L, Bianchi F, Andreassi MG. Arsenic-induced genotoxicity and genetic susceptibility to Arsenic-related pathologies. Int J Environ Res Public Health. 2013;10:1527–46. https://doi.org/10.3390/ijerph10041527.

Flora SJS. Arsenic-induced oxidative stress and its reversibility following combined administration of N-acetylcysteine and meso 2,3-dimercaptosuccinic acid in rats. Clin Exp Pharmacol Physiol. 1999;26:865–9. https://doi.org/10.1046/j.1440-1681.1999.03157.x.

Förstermann U, Münzel T. Endothelial nitric oxide synthase in vascular disease: from marvel to menace. Circulation. 2006; https://doi.org/10.1161/CIRCULATIONAHA.105.602532.

Fowler BA, Selene C-H, Chou R, Jones J, Dexter L, Sullivan W Jr, Chen C-J. Chapter 28 – Arsenic. In: Nordberg GF, Fowler BA, Nordberg M, editors. Handbook on the toxicology of metals. 4th ed. San Diego: Academic; 2015. p. 581–624. https://doi.org/10.1016/B978-0-444-594 53-2.00028-7.

Frankenberg E. 基因的改变NIH public access. Bone. 2012;23:1–7. https://doi.org/10.1007/978-1-59745-416-2.

Gonzalez H, Lema C, Kirken R, Maldonado A, Varela-Ramirez A, Aguilera J. Arsenic-exposed keratinocytes exhibit differential microRNAs expression profile; potential implication of miR-21, miR-200a and miR-141 in melanoma pathway. Clin Cancer Drugs. 2015;2:138–47. https://doi.org/10.2174/2212697x02666150629174704.

Gozin A, Franzini E, Andrieu V, Da Costa L, Rollet-Labelle E, Pasquier C. Reactive oxygen species activate focal adhesion kinase, paxillin and P130CAS tyrosine phosphorylation in endothelial cells. Free Radic Biol Med. 1998; https://doi.org/10.1016/S0891-5849(98)00134-8.

Graves P, Zeng Y. Biogenesis of mammalian microRNAs: a global view. genomics. Proteomics Bioinf. 2012;10:239–45. https://doi.org/10.1016/j.gpb.2012.06.004.

Green DR, Kroemer G. Cytoplasmic functions of the tumour suppressor p53. Nature. 2009; https://doi.org/10.1038/nature07986.

Hegde ML, Hazra TK, Mitra S. Early steps in the DNA base excision/single-strand interruption repair pathway in mammalian cells. Cell Res. 2008;18:27–47. https://doi.org/10.1038/cr.2008.8.

Hei TK, Liu SUX, Waldren C. Mutagenicity of Arsenic in mammalian cells: role of reactive oxygen species. Proc Natl Acad Sci U S A. 1998; https://doi.org/10.1073/pnas.95.14.8103.

Heikens A. Arsenic contamination of irrigation water, soil and crops in Bangladesh: risk implications for sustainable agriculture and food safety in Asia. FAO – RAP Publications. 2006/20 20, 2; 2006.

Hinhumpatch P, Navasumrit P, Chaisatra K, Promvijit J, Mahidol C, Ruchirawat M. Oxidative DNA damage and repair in children exposed to low levels of Arsenic in utero and during early childhood: application of salivary and urinary biomarkers. Toxicol Appl Pharmacol. 2013;273 https://doi.org/10.1016/j.taap.2013.10.002.

Huang HW, Lee CH, Yu HS. Arsenic-induced carcinogenesis and immune dysregulation. Int J Environ Res Public Health. 2019:16. https://doi.org/10.3390/ijerph16152746.

Hug SJ, Leupin O. Iron-catalyzed oxidation of Arsenic(III) by oxygen and by hydrogen peroxide: pH-dependent formation of oxidants in the Fenton reaction. Environ Sci Technol. 2003;37:2734–42. https://doi.org/10.1021/es026208x.

Humphries B, Yang C. The microRNA-200 family: small molecules with novel roles in cancer development, progression and therapy. Oncotarget. 2015;6:6472–98. https://doi.org/10.18632/oncotarget.3052.

Kato K, Hayashi H, Hasegawa A, Yamanaka K, Okada S. DNA damage induced in cultured human alveolar (L-132) cells by exposure to dimethylarsinic acid. Environ Health Perspect. 1994; https://doi.org/10.1289/ehp.94102s3285.

Khairul I, Wang QQ, Jiang YH, Wang C, Naranmandura H. Metabolism, toxicity and anticancer activities of Arsenic compounds. Oncotarget. 2017;8:23905–26. https://doi.org/10.18632/oncotarget.14733.

Krammer PH. CD95(APO-1/Fas)-mediated apoptosis: live and let die. Adv Immunol. 1998; https://doi.org/10.1016/s0065-2776(08)60402-2.

Ledgerwood LG, Kumar D, Eterovic AK, Wick J, Chen K, Zhao H, Tazi L, Manna P, Kerley S, Joshi R, Wang L, Chiosea SI, Garnett JD, Tsue TT, Chien J, Mills GB, Grandis JR, Thomas SM. The degree of intratumor mutational heterogeneity varies by primary tumor sub-site. Oncotarget. 2016;7:27185–98. https://doi.org/10.18632/oncotarget.8448.

Lee TC, Tanaka N, Lamb PW, Gilmer TM, Barrett JC. Induction of gene amplification by Arsenic. Science. 1988;80 https://doi.org/10.1126/science.3388020.

Liao W-T, Lu J-H, Lee C-H, Lan C-CE, Chang J-G, Chai C-Y, Yu H-S. An interaction between Arsenic-induced epigenetic modification and inflammatory promotion in a skin equivalent during Arsenic carcinogenesis. J Invest Dermatol. 2017;137:187–96. https://doi.org/10.1016/j.jid.2016.08.017.

Liu J, Waalkes MP. Liver is a target of Arsenic carcinogenesis. Toxicol Sci. 2008;105:24–32. https://doi.org/10.1093/toxsci/kfn120.

Liu SX, Athar M, Lippai I, Waldren C, Hei TK. Induction of oxyradicals by Arsenic: implication for mechanism of genotoxicity. Proc Natl Acad Sci U S A. 2001; https://doi.org/10.1073/pnas.98.4.1643.

Ludwig S, Hoffmeyer A, Goebeler M, Kilian K, Häfner H, Neufeld B, Han J, Rapp UR. The stress inducer arsenite activates mitogen-activated protein kinases extracellular signal-regulated kinases 1 and 2 via a MAPK kinase 6/p38- dependent pathway. J Biol Chem. 1998;273:1917–22. https://doi.org/10.1074/jbc.273.4.1917.

Martindale JL, Holbrook NJ. Cellular response to oxidative stress: signaling for suicide and survival. J Cell Physiol. 2002; https://doi.org/10.1002/jcp.10119.

Martinez VD, Becker-Santos DD, Vucic EA, Lam S, Lam WL. Induction of human squamous cell-type carcinomas by Arsenic. J Skin Cancer. 2011a;2011:454157. https://doi.org/10.1155/2011/454157.

Martinez VD, Vucic EA, Adonis M, Gil L, Lam WL. Arsenic biotransformation as a cancer promoting factor by inducing DNA damage and disruption of repair mechanisms. Mol Biol Int. 2011b;2011:1–11. https://doi.org/10.4061/2011/718974.

Martinez VD, Vucic EA, Becker-Santos DD, Gil L, Lam WL. Arsenic exposure and the induction of human cancers. J Toxicol. 2011c;2011 https://doi.org/10.1155/2011/431287.

Masuda T, Ishii K, Morishita Y, Iwasaki N, Shibata Y, Tamaoka A. Hepatic histopathological changes and dysfunction in primates following exposure to organic Arsenic diphenylarsinic acid. J Toxicol Sci. 2018; https://doi.org/10.2131/jts.43.291.

Menzel DB, Rasmussen RE, Lee E, Meacher DM, Said B, Hamadeh H, Vargas M, Greene H, Roth RN. Human lymphocyte home oxygenase 1 as a response biomarker to inorganic Arsenic. Biochem Biophys Res Commun. 1998; https://doi.org/10.1006/bbrc.1998.9363.

Moore LE, Smith AH, Eng C, Kalman D, DeVries S, Bhargava V, Chew K, Moore D, Ferreccio C, Rey OA, Waldman FM. Arsenic-related chromosomal alterations in bladder cancer. J Natl Cancer Inst. 2002;94:1688–96. https://doi.org/10.1093/jnci/94.22.1688.

Muenyi CS, Ljungman M, States JC. Arsenic disruption of DNA damage responses-potential role in carcinogenesis and chemotherapy. Biomolecules. 2015;5:2184–93. https://doi.org/10.3390/biom5042184.

Naranmandura H, Xu S, Koike S, Pan LQ, Chen B, Wang YW, Rehman K, Wu B, Chen Z, Suzuki N. The endoplasmic reticulum is a target organelle for trivalent dimethylarsinic acid (DMA III)-induced cytotoxicity. Toxicol Appl Pharmacol. 2012;260:241–9. https://doi.org/10.1016/j.taap.2012.02.017.

Nordenson I, Beckman L. Is the genotoxic effect of Arsenic mediated by oxygen free radicals? Hum Hered. 1991;41:71–3. https://doi.org/10.1159/000153979.

O'Brien J, Hayder H, Zayed Y, Peng C. Overview of microRNA biogenesis, mechanisms of actions, and circulation. Front Endocrinol (Lausanne). 2018;9:402. https://doi.org/10.3389/fendo.2018.00402.

Oketani M, Kohara K, Tuvdendorj D, Ishitsuka K, Komorizono Y, Ishibashi K, Arima T. Inhibition by Arsenic trioxide of human hepatoma cell growth. Cancer Lett. 2002; https://doi.org/10.1016/S0304-3835(01)00800-X.

Page C. Arsenic toxicity. Encycl Met. 2013;143–143 https://doi.org/10.1007/978-1-4614-1533-6_100120.

Porter AC, Fanger GR, Vaillancourt RR. Signal transduction pathways regulated by arsenate and arsenite. Oncogene. 1999; https://doi.org/10.1038/sj.onc.1203214.

Rahman M, Vahter M, Wahed MA, Sohel N, Yunus M, Streatfield PK, El Arifeen S, Bhuiya A, Zaman K, Chowdhury AMR, Ekström EC, Persson LÅ. Prevalence of Arsenic exposure and skin lesions. A population based survey in Matlab, Bangladesh. J Epidemiol Community Health. 2006;60:242–8. https://doi.org/10.1136/jech.2005.040212.

Rahman MS, Kumar A, Kumar R, Ali M, Ghosh AK, Singh SK. Comparative quantification study of Arsenic in the groundwater and biological samples of Simri Village of Buxar District, Bihar, India. Indian J Occup Environ Med. 2019;23:126–32. https://doi.org/10.4103/ijoem.IJOEM_240_18.

Reichard JF, Puga A. Effects of Arsenic exposure on DNA methylation and epigenetic gene regulation. Epigenomics. 2010;2:87–104. https://doi.org/10.2217/epi.09.45.

Ren X, Gaile DP, Gong Z, Qiu W, Ge Y, Huang C, Yan H, Olson JR, Kavanagh TJ, Wu H, Professions H, Professions H, Sciences OH. HHS Public Access. 2016;283:198–209. https://doi.org/10.1016/j.taap.2015.01.014.Arsenic.

Riley T, Sontag E, Chen P, Levine A. Transcriptional control of human p53-regulated genes. Nat Rev Mol Cell Biol. 2008; https://doi.org/10.1038/nrm2395.

Rousseau MC, Straif K, Siemiatycki J. IARC carcinogen update [1]. Environ Health Perspect. 2005;113:580–3. https://doi.org/10.1289/ehp.113-a580.

Sablina AA, Budanov AV, Ilyinskaya GV, Agapova LS, Kravchenko JE, Chumakov PM. The antioxidant function of the p53 tumor suppressor. Nat Med. 2005; https://doi.org/10.1038/nm1320.

Shahid M, Dumat C, Niazi N, Khalid S, Natasha N. Global scale Arsenic pollution: increase the scientific knowledge to reduce human exposure. VertigO. 2018; https://doi.org/10.4000/vertigo.21331.

Tanriover B. Renal cell cancer, environmental Arsenic exposure and carcinogenic mutations. UHOD Uluslararasi Hematol Derg. 2012;22:62–6. https://doi.org/10.4999/uhod.11079.

Tchounwou P, Patlolla A, Centeno J. Carcinogenic and Systemic health effects associated with Arsenic exposure – a critical review. Toxicol Pathol. 2003;31:575–88. https://doi.org/10.1080/714044691.

Vahter M. Mechanisms of Arsenic biotransformation. Toxicology. 2002;181–182:211–7. https://doi.org/10.1016/s0300-483x(02)00285-8.

Vahter M, Concha G. Role of metabolism in Arsenic toxicity. Pharmacol Toxicol. 2001;89:1–5. https://doi.org/10.1034/j.1600-0773.2001.d01-128.x.

Vélez-Cruz R, Johnson DG. The retinoblastoma (RB) tumor suppressor: pushing back against genome instability on multiple fronts. Int J Mol Sci. 2017;18 https://doi.org/10.3390/ijms18081776.

Wu W, Jaspers I, Zhang W, Graves LM, Samet JM. Role of Ras in metal-induced EGF receptor signaling and NF-κB activation in human airway epithelial cells. Am J Phys Lung Cell Mol Phys. 2002:282. https://doi.org/10.1152/ajplung.00390.2001.

Yamamura S. Drinking water guidelines and standards. United Nations Synth Rep Arsen Drink Water. 2001:18.

Yamanaka K, Okada S. Induction of lung-specific DNA damage by metabolically methylated Arsenics via the production of free radicals. Environ Health Perspect. 1994; https://doi.org/10.1289/ehp.94102s337.

Yamanaka K, Hayashi H, Tachikawa M, Kato K, Hasegawa A, Oku N, Okada S. Metabolic methylation is a possible genotoxicity-enhancing process of inorganic Arsenics. Mutat Res Genet Toxicol Environ Mutagen. 1997;394:95–101. https://doi.org/10.1016/S1383-5718(97)00130-7.

Yamauchi H, Aminaka Y, Yoshida K, Sun G, Pi J, Waalkes M. Evaluation of DNA damage in patients with Arsenic poisoning: urinary 8-hydroxydeoxyguanine. Toxicol Appl Pharmacol. 2004;198:291–6. https://doi.org/10.1016/j.taap.2003.10.021.

Yeh S, How SW, Lin CS. Arsenical cancer of skin. Histologic study with special reference to Bowen's disease. Cancer. 1968;21:312–39. https://doi.org/10.1002/1097-0142(196802)21:2<312::aid-cncr2820210222>3.0.co;2-k.

Yu H-S, Liao W-T, Chai C-Y. Arsenic carcinogenesis in the skin. J Biomed Sci. 2006;13:657–66. https://doi.org/10.1007/s11373-006-9092-8.

Zeng K, Zhong B, Fang M, Shen X-L, Huang L-N. Common polymorphisms of the hOGG1, APE1 and XRCC1 genes correlate with the susceptibility and clinicopathological features of primary angle-closure glaucoma. Biosci Rep. 2017;37 https://doi.org/10.1042/BSR20160644.

Zhang T, Yang Z, Kusumanchi P, Han S, Liangpunsakul S. Critical role of microRNA-21 in the pathogenesis of liver diseases. Front Med. 2020;7:1–7. https://doi.org/10.3389/fmed.2020.00007.

Role of Mitochondrial Oxidative Stress in Pathophysiology of Lung Cancer

Archana Sharma, Almaz Zaki, Gulnaz Tabassum, Salman Khan, Mohd Mohsin, and Syed Mansoor Ali

Contents

Abstract Mitochondria are key role players in metabolism, bioenergetics, biosynthesis, and cell survival or death functions. Reactive oxygen species (ROS) function as key mediators of cellular signaling pathways involved in proliferation, survival, apoptosis, and immune response. Mitochondria are critically involved in ROS-dependent lung disorders like lung cancer. Mitochondrial metabolism is a crucial event for tumor survival, proliferation, and metastasis. Besides the mitochondrial metabolic changes, mitochondrial dynamics such as fission and fusion events are also affected in cancer cells. ROS are the key molecules regulating the molecular signaling alterations which lead to lung cancer progression. These mitochondrial events leading to ROS-dependent oxidative stress are in turn also regulated by various microRNAs and vice versa.

A. Sharma · A. Zaki · G. Tabassum · S. Khan · M. Mohsin · S. Mansoor Ali (✉)
Department of Biotechnology, Faculty of Natural Sciences, Jamia Millia Islamia,
New Delhi, India
e-mail: smansoor@jmi.ac.in

© The Author(s), under exclusive license to Springer Nature Switzerland AG 2021   171
K. K. Kesari, N. K. Jha (eds.), *Free Radical Biology and Environmental Toxicity*,
Molecular and Integrative Toxicology, https://doi.org/10.1007/978-3-030-83446-3_8

Keywords Reactive oxygen species · Mitochondria · Apoptosis · Oxidative stress · Lung cancer

1   Introduction

Mitochondria are double membrane bound organelle present in nearly all eukaryotic cells and are essential for their survival. For proper functioning, cell needs energy in the form of ATP which is provided by mitochondria, because of this mitochondria is also known as "power house" of the cell (Liu and Chen 2017).

Depending upon the cell, tissue types, and their requirement, its number varies from one to several thousands. Apart from producing energy, it is involved in various other functions such as production of various intermediate for macromolecule synthesis, maintaining calcium homeostasis, production of reactive oxygen species (ROS), regulation of various cell signaling pathways, and regulation of cell death mechanism by apoptosis (Sreedhar and Zhao 2018).

It comprises several compartments including an inner and outer membrane separated by intermembrane space, the matrix, and the cristae. Each compartment performs different function.

The inner mitochondrial membrane is the site where all the five complexes of electron transport chain (ETC) are located and production of ATP takes place. The outer mitochondrial membrane protects the organelle and involved in regulation of immune response through activation of mitochondrial antiviral signaling proteins whenever virus infection takes place (Mohanty et al. 2019).

The exchange of small molecules, ions, and protein and important signaling event take place because of mitochondrial porins or voltage-dependent anion selective channels (VDAC) which is present on OMM. Both glycolysis and mitochondrial respiration are regulated by VDAC.

Lung cancer of respiratory epithelium origin can be classified into two major histologic groups, i.e., small cell lung cancer (SCLC) and non-small cell lung cancer (NSCLC). Small cell lung cancer that accounts for 15% of lung cancer cases is known to be derived from cells exhibiting neuroendocrine features. NSCLC is a major type of lung carcinoma and found among 85% lung cancer cases, and it is further grouped into three major histologic subcategories: adenocarcinoma (AD), squamous cell carcinoma (SCC), and large cell carcinoma (Herbst et al. 2008; Dela Cruz et al. 2011).

2   Reactive Oxygen Species

The term "reactive oxygen species" includes the most compendious molecules originating from oxygen that have taken extra electrons and capable of oxidizing other subsidiaries molecules (Martínez-Cayuela 1995). Single electron reduction of oxygen produces superoxide that subsequently combine with another superoxide radical by the superoxide dismutase enzyme to form hydrogen peroxide, non-radical but highly reactive molecule and releasing one water molecule. Hydrogen peroxide can further gain the electron from Fe^{2+} Fenton reaction to produce hydroxyl radical. These three forms of ROS have adverse effect on cellular physiology.

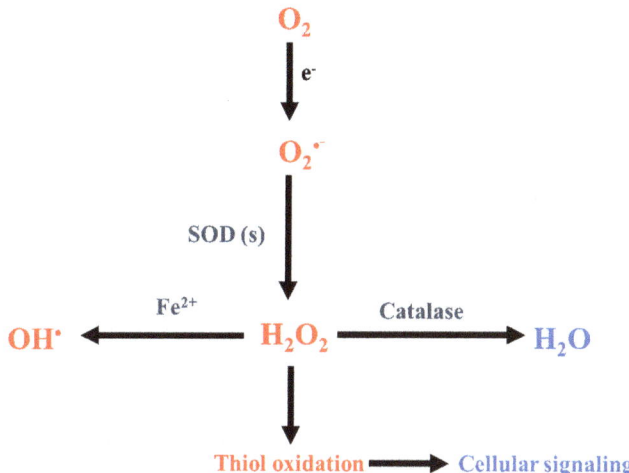

3   Mitochondria as Source of Reactive Oxygen Species in Lung Cancer

In comparison to ROS production, mitochondria are the largest contributor. Emerging evidences indicate that cancer is basically a mitochondrial metabolic disease. It contains eight known superoxide producing sites consuming almost 1% mitochondrial oxygen only to generate superoxide radical. Relative contribution of each site to total cellular ROS is ambiguous. Complexes I, II, and III are well elucidating sites contributing ROS production while other sites contribution in ROS production and signaling is unknown (Quinlan et al. 2012; Handy and Loscalzo 2012; Murphy 2009). Another major source of ROS is NADPH oxidases. Initially, these enzymes were described in phagocytosis where they create localized oxidative stress to kill engulfed pathogens (Babior 1999). Later they were discovered in different tissues with adverse non-immune functions positioned to different organelles like plasma membrane, nucleus, mitochondria, and endoplasmic reticulum (Wang et al. 2014a).

Chronic inflammation and oxidative stress are major factors responsible for rise of lung cancer (Nam et al. 2017). Mitochondria, an important subcellular organelle, is responsible for ATP synthesis and maintaining cellular homeostasis through mechanisms like signaling through ROS, apoptosis regulation, calcium signaling, cellular differentiation and growth, and cell cycle control. Dysfunction in mitochondrial complex and dynamic function has been identified as a potent cause of NSCLC (Archer 2013).

Cancer cells are more proliferative and required more ATP for their energy need. The free energy formed during electron transport is used for ATP generation. Cancer cells used lactate as a fuel source than glucose and generate more ROS, which led to metabolism change from oxidative metabolism to glycolytic metabolism in the fibroblast cancer cells (Ghosh et al. 2020). The properties of cancer cells are alarming due to elevated ROS levels in them. In normal cells, ROS generation is administered and directed by the transcription factor NRF2 and its suppressor protein KEAP1, which is an inducible antioxidant pathway that can activate upon cellular stresses. There are certain genes such as MYC, Braf, and KRAS which can restrain the ROS generation. Compounds that are organic or inorganic and having one or more unpaired electrons are known as free radicals, such as the superoxide radical ($O_2^{-\cdot}$) and hydroxyl radical (.OH), which are both produced endogenously by chemical reactions. The $O_2^{-\cdot}$ radical can bind with NO resulting in peroxynitrite molecule (ONOO–) which is quickly converted to hydrogen peroxidase (H_2O_2) by superoxide dismutase. Hydroxyl radicals can also generate by hydrogen peroxide and peroxyl radicals in enzymatic pathways of xanthine oxidase, NADPH oxidase, and oxidoreductases (Domej et al. 2014). Hydrogen peroxide can convert into hypochlorous acid and hypobromous acid with the help of heme peroxidases or myeloperoxidase in the presence of chloride ions and bromide ions respectively. Exogenous ROS sources are environmental gases NO_2, SO_2, CO, –CHO, cigarette smoke and airborne particulate matters which may cause inflammatory responses (Ritz et al. 2019).

3.1   Mitochondrial Metabolic Cycles as Source of ROS

Metabolic alterations in mitochondria have been observed in cancer cells of NSCLC. Cancer cell shows Warburg effect, i.e., excessive glucose consumption and excretion of lactic acid in the presence of oxygen (aerobic glycolysis) is high in comparison to normal cell as well as inhibition of mitochondrial respiration is observed. Although oxygen is available, unlike normal cells, lung cancer cells prefer glycolysis for the generation of ATP (Vanhove et al. 2019). Even though the outcome of ATP from this process is much lower than oxidative phosphorylation, it facilitates the increased cellular growth rates and proliferation (Roberts and Thomas 2013; Plas and Thompson 2005).

The electron transport chain complexes I, II, and III are involved in the production of ROS during OXPHOS and oxidative stress. This occurs due to loss of homeostatic balance between generation of ROS and antioxidant action. Elevated radical generation or lower antioxidant action leads to ROS production which further affects cell integrity, and results in damage of macromolecules. The complex I subunit plays a significantly important role in tumor progression and metastasis. Mutation in complex I lead to its loss of interaction with complex III in supercomplexes, which causes defect in electron transfer thus cells become unable to identify ROS production. This irregularity causes excessive production of ROS, escalated oxidative stress, and heavy energy loss (Solaini et al. 1807). Complex IV is made up of 13–14 subunits which comprise developmentally regulated and tissue-specific isoforms of OXPHOS complex. Reduction of complex IV function has been observed in lung cancer. Complex V is also called as ATP synthase because of its ability to catalyze ATP synthesis. An over expression of ATP synthase subunit d (ATP5D) has been observed in cancerous tissues than in normal tissues (Kalainayakan et al. 2018). Complex I is seen to be altered in case of lung cancer, specifically in patients that have never smoked.

Mitochondria organelle is one of the most intracellular ROS producing organelle. Complexes I, II, and III of electron transport chain (ETC) are responsible for generating ROS. It was reported that 2% of the consumed oxygen by the mitochondria is entangled to generate ROS (Chance et al. 1979). H_2O_2 is found only when functions of complex I and complex III of ETC are inhibited by rotenone and antimycin A. During electron transport, the flavin groups, quinones, heme moieties, and metalloproteins help in the movement of electrons where oxygen serves as terminal electron acceptor. Free radicals or superoxide are generated by when an electron's movement is stuck. Electrons run in a step-by-step manner, a short delay of electrons at any site of ETC, upsurge the electron on the previous site which produces a chance to meet electron with oxygen to form superoxide and production of ROS. Ubisemiquinone is a free radical formed after the transport of the first electron to complex III, if lag in the discharge of the second electron, the first electron might be trapped by oxygen to form superoxide (Guzy and Schumacker 2006; Sabharwal and Schumacker 2014). The T8993G mutation in mtDNA reduced the activity of complex V and produced more ROS, and these mutations also contribute to tumorgenicity. The enzymes include pyruvate dehydrogenase, alpha-ketoglutarate, succinate dehydrogenase, and NADH dehydrogenase, which are involved in the TCA cycle using flavin-containing prosthetic groups such as flavin adenine dinucleotide (FAD) and flavin mononucleotide (FMN) that can generate ROS (Quinlan et al. 2014). Metal containing enzymes, such as mitochondrial aconitase, also potentially generate ROS (Gardner 1997). The strong electric field generated by the movement of electrons causes the translocation of superoxide to the cytosol. ROS reside in the matrix scavenged by antioxidant but overproduction of these ROS comes into intermembrane space and cytosol and stimulate cytochrome c expression, which mediates the apoptosis (Sabharwal and Schumacker 2014). ROS also cause mtDNA damage, unnecessarily bind with intracellular or surface receptors, and disturb signaling pathway of cancer cells. Mitochondrial ROS is controlled by oxidative stress, mitochondrial membrane potential, and the metabolic state of mitochondria.

3.2 Hypoxia and ROS in Cancer

An imbalance between glycolysis and oxidative phosphorylation leads to the accumulation of mitochondrial metabolites in the cytosol and hence influence metabolism. Hypoxia or low oxygen level is one of the side effects of the above imbalance.

Uncontrolled cell proliferation in tumor, which later outgrows its blood supply, results in lowered availability of oxygen in the cells and thus producing a hypoxic state. The hypoxic conditions promote the genetic instability which results into the cancer progression.

Transcription factors known as hypoxia inducible factors (HIF) mediates the adaptive response during hypoxic conditions in cells leading to loss in inner mitochondrial membrane potential and release of more ROS. Hypoxia elevates the RAS and c-MYC gene expression due to HIF-2alpha. The activation of NADPH oxidase stimulates the superoxide in hypoxia condition (Pourahmad et al. 2016). Complex III of ETC is responsible to generate ROS in hypoxia conditions. Increased level of ROS can oxidize lipids, proteins, and DNA which causes tissue injury and inflammatory responses (Kirkham and Rahman 2006).

3.3   Oncogenes as Source of ROS

Oncogenes such as Bcr/Abl, RAS, c-MYC, and FTL3 increase the ROS generation in cancer cells, due to changing the pro-oxidant and anti-oxidant regulation. Oxidative stress due to exogenous ROS in cancer cell is fatal when compared to the normal cells (Suzuki-Karasaki et al. 2014). The c-MYC expression is higher and increased the ROS level and fluctuate the metabolic rate, disturb the scavenging process, and activate the intracellular oxidases. It also stimulates apoptosis. Caveolin 1 is the part of the membrane caveolae, ROS production in the cancer-associated fibroblast is affected by mitochondrial damage, and dysfunctional mitochondria is stimulated by loss of caveolin 1, which also assists in genomic instability in nearby cells (Martinez-Outschoorn et al. 2010). Paradoxically, arsenic trioxide (ATO) is used in blood cancer treatment to initiate cell death by leakage of electrons through the ETC which generate ROS.

4   ROS-Induced Mitochondrial DNA (mtDNA) Mutations in Cancer

The free radical generated during OXPHOS affects mitochondria and induces oxidative stress, consequently leading to alteration of the structure of respiratory chain, leakage of proton mitochondrial uncoupling, and damage to mitochondrial DNA (Murphy et al. 2011). Lack of protective histones protein, intron, and proper DNA repair mechanism system makes mitochondrial DNA more prone to mutation than nuclear genomic DNA. Tumor cells are deficient in antioxidant defense mechanism (Roberts and Thomas 2013). Hence, continuous oxidative stress results in more DNA damage in the cells and subsequent progression and metastasis of cancer. Mutation of mitochondrial DNA and mtDNA content has positive association with the risk of cancer in lungs (Bonner et al. 2009; Hosgood et al. 2010).

Cancerous cells consume higher amount of ATP to support their growth and proliferation. To fulfill this need, free energy conserved in the processes, such as oxidative phosphorylation (OXPHOS) of lipids, amino acids, and glucose and by transferring electron to electron transport system (ETS), is harnessed to generate ATP. These systems work sequentially and therefore a delay at one location causes electron traffic at earlier sites. This delay provides the chances of electron to escape generating reactive oxygen species (ROS) such as super oxide, hydrogen peroxide, and hydroxyl radical. Mitochondria generate ROS for cellular signaling but excess ROS causes exhaustion of antioxidants further adding damage to lipids, proteins, and DNA by oxidation (Murphy et al. 2011). Previous studies revealed diverse role of mitochondria in regulating inflammation, and innate and adaptive immune responses. However, in account of cancer, uncontrolled chronic inflammation and mutation or damage to mitochondrial DNA end up with tumorigenesis (Brandon et al. 2006; Ishikawa et al. 2008; Neagu et al. 2019).

The tricarboxylic acid cycle (TCA cycle) is summed up by a sequence of enzymes located in the mitochondrial matrix playing a key role in removal of electrons from intermediary metabolites and transferring them to the ETC. The abnormal accumulation of TCA cycle intermediates as well as mutations in genes encoding TCA cycle enzymes can promote carcinogenesis (Raimundo et al. 2011). The TCA cycle has eight steps that are catalyzed by eight different enzymes. The key enzymes that were altered in cancer cells are succinate dehydrogenase (SDH), fumarate hydratase (FH), isocitrate dehydrogenase (IDH), and citrate synthase (CS). Mitochondrial abnormalities induce metabolic reprogramming within cells that are highly dependent on glycolysis, which further facilitates the tumorigenic process (Zong et al. 2016; Ahn and Metallo 2015). Inactivation of any site in the TCA cycle can augment ROS generation, and further ROS can inhibit aconitase, an important enzyme involved in the TCA cycle. Lactate dehydrogenase, pyruvate kinase, phosphofructokinase, and hexokinase are crucial enzymes which are reported with an increased expression in case of lung cancer (Chang et al. 2020).

ROS are well-known genotoxic entities and mitochondrion contains hundreds of identical copies of mtDNA (mitochondrial DNA) which are 10 times more susceptible to mutation than the nuclear DNA due to lack of protective histones. MtDNA can be served as a marker in lung cancer prognosis, stage, and severity (Liu et al. 2017). Mitochondrial complex I was among the most altered protein complex due to mtDNA mutation in non-smoke lung cancer patients (Dasgupta et al. 2012). Pro-inflammatory microenvironment disrupts the mitochondrial homeostasis by generating wide range of mitochondrial damage associated molecular patterns (mitoDAMPs) through mtDNA, mtROS, cardiolipin, ATP, calreticulin, and *N*-formyl peptide (NFP) fMet-Leu-Phy. Among them, ROS produced by mitochondria have detrimental effect on cellular pathophysiology which has been observed as a hallmark for many cancers (Kopecka et al. 2020).

5   Molecular Signaling Regulated by ROS and Affected Pathways in Lung Cancer Cells

Exogenous carcinogens such as arsenic, cadmium, asbestos, beryllium, chromium, nickel, diesel, silica, radon, vanadium, and cigarette smoke are directly connected with ROS production, inflammation, and carcinogenesis. Transcription factors of inflammatory pathways such as AP-1, NF-κB, PI3K (phosphoinositide kinase), and hypoxia inducible factors (HIF) can be activated by oxidants directly involved in carcinogenesis. Figure depicts the general signaling scheme of some key molecular pathways activated in response to inhaled toxic agents resulting in ROS generation leading to chronic inflammation and lung cancer (Azad et al. 2008). AP-1 and NF-kB are two prominent transcription factors that are sensitive to ROS. It is reported that ROS can activate TNF-alpha that induces NF-kB and API-1 expression to modulate proinflammatory genes (Rahman et al. 2002).

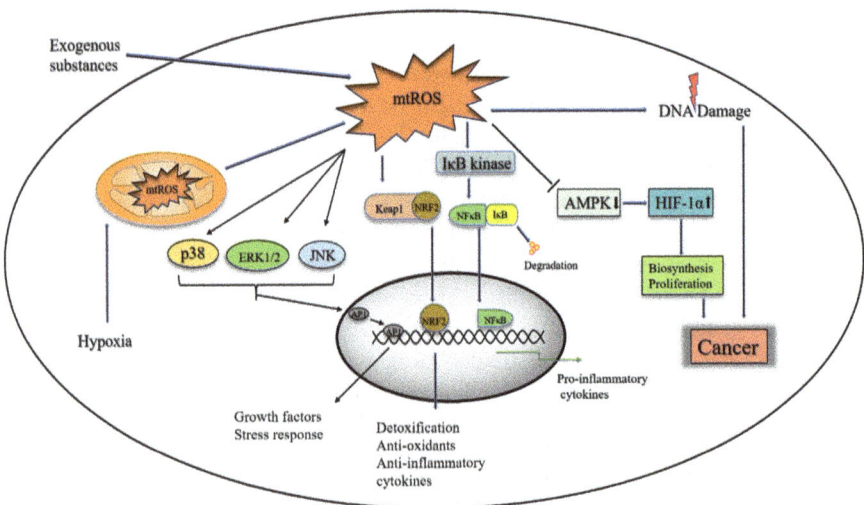

Nuclear factor (erythroid-derived 2)-like 2 (Nrf2) is known as prime regulator of cellular oxidative stress. Under normal circumstances, Nrf2 is degraded by the ubiquitous activity of keap1 ((Kelch-like erythroid cell-derived protein with CNC homology-associated protein 1). Exogenous stimulation, such as oxidants or electrophiles, and mitochondrial ROS (Kasai et al. 2020), cause keap1 modification resulting Nrf2 liberation and translocation to nucleus from the cytosol that initiate expression of broad range of genes involved in detoxification and protection from chemical toxicants (cytoprotection) and free radicals (Ježek et al. 2018). Nrf2-keap1 dysregulation is linked to various tumors, including lung cancer, and its role in cancer chemoprevention and therapy had been observed. Although Nrf2 role in mitochondrial dynamics was reported, its mechanism is poorly understood (Dinkova-Kostova and Abramov 2015). Data released by The Cancer Genome Atlas (TCGA) revealed mutations or deletions in Nrf2, Keap1, and CUL3 occurred in

34% of lung squamous cell carcinomas (Hammerman et al. 2012). Genomic alteration in keap1 leads to stabilization of Nrf2 protein for downstream processing causing cancer progression in unspecified manner (Singh et al. 2020).

Hypoxia or low oxygen level (Warburg effect) can alter mitochondrial metabolism that causes metabolites accumulation inside the cytoplasm that further adds in
genomic instability which drives cancer development and progression (Nelson et al.
2004). Enlarged mitochondria was seen in A549 cell line in hypoxic condition
which are mostly resistant to apoptosis and allow continuous proliferation (Chiche
et al. 2010). Cells during stressful environment, such as hypoxia, stabilize and activate the hypoxia-inducible transcription factor (HIF-α) which initiate a cascade of
transcription of more than 70 hypoxia-related factors to thrive the stress. Stressed
environment causes cellular energy exhaustion or increase in AMP/ATP ratio (Bauer
et al. 2004; El Mjiyad et al. 2011; Ramanathan et al. 2005; Yeung et al. 2008). HIF-α
induced factors help to compensate for low yield by regulating glycolytic conversion. ATP depletion during stress activates 5′ adenosine monophosphate-activated
protein kinase (AMPK) which impedes cell proliferation. But in case of cancerous
cells, AMPK signaling suppression allows tumor cell to proliferate even in nutrient-
depleted environment by suppressing tumor suppressor gene such as p53
(Shackelford and Shaw 2009). The loss of AMPK signaling in cancer cells shifts the
energy production toward glycolysis.

Oxidative stress induced by cigarette smoking, air pollution, and heavy metals is
known to cause lung cancer by generating ROS (Pizzino et al. 2017). Oxidative
stress inducible gene, *CISD2*, is associated with lung cancer. CISD2 protein mainly
localized to mitochondria is known for its role in maintaining mitochondrial integrity and cell proliferation. CISD2 protein is also known for its role in ROS neutralizing efficacy, lung adenocarcinoma development, and increased apoptosis in A549
cell line (Li et al. 2017a). Parkinson's disease gene family was found to be involved
in tumorigenesis where *PINK1* upregulation was strongly found to be associated
with lung cancer occurrence and development with increased ROS production and
dysfunctional mitochondria accumulation (Lu et al. 2020).

Migration is one of the characteristic properties of cancerous cells which can be
regulated by endogenous ROS. Dayoung et al. demonstrated enhanced cell migration in TGF-β treated Nrf2 knockdown A549 cell lines (Ryu et al. 2020). Iron acts
as a cofactor for several mitochondrial enzymes such as catalase and glutathione
peroxidase. A study found intricate role of ferrous ion-dependent hydroxyl radical
production in invasion and migration of lung cancer cells which is directly linked to
tumor progression and metastasis (Luanpitpong et al. 2010).

6   ROS and Apoptosis in Lung Cancer

Apoptosis or programmed cell death tightly regulated by various caspases is one of
the key functions of mitochondria. Defects in complex I lead to alteration of the
apoptosis pathway. Cancer cells may achieve resistance to various apoptotic stimuli

either by increased expression of antiapoptotic proteins and/or functional loss of proapoptotic proteins.

Certain stimulus like excessive ROS generation or hypoxia state causes permeabilization of inner mitochondrial membrane or mitochondrial permeability transition (MPT), hence results in swelling of mitochondria and cell death by apoptotic pathway (Kwong et al. 2007).

Tumorigenesis display defective apoptosis. Sensitivity to apoptosis varies from one type to another type of cancer. For example, small cell lung carcinoma (SCLC) type was more prone to spontaneous apoptosis than NSCLC in case of lung cancer (Joseph et al. 1999).

In case of non-small cell lung cancer, Bcl2 (an anti-apoptotic protein) comparatively shows higher expression in squamous cell carcinoma than adenocarcinoma (Joseph et al. 2000). The function of PTPC is controlled by Bcl-2 family members (Tsujimoto and Shimizu 2000). Bcl-2 through interaction with VDAC controls PTPC and keeps it inactive. Active state of PTPC causes association of Bax with VDAC allowing influx of solutes from matrix such as calcium and other ions, which through depolarization of mitochondrial membrane causes MPT. MOMP subsequently causes release of cytochrome-c, an intermembrane space protein. Cytochrome c release triggers next step of apoptosis, i.e., apoptosome formation. But lung cancer cells display the defect in the formation of apoptosome. Lung cancer and other tumor displays an activated anti-apoptotic protein in early stage of tumor formation while the pro-apoptotic protein is shown to be inhibited in them (Roberts and Thomas 2013).

P53, a tumor suppressor gene, has a key role not only in cell cycle check points, recombination, and DNA repair mechanism but also in regulating the apoptotic pathway. Lung cancer shows defect in P53 pathway. Dysfunctional p53 protein is seen in almost 50% of all lung cancers with 50% of detected mutation of p53 is present in NSCLC (Viktorsson et al. 2005; Hasty and Christy 2013) and > 70% of SCLC. In most of the human cancers, p53 gene is present in an inactive state hence allowing the tumor cell to escape cell senescence and their continuous proliferation. Moreover, p53 upregulated modulator of apoptosis (PUMA), an apoptotic protein plays an important function during p53 dependent apoptosis and cells lacking PUMA are resistance to apoptosis, hence increased carcinogenesis is seen in them and majority of lung cancer cells shows defect in p53 pathway (Yu et al. 2006).

7 ROS and Mitochondrial Dynamics in Lung Cancer

Mitochondrial dynamics involves the process of critical balance between mitochondrial fusion and fission events that regulate the structure, function, and shape of the mitochondrion. For mitochondria to function normally, the regulation of dynamic process is very important. Various proteins play important roles in controlling the fission and fusion process, and among them Mfn1 and Mfn2 (mitofusin) on the outer membrane and OPA1 (optic atropy 1) on the inner membrane maintain the normal level of mitochondrial fusion, and fission is controlled by dynamin related protein (Drp1) (Berman et al. 2008). High expression of Opa 1 is seen in case of lung adenocarcinoma (Fang et al. 2012). High expression of Opa 1 leads to prevention of apoptosis by intrinsic pathway while it does not block the apoptosis mediated by extrinsic pathway (Frezza et al. 2006).

Adenocarcinomic alveolar epithelial cells displays downregulated expression of Drp protein, which further limit mitochondrial fission and also inhibit the downstream processes involved in apoptosis activation (Thomas and Jacobson 2012) as it is known that mitochondria plays an important role in controlling apoptosis by activating caspases. Fission inhibition also alters other mitochondrial function, including excess ROS generation, loss of mtDNA, and depletion of ATP (Parone et al. 2008). Furthermore, in cancerous cells, dysfunctioning of mitochondrial fission directly leads to induction of chromosomal instability and overamplification of centrosome, which further promote cell survival through initiation of the DNA damage response via cell cycle modification (Qian et al. 2012).

8 ROS in Treatment of Lung Cancer

Higher levels of ROS can cause cell senescence, death of cancer cell, and arrest of cell cycle. The findings suggested that ASK1/JNK and ASK1/p38 pathways are involved in ROS-mediated cancer cell death. p38 and JNK pathway mutations are found in various cancers indicating that these alterations can cause cancer cell death. Experimental results showed that p38 and JNK pathway activation through ROS can promote prevention of tumor growth and metastasis. An urethane-induced model of lung carcinoma was used to study the role of macrophage polarization in 6-gingerol (6-G)-associated anti-cancerous effects. The findings revealed that 6-G acts as an arginase inhibitor on tumor supporting macrophages. In the control group, lung carcinogenesis was shown to be positively associated with macrophage (F4/80+) lung interstitial infiltration. After 6-G therapy, lung carcinogenesis was enhanced with increased M1 macrophages and decreased M2 macrophages in the lung interstitial fluid. Increased levels of IL-12 and IFN-γ and decreased levels of TGF-β1 and IL-10 were observed in the alveolar cavity in comparison to control group. LEC-treated mice promoted carcinogenesis-preventing efficacy of 6-G which could be aborted completely in mice treated with pexidartinib (Yao et al. 2018).

Another research shows the connection between antiapoptotic effects of caspase inhibitors and suppression of H_2O_2-induced oxidative stress and GSH depletion. The study was conducted on H_2O_2-treated Calu-6 and A549 lung cancer cells. It was observed that H_2O_2-treated cells showed inhibited growth and induced apoptosis while caspase inhibitors (caspase-3, −8, or −9 inhibitors) averted cell death in H_2O_2-treated lung cancer cells. H_2O_2 increased intracellular ROS levels and catalase activity but triggered GSH deletion. Caspase inhibitors decreased ROS levels but did not affect GSH deletion (Park 2018). The anticancer effect of limonin on A549 cells was assessed by cell viability, development of ROS, and morphological apoptotic changes by staining with AO/EtBr. Expression levels of apoptosis proteins (caspase-3 and caspase-9) were also analyzed. B(a)P-induced mice showed increased lipid peroxidation and inflammatory cytokines levels along with simultaneous decrease in the enzymatic and nonenzymatic antioxidants levels. The results indicated that limoni protects lungs against oxidative damage both in vivo and in vitro (Gong et al. 2019). The influence of combination therapy of hydrogen-rich saline along with the PI3K inhibitor, LY294002, on the proliferation, oxidative stress and apoptosis of NSCLC A549 cells showed enhanced anti-proliferation efficacy and apoptosis. It also showed reduced expression of nuclear factor-κB p65 and heme oxygenase-1 and inhibited phospho-Akt activity.

The results further indicated that reduction in cell proliferation and induced cell apoptosis may be achieved by combination therapy by inhibiting the PI3K pathway and downregulating Akt phosphorylation in NSCLC cell lines (Jiang et al. 2018). Another report revealed the effect of suberoylanilide hydroxamic acid (SAHA); a histone deacetylase (HDAC) inhibitor on apoptosis of lung cancer cells and its association with ROS, glutathione (GSH), and thioredoxin1 (Trx1) levels. Changes in the levels of GSH and ROS in SAHA-treated lung cancer cells have been shown to be partly co-related with cell death (You and Park 2017).

Another study showed that SZC015 exerts its inhibitory effect against H322 cells by suppressing Akt/NF-κB signaling pathway mediated by excessive ROS generation leading to autophagic and apoptotic cell death. It was observed that lung cancer cells treated with SZC015 had decreased cell viability because of induction of autophagy and apoptosis, also H322 cell treated with SZC015 got G0/G1 phase cell cycle arrest. SZC015 decreased levels of total p65, the p-p65, Akt, p-Akt, and p-IκBα in the cytoplasm and nucleus in H322 cells, respectively. High levels of intracellular ROS were observed, which could be prevented by ROS scavenger such as N-acetyl l-cysteine (Sun et al. 2016a).

In another study, it was shown that cholesterol oxidase (COD) from Bordetella species (COD-B) decreases cholesterol content and increases ROS levels, thus contributing to irreversible lung cancer cell apoptosis both in vitro and in vivo. Treatment with COD-B irreversibly inhibited Akt and ERK1/2 phosphorylation while on the other hand it was exacerbated by addition of cholesterol. In addition, COD-B treatment resulted in downregulation of Bcl-2, upregulation of Bax, phosphorylation of JNK and p38, activated caspase-3, and release of cytochrome (Liu et al. 2014). In a study, the anti-growth effect of Zebularine (Zeb), a DNA methyltransferase (DNMT) inhibitor was observed on A549 lung cancer cells along with its relation to ROS levels. It was shown that Zeb leads to inhibition of the growth of A549 lung cancer cells via cell cycle arrest and apoptosis mediated by its influence on ROS and TrxR1 levels. Zeb induced an S phase arrest in A549 cells to inhibit growth as revealed by cell cycle analysis. Zeb also induced ROS levels and glutathione (GSH) depletion in A549 cells. While overexpressed TrxR1 was reported to attenuate cell death and ROS level in Zeb-treated A549 cells, the downregulation of TrxR1 was shown to further intensify ROS levels and cell death in these cells (You and Park 2014).

Sulforaphane (SFN) can promote ROS and prevent EGFR-mediated signaling in NSCLC. Researchers showed that in vivo experimental results suggested that despite resistance to SFN-induced apoptosis of high EGFR levels cells, SFN maintained inhibition of NSCLC tumor growth. This effects on NSCLC by SFN are mainly due to SFN-mediated ROS production (Wang et al. 2019).

Experimental studies on use of levofloxacin for treatment of lung cancer showed that it can be effective against lung cancer cells through inhibition of proliferation and apoptosis induction in in vivo and in vitro models of lung cancer. At molecular level, it prevents ETC complex activity, which leads to mitochondrial respiration inhibition and ATP reduction. Additionally, it is increasing ROS levels, superoxide, and hydrogen peroxide levels of mitochondria (Song et al. 2016).

An alkylating agent named methyl methanesulfonate (MMS) can cause cellular death via necroptosis and apoptosis. A group of researchers established MMS-mediated necroptosis-based lung carcinoma model. Their experimental results confirmed necroptosis induction in A549 cells through MMS, possibly via PIG-3/ROS axis (Jiang et al. 2016).

An experimental study for the investigation of apoptotic effect of a medicinal herb named corosolic acid (CRA) on A549 cells suggested CRA can considerably decrease cell viability of A549 cells. In addition, CRA-induced caspases and ameliorate ROS levels in A549 cells. Furthermore, *N*-acetylcysteine (NAC), a ROS scavenger exposure to cells inhibits CRA-mediated apoptosis, signifying ROS role in CRA-induced apoptosis (Nho et al. 2013).

A thioredoxin (Trx-1) inhibitor, named PX-12 (1-methylpropyl 2-imidazolyl disulfide), has toxicological effects on tumor, but antitumor effects on lung cancer are relatively unknown. An experimental investigation involving PX-12 effects on A549 lung cancer cells showed that PX-12 induces cell cycle arrest and prevents A549 cells growth. In addition, PX-12 significantly increased levels of ROS and deplete GSH levels. Above finding is the first such report, which shows that PX-12 prevents A549 cells growth through cell cycle arrest and ROS-mediated apoptosis (You et al. 2014).

9 Role of MicroRNA in Mitochondrial ROS-Mediated Lung Cancer Pathophysiology

MicroRNAs (miRNAs) are small non-coding RNAs which are involved in physiological regulation of various oxidative stress genes at the post-transcriptional level. miRNAs can regulate the expression of various redox sensors and alter key gene and protein components of the cellular antioxidant machinery and also other way round, oxidative stress may affect the expression levels of multiple miRNAs (Bartel 2018). This complex network between miRNAs and oxidative stress leads to modulation of cellular homeostasis. Moreover, several other cell regulatory factors including phosphorylation (Paroo et al. 2009; Su et al. 2017), ubiquitylation (Kao et al. 2019), deacetylation (Wada et al. 2012), and SUMOylating (Yuan et al. 2017) link the biogenesis of miRNAs biogenesis to different cell signaling pathways involved in lung cancer. Various studies have also revealed that some miRNAs-transcription

Table 1 microRNA (miR) involved in mitochondrial oxidative stress pathways in lung cancer

miRNA	Target pathway or molecule	References
miR-200c	Suppression of Nrf2 expression, overexpression of miR-200c augments ROS and p21 expression levels leading to enhanced radiosensitivity in lung cancer cells.	Cortez et al. (2014)
miR-155	Represses cancer cell death as well as promotes tumor cell colony formation and migration, mediated by upregulation of Nrf2/Keap1 signaling pathway. Downregulation of miR-155 was associated with reduced cellular levels of Nrf2, heme oxygenase-1 (HO-1), NAD(P)H quinone oxidoreductase 1 leading to suppressed cancer cell survival and migration in lung cancer.	Gu et al. (2017)
miR-144	Regulate the cisplatin resistance of lung cancer cells via Nrf2.	Yin et al. (2018)
miR-663	Mitochondrial pathway, targeting BBC3 and BTG2; regulating apoptosis by controlling MOMP.	Fiori et al. (2018)
miR-146a	Downregulates SOD2 and enhances ROS generation; Increases apoptosis, inhibits cell proliferation in lung cancer.	Wang et al. (2014b)
miR-206	Decreasing the angiogenesis by targeting 14–3-3ζ and inhibiting the STAT3/HIF-1α/VEGF pathway.	Xue et al. (2016)
miR-551b	It targeted catalase leading to increased survival and chemoresistance in lung cancer.	Xu et al. (2014)
miR-99a	Downregulation of miR-99a favors metastasis by increasing ROS through the targeting of NOX4.	Sun et al. (2016b)
miR-182	It causes increment of ROS, which further induces tumorigenesis through the JNK pathway via PDK4 targeting-dependent lipogenesis.	Li et al. (2017b)
miR-506	miR-506 negatively regulates NF-κB p65 expression and thus increases ROS production, which, in turn, activates p53 to kill cancer cells through increased apoptosis.	Yin et al. (2015)
miR-210	miR-210-3p directly targets NADH dehydrogenase (ubiquinone) 1 alpha subcomplex 4 (NDUFA4), and succinate dehydrogenase complex subunit D (SDHD), leading to altered physical structure of the mitochondria as well as mitochondrial membrane potential.	Puisségur et al. (2011)

factors are redox-sensitive (Simone et al. 2009; Pradhan et al. 2019). Table 1 enlists the microRNAs which are involved in mitochondrial signaling playing critical role in oxidative stress-mediated lung cancer pathophysiology.

10   Conclusion

Oxidative stress is a key contributing factor to many diseases, including lung cancer. Recent researches emphasize the role of mitochondrial metabolic alterations leading to oxidative stress and ROS-mediated molecular signaling pathway alterations resulting in lung cancer. Various studies also identified new redox-sensitive miR-NAs, as well as the roles of these miRNAs in pathophysiology of lung cancer progression and establishment. Further researches are needed to determine the cellular crosstalk between ROS and cancer cells which are regulated by miRNA or conversely regulate miRNA biogenesis and signaling.

References

Ahn CS, Metallo CM. Mitochondria as biosynthetic factories for cancer proliferation. Cancer Metab. 2015;3:1.

Archer SL. Mitochondrial dynamics – mitochondrial fission and fusion in human diseases. N Engl J Med. 2013;369:2236–51.

Azad N, Rojanasakul Y, Vallyathan V. Inflammation and lung cancer: roles of reactive oxygen/nitrogen species. J Toxicol Environ Heal – Part B Crit Rev. 2008; https://doi.org/10.1080/10937400701436460.

Babior BM. NADPH oxidase: an update. Blood. 1999; https://doi.org/10.1182/blood.v93.5.1464.

Bartel DP. Metazoan microRNAs. Cell. 2018;173:20–51.

Bauer DE, Harris MH, Plas DR, et al. Cytokine stimulation of aerobic glycolysis in hematopoietic cells exceeds proliferative demand. FASEB J. 2004; https://doi.org/10.1096/fj.03-1001fje.

Berman SB, Pineda FJ, Hardwick JM. Mitochondrial fission and fusion dynamics: the long and short of it. Cell Death Differ. 2008;15:1147–52.

Bonner MR, Shen M, Liu C-S, et al. Mitochondrial DNA content and lung cancer risk in Xuan Wei, China. Lung Cancer. 2009;63:331–4.

Brandon M, Baldi P, Wallace DC. Mitochondrial mutations in cancer. Oncogene. 2006; https://doi.org/10.1038/sj.onc.1209607.

Chance B, Sies H, Boveris A. Hydroperoxide metabolism in mammalian organs. Physiol Rev. 1979;59:527–605.

Chang L, Fang S, Gu W. The molecular mechanism of metabolic remodeling in lung cancer. J Cancer. 2020;11:1403–11.

Chiche J, Rouleau M, Gounon P, et al. Hypoxic enlarged mitochondria protect cancer cells from apoptotic stimuli. J Cell Physiol. 2010; https://doi.org/10.1002/jcp.21984.

Cortez MA, Valdecanas D, Zhang X, et al. Therapeutic delivery of miR-200c enhances radiosensitivity in lung cancer. Mol Ther. 2014;22:1494–503.

Dasgupta S, Soudry E, Mukhopadhyay N, et al. Mitochondrial DNA mutations in respiratory complex-I in never-smoker lung cancer patients contribute to lung cancer progression and associated with EGFR gene mutation. J Cell Physiol. 2012; https://doi.org/10.1002/jcp.22980.

Dela Cruz CS, Tanoue LT, Matthay RA. Lung cancer: epidemiology, etiology, and prevention. Clin Chest Med. 2011; https://doi.org/10.1016/j.ccm.2011.09.001.

Dinkova-Kostova AT, Abramov AY. The emerging role of Nrf2 in mitochondrial function. Free Radic Biol Med. 2015; https://doi.org/10.1016/j.freeradbiomed.2015.04.036.

Domej W, Oettl K, Renner W. Oxidative stress and free radicals in COPD – implications and relevance for treatment. Int J Chron Obstruct Pulmon Dis. 2014;9:1207–24.

El Mjiyad N, Caro-Maldonado A, Ramírez-Peinado S, et al. Sugar-free approaches to cancer cell killing. Oncogene. 2011; https://doi.org/10.1038/onc.2010.466.

Fang H-Y, Chen C-Y, Chiou S-H, et al. Overexpression of optic atrophy 1 protein increases cisplatin resistance via inactivation of caspase-dependent apoptosis in lung adenocarcinoma cells. Hum Pathol. 2012;43:105–14.

Fiori ME, Villanova L, Barbini C, et al. miR-663 sustains NSCLC by inhibiting mitochondrial outer membrane permeabilization (MOMP) through PUMA/BBC3 and BTG2 article. Cell Death Dis. 2018; https://doi.org/10.1038/s41419-017-0080-x.

Frezza C, Cipolat S, Martins de Brito O, et al. OPA1 controls apoptotic cristae remodeling independently from mitochondrial fusion. Cell. 2006;126:177–89.

Gardner PR. Superoxide-driven aconitase FE-S center cycling. Biosci Rep. 1997;17:33–42.

Ghosh P, Vidal C, Dey S, et al. Mitochondria targeting as an effective strategy for cancer therapy. Int J Mol Sci. 2020:21.

Gong C, Qi L, Huo Y, et al. Anticancer effect of Limonin against benzo(a)pyrene-induced lung carcinogenesis in Swiss albino mice and the inhibition of A549 cell proliferation through apoptotic pathway. J Biochem Mol Toxicol. 2019;33:e22374.

Gu S, Lai Y, Chen H, et al. MiR-155 mediates arsenic trioxide resistance by activating Nrf2 and suppressing apoptosis in lung cancer cells. Sci Rep. 2017; https://doi.org/10.1038/s41598-017-06061-x.

Guzy RD, Schumacker PT. Oxygen sensing by mitochondria at complex III: the paradox of increased reactive oxygen species during hypoxia. Exp Physiol. 2006;91:807–19.

Hammerman PS, Voet D, Lawrence MS, et al. Comprehensive genomic characterization of squamous cell lung cancers. Nature. 2012; https://doi.org/10.1038/nature11404.

Handy DE, Loscalzo J. Redox regulation of mitochondrial function. Antioxid Redox Signal. 2012; https://doi.org/10.1089/ars.2011.4123.

Hasty P, Christy BA. p53 as an intervention target for cancer and aging. Pathobiol Aging Age Relat Dis. 2013;3 https://doi.org/10.3402/pba.v3i0.22702.

Herbst RS, Heymach JV, Lippman SM. Lung cancer. N Engl J Med. 2008;359:1367–80.

Hosgood HD, Liu C-S, Rothman N, et al. Mitochondrial DNA copy number and lung cancer risk in a prospective cohort study. Carcinogenesis. 2010;31:847–9.

Ishikawa K, Takenaga K, Akimoto M, et al. ROS-generating mitochondrial DNA mutations can regulate tumor cell metastasis. Science (80-). 2008; https://doi.org/10.1126/science.1156906.

Ježek J, Cooper KF, Strich R. Reactive oxygen species and mitochondrial dynamics: the Yin and Yang of mitochondrial dysfunction and cancer progression. Antioxidants. 2018;7(1):13.

Jiang Y, Shan S, Chi L, et al. Methyl methanesulfonate induces necroptosis in human lung adenoma A549 cells through the PIG-3-reactive oxygen species pathway. Tumour Biol. 2016;37:3785–95.

Jiang Y, Liu G, Zhang L, et al. Therapeutic efficacy of hydrogen-rich saline alone and in combination with PI3K inhibitor in non-small cell lung cancer. Mol Med Rep. 2018;18:2182–90.

Joseph B, Ekedahl J, Sirzen F, et al. Differences in expression of pro-caspases in small cell and non-small cell lung carcinoma. Biochem Biophys Res Commun. 1999;262:381–7.

Joseph B, Lewensohn R, Zhivotovsky B. Role of apoptosis in the response of lung carcinomas to anti-cancer treatment. Ann N Y Acad Sci. 2000;926:204–16.

Kalainayakan SP, FitzGerald KE, Konduri PC, et al. Essential roles of mitochondrial and heme function in lung cancer bioenergetics and tumorigenesis. Cell Biosci. 2018;8:56.

Kao S-H, Cheng W-C, Wang Y-T, et al. Regulation of miRNA biogenesis and histone modification by K63-polyubiquitinated DDX17 controls cancer stem-like features. Cancer Res. 2019:canres.2376.2018.

Kasai S, Shimizu S, Tatara Y, et al. Regulation of Nrf2 by mitochondrial reactive oxygen species in physiology and pathology. Biomol Ther. 2020; https://doi.org/10.3390/biom10020320.

Kirkham P, Rahman I. Oxidative stress in asthma and COPD: antioxidants as a therapeutic strategy. Pharmacol Ther. 2006;111:476–94.

Kopecka J, Gazzano E, Castella B, et al. Mitochondrial metabolism: inducer or therapeutic target in tumor immune-resistance? Semin Cell Dev Biol. 2020; https://doi.org/10.1016/j.semcdb.2019.05.008.

Kwong JQ, Henning MS, Starkov AA, et al. The mitochondrial respiratory chain is a modulator of apoptosis. J Cell Biol. 2007;179:1163–77.

Li SM, Chen CH, Chen YW, et al. Upregulation of CISD2 augments ROS homeostasis and contributes to tumorigenesis and poor prognosis of lung adenocarcinoma. Sci Rep. 2017a; https://doi.org/10.1038/s41598-017-12131-x.

Li G, Li M, Hu J, et al. The microRNA-182-PDK4 axis regulates lung tumorigenesis by modulating pyruvate dehydrogenase and lipogenesis. Oncogene. 2017b; https://doi.org/10.1038/onc.2016.265.

Liu X, Chen Z. The pathophysiological role of mitochondrial oxidative stress in lung diseases. J Transl Med. 2017;15:207.

Liu J, Xian G, Li M, et al. Cholesterol oxidase from Bordetella species promotes irreversible cell apoptosis in lung adenocarcinoma by cholesterol oxidation. Cell Death Dis. 2014;5:e1372.

Liu F, Sanin DE, Wang X. Mitochondrial DNA in lung cancer. Adv Exp Med Biol. 2017;1038:9–22.

Lu X, Liu QX, Zhang J, et al. PINK1 overexpression promotes cell migration and proliferation via regulation of autophagy and predicts a poor prognosis in lung cancer cases. Cancer Manag Res. 2020; https://doi.org/10.2147/CMAR.S262466.

Luanpitpong S, Talbott SJ, Rojanasakul Y, et al. Regulation of lung cancer cell migration and invasion by reactive oxygen species and caveolin-1. J Biol Chem. 2010; https://doi.org/10.1074/jbc.M110.124958.

Martínez-Cayuela M. Oxygen free radicals and human disease. Biochimie. 1995; https://doi.org/10.1016/0300-9084(96)88119-3.

Martinez-Outschoorn UE, Balliet RM, Rivadeneira DB, et al. Oxidative stress in cancer associated fibroblasts drives tumor-stroma co-evolution: a new paradigm for understanding tumor metabolism, the field effect and genomic instability in cancer cells. Cell Cycle. 2010;9:3256–76.

Mohanty A, Tiwari-Pandey R, Pandey NR. Mitochondria: the indispensable players in innate immunity and guardians of the inflammatory response. J Cell Commun Signal. 2019;13:303–18.

Murphy MP. How mitochondria produce reactive oxygen species. Biochem J. 2009; https://doi.org/10.1042/BJ20081386.

Murphy MP, Holmgren A, Larsson NG, et al. Unraveling the biological roles of reactive oxygen species. Cell Metab. 2011; https://doi.org/10.1016/j.cmet.2011.03.010.

Nam H-S, Izumchenko E, Dasgupta S, et al. Mitochondria in chronic obstructive pulmonary disease and lung cancer: where are we now? Biomark Med. 2017;11:475–89.

Neagu M, Constantin C, Popescu ID, et al. Inflammation and metabolism in cancer cell – mitochondria key player. Front Oncol. 2019; https://doi.org/10.3389/fonc.2019.00348.

Nelson DA, Tan TT, Rabson AB, et al. Hypoxia and defective apoptosis drive genomic instability and tumorigenesis. Genes Dev. 2004; https://doi.org/10.1101/gad.1204904.

Nho KJ, Chun JM, Kim HK. Corosolic acid induces apoptotic cell death in human lung adenocarcinoma A549 cells in vitro. Food Chem Toxicol. 2013;56:8–17.

Park WH. Antiapoptotic effects of caspase inhibitors on H2O2-treated lung cancer cells concerning oxidative stress and GSH. Mol Cell Biochem. 2018;441:125–34.

Parone PA, Da Cruz S, Tondera D, et al. Preventing mitochondrial fission impairs mitochondrial function and leads to loss of mitochondrial DNA. PLoS One. 2008;3:e3257.

Paroo Z, Ye X, Chen S, et al. Phosphorylation of the human microRNA-generating complex mediates MAPK/Erk signaling. Cell. 2009;139:112–22.

Pizzino G, Irrera N, Cucinotta M, et al. Oxidative stress: harms and benefits for human health. Oxidative Med Cell Longev. 2017; https://doi.org/10.1155/2017/8416763.

Plas DR, Thompson CB. Akt-dependent transformation: there is more to growth than just surviving. Oncogene. 2005;24:7435–42.

Pourahmad J, Salimi A, Seydi E. Role of oxygen free radicals in cancer development and treatment. Free Radicals Dis., InTech; 2016.

Pradhan AK, Bhoopathi P, Talukdar S, et al. MDA-7/IL-24 regulates the miRNA processing enzyme DICER through downregulation of MITF. Proc Natl Acad Sci. 2019;116:5687–92.

Puisségur MP, Mazure NM, Bertero T, et al. MiR-210 is overexpressed in late stages of lung cancer and mediates mitochondrial alterations associated with modulation of HIF-1 activity. Cell Death Differ. 2011; https://doi.org/10.1038/cdd.2010.119.

Qian W, Choi S, Gibson GA, et al. Mitochondrial hyperfusion induced by loss of the fission protein Drp1 causes ATM-dependent G2/M arrest and aneuploidy through DNA replication stress. J Cell Sci. 2012;125:5745–57.

Quinlan CL, Treberg JR, Perevoshchikova IV, et al. Native rates of superoxide production from multiple sites in isolated mitochondria measured using endogenous reporters. Free Radic Biol Med. 2012; https://doi.org/10.1016/j.freeradbiomed.2012.08.015.

Quinlan CL, Goncalves RLS, Hey-Mogensen M, et al. The 2-oxoacid dehydrogenase complexes in mitochondria can produce superoxide/hydrogen peroxide at much higher rates than complex I. J Biol Chem. 2014;289:8312–25.

Rahman I, Gilmour PS, Jimenez LA, et al. Oxidative stress an TNF-α induce histone acetylation and NF-κB/AP-1 activation in alveolar epithelial cells: potential mechanism in gene transcription in lung inflammation. Mol Cell Biochem. 2002; https://doi.org/10.1023/A:1015905010086.

Raimundo N, Baysal BE, Shadel GS. Revisiting the TCA cycle: signaling to tumor formation. Trends Mol Med. 2011;17:641–9.

Ramanathan A, Wang C, Schreiber SL. Perturbational profiling of a cell-line model of tumorigenesis by using metabolic measurements. Proc Natl Acad Sci U S A. 2005; https://doi.org/10.1073/pnas.0502267102.

Ritz B, Hoffmann B, Peters A. The effects of fine dust, ozone, and nitrogen dioxide on health. Dtsch Arztebl Int. 2019;51–52:881–6.

Roberts ER, Thomas KJ. The role of mitochondria in the development and progression of lung cancer. Comput Struct Biotechnol J. 2013;6:e201303019.

Ryu D, Lee JH, Kwak MK. NRF2 level is negatively correlated with TGF-β1-induced lung cancer motility and migration via NOX4-ROS signaling. Arch Pharm Res. 2020; https://doi.org/10.1007/s12272-020-01298-z.

Sabharwal SS, Schumacker PT. Mitochondrial ROS in cancer: initiators, amplifiers or an Achilles' heel? Nat Rev Cancer. 2014;14:709–21.

Shackelford DB, Shaw RJ. The LKB1–AMPK pathway: metabolism and growth control in tumour suppression. Nat Rev Cancer. 2009;9:563–75.

Simone NL, Soule BP, Ly D, et al. Ionizing radiation-induced oxidative stress alters miRNA expression. PLoS One. 2009;4:e6377.

Singh A, Daemen A, Nickles D, et al. NRF2 activation promotes aggressive lung cancer and associates with poor clinical outcomes. Clin Cancer Res. 2020; https://doi.org/10.1158/1078-0432.ccr-20-1985.

Solaini G, Sgarbi G, Baracca A. Oxidative phosphorylation in cancer cells. Biochim Biophys Acta. 1807;2011:534–42.

Song M, Wu H, Wu S, et al. Antibiotic drug levofloxacin inhibits proliferation and induces apoptosis of lung cancer cells through inducing mitochondrial dysfunction and oxidative damage. Biomed Pharmacother. 2016;84:1137–43.

Sreedhar A, Zhao Y. Dysregulated metabolic enzymes and metabolic reprogramming in cancer cells. Biomed Rep. 2018;8:3–10.

Su C, Li Z, Cheng J, et al. The protein phosphatase 4 and SMEK1 complex dephosphorylates HYL1 to promote miRNA biogenesis by antagonizing the MAPK cascade in arabidopsis. Dev Cell. 2017;41:527–539.e5.

Sun B, Gao L, Ahsan A, et al. Anticancer effect of SZC015 on lung cancer cells through ROS-dependent apoptosis and autophagy induction mechanisms in vitro. Int Immunopharmacol. 2016a;40:400–9.

Sun M, Hong S, Li W, et al. MIR-99a regulates ROS-mediated invasion and migration of lung adenocarcinoma cells by targeting NOX4. Oncol Rep. 2016b; https://doi.org/10.3892/or.2016.4672.

Suzuki-Karasaki Y, Suzuki-Karasaki M, Uchida M, et al. Depolarization controls TRAIL-sensitization and tumor-selective killing of cancer cells: crosstalk with ROS. Front Oncol. 2014;4:128.

Thomas KJ, Jacobson MR. Defects in mitochondrial fission protein dynamin-related protein 1 are linked to apoptotic resistance and autophagy in a lung cancer model. PLoS One. 2012;7:e45319.

Tsujimoto Y, Shimizu S. VDAC regulation by the Bcl-2 family of proteins. Cell Death Differ. 2000;7:1174–81.

Vanhove K, Graulus G-J, Mesotten L, et al. The metabolic landscape of lung cancer: new insights in a disturbed glucose metabolism. Front Oncol. 2019;9:1215.

Viktorsson K, De Petris L, Lewensohn R. The role of p53 in treatment responses of lung cancer. Biochem Biophys Res Commun. 2005;331:868–80.

Wada T, Kikuchi J, Furukawa Y. Histone deacetylase 1 enhances microRNA processing via deacetylation of DGCR8. EMBO Rep. 2012;13:142–9.

Wang J, Pareja KA, Kaiser CA, et al. Redox signaling via the molecular chaperone BiP protects cells against endoplasmic reticulum-derived oxidative stress. elife. 2014a; https://doi.org/10.7554/eLife.03496.

Wang Q, Chen W, Bai L, et al. Receptor-interacting protein 1 increases chemoresistance by maintaining inhibitor of apoptosis protein levels and reducing reactive oxygen species through a microRNA-146a-mediated catalase pathway. J Biol Chem. 2014b; https://doi.org/10.1074/jbc.M113.526152.

Wang T-H, Chen C-C, Huang K-Y, et al. High levels of EGFR prevent sulforaphane-induced reactive oxygen species-mediated apoptosis in non-small-cell lung cancer cells. Phytomedicine. 2019;64:152926.

Xu X, Wells A, Padilla MT, et al. A signaling pathway consisting of miR-551b, catalase and MUC1 contributes to acquired apoptosis resistance and chemoresistance. Carcinogenesis. 2014; https://doi.org/10.1093/carcin/bgu159.

Xue D, Yang Y, Liu Y, et al. MicroRNA-206 attenuates the growth and angiogenesis in non-small cell lung cancer cells by blocking the 14-3-3z/STAT3/HIF-1a/VEGF signaling. Oncotarget. 2016; https://doi.org/10.18632/oncotarget.12972.

Yao J, Du Z, Li Z, et al. 6-Gingerol as an arginase inhibitor prevents urethane-induced lung carcinogenesis by reprogramming tumor supporting M2 macrophages to M1 phenotype. Food Funct. 2018;9:4611–20.

Yeung SJ, Pan J, Lee MH. Roles of p53, MYC and HIF-1 in regulating glycolysis – the seventh hallmark of cancer. Cell Mol Life Sci. 2008; https://doi.org/10.1007/s00018-008-8224-x.

Yin M, Ren X, Zhang X, et al. Selective killing of lung cancer cells by miRNA-506 molecule through inhibiting NF-κB p65 to evoke reactive oxygen species generation and p53 activation. Oncogene. 2015; https://doi.org/10.1038/onc.2013.597.

Yin Y, Liu H, Xu J, et al. miR-144-3p regulates the resistance of lung cancer to cisplatin by targeting Nrf2. Oncol Rep. 2018; https://doi.org/10.3892/or.2018.6772.

You BR, Park WH. Zebularine inhibits the growth of A549 lung cancer cells via cell cycle arrest and apoptosis. Mol Carcinog. 2014;53:847–57.

You BR, Park WH. Suberoylanilide hydroxamic acid induces thioredoxin1-mediated apoptosis in lung cancer cells via up-regulation of miR-129-5p. Mol Carcinog. 2017;56:2566–77.

You BR, Shin HR, Park WH. PX-12 inhibits the growth of A549 lung cancer cells via G2/M phase arrest and ROS-dependent apoptosis. Int J Oncol. 2014,44.301–8.

Yu J, Yue W, Wu B, et al. PUMA sensitizes lung cancer cells to chemotherapeutic agents and irradiation. Clin Cancer Res. 2006;12:2928–36.

Yuan H, Deng R, Zhao X, et al. SUMO1 modification of KHSRP regulates tumorigenesis by preventing the TL-G-Rich miRNA biogenesis. Mol Cancer. 2017;16:157.

Zong W-X, Rabinowitz JD, White E. Mitochondria and cancer. Mol Cell. 2016;61:667–76.

Regulation of Glucose Transporters in Cancer Progression

Sibi Raj, Manish K. Mishra, Sitaram Harihar, Ashok Kumar, Shubhadeep Roychoudhury, Arun Kumar, Brijesh Rathi, and Dhruv Kumar

Contents

S. Raj · D. Kumar (✉)
Amity Institute of Molecular Medicine and Stem Cell Research (AIMMSCR), Amity University, Noida, Uttar Pradesh, India
e-mail: dkumar13@amity.edu

M. K. Mishra
Environmental Radioactivity Measurement Section, Environmental Monitoring & Assessment Division, Bhabha Atomic Research Centre, Mumbai, India

S. Harihar
Department of Genetic Engineering, SRM Institute of Science and Technology, Chennai, Tamil Nadu, India

A. Kumar
Department of Biochemistry, All India Institute of Medical Sciences (AIIMS), Bhopal, Bhopal, Madhya Pradesh, India

S. Roychoudhury
Department of Life Science and Bioinformatics, Assam University, Silchar, India

A. Kumar
Mahavir Cancer Institute & Research Centre, Patna, Bihar, India

B. Rathi
Laboratory for Translational Chemistry and Drug Discovery, Hansraj College University of Delhi, New Delhi, India

© The Author(s), under exclusive license to Springer Nature Switzerland AG 2021
K. K. Kesari, N. K. Jha (eds.), *Free Radical Biology and Environmental Toxicity*, Molecular and Integrative Toxicology, https://doi.org/10.1007/978-3-030-83446-3_9

Abstract Glycolysis has been one of the major hallmarks of cancer, as cancer cells exhibit a high rate of glucose consumption beyond that required for energy production. Glucose acts as a substrate to generate biomass and regulates cell signaling that is required for the cancer progression. Glucose transporters (GLUTs) have been majorly playing their role in glucose transport across the cell membrane to meet their metabolic demands. Overexpression of GLUTs has been reported in various cancer types. Also, activation of certain oncogenes such as c-myc, ras, and src and several other transcription factors such as hypoxia inducible factor-1α induces the overexpression of GLUTs in cancer cells. GLUT regulation has been studied on the epigenetic level as well to provide fundamental information on its regulation. Targeting and understanding GLUTs function in cancer cells could pave novel path for therapeutic strategies against cancer.

Keywords GLUTs · Glycolysis · Warburg effect · Epigenetics · c-myc · Mtor · Cancer

Graphical Abstract

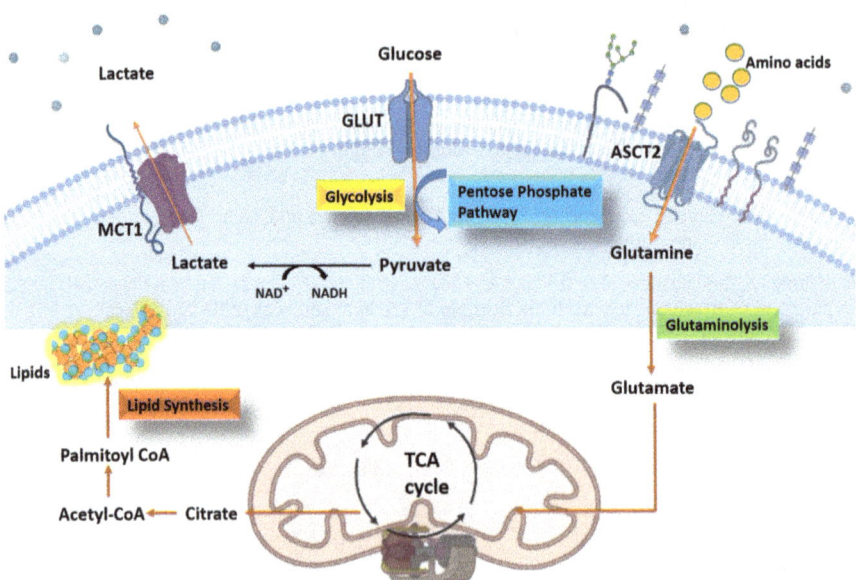

1 Introduction

Tumorigenesis is often characterized by changes at cellular, genetic, and epigenetic levels leading to abnormal growth of the cells. There are several factors that contribute to this abnormal and uncontrolled growth of cells, one among which is metabolic regulations. The changes in intracellular and extracellular metabolites often contribute to metabolic reprogramming which have profound effects on gene expression, cellular differentiation, and cellular microenvironment. Glycolysis is often the preferred metabolic pathway for cancer cells that enables them to acquire energy and other metabolites for their growth and survival (Mathupala et al. 1797). A remarkable observation of increased glucose uptake in cancer cells was observed by a German physiologist Otto Warburg in 1920s (Warburg et al. 1927). Cancer cells proliferate at a very high rate as compared to normal cells; this phenomenon makes the necessity for cancer cells to have more fuel to grow. Consumption of glucose by cancer cells provide them the mechanism to generate energy as well as to obtain other metabolic requirements to proliferate and survive. This reprogrammed metabolism in cancer cells is considered as one of the hallmarks of cancer. The increase in glycolysis allows the glycolytic intermediates to fulfill the metabolic demands of proliferating cells. Cancer cells thus rely on inefficient glycolytic mode of ATP synthesis (2 ATP/glucose), rather than mitochondrial respiration that comparatively produces more ATPs (36 ATP/glucose). This increase in glucose transport is supported by glucose transporters (Glut1 and Glut3) which are overexpressed in the plasma membrane (Krzeslak et al. 2012). Also, increase in glucose consumption is used as a carbon source for other biosynthetic processes to support cell proliferation. Major biosynthetic pathways lead to the production of nucleotides, lipids, and proteins. The diversion of glycolytic flux toward serine biosynthesis with the help of glycolytic enzyme phosphoglycerate dehydrogenase (PHGDH) is one such example (Locasale et al. 2011). Tumor cells undergo a metabolic shift from mitochondrial respiration to glycolysis generating two molecules of pyruvate. Pyruvate generated is converted to lactate by the enzyme lactate dehydrogenase-A (LDH-A). The NAD+ generated through the conversion of pyruvate to lactate helps to maintain the glycolytic flux in cancer cells. The possible reasons why cancer cells highly depend on glycolysis are: (1) ATP production via glycolysis is much faster than OXPHOS, (2) increased glycolytic flux produces enough intermediates for the biosynthesis of amino acids, nucleic acids, lipid bilayer, and fatty acid for the cell growth, and (3) NADPH produced by pentose phosphate pathway (PPP) maintains glutathione levels in cells, resulting in the resistance of cancer cells to chemotherapeutic molecules (Vander Heiden et al. 2009). The enzymatic conversion of glucose into pyruvate in HNSCC is regulated to meet major cellular needs by the production of intermediates for synthetic reactions, such as the formation of fatty acids, amino acids, and nucleotides. In glycolysis, the reactions catalyzed by hexokinase, phosphofructokinase, and pyruvate kinase are majorly upregulated in cancer cells leading to the generation of intermediate metabolites for the biosynthetic pathways such as pentose phosphate pathway, serine biosynthesis, glutaminolysis, and glyceraldehyde 3-phosphate pathway for the production of amino acids, nucleotides, and lipids for the growth and survival of cancer cells.

There are various forms of glucose transporters (GLUTs) that have been described and their expression is cell specific. The major function of GLUTs is to ensure the continuous availability of glucose for metabolism via controlling the movement of glucose between intracellular and extracellular compartments. The GLUT gene family belongs to the 2A solute carrier family and is recognized by the gene symbol SLC2A (Chen et al. 2010). GLUT genes are described into 14 isoforms and are further classified into three groups based on their primary sequences. GLUTs are intrinsic membrane proteins which differ in tissue-specific expression and majorly functions depending upon the metabolic and hormonal regulation. In note to this, the brain heavily relies on glucose to support metabolic demands and specifically shows affinity toward GLUT1 and GLUT3 (Mueckler and Thorens 2013). Also, the muscle tissues express multiple GLUTs such as GLUT3, GLUT-5, GLUT-10, and GLUT-11. The various types of GLUTs also play a role in embryo development and express different affinities for glucose and other sugar molecules such as fructose or mannose. Increased expression of GLUT proteins has been observed in several types of cancer. GLUT1 and GLUT3 are the major isoforms seen to be overexpressed in cancer (Kim et al. 2017; Han et al. 2017). The level of GLUT1 expression is considered to be a suitable marker of hypoxia and glucose metabolism and can be measured in easy ways through routine histologic assessment of tumors. The expression pattern of GLUT1 has been reportedly varied in different tumor types and higher levels of GLUT1 indicate a poor prognosis.

2 Glucose Transporters

The major function of GLUTs is to ensure the continuous glucose availability for the metabolism to generate energy and this mechanism is achieved by regulating the glucose movement between intracellular and extracellular compartments. There are 14 isoforms of glucose transporters and are classified into three main categories based on their primary sequences (Macheda et al. 2005). The first class of glucose transporters is known as classical transporters and involves GLUT1-4 and GLUT14. GLUT14 is considered to be a duplicate of GLUT3. Class two group of genes involves GLUT5, 7, 9, and 11. Lastly, class three includes GLUT 6, 8, 10, 12, and 13. Glut 13 is also known as the H^+ myoinositol transporter (HMIT) (Macheda et al. 2005). All the various GLUT isoforms show a same transmembrane anatomy pattern having large highly conserved transmembrane domain and an asymmetric less conserved cytoplasmic and exoplasmic domain. However, each of the transporters has different functions depending on their localization, protein interactions, and transport kinetics. 13-14% of the protein sequences across the GLUTs are identical with 30–70% of them being conserved (Augustin 2010). They have been variedly distributed across the tissue and are also cell specific with having different affinities for glucose and other sugar molecules such as fructose and galactose.

3   Role for GLUTs in Cellular Level

GLUT1 has been discussed for their various other roles in cellular level. Beyond playing a role in glycolysis the function of glucose carriers is been explored in several other terms. Protein–protein interactions and glucose dependent signaling are the other sources of functions that have been largely studied.

GLUTs are known to interact with other membrane proteins which consequently would disrupt the activity of glucose transporter or the interacting proteins. This was in note demonstrated between the secondary active Na^+/glucose cotransporter SGLT1 and the plasma membrane receptor tyrosine kinase EGFR. The gain-of-function activity in EGFR through gene mutation or amplification facilitated tumor progression in various cancer subtypes such as glioblastoma, LUAD, and CRC (Weihua et al. 2008). EGFR has been widely reported to be an essential factor that plays a critical role in cancer progression and is been targeted using many small molecule receptor kinase inhibitors that inhibit the cancer progression and survival. In contrast, a kinase-independent prosurvival function was observed where EGFR prevents autophagy-associated cell death in cancer cells through sustained glucose uptake. This was due to the binding of EGFR with SGLT1 and its stabilization with the molecule (Dittmann et al. 2013). The coexpression of both EGFR and SGLT1 leads to the survival of cancer cells in vitro even under less glucose concentration. This interaction has also led to a finding where it was seen to be highly increased in cancer cells treated with radiation therapy and also had a role in recovery from radiation therapy-mediated cellular ATP reduction. This interaction might be dynamically regulated, as it was reported to be increased in cancer cells treated with radiation therapy, and to contribute to the recovery from radiation therapy-mediated cellular ATP depletion. However, in that situation SGLT1 was phosphorylated and stabilized in an EGFR tyrosine kinase activity-dependent manner. The role of GLUT transporters in association with EGFR signaling has not been extensively studied, although there is a possibility that other glucose transporters also interact with membrane-associated proteins that have signaling properties. Immunoprecipitation experiments followed by mass spectrometry analyses might help to uncover physically and functionally linked GLUT interacting proteins.

The intermediate glycolytic molecules also act as signaling molecules to deviate the glycolytic cycle to several other biosynthesis required for cancer cell proliferation and survival. The process of glycolysis generates various other metabolic intermediates that could possibly serve as substrates for enzymes responsible for chromatin remodeling which could have potential impact in cell signaling. For example, alpha-ketoglutarate acts as a cosubstrate for 10–11 translocation (TET) enzymes which are major players in DNA demethylation (Ito et al. 2010; Loenarz and Schofield 2011). Also, studies in mice with *supra*-physiological administration of glucose elevates the production of α-KG that is correlated with a higher global hydroxymethylation in various tissues as assessed by GC-MS analysis (Yang et al. 2014). Also, this treatment also led to the alteration of 732 genes among which 89 genes showed higher hydroxymethylation level. Adding to this complexity, the oncometabolite 2-hydroxyglutarate competes with α-KG for dioxygenase (including TET) enzyme regulation (Xu et al. 2011).

4   Mechanisms of GLUT1 and GLUT3 Regulation

Cancer cells are reportedly known to have accelerated metabolic rates and high glucose demand to support their growth and proliferation. Increased glycolysis hugely is associated with deregulated expression of glucose carriers, particularly GLUT1 and GLUT3. Both of these transporters show an amino acid sequence similarity of about 64% and portray high affinity and turnover number. Several studies exploring publicly available gene expression datasets such as The Cancer Genome Atlas (TCGA) or oncomine have highlighted the increased expression of GLUT receptors across the cancer types. GLUT2 expression is observed to be very high in hepatocellular carcinoma and correlates with poor survival (Kim et al. 2017), whereas GLUT3 is overexpressed in muscle invasive bladder cancer. Also, GLUT-3 expression is high in aggressive glioblastoma as compared to grade lesions and is associated with poor survival (Han et al. 2017). Concluding with that overexpression of GLUT1 and GLUT3 is associated with poor survival in several cancer types such as colorectal cancer, breast cancer, lung cancer, squamous cell carcinoma, ovarian cancer, and glioblastoma cancer (Flavahan et al. 2013; Chai et al. 2017).

GLUT1 and GLUT3 have known to share similar mechanistic features as they respond to HIF-1α or p53, but also have been reported to respond to many different stimuli responses. GLUT-1 has been reported to be activated in response to HIF-1α in glioblastoma stem cells (Li et al. 2009). Also, other factors associated with hypoxia such as VEGF receptor and calcium channel transactivation reportedly upregulated GLUT1 synthesis and trafficking to the cell membrane (Suh and Han 2013). Whereas at the gene level, GLUT1 is controlled by a different mechanism. This involves the transcription factors c-MYC and sine oculis homeobox 1 (SIX1) which promotes glycolysis via direct transactivation of GLUT1 and other related genes (Osthus et al. 2000; Li et al. 2018). Overexpression of GLUT1 in Burkitt's lymphoma cell lines were characterized by chromosomal translocations of MYC. Whereas in nonsmall cell lung cancer (NSCLC), overexpression of GLUT1 was correlated with mutation in oncogene EGFR and KRAS (Sasaki et al. 2012). Studies have also reported that the oncogenic mutant Kras copy number is a critical determinant of metabolic shift (Kerr et al. 2016). Also, evidences suggest that upregulation of GLUT1 takes place upon Kras activation in brochoalveolar stem cells, and lung tumor formation was accelerated in chronic hyperglycemia under the influence of streptozotocin (Micucci et al. 2014). GLUT1 expression was shown to be higher in colorectal cancer with KRAS or BRAF mutation giving them an opportunity to survive even at low glucose conditions. Surprisingly, when cells with wild type KRAS were exposed to low glucose, few cells showed increase in GLUT1 expression and few others acquired KRAS mutation which highlighted the selection pressure imposed by glucose deprivation (Yun et al. 2009). Thioredoxin-interacting protein (TXNIP) has been reported to negatively regulate GLUT1 expression. This is a multifunctional protein that represses GLUT1 function by two ways, either by binding to GLUT1 or leading to a clathrin dependent endocytosis thereby reducing GLUT1 mRNA expression. Upon stress conditions or in the presence of insulin

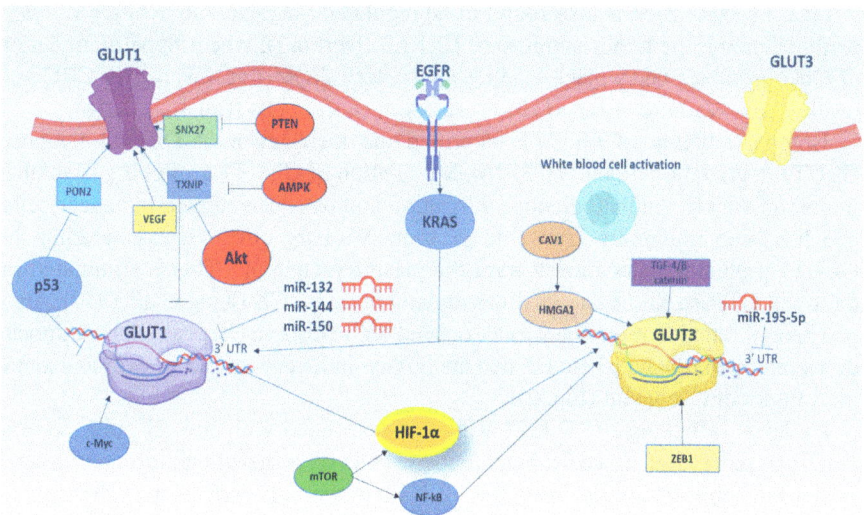

Fig. 1 *Deregulation of glucose transporters in cancer.* Various factors has been reported to down-regulate as well as upregulate GLUT1 and GLUT3. GLUT-1 is directly regulated via multiple transcriptional regulators (TXNIP, p53, c-Myc, SIX1, HIF-1a) and chromatin modifier (HDAC2). Also, microRNAs (miR-132, miR-144, miR-150) specifically target GLUT1. GLUT1 regulation also takes place through regulated protein levels that functions in trafficking or degradation of proteins (PTEN, DERL3, VEGF, PON2, Akt). The regulation of GLUT3 is performed by factors such as (NF-jB, HIF-1a, HMGA1, ZBTB7A) or 30 untranslated regions (30 UTR) of the mRNA (miR-195-5p)

TXNIP is phosphorylated via AMPK and AKT which degrades the multifunctional protein and increases the GLUT1 dependent glucose uptake and restores energy homeostasis (Wu et al. 2013). p53 is another major tumor suppressor gene which negatively regulates GLUT1 and GLUT3 transcription. Also, studies in pancreatic ductal carcinoma p53 was reported to repress the transcription of paraoxonase 2 (PON2) which promotes higher interaction between GLUT1 and the membrane protein stomatin which reduces the glucose uptake (Waldhart et al. 2017; Schwartzenberg-Bar-Yoseph et al. 2004) (Fig. 1).

GLUT1 regulation was also explored on the level of epigenetics, where in the mouse model it was observed that upon fasting the release of ketone body beta-hydroxybutane increases the expression of GLUT1. This was associated with the increase in H3K9 acetylation at a cis-regulatory region of Glut1 gene. This was further explored through CRISPR/Cas9-mediated disruption of *Hdac2*, which increased *Glut1* expression (Kawauchi et al. 2008). MicroRNAs have also been reported to regulate GLUT1 expression, decreased expression of miR-144 and miR-132 has been associated to increase GLUT1 expression and glucose uptake in tumors as compared to healthy tissue in lung and prostate cancer (Nagarajan et al. 2017; Zhang et al. 2001). In CD46 costimulated CD4[+] T cells, downregulation of miR-150 and *GLUT1* upregulation were observed, identifying *GLUT1* as a target of this microRNA in T cells (Tanegashima et al. 2017) (Fig. 1).

GLUT1 expression is also found to be regulated via DNA methylation by indirectly silencing the tumor suppressor DERL3 (Derlin-3), where hypermethylation of CpG promoter takes place leading to shorter relapse free survival in CRC and resulted in a Warburg effect in HCT-116 colon cancer cells (Liu et al. 2016).

Isotopic labeling of GLUT1 identified the mechanism of DERL3-mediated GLUT1 degradation. (Qu et al. 2016). Modulation of GLUT1 trafficking is another important way to regulate glucose utilization and has been studied in cancer cells. p53 has been reportedly known to promote Warburg effect by accelerating the GLUT1 protein translocation toward the plasma membrane. This is stimulated by the small GTPase RhoA and its downstream kinase ROCK (King et al. 1950). Also, p53 gain of function mutation was associated with Warburg effect as GLUT1 knockdown inhibited mutant p53-mediated anchorage-independent cell growth and xenograft tumor development (Fig. 2).

Cellular trafficking of GLUT1 was also associated with several other factors. One such report was based on hematopoetic cells were growth factor interleukin-3 was responsible for maintaining the GLUT1 at the cell surface with the help of small GTPase Rab11a-dependent recycling of intracellular GLUT1 (Lopez-Serra et al. 2014). Also, PI3-K effector Akt pathway was also enhanced upon

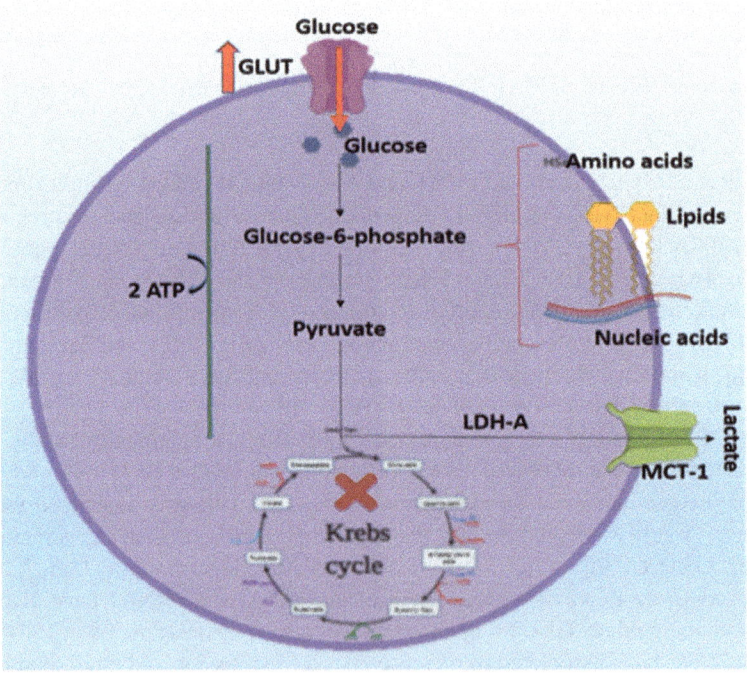

Fig. 2 Upon stress conditions such as limited O_2 availability, pyruvate is metabolized to lactate. GLUT receptors are upregulated in cancer cells and metabolize mostly glucose to lactate despite the availability of O_2 (the Warburg effect), forwarding glucose metabolites from energy production to anabolic process to accelerate cell proliferation, at the expense of generating only two ATPs per glucose

interleukin-3 stimulation, which further enhanced both GLUT1 gene transcription and GLUT1 protein trafficking to the cell membrane (Zhang et al. 2013; Wieman et al. 2007; Shinde and Maddika 2017; Baldwin et al. 1995). Transgenic expression of Glut1 promoted T cell activation thereby accumulating activated memory-phenotype T cells with signs of autoimmunity in aged mice (Barthel et al. 1999). A tumor suppressor gene PTEN catalyzed by PI3-K interacts with SNX27 and effectively prevents the binding between SNX27 and VPS26 disrupting the formation of a functional retromer complex, and in turn inhibiting the GLUT1 recycling to the plasma membrane. Surprisingly, this effect was independent on its phosphatase activity (Wieman et al. 2007). Altogether, these studies highlight a tight control of GLUT1 expression and localization by growth factor-mediated PI3K-Akt signaling, in diverse cell types including hematopoietic and cancer cells (Jacobs et al. 2008).

5   GLUT3 Regulation

GLUT3 is yet another factor that majorly has a role in glycolysis in cancer cells. There have been studies that report the involvement of different factors in the regulation of GLUT-3. With the help of actinomycin-D and the cloning of GLUT3 promoter enabled the identification of the second messenger cAMP synthetic analogue 8-br-cAMP and its role in the transcriptional regulation of *GLUT3* in a breast cancer cell line (Meneses et al. 2008). In colon cancer cell lines, caveolin-1 (CAV1) triggers a nuclear localization of HMGA1, which stimulates *GLUT3* transcription by binding to a DNA sequence located within the *GLUT3* promoter (Ha et al. 2012). Studies based on Tsc1- or Tsc2-null mouse embryonic mouse fibroblasts and human cancer lines showed the impact of cancer-related signaling pathways such as mTOR complex having a role in GLUT3 upregulation. This mechanism of mTOR association with increased GLUT-3 takes place via NF-κB pathway stimulation (Zha et al. 2015). Additionally, increased NF-κB signaling, resulting from *p53* deletion in mouse embryonic fibroblasts, stimulates *Glut3* transcription by binding to a specific NF-κB binding element located within the first intron of the mouse *Glut3* gene. Also, studies have shown the association of EGFR signaling along with GLUT-3 in lung adenocarcinoma (LUAD) cells which happens through the gain of function of the receptor kinase which enhances the glucose uptake and glycolysis in cancer cells. A study of drug sensitive in lung tumor cells, the inhibition of EGFR tyrosine kinase decreased the utilization of glucose and lactate production and also inhibited GLUT3 expression (Makinoshima et al. 2014). Activation of the PI3K-Akt-mTOR signaling pathway by IGF-I leads to GLUT3 induction through HIF-1α in neuronal PC12 cells, involving a hypoxia-response element (HRE) in the *GLUT3* promoter (Yu et al. 2012). OCT4 protein has been reportedly shown to exhibit cooperative interactions with HIF-1α and has also been associated with GLUT3 expression; it is also a pluripotency marker (Christensen et al. 2015). In cancer cells, an enhancer located at the second intron of human GLUT3 is necessary for gene induction in association with various other transcription factor activities. Also, ZEB1 which is a epithelial-mesenchymal transition (EMT) transcription factor is responsible for the activation of this intronic enhancer in a panel of human NSCLC and hepatocellular carcinoma cell lines. Thus, inhibition of GLUT3 expression consequently decreases the uptake of glucose and growth of the mesenchymal lung tumor cells, whereas its ectopic expression in epithelial cells sustains their proliferation in low glucose (Apostolou et al. 2013).

Role of epigenetics has always been crucial in cancer initiation and progression. GLUT3 has been very less studied on the level of epigenetics. Inhibition of GLUT3 with epigenetic targeting agents such as histone deacetylase inhibitors decreased the GLUT3 expression induced in NSCLC cell lines, with minimal effect in non-transformed immortalized human bronchial epithelial cells. In bladder cancer cells, miR-195-5p targets the GLUT3 3'-untranslated region (UTR). Small interfering RNA (siRNA)- and miR-195-5p-mediated GLUT3 knockdown experiments revealed that miR-195-5p decreases glucose uptake, inhibits cell growth, and promotes apoptosis by suppressing GLUT3 (Masin et al. 2014). GLUT3 regulation is been studied at protein level via the membrane localization that initiates through immune cell activation. Insulin has been reported to stimulate GLUT3 to GLUT4 translocation in resting immune cells to the plasma membrane in B-lymphocytes and in monocytes. Whereas, upon the activation of immune cell along with phorbol myristate acetate and lipopolysaccharide, GLUT3 expression was seen to elevate at the plasma membrane in B- and T-lymphocytes, as well as monocytes and neutrophils (Fei et al. 2012; Estrada et al. 1994; Fu et al. 2004; Maratou et al. 2007). Molecular mechanisms involved in GLUT3 re-localization have been reviewed (Simpson et al. 2008).

The epigenome-based study of glucose regulation was also observed at the level of histone where several studies observed an association between glycolysis and histone acetylation where acetyl-CoA has a major role. The acetylation of the lysine residues in histones promotes the relaxing of the chromatin, necessary for gene transcription. Upon the addition of inhibitors for glycolysis a significant observation was highlighted where alteration of histone acetylation takes place and has been correlated with the levels of acetyl-CoA. This link was further shown in a study involving the differentiation of embryonic stem cell, where glycolysis leads to the production of a large proportion of acetyl-CoA produced by pluripotent cells. Further, functional assays reported the association of cell differentiation with acetyl-CoA and histone acetylation levels.

6 Glucose Transporters in Cancer Therapies

The detection of tumor at early stage has always been a challenge to cure the disease. Several techniques have been already used widely to treat this disease such as X-ray imaging, computed tomography scan (CT scan), or positron emission tomography (PET) scan, which essentially provide useful information about tumor volumes, anatomy, location, function, or metabolism. PET scan is one such technique which heavily relies on the glucose uptake level of cancer cells. It uses radiolabeled glucose analog 2-deoxy-2-[fluorine-18] fluoro-D-glucose (^{18}F-FDG) (Fass 2008; Zhu et al. 2011). GLUTs help this radiotracer transportation across cell membranes, which is then phosphorylated by hexokinases into FDG-6-phosphate and restricts further metabolization. This led to the accumulation of FDG-6-phosphate in the cytoplasm making its cell tracing activity increased (Plathow and Weber 2008). Although PET is a standard technique to detect the increased tumor metabolism, the sensitivity of this technique differs among the different cancer types (Wuest et al. 2018). This heterogeneity has been particularly associated with GLUT1 or GLUT3 tumor expression since ^{18}F-FDG has a high affinity for glucose transporters, and especially for these two uniporters. This positive association between GLUT1 over-expression and ^{18}F-FDG uptake has been reported in cervical cancer, thymic epithe-lial tumors, primary gastric lymphoma (PGL), as well as metastatic pulmonary tumors (Kaira et al. 2010; Watanabe et al. 2013; Kaira et al. 2011; Takahashi et al. 2014). Similarly, GLUT3 expression and ^{18}F-FDG accumulation have been corre-lated in primary central nervous system lymphoma (PCNSL) (Hong and Lim 2012). But surprisingly this correlation of GLUT1 expression and ^{18}F-FDG was not seen in adenocarcinoma (Yen et al. 2004). These highlights the limitations and the necessity to utilize alternative carbon sources that can be used other than glucose such as lactate. Usage of lactate has made it a predominant carbon fuel in lung tumors. Glutamine is also considered one other carbon source that contributes to Krebs cycle in pancreatic cancer (Hui et al. 2017; Housman et al. 2014; Kuang et al. 2017).

Many of the cancer cells initially are sensitive to chemotherapy but at later stages most of the cancer types acquire the ability to relapse and resist the therapeutic strategies. This resistance is usually acquired by variety of mechanisms, one among which largely involved is metabolic regulations that promote drug inhibition, degra-dation, or cellular export (Tsukioka et al. 2007). Among the different possibilities that enable cancer cells to resist therapy, glucose transporters are reported to be involved. As an example, the antiangiogenic VEGF-neutralizing antibody called bevacizumab, used in the treatment of glioblastoma, has a limited response dura-tion. A recent study comparing bevacizumab-responsive to resistant patient-derived tumor xenografts identified higher glucose uptake, glycolysis, and survival in low glucose conditions in drug-resistant tumors, phenotypes that were recapitulated upon GLUT3 overexpression (Huang et al. 2014). Accordingly, GLUT3 was upreg-ulated in bevacizumab-resistant versus sensitive tumors. Studies in ovarian cancer also showed the association between glucose transporter expression and tumor angiogenesis using immunohistochemistry analysis (Kunkel et al. 2007). A study was done with patients having advanced cervical squamous cell carcinoma to observe the role of GLUT1 in response to radiotherapy. Of the patients 53% report-edly showed overexpression of GLUT1 who were radiation sensitive. Patients with increased GLUT1 expression in tumors showed an increased resistance to radio-therapy and shorter progression-free survival than those with low GLUT1. GLUT1 has been reportedly known to be a biomarker in oral squamous cell carcinoma which was proclaimed by performing immunohistochemistry from 40 pretreated tumor biopsies. GLUT1 was seen to be a prominent factor in chemoresistance and was confirmed by a study done with the NCI-60 panel of human cancer cell lines. The underlying mechanisms by which GLUTs are involved in chemo- or radio-resistance remain largely unknown. However, a few studies suggest a connection between glycolysis and the anti-apoptotic gene *MCL1*, whereby glycolysis inhibition blocks its translation. Importantly, glycolysis inhibition used in combination with ABT-737, a pro-apoptotic BH3-mimetic compound, increases the sensitivity of lymphoma cells to this drug (Evans et al. 2008).

7   *Glut* Gene Targeting to Interrogate Their Function in Tumors

Targeting of glut gene was done at each level of gametogenesis, embryogenesis, or specific cell types including pancreatic β-cells or neurons to know its consequence (Meynet et al. 2012; Pradelli et al. 2010; Guillam et al. 1997; Schmidt et al. 2008, 2009). Knockdown of Glut1 and Glut3 has reported to lead embryonic lethality. This has led to the successful development of conditional mouse models (Pradelli et al. 2010; Wang et al. 2006). This issue was solved by developing mice with spatiotemporal control of Glut deletion. Although all Glut genes from class 1 have been engineered to enable Cre/LoxP-mediated recombination ($Glut1^{Flox/Flox}$, $Glut2^{Flox/Flox}$, $Glut3^{Flox/Flox}$ and $Glut4^{Flox/Flox}$ mice) (Young et al. 2011; Seyer et al. 2013; Fidler et al. 2017), only *Glut1* conditional gene deletion has been studied in the context of tumor development, in breast cancer. First, isolated mammary cells from $Glut1^{Flox/Flox}$ mice were transformed *in vitro* with polyomavirus middle T antigen (PyMT), followed by recombination using adenovirus-Cre (Ad-Cre) or Ad-GFP as control. This made the cells to uptake less glucose, synthesize lipids, and proliferate. Implantation of Glut-1 deficient cells into the mammary fat pads of immunodeficient mice formed tumors that grew slower and had a reduced cell proliferation compared to control tumors This led to an elegant model of breast cancer, where immunocompetent mice express or not Glut1 (NIC-Glut1$^{+/+}$, NIC-Glut1$^{Flox/+}$, and NIC-Glut1$^{Flox/Flox}$) in the mammary epithelial compartment and the tumor epithelial cells (Abel et al. 1999). At 200 days of life, all NIC-Glut1$^{+/+}$ mice had tumors, whereas almost all NIC-Glut1$^{Flox/Flox}$ mice remained tumor-free even at >500 days. Strikingly, NIC-Glut1$^{Flox/+}$ heterozygous mice behaved like NIC-Glut1$^{Flox/Flox}$ mice, with most animals having no tumor in the long term, indicating that the loss of even a single *Glut1* allele is sufficient to impose a strong break to breast tumor development in this model. Upon the deletion of one or both alleles of *Glut1* the epithelial cells from mammary glands showed less proliferation, but the appearance of the mammary ductal was overall normal. The deletion of *Glut1* allele deletion leads to the elimination of tumorigenic mammary cells providing an inability to form tumors (Wellberg et al. 2016). In conclusion, similar studies and investigations using cancer mouse models with *Glut* deletion can provide important biological information on their role in cancer development. This understanding of complete mechanism can also enable the finding of metabolic targets to cure the disease. Undoubtedly, additional investigations from this and other mouse models of cancer with *Glut* conditional deletion will continue to provide important biological information on their role in cancer development.

8 Conclusions

Glycolysis has been a very major area of study related to cancer progression. The metabolic regulation serves as one of the key hallmarks of cancer. So understanding and studying the different players in this process have been a necessity to bring up novel therapeutic strategies to treat cancer. Among the several major players, GLUTs have been a subject of multiple investigations. The major step in glycolysis is the consumption of glucose to generate energy and lead to multiple other signaling pathways to generate essential metabolites to enhance the cancer progression. The overexpression of glucose transporters in cancer has set a benchmark to target the glucose transporters and halt the cancer progression. Several in vitro studies have highlighted the importance of glucose transporters in cancer progression which makes it critical to be studied in in vivo system as well. Additionally, the impact of GLUT proteins outside the cancer cells also has to be explored largely to understand the fundamental knowledge on their regulation and deregulation of glucose transporters. Studies on GLUT1 regulation has been explored on the level of genetics also using the cre recombinase strategy with the help of mice models of cancer that highlights the importance of GLUTs in distinct cell types of the tumor mass. Although GLUTs' role has been widely studied in cancer progression, there are no widely available inhibitors against it. So, addressing the chemical inhibition of GLUTs is the need of the hour to come up with effective chemotherapeutic strategy against cancer cells.

References

Abel ED, Kaulbach HC, Tian R, Hopkins JCA, Duffy J, Doetschman T, Minnemann T, Boers M-E, Hadro E, Oberste-Berghaus C, et al. Cardiac hypertrophy with preserved contractile function after selective deletion of GLUT4 from the heart. J Clin Invest. 1999;104:1703–14.

Apostolou E, Ferrari F, Walsh RM, Bar-Nur O, Stadtfeld M, Cheloufi S, Stuart HT, Polo JM, Ohsumi TK, Borowsky ML, et al. Genome-wide chromatin interactions of the Nanog locus in pluripotency, differentiation, and reprogramming. Cell Stem Cell. 2013;12:699–712.

Augustin R. The protein family of glucose transport facilitators: it's not only about glucose after all. IUBMB Life. 2010;62:315–33.

Baldwin SA, Barros LF, Griffiths M. Trafficking of glucose transporters–signals and mechanisms. Biosci Rep. 1995;15:419–26.

Barthel A, Okino ST, Liao J, Nakatani K, Li J, Whitlock JP, Roth RA. Regulation of GLUT1 gene transcription by the serine/threonine kinase Akt1. J Biol Chem. 1999;274:20281–6.

Chai YJ, Yi JW, Oh SW, Kim YA, Yi KH, Kim JH, Lee KE. Upregulation of SLC2 (GLUT) family genes is related to poor survival outcomes in papillary thyroid carcinoma: analysis of data from The Cancer Genome Atlas. Surgery. 2017;161:188–94.

Chen L-Q, Hou B-H, Lalonde S, Takanaga H, Hartung ML, Qu X-Q, Guo W-J, Kim J-G, Underwood W, Chaudhuri B, et al. Sugar transporters for intercellular exchange and nutrition of pathogens. Nature. 2010;468:527–32.

Christensen DR, Calder PC, Houghton FD. GLUT3 and PKM2 regulate OCT4 expression and support the hypoxic culture of human embryonic stem cells. Sci Rep. 2015;5:17500.

Dittmann K, Mayer C, Rodemann HP, Huber SM. EGFR cooperates with glucose transporter SGLT1 to enable chromatin remodeling in response to ionizing radiation. Radiother Oncol. 2013;107:247–51.

Estrada DE, Elliott E, Zinman B, Poon I, Liu Z, Klip A, Daneman D. Regulation of glucose transport and expression of GLUT3 transporters in human circulating mononuclear cells: studies in cells from insulin-dependent diabetic and nondiabetic individuals. Metabolism. 1994;43:591–8.

Evans A, Bates V, Troy H, Hewitt S, Holbeck S, Chung Y-L, Phillips R, Stubbs M, Griffiths J, Airley R. Glut-1 as a therapeutic target: increased chemoresistance and HIF-1-independent link with cell turnover is revealed through COMPARE analysis and metabolomic studies. Cancer Chemother Pharmacol. 2008;61:377–93.

Fass L. Imaging and cancer: a review. Mol Oncol. 2008;2:115–52.

Fei X, Qi M, Wu B, Song Y, Wang Y, Li T. MicroRNA-195-5p suppresses glucose uptake and proliferation of human bladder cancer T24 cells by regulating GLUT3 expression. FEBS Lett. 2012;586:392–7.

Fidler TP, Campbell RA, Funari T, Dunne N, Balderas Angeles E, Middleton EA, Chaudhuri D, Weyrich AS, Abel ED. Deletion of GLUT1 and GLUT3 reveals multiple roles for glucose metabolism in platelet and megakaryocyte function. Cell Rep. 2017;20:881–94.

Flavahan WA, Wu Q, Hitomi M, Rahim N, Kim Y, Sloan AE, Weil RJ, Nakano I, Sarkaria JN, Stringer BW, et al. Brain tumor initiating cells adapt to restricted nutrition through preferential glucose uptake. Nat Neurosci. 2013;16:1373–82.

Fu Y, Maianu L, Melbert BR, Garvey WT. Facilitative glucose transporter gene expression in human lymphocytes, monocytes, and macrophages: a role for GLUT isoforms 1, 3, and 5 in the immune response and foam cell formation. Blood Cells Mol Dis. 2004;32:182–90.

Guillam M-T, Hummler E, Schaerer E, Wu J-Y, Birnbaum MJ, Beermann F, Schmidt A, Deriaz N, Thorens B. Early diabetes and abnormal postnatal pancreatic islet development in mice lacking Glut-2. Nat Genet. 1997;17:327.

Ha T-K, Her N-G, Lee M-G, Ryu B-K, Lee J-H, Han J, Jeong S-I, Kang M-J, Kim N-H, Kim H-J, et al. Caveolin-1 increases aerobic glycolysis in colorectal cancers by stimulating HMGA1-mediated GLUT3 transcription. Cancer Res. 2012;72:4097–109.

Han AL, Veeneman BA, El-Sawy L, Day KC, Day ML, Tomlins SA, Keller ET. Fibulin-3 promotes muscle-invasive bladder cancer. Oncogene. 2017;36:5243–51.

Hong R, Lim S-C. 18F-fluoro-2-deoxyglucose uptake on PET CT and glucose transporter 1 expression in colorectal adenocarcinoma. World J Gastroenterol. 2012;18:168–74.

Housman G, Byler S, Heerboth S, Lapinska K, Longacre M, Snyder N, Sarkar S. Drug resistance in cancer: an overview. Cancer. 2014;6:1769–92.

Huang X-Q, Chen X, Xie X-X, Zhou Q, Li K, Li S, Shen L-F, Su J. Co-expression of CD147 and GLUT-1 indicates radiation resistance and poor prognosis in cervical squamous cell carcinoma. Int J Clin Exp Pathol. 2014;7:1651–66.

Hui S, Ghergurovich JM, Morscher RJ, Jang C, Teng X, Lu W, Esparza LA, Reya T, Zhan L, Yanxiang Guo J, et al. Glucose feeds the TCA cycle via circulating lactate. Nature. 2017;551:115–8.

Ito S, D'Alessio AC, Taranova OV, Hong K, Sowers LC, Zhang Y. Role of Tet proteins in 5mC to 5hmC conversion, ES-cell self-renewal and inner cell mass specification. Nature. 2010;466:1129–33.

Jacobs SR, Herman CE, MacIver NJ, Wofford JA, Wieman HL, Hammen JJ, Rathmell JC. Glucose uptake is limiting in T cell activation and requires CD28-mediated Akt-dependent and independent pathways. J Immunol. 2008;180:4476–86.

Kaira K, Endo M, Abe M, Nakagawa K, Ohde Y, Okumura T, Takahashi T, Murakami H, Tsuya A, Nakamura Y, et al. Biologic correlation of 2- [18 F]-Fluoro-2-Deoxy-D-glucose uptake on positron emission tomography in thymic epithelial tumors. J Clin Oncol. 2010;28:3746–53.

Kaira K, Okumura T, Ohde Y, Takahashi T, Murakami H, Oriuchi N, Endo M, Kondo H, Nakajima T, Yamamoto N. Correlation between 18F-FDG uptake on PET and molecular biology in metastatic pulmonary tumors. J Nucl Med. 2011;52:705–11.

Kawauchi K, Araki K, Tobiume K, Tanaka N. p53 regulates glucose metabolism through an IKK-NF-kappaB pathway and inhibits cell transformation. Nat Cell Biol. 2008;10:611–8.

Kerr EM, Gaude E, Turrell FK, Frezza C, Martins CP. Mutant Kras copy number defines metabolic reprogramming and therapeutic susceptibilities. Nature. 2016;531:110–3.

Kim YH, Jeong DC, Pak K, Han M-E, Kim J-Y, Liangwen L, Kim HJ, Kim TW, Kim TH, Hyun DW, et al. SLC2A2 (GLUT2) as a novel prognostic factor for hepatocellular carcinoma. Oncotarget. 2017;8:68381–92.

King BC, Esguerra JLS, Golec E, Eliasson L. CD46 activation regulates miR-150-mediated control of GLUT1 expression and cytokine secretion in human CD4 + T cells. J Immunol. 1950;196:1636–45.

Krzeslak A, Wojcik-Krowiranda K, Forma E, et al. Expression of glut1 and glut3 glucose transporters in endometrial and breast cancers. Pathol Oncol Res. 2012;18:721–8.

Kuang R, Jahangiri A, Mascharak S, Nguyen A, Chandra A, Flanigan PM, Yagnik G, Wagner JR, De Lay M, Carrera D, et al. GLUT3 upregulation promotes metabolic reprogramming associated with antiangiogenic therapy resistance. JCI Insight. 2017;2:e88815.

Kunkel M, Moergel M, Stockinger M, Jeong J-H, Fritz G, Lehr H-A, Whiteside TL. Overexpression of GLUT-1 is associated with resistance to radiotherapy and adverse prognosis in squamous cell carcinoma of the oral cavity. Oral Oncol. 2007;43:796–803.

Li Z, Bao S, Wu Q, Wang H, Eyler C, Sathornsumetee S, Shi Q, Cao Y, Lathia J, McLendon RE, et al. Hypoxia-inducible factors regulate tumorigenic capacity of glioma stem cells. Cancer Cell. 2009;15:501–13.

Li L, Liang Y, Kang L, Liu Y, Gao S, Chen S, Li Y, You W, Dong Q, Hong T, et al. Transcriptional regulation of the Warburg effect in cancer by SIX1. Cancer Cell. 2018;33:368–385. e7.

Liu M, Gao J, Huang Q, Jin Y, Wei Z. Downregulating microRNA-144 mediates a metabolic shift in lung cancer cells by regulating GLUT1 expression. Oncol Lett. 2016;11:3772–6.

Locasale JW, et al. Phosphoglycerate dehydrogenase diverts glycolytic flux and contributes to oncogenesis. Nat Genet. 2011;43(9):869–74.

Loenarz C, Schofield CJ. Physiological and biochemical aspects of hydroxylations and demethylations catalyzed by human 2-oxoglutarate oxygenases. Trends Biochem Sci. 2011;36:7–18.

Lopez-Serra P, Marcilla M, Villanueva A, RamosFernandez A, Palau A, Leal L, Wahi JE, SetienBaranda F, Szczesna K, Moutinho C, et al. A DERL3-associated defect in the degradation of SLC2A1 mediates the Warburg effect. Nat Commun. 2014;5:3608.

Macheda ML, Rogers S, Best JD. Molecular and cellular regulation of glucose transporter (glut) proteins in cancer. J Cell Physiol. 2005;202:654–62.

Makinoshima H, Takita M, Matsumoto S, Yagishita A, Owada S, Esumi H, Tsuchihara K. Epidermal growth factor receptor (EGFR) signaling regulates global metabolic pathways in EGFR-mutated lung adenocarcinoma. J Biol Chem. 2014;289:20813–23.

Maratou E, Dimitriadis G, Kollias A, Boutati E, Lambadiari V, Mitrou P, Raptis SA. Glucose transporter expression on the plasma membrane of resting and activated white blood cells. Eur J Clin Investig. 2007;37:282–90.

Masin M, Vazquez J, Rossi S, Groeneveld S, Samson N, Schwalie PC, Deplancke B, Frawley LE, Gouttenoire J, Moradpour D, et al. GLUT3 is induced during epithelial-mesenchymal transition and promotes tumor cell proliferation in non-small cell lung cancer. Cancer Metab. 2014;2:11.

Mathupala SP, Ko YH, Pedersen PL. The pivotal roles of mitochondria in cancer: Warburg and beyond and encouraging prospects for effective therapies. Biochim Biophys Acta. 1797;2010:1225–30.

Meneses AM, Medina RA, Kato S, Pinto M, Jaque MP, Lizama I, García Mde los A, Nualart F, Owen GI. Regulation of GLUT3 and glucose uptake by the cAMP signalling pathway in the breast cancer cell line ZR-75. J Cell Physiol. 2008;214:110–6.

Meynet O, Beneteau M, Jacquin MA, Pradelli LA, Cornille A, Carles M, Ricci J-E. Glycolysis inhibition targets Mcl-1 to restore sensitivity of lymphoma cells to ABT-737-induced apoptosis. Leukemia. 2012;26:1145–7.

Micucci C, Orciari S, Catalano A. Hyperglycemia promotes K-Ras-induced lung tumorigenesis through BASCs amplification. PLoS One. 2014;9:e105550.

Mueckler M, Thorens B. The SLC2 (GLUT) family of membrane transporters. Mol Asp Med. 2013;34:121–38.

Nagarajan A, Dogra SK, Sun L, Gandotra N, Ho T, Cai G, Cline G, Kumar P, Cowles RA, Wajapeyee N. Paraoxonase 2 facilitates pancreatic cancer growth and metastasis by stimulating GLUT1-mediated glucose transport. Mol Cell. 2017;67:685–701. e6.

Osthus RC, Shim H, Kim S, Li Q, Reddy R, Mukherjee M, Xu Y, Wonsey D, Lee LA, Dang CV. Deregulation of glucose transporter 1 and glycolytic gene expression by c-Myc. J Biol Chem. 2000;275:21797–800.

Plathow C, Weber WA. Tumor cell metabolism imaging. J Nucl Med. 2008;49:43S–63S.

Pradelli LA, Beneteau M, Chauvin C, Jacquin MA, Marchetti S, Munoz-Pinedo C, Auberger P, Pende M, Ricci J-E. Glycolysis inhibition sensitizes tumor cells to death receptors-induced apoptosis by AMP kinase activation leading to Mcl-1 block in translation. Oncogene. 2010;29:1641–52.

Qu W, Ding S, Cao G, Wang S, Zheng X, Li G. miR-132 mediates a metabolic shift in prostate cancer cells by targeting Glut1. FEBS Open Bio. 2016;6:735–41.

Sasaki H, Shitara M, Yokota K, Hikosaka Y, Moriyama S, Yano M, Fujii Y. Overexpression of GLUT1 correlates with Kras mutations in lung carcinomas. Mol Med Rep. 2012;5:599–602.

Schmidt S, Gawlik V, Holter SM, Augustin R, Scheepers A, Behrens M, Wurst W, Gailus-Durner V, Fuchs H, de Angelis MH, et al. Deletion of glucose transporter GLUT8 in mice increases locomotor activity. Behav Genet. 2008;38:396.

Schmidt S, Hommel A, Gawlik V, Augustin R, Junicke N, Florian S, Richter M, Walther DJ, Montag D, Joost H-G, et al. Essential role of glucose transporter GLUT3 for post-implantation embryonic development. J Endocrinol. 2009;200:23–33.

Schwartzenberg-Bar-Yoseph F, Armoni M, Karnieli E. The tumor suppressor p53 down-regulates glucose transporters GLUT1 and GLUT4 gene expression. Cancer Res. 2004;64:2627–33.

Seyer P, Vallois D, Poitry-Yamate C, Schutz F, Metref S, Tarussio D, Maechler P, Staels B, Lanz B, Grueter R, et al. Hepatic glucose sensing is required to preserve b cell glucose competence. J Clin Invest. 2013;123:1662–76.

Shinde SR, Maddika S. PTEN regulates glucose transporter recycling by impairing SNX27 retromer assembly. Cell Rep. 2017;21:1655–66.

Simpson IA, Dwyer D, Malide D, Moley KH, Travis A, Vannucci SJ. The facilitative glucose transporter GLUT3: 20 years of distinction. Am J Physiol Endocrinol Metab. 2008;295:E242–53.

Suh HN, Han HJ. Fibronectin-induced VEGF receptor and calcium channel transactivation stimulate GLUT-1 synthesis and trafficking through PPARc and TC10 in mouse embryonic stem cells. Stem Cell Res. 2013;10:371–86.

Takahashi Y, Akahane T, Yamamoto D, Nakamura H, Sawa H, Nitta K, Ide W, Hashimoto I, Kamada H. Correlation between positron emission tomography findings and glucose transporter 1, 3 and L-type amino acid transporter 1 mRNA expression in primary central nervous system lymphomas. Mol Clin Oncol. 2014;2:525–9.

Tanegashima K, Sato-Miyata Y, Funakoshi M, Nishito Y, Aigaki T, Hara T. Epigenetic regulation of the glucose transporter gene Slc2a1 by β-hydroxybutyrate underlies preferential glucose supply to the brain of fasted mice. Genes Cells. 2017;22:71–83.

Tsukioka M, Matsumoto Y, Noriyuki M, Yoshida C, Nobeyama H, Yoshida H, Yasui T, Sumi T, Honda K-I, Ishiko O. Expression of glucose transporters in epithelial ovarian carcinoma: correlation with clinical characteristics and tumor angiogenesis. Oncol Rep. 2007;18:361–7.

Vander Heiden MG, Cantley LC, Thompson CB. Understanding the Warburg effect: the metabolic requirements of cell proliferation. Science. 2009;324:1029–33.

Waldhart AN, Dykstra H, Peck AS, Boguslawski EA, Madaj ZB, Wen J, Veldkamp K, Hollowell M, Zheng B, Cantley LC, et al. Phosphorylation of TXNIP by AKT mediates acute influx of glucose in response to insulin. Cell Rep. 2017;19:2005–13.

Wang D, Pascual JM, Yang H, Engelstad K, Mao X, Cheng J, Yoo J, Noebels JL, De Vivo CD. A mouse model for Glut-1 haploinsufficiency. Hum Mol Genet. 2006;15:1169–79.

Warburg O, Wind F, Negelein E. The metabolism of tumors in the body. J Gen Physiol. 1927;8(6):519–30.

Watanabe Y, Suefuji H, Hirose Y, Kaida H, Suzuki G, Uozumi J, Ogo E, Miura M, Takasu K, Miyazaki K, et al. 18F-FDG uptake in primary gastric malignant lymphoma correlates with glucose transporter 1 expression and histologic malignant potential. Int J Hematol. 2013;97:43–9.

Weihua Z, Tsan R, Huang W-C, Wu Q, Chiu C-H, Fidler IJ, Hung M-C. Survival of cancer cells is maintained by EGFR independent of its kinase activity. Cancer Cell. 2008;13:385–93.

Wellberg EA, Johnson S, Finlay-Schultz J, Lewis AS, Terrell KL, Sartorius CA, Abel ED, Muller WJ, Anderson SM. The glucose transporter GLUT1 is required for ErbB2-induced mammary tumorigenesis. Breast Cancer Res. 2016;18:131.

Wieman HL, Wofford JA, Rathmell JC. Cytokine stimulation promotes glucose uptake via phosphatidylinositol-3 Kinase/Akt regulation of Glut1 activity and trafficking. Mol Biol Cell. 2007;18:1437–46.

Wu N, Zheng B, Shaywitz A, Dagon Y, Tower C, Bellinger G, Shen C-H, Wen J, Asara J, McGraw TE, et al. AMPK-dependent degradation of TXNIP upon energy stress leads to enhanced glucose uptake via GLUT1. Mol Cell. 2013;49:1167–75.

Wuest M, Hamann I, Bouvet V, Glubrecht D, Marshall A, Trayner B, Soueidan O-M, Krys D, Wagner M, Cheeseman C, et al. Molecular imaging of GLUT1 and GLUT5 in breast cancer: a multitracer positron emission tomography imaging study in mice. Mol Pharmacol. 2018;93:79–89.

Xu W, Yang H, Liu Y, Yang Y, Wang P, Kim S-H, Ito S, Yang C, Wang P, Xiao M-T, et al. Oncometabolite 2-hydroxyglutarate is a competitive inhibitor of a-ketoglutarate-dependent dioxygenases. Cancer Cell. 2011;19:17–30.

Yang H, Lin H, Xu H, Zhang L, Cheng L, Wen B, Shou J, Guan K, Xiong Y, Ye D. TETcatalyzed 5-methylcytosine hydroxylation is dynamically regulated by metabolites. Cell Res. 2014;24:1017–20.

Yen T-C, See L-C, Lai C-H, Yah-Huei CW, Ng K-K, Ma S-Y, Lin W-J, Chen J-T, Chen W-J, Lai C-R, et al. 18F-FDG uptake in squamous cell carcinoma of the cervix is correlated with glucose transporter 1 expression. J Nucl Med. 2004;45:22–9.

Young CD, Lewis AS, Rudolph MC, Ruehle MD, Jackman MR, Yun UJ, Ilkun O, Pereira R, Abel ED, Anderson SM. Modulation of glucose transporter 1 (GLUT1) expression levels alters mouse mammary tumor cell growth in vitro and in vivo. PLoS One. 2011;6:e23205.

Yu J, Li J, Zhang S, Xu X, Zheng M, Jiang G, Li F. IGF-1 induces hypoxia-inducible factor 1amediated GLUT3 expression through PI3K/Akt/ mTOR dependent pathways in PC12 cells. Brain Res. 2012;1430:18–24.

Yun J, Rago C, Cheong I, Pagliarini R, Angenendt P, Rajagopalan H, Schmidt K, Willson JKV, Markowitz S, Zhou S, et al. Glucose deprivation contributes to the development of KRAS pathway mutations in tumor cells. Science. 2009;325:1555–9.

Zha X, Hu Z, Ji S, Jin F, Jiang K, Li C, Zhao P, Tu Z, Chen X, Di L, et al. NFjB up-regulation of glucose transporter 3 is essential for hyperactive mammalian target of rapamycin-induced aerobic glycolysis and tumor growth. Cancer Lett. 2015;359:97–106.

Zhang JZ, Abbud W, Prohaska R, Ismail-Beigi F. Overexpression of stomatin depresses GLUT-1 glucose transporter activity. Am J Phys Cell Physiol. 2001;280:C1277–83.

Zhang C, Liu J, Liang Y, Wu R, Zhao Y, Hong X, Lin M, Yu H, Liu L, Levine AJ, et al. Tumourassociated mutant p53 drives the Warburg effect. Nat Commun. 2013;4:2935.

Zhu A, Lee D, Shim H. Metabolic PET imaging in cancer detection and therapy response. Semin Oncol. 2011;38:55–69.

Oxidative Stress: A Potential Link Between Pesticide Exposure and Early-Life Neurological Disorders

Shalini Mani, Anvi Jain, Aaru Gulati, Sakshi Tyagi, Km Vaishali Pal, Himanshi Jaiswal, and Manisha Singh

Contents

Abstract Since the past few decades, various chemicals and termed pesticides have been used to protect the crops from pests. The pesticides are mostly used in large quantities in agriculture and toxic exposure to such chemicals is reported to impact human health by causing morbidity and severe disabilities. The indiscriminate use of different pesticides is reported to cause neurotoxicological problems at different stages of human lives. Most of the pesticides are reported to raise the oxidative stress level in the human system. This elevated level of reactive oxygen species is known to exhibit severe damage to different cellular systems. Neuron cells are highly sensitive toward the elevation of oxidative stress and thus pesticide-mediated toxicity leads to different neurological disorders in humans. This chapter elaborates upon different classes of pesticides and their effect on oxidative stress. Following this, the association between pesticide exposure and the development of different neurological disorders are also discussed in this chapter.

Authors Anvi Jain and Aaru Gulati have equally contributed to this chapter.

S. Mani (✉) · A. Jain · A. Gulati · S. Tyagi · K. V. Pal · H. Jaiswal · M. Singh
Centre for Emerging Diseases, Department of Biotechnology, Jaypee Institute of Information Technology, Noida, Uttar Pradesh, India
e-mail: shalini.mani@jiit.ac.in

Keywords Pesticides · Oxidative stress · Neurological disorders · Reactive oxygen species · Nervous system

1 Introduction

Pesticides are a subcategory of agrochemicals that are a collective term for various chemicals that are used in agriculture or in public health protection programs to kill unwanted plants, insects, rodents, and molds (Fianko et al. 2011; Gilden et al. 2010). Acephate, glyphosate, Deet, metaldehyde, propoxur, boric acid, dursban, diazinon, malathion, and DDT are some specific examples of synthetic chemical pesticides.

Pesticides are classified into insecticides, herbicides, fungicides, and rodenticides based on the type of pests they kill (Aktar et al. 2009; Gilden et al. 2010). Another way of grouping pesticide is by considering their derivatives or chemical forms from a particular source or manufacturing process, namely organophosphates (OPs), bipyridyl, organochlorine (OCs) insecticides, carbamates, chlorophenoxy pyrethroid, neonicotinoid, and phosphonate. The drastic usage of pesticides is increasing day by day globally. It is estimated that pesticide usage will increase to more than 3.6 million tonnes globally. Even though pesticide utilization is immensely crucial for the agricultural industry, its dire consequences cannot be neglected. Pesticides cause grave health threats to living beings especially to infants and children because of their known biomagnification and persistent nature. Worldwide estimates of acute pesticide poisoning ranged from around 1–41 million people being affected yearly as per "The Pesticide Action Network" in 2010 (Pesticide Action Network 2010). The World Health Organization (WHO 2006) has estimated that more than 30% of the total overall global risk of diseases in children can be traced to environmental toxicants and causes, the major pesticides are prime contributors (Lekei et al. 2017). There are several different ways by which pesticides can find their way into the human body as shown in Fig. 1.

Pesticides have been found in the amniotic fluid, body tissues of humans, and fetuses as they can reach the placenta even during the early stages of fetal life. This poses a higher risk for the child's natural development and growth, and the child appears to be infected by breastfed after birth as breast milk can be tainted with pesticides and is at the very top of the food chain. Meconium has also been found to contain pesticides. Placentas, amniotic fluid, blood, umbilical cord, fetuses, and breast milk (including HCB, DDT, and HCH isomers) of women in polluted areas have been shown to contain traces of organochlorine pesticides (UNICEF 2018). In particular, babies and young children are more exposed and more vulnerable to the toxic effects of pesticides, especially in rural areas. Children are subjected to toxic contaminants from the widespread use of pesticides in crops and households, gardens, food processing, public areas, and schools, and are thus subject to chronic health issues. According to a study done in Egypt, children who were aged between 9 and 15 years and seasonally worked in the fields of cotton to apply pesticides were reported to have remarkably reduced cholinesterase levels and drastically more

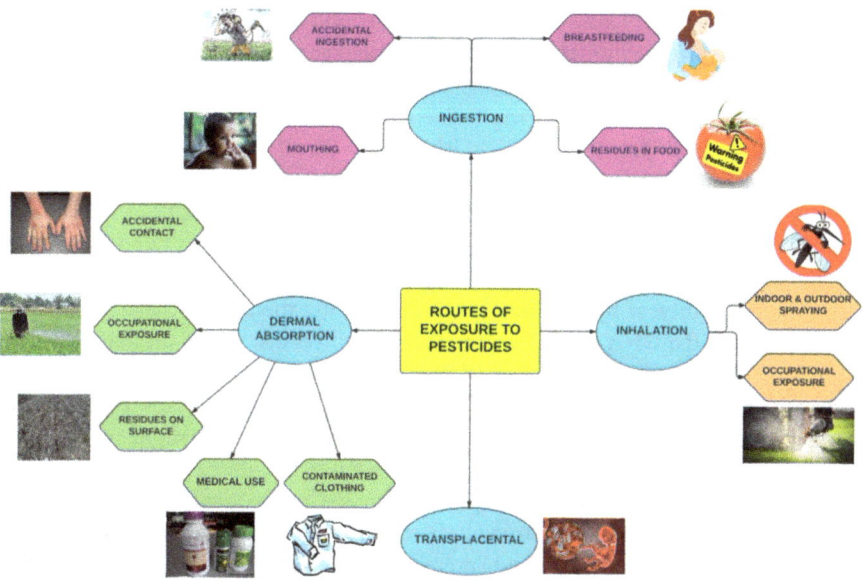

Fig. 1 Schematic representation of various routes of exposure to pesticides

neurological symptoms compared to the controls (Abdel Rasoul et al. 2008). In adulthood, fetal exposure to pesticides can also cause some forms of lethal diseases. Exposure to early life will affect their developing brain, other body nervous systems, and interrupt their physiological and mental development, resulting in a variety of disorders and diseases. Any of the diseases may occur later in adulthood, and by genetic abnormalities or defects, especially changes in the reproductive system and cancers, some may be passed on to successive generations. Scientific research indicates that even low exposure to pesticides in the womb and early childhood, which is well below what is commonly considered healthy, presents a significant danger to the natural development and development of children. Pesticides have been linked to endocrine disruption, birth defects, damage to the nervous system, cancer, breathing difficulties, and deeply impact behavioral and intellectual development.

In certain parts of the brain, nerve cells which will eventually be close to the surface of the brain move from more central locations at the time of early infant and late fetal life. Throughout childhood, the activity of myelination of nerve tracts continues, but the incomplete myelination of nerve fibers modifies the responses to neurotoxins. Not only these two developmental processes, but many developmental processes take place in the brain which can be disrupted by neurotoxic pesticides. Unlike other organs of the body, the brain cannot repair damaged cells (Lekei et al. 2017). There have been various studies relating prenatal or early exposure to pesticides with postnatal neuronal diseases, such as attention deficit hyperactivity disorder (ADHD), cerebral palsy, epilepsy, autism spectrum disorder (ASD), multiple sclerosis. Also, early exposure to neurotoxic pesticides can later in life increase the risk of chronic neurologic diseases like Parkinson's disease, dementia, Alzheimer's, and amyotrophic lateral sclerosis (ALS). It is understood that pesticides activate certain pathways responsible for free radical formation and modification of antioxidant defense mechanisms. In exchange, this induces an imbalance in the cell's redox state which causes oxidative stress. Progressive oxidative damage can be responsible for the excessive development of ROS and RNS. As a second messenger, reactive oxygen species (ROS) at low concentrations play an incredibly critical role in cell signaling.

But ROS can injure cellular macromolecules such as proteins, deoxyribonucleic acid (DNA), and lipids at higher concentrations and long-term exposure, which eventually makes way to necrotic and apoptotic cell death (Chen et al. 2012). ROS have the capability of lipids denaturation, producing membrane damage, structurally damaging the DNA, and changing the inner proteins, which ultimately affects their structure and function. When ROS is in excessive amounts, these symptoms occur in the brain and the capacity of neuronal antioxidant processes to counterbalance the destructive reactions is then severely diminished (Popa-Wagner et al. 2013).

The brain, because of its strong need for oxygen, is vulnerable to the effects of ROS and also has comparatively poor defensive antioxidant mechanisms. The human brain has a wealth of extremely peroxidizable substrates that serve as a prooxidant in hemorrhagic stroke and sometimes cause autophagic cell death. In several neurodegenerative disorders, oxidative stress has been observed. This chapter aims to understand the crucial aspect of the detailed association of pesticides, neuronal disorders, and the involvement of oxidative stress in such cases.

2 Major Pesticides and Their Effect on Brain During Early Life

Pesticides are usually classified based on applications, target organism, and chemical nature. They involve a broad range of compounds, including insecticides, herbicides, fungicides, rodenticides, and fumigants, intended to prevent, destroy, or repel pests. The primary focus is on those insecticides, herbicides, and specific chemical classes that are likely to cause more acute and chronic neurotoxicity among children. Among various types of pesticides, fumigants and fungicides are potentially less noxious and less active in causing acute toxicity and poisoning in early life or childhood. Various classes of pesticide and their effect on neuronal functioning are discussed in Table 1.

2.1 Organophosphates

Organophosphates (OPs) are widely used in the world as a major insecticide class that blocks insects, via suppressing the acetylcholinesterase (AChE) enzyme activity that hydrolyzes the neurotransmitter acetylcholine in both pest and other off-target organisms in the peripheral and central nervous system (Bjørling-Poulsen

Table 1 Summary of different classes of pesticides, their effect on the biological system of humans, and associated early life neurological problems

Pesticide class	Pesticide example	Use	Effect of pesticides	Associated neurological disorder	References
Organophosphate	Chlorpyrifos, malathion, diazinon, parathion, dimethoate	Insecticide	Can cause reduced IQ levels and cognitive and psychomotor development Inhibit AChE enzyme Altered neuronal-differentiation Disruption of neurotransmitter signals Altered brain development Behavioral impairments, including motor development/coordination deficits	ADHD, seizures, ASD, memory loss, cerebral palsy, pervasive developmental disorders	Abreu-Villaça and Levin (2018), Voorhees et al. (2017), Bjørling-Poulsen et al. (2008), Miodovnik (2011), and Liew et al. (2020)
Organochlorine	DDT, lindane, aldrin, dieldrin, chlordane, endrin	Insecticide	Altered neuronal differentiation and survival Low cognitive development, hampered motor functions, quantitative, verbal, memory ability Inhibition of voltage-gated calcium channels Altered neurotransmitter levels Oxidative stress and mitochondrial damage	Impaired psychomotor and cognitive development, ASD, ADHD, memory loss, seizures, anxiety disorders, multiple sclerosis	Abreu-Villaça and Levin (2018), Saeedi Saraviand Dehpour (2016), and Yamazaki et al. (2017)

(continued)

Table 1 (continued)

Pesticide class	Pesticide example	Use	Effect of pesticides	Associated neurological disorder	References
Pyrethroid	Pyrethrins, deltamethrin, cypermethrin, fenvalerate, allethrin, bifenthrin	Insecticide	Altered levels of neurotransmitters and metabolites of monoamine neurotransmitters Increase oxidative stress Increases blood–brain barrier permeability Imbalances between pro- and anti-inflammatory responses Altered brain vascular formation	ASD, ADHD, learning and memory disorder	Mohammadi et al. (2019), Abreu-Villaça and Levin (2018), and Richardson et al. (2015)
Carbamates	Carbaryl, carbofuran, propoxur, bendiocarb, methomyl, aldicarb	Insecticide	Decreased neurogenesis Altered neuronal–differentiation Decreased survival of progenitor cells Delayed motor and sensorimotor development and activity Increased oxidative stress	The cognitive deficit, anxiety and depression disorder, hypotonia, cerebral palsy, epilepsy	Abreu-Villaça and Levin (2018), Liew et al. (2020), and Bjørling-Poulsen et al. (2008)
Neonicotinoids	Imidacloprid, acetamiprid, thiacloprid	Insecticide	Increased brain AChE activity Abnormal social interactions	Sensorimotor impairment, ADHD	Abreu-Villaça and Levin (2018) and Roberts et al. (2019)

Pesticide class	Pesticide example	Use	Effect of pesticides	Associated neurological disorder	References
Bipyridyl	Paraquat, diquat	Herbicide	Reduced levels of locomotor activity Decreased cognitive function Loss of dopaminergic neurons	Behavioral disorders, learning deficits, and memory deficits	Saeedi Saravi and Dehpour (2016) and Konthonbut et al. (2018)
Chlorophenoxy	2,4-Dichlorophenoxyacetic acid (2,4-D)	Herbicide	Increased dopamine D2-type receptor Decreased AChE activity	Aggression, hypertonia, ataxia, nystagmus, hallucinations, convulsions	Bjørling-Poulsen et al. (2008)
Phosphonate	Glyphosate	Herbicide	Delayed developmental reflexes, sensorimotor, vestibular, and/or proprioceptive function Reduces locomotor activity Loss of dopaminergic neurons Loss of neurotransmitters	ASD, dementia, anxiety and depression disorder, ADHD	Ait-Bali et al. (2020), Pu et al. (2020), and Seneff et al. (2015)

et al. 2008). It's being used for both agricultural and residential applications and is the major class of toxic compounds. Several studies have indicated that OPs (e.g., malathion, parathion, diazinon, chlorpyrifos (CPF), Phosmet) as one of the environmental pollutant threats to human health. OPs do not bioaccumulate and therefore are more advantageous as compared to organochlorines (Abreu-Villaça and Levin 2017; Roberts and Karr 2012). Mostly, the OPs are absorbed through the respiratory system, gastrointestinal tract, and skin. OPs are further reported to cause neurotoxicity as well as neurophysical, neurobehavioral, and neurochemical alterations in the brain. Among all the OPs, CPF is perhaps the mainly reported in studies, using both *in vitro* and animal models. Some of the adverse consequences of CPF are impaired cognition and motor function, attention deficit hyperactivity disorder and developmental problems, altered brain development, and altered neural differentiation (Bjørling-Poulsen et al. 2008). Another effect of CPF is inhibition of DNA synthesis, which has been related to the anticholinesterase pathway of developmental neurotoxicity. This effect was apparent before the emergence of the cholinergic synapses (Abreu-Villaça and Levin 2018).

Human exposure to different organophosphate during gestation is linked with poor neurodevelopment and disorders in children as mentioned in Table 1. Epidemiological studies have shown that OP is correlated with learning and memory deficits (Voorhees et al. 2017). OPs exposures in the third trimester of gestation are consistent in causing ADHD, poor motor and visuospatial functions, visuospatial memory and ASD, in the early age of the child (Abreu-Villaça and Levin 2018). Within mothers, OPs were linked to peculiar reflexes and signs of pervasive developmental disorder at age of 2 years (Miodovnik 2011). Many case studies indicated a high possibility of seizures because of OPs exposure (Roberts and Karr 2012). Dimethoate exposure demonstrated a consistent association with cerebral palsy (Liew et al. 2020).

2.2 Organochlorines

Organochlorines (OCs) insecticides are persistent environmental contaminants that bioaccumulate in humans through the food chain and have elevated stability and lipophilicity, and sluggish rate of removal from the body, hence they are restricted and banned in most countries (Abreu-Villaça and Levin 2018; Arora et al. 2013). OCs include DDT (dichlorodiphenyltrichloroethane), hexachlorocyclohexane, and cyclodienes. By modifying energy metabolism, a constant widening of the sodium channels, and association with gamma-aminobutyric acid (GABA) receptors, OCs are known to have their insecticidal property, resulting in the disturbance of the nerve fibers' sodium and potassium currents. Because of their excessive lipid solubility, they can penetrate the breast milk through the blood, and also capable of going through the placenta and blood–brain barrier (BBB) (Abreu-Villaça and Levin 2018). Therefore, the chronic exposure of OCs during pregnancy is the main concern for developmental health effects, particularly poor neurodevelopmental and perinatal neuropsychological deficits like low cognitive development, diminished motor functions and psychomotor growth, loss of memory, anxiety, ASD, ADHD, and altered behavioral activity as summarized in Table 1. Also, usually used OCs can have irreversible effects on the CNS and peripheral nervous systems as well (Saeedi Saravi and Dehpour 2016).

It has been found that endosulfan is possibly associated with idiopathic seizures in children (Arora et al. 2013). Aldrin, dieldrin, DDT, and toxaphene, which are some well-known organochlorines, are correlated with ALS (Kamel et al. 2012). Research on animal models evidenced that exposed organochlorines during gestation and lactation induces altered neurotransmitter level and prevents the voltage-gated calcium channels (Abreu-Villaça and Levin 2018).

2.3 Pyrethroid

Pyrethroids are a relatively more recent group of insecticides obtained from natural sources, i.e., pyrethrins from *Chrysanthemum* plants. Pyrethroids are composed of an acid moiety, ester, and alcohol moiety. There are two chiral carbons (trans and cis) in the acid moiety, which render stereoisomeric and cis isomers stronger in comparison to the trans isomers (Bjørling-Poulsen et al. 2008; Mohammadi et al. 2019). Synthetic pyrethroids are not much liable to hydrolysis and photodegradation than esters which have moderate insecticidal intensity, and significantly lower mammalian toxic effects (Abreu-Villaça and Levin 2017). Pyrethrins don't have lasting effects in heat or sunlight and, hence, usually have indoor utilization. Therefore, these insecticides are synthetically obtained compounds that are adjusted to be much more stable to the sun rays or heat and are used for pest management, mainly outdoors (Roberts and Karr 2012). Pyrethroid has been divided into types: the type I pyrethroids such as allethrin, bifenthrin, and permethrin don't have any cyano-group, while the type II pyrethroids such as deltamethrin, cypermethrin, and fenvalerate have a cyano-group in the α-position (Abreu-Villaça and Levin 2018).

The primary target of both types of pyrethroid in pest and off-target organisms is slowing down the regulation of voltage-gated sodium channels (Mohammadi et al. 2019). Therefore, the channel stays persistent for a prolonged period, letting sodium ions pass, and resulting in a reduction in the polarization of the neuronal membrane. Secondly, pyrethroid neurotoxicity is also caused by altering the kinetics of calcium and chloride channels and hence, leading to an alteration in the degree of neurotransmitter in the synaptic cleft. Type II pyrethroids bind to GABA receptors and block them in the human brain, triggering a reduction in the chlorine influx and thus causing hyperstimulation. Early exposure during the gestation period enhances the BBB permeability, triggers oxidative stress, and adversely affects the dopaminergic and noradrenergic systems and similarly, during the postnatal exposure, they imbalance pro- and anti-inflammatory actions; alteration in the serotonergic muscarinic system (Abreu-Villaça and Levin 2018). A study recently has manifested that pyrethroid intake during pregnancy has resulted in developmental neurotoxicity as shown by delayed or slow neurocognitive growth in 6-year-old kids (Viel et al. 2015) and also delayed mental development with poor motor skills, visual ability, impaired learning and memory, and attention problems as presented by reports. An experimental study in mice demonstrated that exposure to permethrin and cypermethrin in the neonatal stage caused severe behavioral activities, the difference in monoamine concentrations in the striatum, and induction of oxidative stress (Bjørling-Poulsen et al. 2008). Type II pyrethroid pesticide during development in mice exhibits ADHD, including hyperactive and impulsive activities (Abreu-Villaça and Levin 2018). Another study showed prenatal exposures to pyrethroid have been associated with ASD in children (Furlong et al. 2017).

2.4   Carbamates

Carbamates originate through carbamic acid and almost all of them are the esters of N-methyl carbamic acid (Abreu-Villaça and Levin 2017). Carbamates insecticides (such as carbaryl, carbofuran, methiocarb, and propoxur) are being utilized in farming as a replacement to the organophosphate. Carbamates, similar to organophosphate insecticides, inhibit AChE in the pests and off-target organisms, resulting in cholinergic toxicity and deposition of ACh in the nervous system (Roberts and Karr 2012). Carbamates induce only reversible inhibition of AChE. The effects of early exposure to these pesticides are more injurious and can linger till adulthood. Carbamates poisoning (mainly by aldicarb or methomyl) in children aged 1–8 years causes CNS depression and hypotonia in children (Bjørling-Poulsen et al. 2008). Exposure during the gestation period of pregnancy elicits neurodegeneration and reduces progenitor cell survival and neurogenesis. It is also reported to increased the oxidative stress in cerebellar Purkinje cells (Abreu-Villaça and Levin 2017). Fenobucarb, exposure to the zebrafish is also observed to induce developmental toxicity, including brain apoptosis, central nerve, and peripheral motor neuron damage, axon and myelin degeneration with reduced motility as well as hypoactivity (Zhu et al. 2020). Moreover, carbamates in lactating rats pointed toward the main reason for the late motor and sensorimotor development and increased level of anxiety in teenage rats treated to carbosulfan during the gestation period (Abreu-Villaça and Levin 2018). In a study, carbamates exposure during the first trimester was positively associated with cerebral palsy in female offspring (Liew et al. 2020). Carbamate-induced seizures have features in common with epileptic seizures which can begin in various brain regions and spread throughout causing lesions in many areas, including the cortex, hippocampus, thalamus, and amygdala (Jett 2012).

2.5   Neonicotinoids

Neonicotinoids (e.g., imidacloprid, acetamiprid, thiacloprid, and clothianidin) are a relatively new class of insecticides and a widely used class of insecticide. They are intended for pest management, in the veterinary market, and also for controlling the pest in fish agriculture. Neonicotinoid insecticides were developed to replace organophosphates and carbamates due to broad-spectrum insect toxicity, a low threat to the environment and off-target organisms, and high potency for insects (Abreu-Villaça and Levin 2018). They are derived from metabolic alterations to the formerly used "nicotinoid" insecticides, like nicotine especially. These insecticides are systemic, hence, they diffuse throughout the plant tissues rendering them poisonous toward insects and potentially vertebrates (Simon-Delso et al. 2015). They target the nicotinic N-acetylcholine receptors (nAChRs) and displace the acetylcholine (Roberts and Karr 2012). In insects, their attachment to nAChRs causes repeated neuronal discharges, leading to dysfunction and failure of cellular energy (Simon-Delso et al. 2015). Studies have characterized human exposure to neonicotinoids are likely to cause developmental disorders. Exposure of fruit flies (Drosophila) to Imidacloprid, showed aberrant social behavior with elevated distance, speed of flight, and agility all suggest changes in behavioral patterns of the fly that would be similar to the ADHD in children (Roberts et al. 2019). Early exposure during pregnancy induces sensorimotor dysfunction and greater brain AChE activity, mAChR binding, and glial fibrillary acidic protein expression in the motor cortex and hippocampus in young offspring (Abreu-Villaça and Levin 2018).

2.6 Bipyridyl

Bipyridyl (or bipyridinium) herbicides resemble the structure of quaternary ammonium compounds. They are non-selective contact herbicides, toxic to a large variety of grasses and broad-leafed crops when applied topically in the presence of sunlight (Eddleston 2016). Paraquat and diquat herbicides have a wide range of applications in agriculture and industries. Paraquat is a widely used herbicide and highly toxic to humans and, therefore, banned in many countries. Diquat, on the other hand, is much less widely used and often formulated in combination with paraquat or other herbicides. They exert their herbicidal activity by undergoing redox processing, resulting in the production of superoxide anions (Bjørling-Poulsen et al. 2008). Hydrogen peroxide and subsequently the highly reactive hydroxyl radical can be generated by the action of these anions formed, which can then cause lipid peroxidation and cell death. Severe intoxication with paraquat and diquat results in neuropsychological side effects because of the brain stem infarction and due to the intracranial hemorrhage (Bjørling-Poulsen et al. 2008). Chronic prenatal exposure of paraquat in mice alters the developing brain and its roles or activities, causing abnormal motor coordination, diminished rate of locomotor activity, and cognitive performance that may become apparent with advancing age (Konthonbut et al. 2018). High doses of paraquat promote the loss of midbrain dopaminergic neurons (Rudyk et al. 2017).

2.7   Chlorophenoxy

The chlorophenoxy herbicides are commonly used for the prevention of broad-leaved weeds/crops. They consist of a particular moiety of aliphatic carboxylic acid that is bound by an ether bond to chlorine- (or methyl-) substituted aromatic ring. One of the widely known chlorophenoxy herbicides is 2,4-dichlorophenoxyacetic acid (2,4-D) (Roberts and Karr 2012). 2,4-D is highly hydrophilic and disperse typically across the body without any selective accumulation in any single body organ.

2,4-D treatment in animal models cause alterations in the quality of CNS neurotransmitters and changes in behavior activity (depression, anxiety, aggression), and effect on blood–brain barrier transport. Exposure of 2,4-D in rats beyond lactation showed an increase in the dopamine D2-type receptor (Bjørling-Poulsen et al. 2008). Moreover, ingestion of 2,4-D is involved in causing severe neurotoxic effects like hypertonia, ataxia, nystagmus, hallucinations, and convulsions (Bjørling-Poulsen et al. 2008).

2.8   Phosphonate (Glyphosate)

Glyphosate is common and widely used worldwide. Glyphosate's herbicidal activity is caused by the inhibition of a vital enzyme of a plant, i.e., 5-enolpyruvyl shikimate-3-phosphate synthase. Also, glyphosate is poorly metabolized and generally eliminated in the urine and feces (Ait-Bali et al. 2020; Roberts and Karr 2012). Despite the facts, experimental studies and reports showed adverse effects of glyphosate on human health. Glyphosate is capable of crossing the placental barrier which could perhaps cause alteration in the developmental process of the fetus. Glyphosate-based herbicide (GBH) exposed offspring showcased retarded developmental responses pointing toward abnormal development of sensorimotor, vestibular, and sense of movement and functions. Another study showed maternal glyphosate exposure, reported onset of ASD-like behaviors (i.e., increased grooming time and social interaction deficits) in murine offspring (Pu et al. 2020). Glyphosate applied to corn and soy crops showed a strong association with ADHD anxiety disorder and dementia (Samsel and Seneff 2015). Furthermore, GBH leads to a significant decrease in AChE activity, which is closely linked to altered cognitive functions and also, loss of serotonin, dopamine, and norepinephrine neurotransmitters, and dopaminergic neurons cause altered locomotor activity (Ait-Bali et al. 2020).

Thus, there are different routes of exposure to pesticides and consequently, the function of neuronal cells may severely get affected leading to different brain disorders in early life as well (Fig. 2).

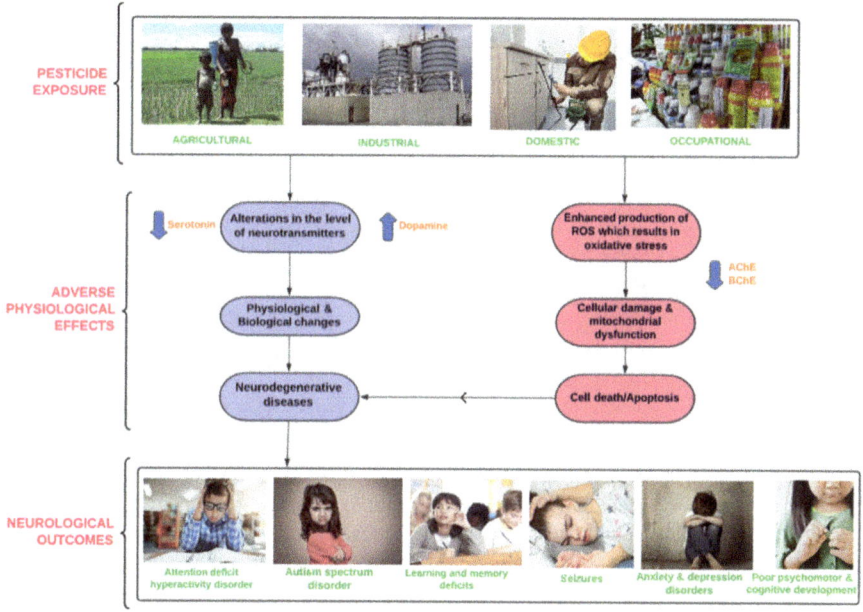

Fig. 2 Pesticidal exposure to humans through occupational sources associated with several catastrophic health effects, altered neurotransmitter and biochemical levels, and neurobehavioral outcomes

3 Pesticide and Oxidative Stress

As discussed, pesticide exposure is revealed to cause various neurological disorders in humans. A large number of studies have further indicated that this effect of pesticide exposure may primarily be due to its ability to generate oxidative stress in the human system. Oxidative stress can be caused by pesticide exposure, by enhancing the production of free radicals that grow more in the cell, by modifying antioxidant defense mechanisms, including scavenging enzymes and detoxification, or by interfering with subcellular or cellular membranes between ROS and resulting in an increase in lipid peroxidation (Abdollahi et al. 2004). Mitochondrial DNA comprises around 1% of the total cellular DNA and is perceived to be especially vulnerable to oxidative stress-related ROS attacks. Oxidative stress is believed to disrupt the pathway of cell signaling because ROS is thought to be the most important redox signaling messengers. The changes brought about by oxidative stress contributes to the production of multiple disorders, such as neurodegenerative diseases, including cerebral palsy, ALS, ADHD, mediating or amplifying the dysfunction of the neurons and inducing neurodegeneration, immunodeficiency syndromes, diabetes, respiratory disorders, schizophrenia, aging, hypertension, and cancer. Even in the early stages of these diseases, oxidative damage has been identified, suggesting that etiologies of theirs are related to free radicals (Díaz-Hung and González Fraguela 2014). The ROS is mostly produced in mitochondria as a by-product of oxidative phosphorylation (OXPHOS) activity (Fig. 3).

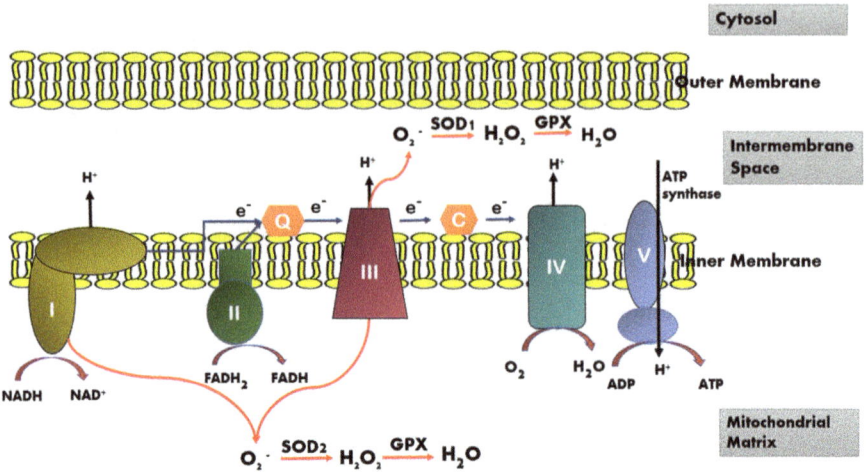

Fig. 3 Schematic representation of oxidative phosphorylation, ATP generation, ROS production, and their neutralization by antioxidant enzymes

3.1 Mitochondria and ATP Generation

Mitochondrion plays a crucial role in human health by performing various yet interconnected functions. They help in ATP (adenosine triphosphate) generation, the production of many biosynthetic intermediates, and are also involved in processes such as autophagy and apoptosis in response to cellular stresses. Some of the important roles of mitochondria in human health have been discussed below.

Mitochondria are regarded as the cellular powerhouse as they play quite a prominent role in energy production. They are the chief source for the production of ATP in eukaryotic cells. ATP is the energy-rich compound that is necessary for simple cell functions, including protein folding, protein degradation, biosynthesis, generation of power (in the case of cell division and muscle contraction), maintenance, and generation of membrane potential (Kühlbrandt 2015). Whenever there is an energy requirement in the cell, the mitochondrial ATP synthase produces ATP from ADP (adenosine diphosphate) and phosphate ions. ATP is generated in mitochondria via the action of OXPHOS. There is the generation of electron carriers (NADH and FADH2) via citric acid cycle enzymes in the matrix. OXPHOS is the procedure of generation of ATP through the transfer of electrons to O_2 via electron carriers like NADH or $FADH_2$. These electron carriers, found in the inner mitochondrial membrane, donate electrons to the electron transport chain (ETC). The ETC comprises four protein complexes (I–IV) (Herst et al. 2017). These complexes undergo conformational changes via sequential redox reactions resulting in the pumping of protons into the space of intermembrane from the matrix. As a consequence, an uneven distribution of protons, a pH gradient, and a transmembrane potential are generated thus creating a proton motive force. This flow back of protons to the mitochondrial matrix via enzyme complex results in the ATP synthesis (Duchen 2004). Hence, mitochondrion through ATP generation helps in various metabolic processes that are essential for cell survival.

3.2 ROS Production and Its Significance

ROS is a term that is generally used for the description of numerous molecules originating from molecular oxygen and free radicals (chemical compounds of one unpaired electron). Through electron transfer reactions or by energy transfer, the ground state oxygen may get converted to much more reactive forms. Under usual circumstances, molecular oxygen in its external orbital can accept up to four electrons which results in two molecules of H_2O. But where there is a state of electron leakage, oxygen by receiving an electron may undergo a univalent reduction, divalent reduction by acceptance of two electrons or either trivalent reduction by acceptance of three electrons, thereby creating the "ROS" (Cortés-Iza and Rodríguez 2018). ROS is produced in various cellular organelles. Mitochondrial metabolism is the primary source of ROS production and it is known to produce 90% of total ROS present inside the cell. Among all four mitochondrial complexes, complex I, II, and III mainly reside within the chain of respiration and are proposed to be the major contributor to the production of ROS species (Lenaz 2001; Dan Dunn et al. 2015). OXPHOS is the dominant process for the generation of unpaired electrons. These unpaired electrons further interact with O_2, thus resulting in the generation of superoxide ions which in turn are converted to other ROS species such as H_2O_2 and hydroxyl ions (–OH). However, free radicals and other reactive oxygen species are continuously formed. Albeit at a low concentration, it is important for various physiological functions and helps in maintaining cellular viability and proliferation too. A greater concentration of ROS species, on the other hand, may result in damage to the cellular structure. These free radicals are not only toxic themselves but are also involved in the pathophysiology of many disorders. The presence of unpaired electrons makes the molecule highly reactive (Kaur and Thakur 2018) and increases the susceptibility of biomolecules like proteinaceous enzymes, lipidic membranes, and DNA to oxidative damage due to ROS. As a result, the main source of production of ROS i.e., the mitochondrion is directly affected (Mittler 2017).

3.3 Neutralization of ROS

The body of a human being has multiple functions to prevent the damage caused by free radicals from happening. Antioxidant agents and pathways are the dominant and most prominent defense mechanisms of the human body. Any material that prevents or slows oxidative harm to a target molecule can be explained by the word antioxidant. By contributing electrons, these molecules are sufficiently stable to neutralize free radicals (Abdollahi et al. 2004). Mitochondria with the help of its antioxidant defenses are usually protected from ROS and oxidative damage (Deavall et al. 2012). Antioxidant mechanisms may be classified into two prominent systems in the body of a human being. The enzymatic mechanism that resists oxidation is the primary protection system against free radical disruption. This multi-layer network of the antioxidant system consists of superoxide dismutase (SODs), glutathione peroxidase (GPX), catalase (CAT), thioredoxin peroxidase (TRXP), and also glutathione reductase. Along with these antioxidant enzymes, mitochondria possess low molecular weight antioxidants like α-tocopherol and ubiquinol (Wei et al. 2001; Espinosa-Diez et al. 2015). These antioxidant molecules serve as lipid peroxyl radical scavengers and thereby deter further free radical chain lipid peroxidation reactions from occurring. But if oxidative stress is much greater than the ability of the enzymatic system, vitamins can play a role in the second line of protection. Vitamins such as vitamin C, vitamin E, vitamin A precursor, and beta-carotene are oxidized and inactivated, but free radicals are scavenged and extinguished. Every antioxidant nutrient has a particular job and always works together to improve the body's total antioxidant ability (Abdollahi et al. 2004). However, as mentioned above the lower concentration of ROS is crucial for various physiological functions such as survival of cells, differentiation of cells, and apoptosis. However, disruption of the balance between antioxidants and free radicals results in deleterious events, causing oxidative stress, which further may cause chronic and permanent damage.

3.4 Cellular Damage Due to ROS Accumulation

While partially or baseline regulated ROS levels have a notable pro-survival effect on cancer cells by acting as essential second intracellular messengers for numerous growth factors and cytokines, elevated ROS levels may result in damage at the cellular level and may even cause apoptosis. Although cancerous cells have evolved adaptive mechanisms to lessen the effects of damage due to oxidation, extreme levels of ROS may interfere with redox homeostasis by irreparable destruction of cellular macromolecules which includes the DNA, proteins, carbohydrates, and lipids, and thus affecting death or survival fate (Dalle-Donne et al. 2003) or at the level of signal transduction or transcriptional regulation by modulating redox-sensitive signaling proteins (Trachootham et al. 2008) or both.

Most of the components of cells are attacked by ROS which results in oxidation and fragmentation of nucleic acids, lipids, and also proteins.

Carbonylation activated due to oxidative stress, the process of S-nitrosylation, glutathionylation, and nitration of tyrosine can significantly harm proteins. The process of carbonylation is significantly crucial among protein oxidative modifications since it takes place at high *in vivo* levels and is also used as an oxidative stress marker. There are several ways of protein carbonyl formation. One way is via a direct oxidative attack on the threonine, proline, arginine, and lysine amino acids' side chains. Another approach is by an indirect reaction of the core amino group of lysine, cysteine, and histidine to reactive carbonyl compounds (deoxyosones, ketoamines, and ketoaldehydes) or through oxidized lipids comprising carbonyl (Nyström 2005). There are two main results of carbonylation. Firstly, heavily carbonylated proteins appear to form high molecular weight aggregates that are degradation resistant. Like degraded or unfolded proteins, they accumulate within the cell, which may further inhibit the activity of the proteasome (Dalle-Donne et al. 2006). Secondly, both the composition and function of various forms of proteins can be strongly modified by carbonylation. Actin carbonylation, for example, contributes to the disruption of the actin cytoskeleton that induces dramatic changes in cellular function (Dalle-Donne et al. 2001). Nevertheless, the inactivation of TRX and TRX reductase is consistent with dysregulation of cellular redox status and stress signaling owing to the carbonylation of their active-site cysteine and selenocysteine residues (Fang and Holmgren 2006). Besides, ROS-induced alterations target several other proteins that are critical in events of cell signaling, including those associated with apoptosis or cell survival, i.e., NFrB, H-Ras, PKC-ε AP-1, MAPK, p53, IP3 kinase, HIF-1, p38 MAPK, Bcl-2, ASK-1, caspases, Ras, and JNK (England and Cotter 2005).

Lipids are vulnerable to oxidative alteration as well. Lipid peroxidation results in the production of lipid radicals and other decomposition products of low molecular weight, such as acrolein, 4-hydroxy-2-nonenal, and malondialdehyde, which are particularly reactive to DNA, proteins, and phospholipids (Niki 2009). In both cell and animal models, these aldehydrous compounds mediate toxic impacts. Oxidized phospholipids also cause signaling pathways that contribute to different inflammatory or apoptosis responses (Fruhwirth et al. 2007). ROS causes cell destruction and impairs mitochondrial respiration by the direct oxidation of cellular macromolecules, inducing apoptotic cell death.

The reduction of molecular oxygen due to pesticide exposure results in the production of highly reactive ROS, which subsequently leads to DNA damage. ROS causes damage to DNA by escaping detoxification. The weakened DNA causes activation of poly (ADP-ribose) enzyme that causes NAD^+ splitting, and the NAD^+ level becomes negligible because of this process, which eventually leads to cell function loss. The damage to DNA also results in forming several modified bases which include glycol, 5,6-dihydroxycytosine, thymine, 2,6-diamino-4-hydroxy-5-formamidopyrimidine, and 8-hydroxyguanine products by the breakdown of sugar (such as, 2-deoxypentose-4-ulose, 2-deoxypentonic acid, erythrose, lactone, strand breaks, and also base-free sites) (Dizdaroglu et al. 2002; Evans et al. 2004). Free radicals attack base moieties and sugars which results in the damage of DNA due to oxidation, this includes changes/losses of nucleic acids. These ROS species are also produced both during exposure to mutagens such as UVA radiation, ionizing radiation, or pesticides, and H_2O_2. These ROS species are also generated as by-products of natural cell metabolism (Lindahl and Barnes 2000). Mitochondrial dysfunction is often identified by a loss of electron transport efficiency in the mitochondrial membrane resulting in reduced energy production (Nicolson 2014). Additionally, it can lead to cell death and impaired cellular functions (Abdollahi et al. 2004; Nicolson 2014). Mitochondrial dysfunction accelerates to an increase of cascades, causing extensive oxidation of macromolecules which are known as damage-associated molecular patterns (DAMPs). Aggregation of DAMPs in the extracellular matrix triggers several agents of chronic inflammatory processes.

These ROS-mediated cellular damage may be responsible for the pathogenesis of various pesticide-mediated neurological disorders in humans.

4 Oxidative Stress: Connection Between Pesticide Exposure Early Life Brain Disorders

4.1 Organophosphates (OPs) and Oxidative Stress

In both the studies, be it *in vitro* or *in vivo*, OPs are known to affect the central and peripheral nervous system by bringing about changes in the redox processes which induces oxidative stress that disrupts the neuronal development and finally causes neuronal disorders. This has been proved by ample case studies. For instance, a notable lower activity of SOD and CAT levels was seen in subjects exposed to OPs relative to those who weren't. A reduction in the activity of these enzymes has also been shown to facilitate the aggregation of free oxygen radicals in erythrocytes and other cells, contributing to increased oxidative stress and, eventually, damage in the tissues. This can be particularly harmful as pesticide use persists for several years and in long run may be responsible for multiple neurological tissues and organ damages (Lopez-Sandoval et al. 2018; Pellegrino et al. 2019; Sheikhansari et al. 2019).

Even in animal studies, OPs induced oxidative stress which caused noticeable DNA disruption and disrupted redox homeostasis in the rat brain (Ojha and Srivastava 2012; Ojha et al. 2013). There is growing evidence that OPs have an effect on neurodevelopment and behavior in children and infants. OPs exposure especially in children has been associated with an array of neurodevelopmental disorders such as poor cognitive and intellectual abilities, including short-term memory, attention problems, mental flexibility (Lizardi et al. 2008; Ruckart et al. 2004; Abdel Rasoul et al. 2008; Yu et al. 2016; Slotkin et al. 2008a) longer reaction time, learning disabilities (Rohlman et al. 2005; Grandjean et al. 2006; Slotkin et al. 2008b), impaired/delayed mental and emotional development, reduced and poor physical abilities which includes coordination problems, reduced psychomotor skills, and poor sensory skills (Keifer et al. 1996; Harari et al. 2010; Handal et al. 2007; Handal et al. 2008; Gonzalez-Alzaga et al. 2015; Liu et al. 2015). Neurobehavioral difficulties (Rohlman et al. 2005) and neuromuscular disorders (Ismail et al. 2010) have also been observed in some studies. Babies and fetuses are highly susceptible to the exposure of OPs, because of the ready placental transmission of OPs, and the primitiveness of metabolic pathways required to process and excrete these compounds. Epidemiological studies of pregnant women exposed to chlorpyrifos proved a correlation between exposure to chlorpyrifos *in utero* and decreased newborn head circumference (Whyatt et al. 2004; Whyatt and Barr 2001;

Berkowitz et al. 2004). A similar study showed that chlorpyrifos is responsible for changing brain development in infants, whose symptoms are evident after 11 years of age (Rauh et al. 2012). Various works have found that prenatal subjection of OPs is interconnected with a wide range of cognitive, neurobehavioral (Yolton et al. 2013; Zhang et al. 2014), and neurodevelopmental outcomes in childhood. Studies have been conducted to explore the association between prenatal OPs exposure and abnormal reflexes in infants. An elevation in the number of irregular reflexes is correlated with increased prenatal levels of urinary OPs metabolites (Young et al. 2005; Engel et al. 2007). Animal experiments have shown that even low toxicity to OPs pesticides to neonatal rats induces irreversible decreases of cholinergic neurons in the brain and is responsible for teenage hyperactivity (Ahlbom et al. 1995). ADHD is a neurodevelopmental disease with a broad spectrum of behavioral symptoms that affects the functions of the brain. Children suffering from ADHD show impulsive actions, hyperactivity, and lack of capacity to control concentration. While ADHD is commonly believed to be of hereditary origin, the evidence is now growing that environmental contaminants like pesticides are also a potential factor (Aguiar et al. 2010). Neurotoxic OPs insecticides are indicated to be one of the key factors in ADHD. For instance, research evinced a correlation between exposure to OPs, the risk of childhood ADHD, and oxidative stress. It also proposed that ADHD might have a relation with the mechanism of lipid peroxidation (Chang et al. 2018). Supporting this, numerous other studies provide strong evidence that exposure to OPs causes ADHD in children or can result in an elevated risk of development of ADHD (Bouchard et al. 2010; Kuehn 2010; Marks et al. 2010; Yu et al. 2016). Similar studies were performed on animals too, and in a study, it was proved that OPs agents induce ADHD similar behaviors in adolescent male mice (Ito et al. 2020). ASD is a series of neurodevelopmental diseases marked by poor speech, decreased social contact, and stereotypical repetitive habits. In about 40–55% of cases, association of mental retardation is present (Newschaffer et al. 2007). It is widely accepted that ASD develops during essential windows of weakness in fetal development due to changes in particular brain structures. With both OPs and OCs listed in the top 10 causes, pesticides are also among the leading causes of autism (Landrigan et al. 2012). Mechanisms implicated in the development of autism include oxidative stress, acetylcholinesterase inhibition during neuronal development, prenatal immune development disturbance, and disruption of GABA signaling pathways (Shelton et al. 2012). A study estimated the exposure of OPs at the time of pregnancy with measurements of dialkyl phosphates (DAP) metabolites in urine, and discovered evidence associating OP pesticide exposures with traits related to ASD which can result in the rise of clinically diagnosed ASD (Sagiv et al. 2018). Findings of a California-based study suggests that prenatal exposure during gestation to ambient pesticides within 2000 m of the residence of their mother, the offspring's ASD risk increases, and infant's exposure could further elevate risks for autism spectrum disorder with comorbid intellectual disability (von Ehrenstein et al. 2019). A similar study found associations with chlorpyrifos as the strongest (Shelton et al. 2014). A related study on mice concluded that chlorpyrifos has long-term effects on traits of ASD (Lan et al. 2017). Another study examined the relationship between maternal blood levels of a diethyl phosphate pesticide at the time of pregnancy and performance of the growth of babies continued for 3 years of age. Three-year-olds with increased prenatal amounts of chlorpyrifos in cord plasma were shown to have displayed substantially more deficits in psychomotor and behavioral growth, and their mothers revealed their babies having issues with concentration and signs of pervasive developmental disorders (Rauh et al. 2006). A similar California-based study found a drastic 230% increased risk of pervasive developmental disorders in prenatally OPs subjected babies with 24 months of age, that also comprised ASD (Eskenazi et al. 2007). Subsequent research in the same population implied exposure to organophosphate pesticides results in ASD (Eskenazi et al. 2010). Cerebral palsy is the most common neuro-motor disability in young children. An association of OPs and cerebral palsy was found by a population-based study in California. The study found that maternal first-trimester subjection to ambient pesticides was correlated with an elevated small-to-moderate threat for cerebral palsy in female offspring (Liew et al. 2020).

4.2 Organochlorines (OCs) and Oxidative Stress

The process of development of the nervous system and the human brain starts soon after fertilization and continues past birth (Rice and Barone 2000). Various works have demonstrated that OCs such as DDT, dichloro-diphenyl-trichloroethane (DDE), hexachlorobenzene (HCB) and others are capable of crossing the placenta and are even expelled along with breast milk, eventually leading to neurodevelopmental toxicity (Rogan and Chen 2005; Fenster et al. 2007). The emerging body of evidence has shown the positive association between prenatal exposure to OCs and the presence of neurodevelopmental disorders in infancy and childhood, including memory loss, longer response time, poor concentration decreased cognitive functioning, impaired/delayed mental and psychomotor development, anxiety, low IQ, and learning disabilities (Eskenazi et al. 2006; Torres-Sánchez et al. 2007; Julvez et al. 2011; Morales et al. 2008; NIOH 2002; Quijano 2002; Bahena-Medina et al. 2011; Ribas-Fitó et al. 2006; Torres-Sánchez et al. 2009). Similar findings were discovered by ample other studies conducted (Jacobson and Jacobson 2003; Sagiv et al. 2008; Forns et al. 2012; Dallaire et al. 2012; Lee et al. 2007; Faroon et al. 1995; Kenet et al. 2007; Shelton et al. 2012; Guzelian 1992). Exposure to OCs in childhood can also lead to similar neurological disorders. For instance, The Netherlands cohort study among 9-year-olds found positive associations between polychlorinated biphenyls (PCBs) and longer response time and increased variations in the response time (Vreugdenhil et al. 2004). Even though OCs have known adverse effects on the neurodevelopment of infants and children, only a handful of studies have examined the association of prenatal exposure to OCs and specifically the risk of ASD. For instance, studies have indicated, the threat of ASD is upto 6.1 times more in young ones whose mothers were exposed to OCs during their first trimester of pregnancy (Ornoy et al. 2015; Roberts et al. 2007). Supporting this, a study suggested that increased levels of a few OC compounds at the time of pregnancy were associated with ASD and intellectual disability without autism (Lyall et al. 2017). This can be aided by numerous other studies and researches (Kim et al. 2009; Kim and Pessah 2011). This hypothesis can also be assisted by a study by Finnish Prenatal Study of Autism which concluded that in 75 ASD cases, HCB and DDE were found (Shen et al. 2005; Cheslack-Postava et al. 2013). In the pathophysiology of autism, mechanisms of action for OC toxins, signaling dysregulation due to the mediation of Ca^2++, mitochondrial ROS development, and oxidative stress are indicated as leading factors for ASD (Mariussen and Fonnum 2006; Shelton et al. 2012; Deth et al. 2008). This can be cited by a study done on the brains of mice that concluded a correlation between oxidative stress, damage due to oxidation, and dysfunction in mitochondrial systems (Schuh et al. 2009). Oxidative stress development is documented as a mechanism that underlies the neuropsychological problems induced by certain OCs such as HCB (Song et al. 2006; Bleavins et al. 1984; Goldey and Taylor 1992). Also, in work by Kumar et al., elevated plasma oxidative stress markers are associated with persistent organic pollutants subjection and may be a source of oxidative stress (Kumar et al. 2014). Parental exposure to OC pesticides such as DDT has a certain role in causing ADHD, though so far just a limited amount of research on this relationship has been conducted. Not long ago, a US research discovered that young ones born with an increased quantity of PCB and DDE in their umbilical cord blood had up to 70% more risk of ADHD (Sagiv et al. 2010). Such work can be aided by other studies (Eubig et al. 2010; Nakagami et al. 2011). A case was reported in Sweden in which a mother and her young one with ADHD were both found to have elevated DDE levels in their body fat (Hardell et al. 2002). Additional studies supported such detections (Sagiv et al. 2012). Positive associations of PCB with ADHD-related behaviors have been observed by certain works on non-human primates such as rodents (Berger et al. 2001; Holene et al. 1998; Rice 2000) and humans (Sagiv et al. 2008) as well.

The most prominent behavioral and neurological problems include anxiety marked by a series of behaviors like nervousness, stress, fear, and unpleasant feelings, affecting an almost one-eighth population of the entire world (Joshi et al. 2015; Workman et al. 2008; Joiner et al. 2005). Exposure to lindane has been observed to cause anxiety, irritability, and nervousness in humans (Llorens et al. 1990; Harvey 1980; Kbare et al. 1977). In animal studies, the panicogenic effects resulting due to exposure to sub-convulsive doses of lindane were indicated in the mice (Llorens et al. 1989). A similar result of work was found to be consistent with the fact that anxiogenic effects are exerted by lindane (Corda and Biggio 1986; File and Lister 1984). Exposure to OCs like chlordecone and PCB has also shown symptoms of anxiety (Faroon et al. 1995; Guzelian 1992; Porta et al. 2007; Colciago et al. 2009). In humans, OCs are widely known as neurogenic compounds. Studies have found that DDT is responsible for nerve impulses discharge repetitively and seizures (Jett 2012; Chang and Dyer 1995). The Food and Drug Administration (FDA) has considered lindane as toxic due to its role in seizures and neurotoxicity (Kaminski et al. 2004; FDA 2003). Howland (1998) observed status epilepticus due to lindane intoxication. A study by Arora et al. (2013) found that endosulfan may have a relation with idiopathic seizures in children. Lindane and endosulfan have been found to induce seizures in animal models as well (Fishman and Gianutsos 1988; Gilbert and Mack 1995; Tochman et al. 2000). Other OCs such as DDT, dieldrin, and aldrin may also induce seizures in mammals (Matin et al. 1981; Albertson et al. 1985; Castro and Palermo-Neto 1989; Kim et al. 2009; Kim and Pessah 2011). Studies have also discovered that OPs subjectivity can even cause epilepsy (Bolton et al. 2011; Tuchman and Cuccaro 2011; NIOH 2002; Quijano 2002), tremors, and motor dysfunction (Gerhart et al. 1985; Benet et al. 1985; Fujimori et al. 1986).

Depression is the leading cause of mental ability worldwide and is considered a significant contributor to the burden of diseases globally. Evidence of works by researchers and scientists have revealed a positive relation between exposure to OCs and depressive disorders (Beard et al. 2013; Amr et al. 1997; Boucher et al. 2013; Cannon et al. 1978; Cohn et al. 1978; Taylor 1982). ALS/Lou Gehrig's disease is a nervous system disease that targets the spinal cord and nerve cells in the brain characterized by loss of muscle control. A meta-analysis study by Kamel et al. pointed out that ALS is predominantly correlated with the use of organochlorines (Kamel et al. 2012). It was reported by Horner et al. that US veterans who participated in the first Persian Gulf war (Horner et al. 2003) had an elevated ALS threat which may be due to the exposure of OCs and OPs (Golomb 2008). Another cohort study by Burns et al. supported the association of OCs and ALS (Burns et al. 2001). Research, works, and case studies point that subjection to endosulfan in the fetal stage has also been known to cause cerebral palsy in children and infants (NIOH 2002; Quijano 2002).

Sinaii et al. concluded that subjection to dioxins, an OP, may cause several autoimmune diseases such as multiple sclerosis (Sinaii et al. 2002).

4.3 Pyrethroid and Oxidative Stress

Pyrethroid induces oxidative stress, which particularly attacks the brain due to its high ability to absorb oxygen and more sensitivity among all other organs, thus considered as the key mechanism of pesticide toxicity (Mohammadi et al. 2019; Ansari et al. 2012). For instance, lambda-cyhalothrin exposure is carried out in rats during the post-lactational and prenatal phase. The results of these showed an increase in oxidative stress, alteration in NMDA receptors in the hippocampus attributed with impairment in learning, memory, and disruption of cholinergic functions associated with a decrease in grip strength indicating the muscle weakness (Ansari et al. 2012; Dhuriya et al. 2017). To support this case, further report of lambda-cyhalothrin exposure in rat showed a decrease in AChE and resulting in rapid, uncontrolled fasciculation of voluntary muscles which eventually leads to cholinergic hyperactivity and also, decrease in the activity of the antioxidant defense enzymes i.e., CAT, SOD (Ali 2012). Permethrin, another pyrethroid insecticide in the zebrafish model induced an increase in ROS, causing an increase in lipid peroxidation, DNA damage. This modulates anxiety-like behavior patterns (Nunes et al.

2019). Neonatal exposure even to a low dose of permethrin or cypermethrin alters the dopaminergic activity, and behavior, striatal monoamine level, and increased oxidative stress (Nasuti et al. 2007). Several studies have shown that pyrethroid exposure during early development may have negative effects on developing nervous systems causing neurodevelopmental deficits (Richardson et al. 2015; Fluegge et al. 2016; Sinha et al. 2006; Shelton et al. 2014). Pyrethroids are commonly used as one of the important compounds in mosquito repellent (MR). A study revealed that exposre of rats to Pyrethroids, during their gestation period, significantly causes the damage to blood–brain barrier, causing abnormalities in to the central nervous system (Sinha et al. 2004). Besides this, the mixture of different pyrethroids in a study demonstrated decreases in motor activity of rats, in addition to this another study showed, impaired motor coordination by developmental exposure to cypermethrin (Wolansky et al. 2009; Gómez-Giménez et al. 2018). Moreover, pyrethroid insecticide metabolites were assessed in urine samples at 6 years of age, indicating poor neurocognitive abilities, low verbal comprehension scores, and working memory scores (Viel et al. 2015). Fenpropathrin (FNP) exposure elicits sensorimotor deficiencies, dysfunction in the nervous system, or skeletal muscles, deterioration of the cells of the Purkinje cell layer in the brain (Abd-Elhakim et al. 2020). The potential neurobehavioral toxicity and cognitive deficits associated with pyrethroid exposure in children have just started to gain attention (Viel et al. 2017; Muñoz-Quezada et al. 2020). Recently, researchers studied the effect of maternal exposure to fenvalerate during pregnancy, delayed the growth and neurobehavioral development, increase anxiety, impaired spatial learning, and memory, also a similar study in rats on other pyrethroid evidenced with poor socio-emotional development in the 1-year-olds (Eskenazi et al. 2018; Liu et al. 2018). During prenatal and in infants, pyrethroid are known to escalate the prevalence of ASD in children and cause severely impaired phenotypes with comorbid intellectual disability (von Ehrenstein et al. 2019). A US study from 2015 reported a high prevalence of ADHD-like symptoms in urinary pyrethroid pesticide biomarkers and this relation was more prominent for boys in comparison to girls. In addition to this, hyperactive-impulsive and inattentive symptoms were also evident in 2–4-year aged children (Wagner-Schuman et al. 2015; Dalsager et al. 2019). Before this, another study found that a commonly used pyrethroid, deltamethrin exposure to causes alterations to the dopamine system, protein expression, and ADHD-like behaviors (Vester et al. 2019).

4.4 Carbamates and Oxidative Stress

From various in vivo and *in vitro* studies of carbamate pesticides (bendiocarb, carbaryl, methomyl, carbofuran, and many others), they have been as neurotoxicants during development (Moser et al. 2012; Rice and Barone 2000; Slotkin et al. 2007; Slotkin and Seidler 2008). The brain is the most susceptible to oxidative stress and hence alteration in oxidative stress can lead to developmental disorders in early life (Atamaniuk et al. 2014). For this relation, a study in rat brain proved that carbofuran exposure is known to significantly reduce the AChE activity, alteration in mitochondrial oxidative stress and further decreasing the glutathione levels, leading to impaired neuronal functioning and eventually causing neurobehavioral deficits, impaired cognitive and motor functions (Leung and Meyer 2019; Kamboj et al. 2008). On the side of this, another evaluation of carbofuran exposure demonstrated higher oxidative stress, lipid peroxidation, and causing neurobehavioral disorders (Rai and Sharma 2007). Gestational exposure to carbofuran leads to a decrease in neuron generation in the brain, disturbing neuronal and glial division, elevated cell dissolution in the hippocampus, and cognitive disorders (like learning and memory problems) in juvenile rodents (Mishra et al. 2012). Carbamate insecticides inhibit AChE, causing a notable surge in the levels of the neurotransmitter acetylcholine in the synaptic cleft, hindering cholinergic neurotransmission (Darvesh et al. 2008; Smulders et al. 2003). This causes adverse effects on the central nervous system, resulting in depression and hypotonia (Lifshitz et al. 1999). By using the zebrafish model, carbofuran and carbaryl exposure trigger mRNA expression and enzyme activity of tyrosine hydroxylase which subsequently induce an increase in levels of dopamine and norepinephrine, causing anxiety-like behavior and decreased locomotion/hypoactivity (Liu et al. 2020; Correia et al. 2019). Likewise, another carbamate i.e., propoxur which is a widely used pesticide in the United States showed an ill effect on motor development at 2 years of age in children and subsequently affecting cognitive functions and language development (Ostrea et al. 2012; Rowe et al. 2016). Another study presented carbaryl to cause learning and memory deficits. The behavioral problems may perhaps be due to the disruption of usual brain development, as the levels of various proteins were shown to be disrupted (Lee et al. 2015). A population-based study of carbofuran phenol level in urine samples of 7-year-old children showed decreased IQ pointing toward the fact that early-life carbamate exposure can negatively affect neurodevelopmental processes (Zhang et al. 2020). Studies indicated that carbaryl exposure during pregnancy showed a high risk of cerebral palsy in children (Liew et al. 2020). Another adverse effect of carbamate exposure is seizures and convulsions as indicated by the studies (Gupta 2004; Jett 2012).

4.5 Neonicotinoids and Oxidative Stress

The EFSA Panel on Plant Protection Products and their Residues in 2013 revealed that there is ample evidence that two neonicotinoids, imidacloprid and acetamiprid, are capable of damaging the evolving human nervous system which includes the cerebrum, and may bring about changes in the development of brain structures and neurons linked with functions like memory and learning and hence classified these pesticides as potential neurodevelopmental toxins (EFSA Panel on Plant Protection Products and their Residues 2013; European Food Safety Authority 2015). This can be supported by an ex vitro study by Kimura-Kuroda et al., which concluded, desensitization and/or excitation of nAChRs by imidacloprid and acetamiprid may be a factor in influencing the development of nervous systems for mammals

(Kimura-Kuroda et al. 2012). A few studies have identified neonicotinoid exposure to humans as potentially toxic to human health. One of them is a Japanese study wherein six patients who munched more than 500 g per day of domestically grown eatables which were sprayed with neonicotinoid pesticides, reported various symptoms, including impaired short-term memory, finger tremor, and muscle weakness/muscle pain//muscle spasm (Taira 2014; Taira et al. 2011). Similar symptoms were observed in volunteers by another research that discovered a correlation between concentrations of *N*-desmethyl-acetamiprid in urine and elevated incidence of neurological symptoms in volunteers (Marfo et al. 2015). The probable health impacts on staff who apply pesticides have also been studied by researchers. Koureas et al. discovered that neonicotinoid treatment of 80 pesticide applicators was consistent with the induction of oxidative damage to DNA in their blood (Koureas et al. 2014). Other studies have supported the hypothesis that neonicotinoid toxicity also induces oxidative stress (Annabi et al. 2015; Kapoor et al. 2010; Lonare et al. 2014; Duzguner and Erdogan 2012; Aydin 2011; El-Gendy et al. 2010). Associations between adverse birth outcomes and exposure of mothers to neonicotinoids during gestation have also been found (Lin et al. 2013). The relation between fetal stage imidacloprid subjection and increased risk for ASD were found among users who frequently used tick and flea medicines that contained imidacloprid (Keil et al. 2014). Similarly, in a study, an association between anencephaly in newborns and residential nearness to agricultural use of imidacloprid was found (Yang et al. 2014). Various similar studies have been done on animals and insects. Findings demonstrate that exposure to thiamethoxam and imidacloprid affect motor activity in rats (Rodrigues et al. 2010; Lonare et al. 2014; Bhardwaj et al. 2010). Tremors were observed in mice that were exposed to neonicotinoids (Chao and Casida 1997). Thiamethoxam was the causative factor for an anxiogenic effect in rats (Rodrigues et al. 2010) and clothianidin in mice was found to elevate anxiety-like behavior (Hirano et al. 2015), affected parameters of neurobehavior in mice (Tanaka 2012), and cognitive activity in mice (Özdemir et al. 2014). Several studies on rats show that after exposure to imidacloprid, decreased sensorimotor performance was observed, and following thiamethoxam exposure, alteration in biochemical processes of rat's cholinergic systems, and behavioral alterations were seen (Abou-Donia et al. 2008; Rodrigues et al. 2010; de Oliveira et al. 2010). Another study after exposing drosophila to imidacloprid observed behavioral deficiency which is a trait of ADHD. The study concluded that when exposed to imidacloprid, behavioral disorders can arise in drosophila (Kim et al. 2017).

4.6   Bipyridyl and Oxidative Stress

Bipyridyl herbicides use is restricted in the United States and banned in Europe but widely used by the developing countries and thus, explored for its developmental neurotoxicity by many experimental reports. One such *in vitro* report demonstrated paraquat exposure in 3D rat brain cell culture systems to cause adverse effects on glutamatergic, GABA, and dopaminergic neurons through oxidative stress (Sandström et al. 2017; Yang and Tiffany-Castiglioni 2005). Another similar study in rats showed depletion in glutathione level (Schmuck et al. 2002). Furthermore, by using *Drosophila melanogaster* brain tissue for studying the paraquat effect on the brain, the study showed significant neuronal damage, decrease in SOD, alteration in oxidative stress which results in DNA damage, and at last, this causes impairment in locomotor functioning (Mehdi and Qamar 2013). The prenatal chronic exposure to the paraquat herbicide results indicated an alteration in the magnitude of synaptic transmission in the developing rat cerebellar cortex (Miranda-Contreras et al. 2005). It also caused motor impairments, degradation of dopaminergic neurons, and astrogliosis in the striatal area (Fernandes et al. 2021). Since the hippocampus is the important structure in memory/learning formation and paraquat exposure causes changes in the hippocampal neuron resulting in impairment in cognitive function like learning and memory (Li et al. 2016). Low levels of paraquat cause neurodevelopmental toxicity in rats, by increasing ROS and alteration in neural stem cells (Colle et al. 2018).

4.7   Chlorophenoxy and Oxidative Stress

Chlorophenoxy herbicides and their derivatives are widely used herbicides globally and since developing the nervous system is the critical period of neuronal development where all vital processes of regional brain structure and function are set up. So, perturbations of these can alter the functions of the nervous system of offspring. As a proof of this, acid 2,4-D exposure to neonates rats through mother's milk induces a significant rise of ROS in the prefrontal cortex and midbrain decreased glutathione and CAT activity (Bongiovanni et al. 2007; Ferri et al. 2007). Apart from this, other neurotoxic effects associated with 2,4-D were decreases in neurite growth and neuronal differentiation (Rosso et al. 2000). Moreover, other studies also evaluated 2,4-D exposure during gestation and lactation period and effect in offspring, increases dopaminergic D2-like receptor levels in all the brain areas, and also a slight alteration in the stability of double-helix DNA, causing DNA damage (Bortolozzi et al. 2004; Benfeito et al. 2014). The 2,4-D herbicide toxicity is explored in the zebrafish model, the results demonstrated altered mitochondrial metabolism and oxidative status, reduced the SOD/CAT, and impaired innate behavior and locomotor activity (Thiel et al. 2020). Various studies also studied the chlorophenoxy herbicide associated with neurobehavioral patterns, hence the newborn rats exhibit impaired behavior patterns, motor anomalies, erect head motion, and impulsiveness due to the serotonergic modification (Bortolozzi et al. 1999; Garcia et al. 2001). The 2,4-D exposure during growth causes damage to the central nervous system (Garcia et al. 2004). Some of the behavioral outcomes associated with the use of 2,4-D were impaired motor coordination and emotional states. However, these behavioral patterns disappear with time (Bortolozzi et al. 1999; Charles et al. 2001).

4.8 Phosphonate and Oxidative Stress

The most widely used pesticides are glyphosate-based herbicides (GBH), mostly because of their use in crops that are genetically modified. Recently, neurotoxic effects induced by GBH are capturing lots of attention due to their potential risk for the development of neurodegenerative disorders, more than 36% of the 271 cases involving acute glyphosate poisoning had neurological effects, according to a survey by the Environmental Protection Agency (EPA), which indicates the presence of glyphosate toxicity in the nervous system and brain (Hawkins 2009). Studies have also found an association between exposure to glyphosate and the presence of ASD. US-based studies have reported that the rise of usage of glyphosate on soy crops and corn in the United States over the years 1995–2010 resulted in elevated ASD rates during the aforesaid time as announced by the US public school system (Samsel and Seneff 2013; Samsel and Seneff 2015; Swanson et al. 2014). This can be supported by other population-based studies that concluded that glyphosate exposure is related to an elevated risk of ASD (Ehrenstein et al. 2019; Pu et al. 2020). The hypothesis stated that exposure to glyphosate in utero might be responsible for developmental defects that lead to autism (Beecham et al. 2016). ADHD has also been proved to be related to glyphosate exposure. It was found that 43% of children suffering from ADHD were fathered by people who were subjected to used glyphosate regularly due to their occupation (Garry et al. 2002). According to some studies in human cell lines, glyphosate elevated the process of apoptosis and necrosis (Mesnage et al. 2013; Gasnier et al. 2009). GBH treatment in multi-age animal models indicates that it may be neurotoxic and may impair the growth of the brain and eventually adult behavior. In the brains of malformed piglets, glyphosate has been detected (Krüger et al. 2014). On performing an animal study, it was figured out that the chronic and sub-chronic exposure of glyphosate increased depressive and anxiety-like behaviors (Aitbali et al. 2018). Likewise, treatment with glyphosate in rats generated oxidative stress (Cattani et al. 2014; El-Shenawy 2009; Astiz et al. 2009) and hence affected the oxidation–antioxidation homeostasis, which promoted cellular death and apoptosis (Yang and Sun 1998; Shimada et al. 1998) thus inducing the depression and anxiety-like phenotype and reduction in the motor development of the GBH-exposed animals. Similarly, Gallegos et al. observed an anxiolytic-like effect when rats were prenatally exposed to GBH (Gallegos et al. 2016). It was revealed that exposure to GBH is a causative factor for neuronal death (Negga et al. 2012).

5   Conclusion and Future Perspectives

Pesticides are a diverse group of chemicals with a wide chronic and acute toxicity spectrum. Poison control centers record lower reports of more serious poisonings, but comparable overall acute toxicity figures for children appear to be published.

There is an increasing body of literature indicating that pesticides, including neuro-developmental or behavioral issues, birth defects, asthma, and cancer, can cause chronic health complications in children. For families and societies, pediatricians are a reliable source of knowledge, while existing instruction on pesticide toxicity and environmental health is minimal in general. Pediatricians should be familiar with the different forms of pesticides, acute toxicity signs and symptoms, and chronic health effects. Efforts should be taken to restrict the exposure of children to the greatest degree practicable and to guarantee that goods put on the market are properly checked for protection to prevent fetuses, babies, and children from adverse effects.

Longitudinal, prospective, and broader trials of children's well-being and pesticide consumption hazards should be used in future studies. To determine modes of behavior and explain some of the less consistent results, further analysis should be undertaken. To establish the exact quantities of toxic pesticides harmful to infants, studies must be underway. Given the proof of the apparent connections between exposure to prenatal and early childhood pesticides and cognitive and behavioral deficits, several future legislation, public health, and individual interventions should be evaluated. To increase knowledge of exposure and effects, it is important to examine how children are exposed, the clinical consequences of exposure, and education. Hospital practice and risk management, curriculum, legislation, and enforcement are important fields.

Acknowledgment The authors acknowledge Jaypee Institute of Information Technology, Noida for providing the entire infrastructure to complete this chapter.

Conflicts of Interest The authors declare no conflict of interest.

References

Abdel Rasoul GM, Abou Salem ME, Mechael AA, Hendy OM, Rohlman DS, Ismail AA. Effects of occupational pesticide exposure on children applying pesticides. Neurotoxicology. 2008;29(5):833–8. https://doi.org/10.1016/j.neuro.2008.06.009.

Abd-Elhakim YM, El-Sharkawy NI, Mohammed HH, Ebraheim LLM, Shalaby MA. Camel milk rescues neurotoxic impairments induced by fenpropathrin via regulating oxidative stress, apoptotic, and inflammatory events in the brain of rats. Food Chem Toxicol. 2020;135 https://doi.org/10.1016/j.fct.2019.111055.

Abdollahi M, Ranjbar A, Shadnia S, Nikfar S, Rezaie A. Pesticides and oxidative stress: a review. Med Sci Monit. 2004;10(6):RA141–7.

Abou-Donia MB, Goldstein LB, Bullman S, Tu T, Khan WA, Dechkovskaia AM, et al. Imidacloprid induces neurobehavioral deficits and increases expression of glial fibrillary acidic protein in the motor cortex and hippocampus in offspring rats following in utero exposure. J Toxicol Environ Health A. 2008;71(2):119–30. https://doi.org/10.1080/15287390701613140.

Abreu-Villaça Y, Levin ED. Developmental neurotoxicity of succeeding generations of insecticides. Environ Int. 2017;99:55–77. https://doi.org/10.1016/j.envint.2016.11.019.

Abreu-Villaça Y, Levin ED. Developmental neurobehavioral neurotoxicity of insecticides. 2018:435–466.

Aguiar A, Eubig PA, Schantz SL. Attention deficit/hyperactivity disorder: a focused overview for children's environmental health researchers. Environ Health Perspect. 2010;118(12):1646–53. https://doi.org/10.1289/ehp.1002326.

Ahlbom J, Fredriksson A, Eriksson P. Exposure to an organophosphate (DFP) during a defined period in neonatal life induces permanent changes in brain muscarinic receptors and behaviour in adult mice. Brain Res. 1995;677(1):13–9. https://doi.org/10.1016/0006-8993(95)00024-k.

Aitbali Y, Ba-M'hamed S, Elhidar N, Nafis A, Soraa N, Bennis M. Glyphosate based- herbicide exposure affects gut microbiota, anxiety and depression-like behaviors in mice. Neurotoxicol Teratol. 2018;67:44–9. https://doi.org/10.1016/j.ntt.2018.04.002.

Ait-Bali Y, Ba-M'hamed S, Gambarotta G, Sassoè-Pognetto M, Giustetto M, Bennis M. Pre- and postnatal exposure to glyphosate-based herbicide causes behavioral and cognitive impairments in adult mice: evidence of cortical ad hippocampal dysfunction. Arch Toxicol. 2020;94(5):1703–23. https://doi.org/10.1007/s00204-020-02677-7.

Aktar MW, Sengupta D, Chowdhury A. Impact of pesticides use in agriculture: their benefits and hazards. Interdiscip Toxicol. 2009;2(1):1–12. 10.2478/v10102-009-0001-7.

Albertson TE, Joy RM, Stark LG. Chlorinated hydrocarbon pesticides and amygdaloid kindling. Neurobehav Toxicol Teratol. 1985;7(3):233–7.

Ali ZY. Neurotoxic effect of lambda-cyhalothrin, a synthetic pyrethroid pesticide: involvement of oxidative stress and protective role of antioxidant mixture. N Y Sci J. 2012;5(9):93–103.

Amr MM, Halim ZS, Moussa SS. Psychiatric disorders among Egyptian pesticide applicators and formulators. Environ Res. 1997;73(1–2):193–9. https://doi.org/10.1006/enrs.1997.3744.

Annabi A, Dhouib IB, Lamine AJ, El Golli N, Gharbi N, El Fazâa S, et al. Recovery by N-acetylcysteine from subchronic exposure to Imidacloprid-induced hypothalamic-pituitary-adrenal (HPA) axis tissues injury in male rats. Toxicol Mech Methods. 2015;25(7):524–31. https://doi.org/10.3109/15376516.2015.1045663.

Ansari RW, Shukla RK, Yadav RS, Seth K, Pant AB, Singh D, et al. Cholinergic dysfunctions and enhanced oxidative stress in the neurobehavioral toxicity of lambda-cyhalothrin in developing rats. Neurotox Res. 2012;22(4):292–309. https://doi.org/10.1007/s12640-012-9313-z.

Arora SK, Batra P, Sharma T, Banerjee BD, Gupta S. Role of organochlorine pesticides in children with idiopathic seizures. Int Sch Res Notices. 2013;2013:849709. https://doi.org/10.1155/2013/849709.

Astiz M, de Alaniz MJ, Marra CA. Effect of pesticides on cell survival in liver and brain rat tissues. Ecotoxicol Environ Saf. 2009;72(7):2025–32. https://doi.org/10.1016/j.ecoenv.2009.05.001.

Atamaniuk TM, Kubrak OI, Husak VV, Storey KB, Lushchak VI. The Mancozeb-containing carbamate fungicide tattoo induces mild oxidative stress in goldfish brain, liver, and kidney. Environ Toxicol. 2014;29(11):1227–35. https://doi.org/10.1002/tox.21853.

Aydin B. Effects of thiacloprid, deltamethrin and their combination on oxidative stress in lymphoid organs, polymorphonuclear leukocytes and plasma of rats. Pestic Biochem Physiol. 2011;100(2):165–71. https://doi.org/10.1016/j.pestbp.2011.03.006.

Bahena-Medina LA, Torres-Sánchez L, Schnaas L, Cebrián ME, Chávez CH, Osorio-Valencia E, et al. Neonatal neurodevelopment and prenatal exposure to dichlorodiphenyldichloroethylene (DDE): a cohort study in Mexico. J Expo Sci Environ Epidemiol. 2011;21(6):609–14. https://doi.org/10.1038/jes.2011.25.

Beard JD, Hoppin JA, Richards M, Alavanja MC, Blair A, Sandler DP, et al. Pesticide exposure and self-reported incident depression among wives in the Agricultural Health Study. Environ Res. 2013;126:31–42. https://doi.org/10.1016/j.envres.2013.06.001.

Beecham JE, Seneff S, et al. J Autism. 2016;3(1) https://doi.org/10.7243/2054-992X-3-1.

Benet H, Fujimori K, Ho IK. The basal ganglia in chlordecone-induced neurotoxicity in the mouse. Neurotoxicology. 1985;6(1):151–8.

Benfeito S, Silva T, Garrido J, Andrade PB, Sottomayor MJ, Borges F, et al. Effects of chlorophenoxy herbicides and their main transformation products on DNA damage and acetylcholinesterase activity. Biomed Res Int. 2014;2014:709036. 10.1155/2014/709036.

Berger DF, Lombardo JP, Jeffers PM, Hunt AE, Bush B, Casey A, et al. Hyperactivity and impulsiveness in rats fed diets supplemented with either Aroclor 1248 or PCB-contaminated St. Lawrence river fish. Behav Brain Res. 2001;126(1–2):1–11. https://doi.org/10.1016/S0166-4328(01)00244-3.

Berkowitz GS, Wetmur JG, Birman-Deych E, Obel J, Lapinski RH, Godbold JH, et al. In utero pesticide exposure, maternal paraoxonase activity, and head circumference. Environ Health Perspect. 2004;112(3):388–91. https://doi.org/10.1289/ehp.6414.

Bhardwaj S, Srivastava MK, Kapoor U, Srivastava LP. A 90 days oral toxicity of imidacloprid in female rats: morphological, biochemical and histopathological evaluations. Food Chem Toxicol. 2010;48(5):1185–90. https://doi.org/10.1016/j.fct.2010.02.009.

Bjørling-Poulsen M, Andersen HR, Grandjean P. Potential developmental neurotoxicity of pesticides used in Europe. Environ Health. 2008:7–50. https://doi.org/10.1186/1476-069X-7-50.

Bleavins MR, Bursian SJ, Brewster JS, Aulerich RJ. Effects of dietary hexachlorobenzene exposure on regional brain biogenic amine concentrations in mink and European ferrets. J Toxicol Environ Health. 1984;14(2–3):363–77. https://doi.org/10.1080/15287398409530586.

Bolton PF, Carcani-Rathwell I, Hutton J, Goode S, Howlin P, Rutter M. Epilepsy in autism: features and correlates. Br J Psychiatry. 2011;198(4):289–94.

Bongiovanni B, De Lorenzi P, Ferri A, Konjuh C, Rassetto M, Evangelista de Duffard AM, et al. Melatonin decreases the oxidative stress produced by 2,4-dichlorophenoxyacetic acid in rat cerebellar granule cells. Neurotox Res. 2007;11(2):93–9. https://doi.org/10.1007/bf03033388.

Bortolozzi AA, Duffard RO, Evangelista De Duffard AM. Behavioral alterations induced in rats by a pre- and postnatal exposure to 2,4-Dichlorophenoxyacetic acid. Neurotoxicol Teratol. 1999;21(4):451–65. https://doi.org/10.1016/S0892-0362(98)00059-2.

Bortolozzi AA, Evangelista De Duffard AM, Duffard RO, Antonelli MC. Effects of 2,4-dichlorophenoxyacetic acid exposure on dopamine D2-like receptors in rat brain. Neurotoxicol Teratol. 2004;26(4):599–605. https://doi.org/10.1016/j.ntt.2004.04.001.

Bouchard MF, Bellinger DC, Wright RO, Weisskopf MG. Attention-deficit/hyperactivity disorder and urinary metabolites of organophosphate pesticides. Pediatrics. 2010;125(6):e1270–7. https://doi.org/10.1542/peds.2009-3058.

Boucher O, Simard MN, Muckle G, Rouget F, Kadhel P, Bataille H, et al. Exposure to an organochlorine pesticide (chlordecone) and development of 18-month-old infants. Neurotoxicology. 2013;35:162–8. https://doi.org/10.1016/j.neuro.2013.01.007.

Burns CJ, Beard KK, Cartmill JB. Mortality in chemical workers potentially exposed to 2,4-dichlorophenoxyacetic acid (2,4-D) 1945-94: an update. Occup Environ Med. 2001;58(1):24–30. https://doi.org/10.1136/oem.58.1.24.

Cannon SB, Veazey JM Jr, Jackson RS, Burse VW, Hayes C, Straub WE, et al. Epidemic kepone poisoning in chemical workers. Am J Epidemiol. 1978;107(6):539–7. https://doi.org/10.1093/oxfordjournals.aje.a112572.

Castro VL, Palermo-Neto J. Effects of long-term aldrin administration on seizure susceptibility of rats. Pharmacol Toxicol. 1989;65(3):204–8. https://doi.org/10.1111/j.1600-0773.1989.tb01157.x.

Cattani D, de Liz Oliveira Cavalli VL, Heinz Rieg CE, Domingues JT, Dal-Cim T, Tasca CI, et al. Mechanisms underlying the neurotoxicity induced by glyphosate-based herbicide in immature rat hippocampus: involvement of glutamate excitotoxicity. Toxicology. 2014;320:34–45. https://doi.org/10.1016/j.tox.2014.03.001.

Chang LW, Dyer RS. Handbook of neurotoxicology. Toxicol Lett. New York: Marcel Dekker Inc. 1995;80(1–3).

Chang CH, Yu CJ, Du JC, Chiou HC, Chen HC, Yang W, et al. The interactions among organophosphate pesticide exposure, oxidative stress, and genetic polymorphisms of dopamine receptor D4 increase the risk of attention deficit/hyperactivity disorder in children. Environ Res. 2018;160:339–46. https://doi.org/10.1016/j.envres.2017.10.011.

Chao SL, Casida JE. Interaction of imidacloprid metabolites and analogs with the nicotinic acetylcholine receptor of mouse brain in relation to toxicity. Pestic Biochem Physiol. 1997;58(1):77–88. https://doi.org/10.1006/pest.1997.2284.

Charles JM, Hanley TR Jr, Wilson RD, van Ravenzwaay B, Bus JS. Developmental toxicity studies in rats and rabbits on 2,4-dichlorophenoxyacetic acid and its forms. Toxicol Sci. 2001;60(1):121–31. https://doi.org/10.1093/toxsci/60.1.121.

Chen X, Guo C, Kong J. Oxidative stress in neurodegenerative diseases. Neural Regen Res. 2012;7(5):376–85. https://doi.org/10.3969/j.issn.1673-5374.2012.05.009.

Cheslack-Postava K, Rantakokko PV, Hinkka-Yli-Salomäki S, Surcel HM, McKeague IW, Kiviranta HA, et al. Maternal serum persistent organic pollutants in the Finnish Prenatal Study of Autism: a pilot study. Neurotoxicol Teratol. 2013;38:1–5. https://doi.org/10.1016/j.ntt.2013.04.001.

Cohn WJ, Boylan JJ, Blanke RV, Fariss MW, Howell JR, Guzelian PS. Treatment of chlordecone (Kepone) toxicity with cholestyramine. Results of a controlled clinical trial. N Engl J Med. 1978;298(5):243–8. https://doi.org/10.1056/NEJM197802022980504.

Colciago A, Casati L, Mornati O, Vergoni AV, Santagostino A, Celotti F, et al. Chronic treatment with polychlorinated biphenyls (PCB) during pregnancy and lactation in the rat Part 2: effects on reproductive parameters, on sex behavior, on memory retention and on hypothalamic expression of aromatase and 5alpha-reductases in the offspring. Toxicol Appl Pharmacol. 2009;239(1):46–54. https://doi.org/10.1016/j.taap.2009.04.023.

Colle D, Farina M, Ceccatelli S, Raciti M. Paraquat and maneb exposure alters rat neural stem cell proliferation by inducing oxidative stress: new insights on pesticide-induced neurodevelopmental toxicity. Neurotox Res. 2018;34:820–33. https://doi.org/10.1007/s12640-018-9916-0.

Corda MG, Biggio G. Proconflict effect of GABA receptor complex antagonists: reversal by diazepam. Neuropharmacology. 1986;25(5):541–4. https://doi.org/10.1016/0028-3908(86)90181-4.

Correia D, Almeida AR, Santos J, Machado AL, Koba Ucun O, Žlábek V, et al. Behavioral effects in adult zebrafish after developmental exposure to carbaryl. Chemosphere. 2019;235:1022–9. https://doi.org/10.1016/j.chemosphere.2019.07.029.

Cortés-Iza SC, Rodríguez AI. Oxidative stress and pesticide disease: a challenge for toxicology. Rev Fac Med. 2018;66(2):261–7. https://doi.org/10.15446/revfacmed.v66n2.60783.

Dallaire R, Muckle G, Rouget F, Kadhel P, Bataille H, Guldner L, et al. Cognitive, visual, and motor development of 7-month-old Guadeloupean infants exposed to chlordecone. Environ Res. 2012;119:79–85. https://doi.org/10.1016/j.envres.2012.07.006.

Dalle-Donne I, Rossi R, Giustarini D, Gagliano N, Lusini L, Milzani A, et al. Actin carbonylation: from a simple marker of protein oxidation to relevant signs of severe functional impairment. Free Radic Biol Med. 2001;31(9):1075–83. https://doi.org/10.1016/S0891-5849(01)00690-6.

Dalle-Donne I, Giustarini D, Colombo R, Rossi R, Milzani A. Protein carbonylation in human diseases. Trends Mol Med. 2003;9(4):169–76. https://doi.org/10.1016/S1471-4914(03)00031-5.

Dalle-Donne I, Aldini G, Carini M, Colombo R, Rossi R, Milzani A. Protein carbonylation, cellular dysfunction, and disease progression. J Cell Mol Med. 2006;10(2):389–406. https://doi.org/10.1111/j.1582-4934.2006.tb00407.x.

Dalsager L, Fage-Larsen B, Bilenberg N, Jensen TK, Nielsen F, Kyhl HB, et al. Maternal urinary concentrations of pyrethroid and chlorpyrifos metabolites and attention deficit hyperactivity disorder (ADHD) symptoms in 2-4-year-old children from the Odense Child Cohort. Environ Res. 2019;176 https://doi.org/10.1016/j.envres.2019.108533.

Dan Dunn J, Alvarez LA, Zhang X, Soldati T. Reactive oxygen species and mitochondria: a nexus of cellular homeostasis. Redox Biol. 2015;6:472–85. https://doi.org/10.1016/j.redox.2015.09.005.

Darvesh S, Darvesh KV, McDonald RS, Mataija D, Walsh R, Mothana S, et al. Carbamates with differential mechanism of inhibition toward acetylcholinesterase and butyrylcholinesterase. J Med Chem. 2008;51(14):4200–12. https://doi.org/10.1021/jm8002075.

de Oliveira IM, Nunes BV, Barbosa DR, Pallares AM, Faro LR. Effects of the neonicotinoids thiametoxam and clothianidin on in vivo dopamine release in rat striatum. Toxicol Lett. 2010;192(3):294–7. https://doi.org/10.1016/j.toxlet.2009.11.005.

Deavall DG, Martin EA, Horner JM, Roberts R. Drug-Induced oxidative stress and toxicity. J Toxicol. 2012;2012:1–13. https://doi.org/10.1155/2012/645460.

Deth R, Muratore C, Benzecry J, Power-Charnitsky VA, Waly M. How environmental and genetic factors combine to cause autism: a redox/methylation hypothesis. Neurotoxicology. 2008;29(1):190–201. https://doi.org/10.1016/j.neuro.2007.09.010.

Dhuriya YK, Srivastava P, Shukla RK, Gupta R, Singh D, Parmar D, et al. Prenatal exposure to lambda-cyhalothrin impairs memory in developing rats: role of NMDA receptor induced post-synaptic signalling in hippocampus. Neurotoxicology. 2017;62:80–91. https://doi.org/10.1016/j.neuro.2017.04.011.

Díaz-Hung ML, González Fraguela ME. Oxidative stress in neurological diseases: cause or effect? Neurologia. 2014;29(8):451–2. https://doi.org/10.1016/j.nrl.2013.06.022.

Dizdaroglu M, Jaruga P, Birincioglu M, Rodriguez H. Free radical-induced damage to DNA: mechanisms and measurement. Free Radic Biol Med. 2002;32(11):1102–15. https://doi.org/10.1016/S0891-5849(02)00826-2.

Duchen MR. Roles of mitochondria in health and disease. Diabetes. 2004;53(suppl 1):S96–S102. https://doi.org/10.2337/diabetes.53.2007.S96.

Duzguner V, Erdogan S. Chronic exposure to imidacloprid induces inflammation and oxidative stress in the liver & central nervous system of rats. Pestic Biochem Physiol. 2012;104(1):58–64. https://doi.org/10.1016/j.pestbp.2012.06.011.

Eddleston M. Bipyridyl herbicides. In: Brent J, Burkhart K, Dargan P, Hatten B, Megarbane B, Palmer R, editors. Critical care toxicology. Cham: Springer; 2016. https://doi.org/10.1007/978-3-319-20790-2_100-1.

EFSA Panel on Plant Protection Products and their Residues. Scientific opinion on the developmental neurotoxicity potential of Acetamiprid and imidacloprid. EFSA J. 2013;11(12). 10.2903/j.efsa.2013.3471.

El-Gendy KS, Aly NM, Mahmoud FH, Kenawy A, El-Sebae AK. The role of vitamin C as antioxidant in protection of oxidative stress induced by imidacloprid. Food Chem Toxicol. 2010;48(1):215–21. https://doi.org/10.1016/j.fct.2009.10.003.

El-Shenawy NS. Oxidative stress responses of rats exposed to Roundup and its active ingredient glyphosate. Environ Toxicol Pharmacol. 2009;28(3):379–85. https://doi.org/10.1016/j.etap.2009.06.001.

Engel SM, Berkowitz GS, Barr DB, Teitelbaum SL, Siskind J, Meisel SJ, et al. Prenatal organophosphate metabolite and organochlorine levels and performance on the Brazelton Neonatal Behavioral Assessment Scale in a multiethnic pregnancy cohort. Am J Epidemiol. 2007;165(12):1397–404. https://doi.org/10.1093/aje/kwm02Ahlbom.

England K, Cotter TG. Direct oxidative modifications of signaling proteins in mammalian cells and their effects on apoptosis. Redox Rep. 2005;10(5):237–45. https://doi.org/10.1179/135100005X70224.

Eskenazi B, Marks AR, Bradman A, Fenster L, Johnson C, Barr DB, et al. In utero exposure to dichlorodiphenyltrichloroethane (DDT) and dichlorodiphenyldichloroethylene (DDE) and neurodevelopment among young Mexican American children. Pediatrics. 2006;118(1):233–41. https://doi.org/10.1542/peds.2005-3117.

Eskenazi B, Marks AR, Bradman A, Harley K, Barr DB, Johnson C, et al. Organophosphate pesticide exposure and neurodevelopment in young Mexican-American children. Environ Health Perspect. 2007;115(5):792–8. https://doi.org/10.1289/ehp.9828.

Eskenazi B, Huen K, Marks A, et al. PON1 and neurodevelopment in children from the CHAMACOS study exposed to organophosphate pesticides in utero. Environ Health Perspect. 2010;118(12):1775–81. https://doi.org/10.1289/ehp.1002234.

Eskenazi B, An S, Rauch SA, Coker ES, Maphula A, Obida M, et al. Prenatal exposure to DDT and pyrethroids for malaria control and child neurodevelopment: the VHEMBE Cohort, South Africa. Environ Health Perspect. 2018;126(4) https://doi.org/10.1289/EHP2129.

Espinosa-Diez C, Miguel V, Mennerich D, Kietzmann T, Sánchez-Pérez P, et al. Antioxidant responses and cellular adjustments to oxidative stress. Redox Biol. 2015;6:183–97. https://doi.org/10.1016/j.redox.2015.07.008.

Eubig PA, Aguiar A, Schantz SL. Lead and PCBs as risk factors for attention deficit/hyperactivity disorder. Environ Health Perspect. 2010;118(12):1654–67. https://doi.org/10.1289/ehp.0901852.

European Food Safety Authority. Conclusion on the peer review of the pesticide risk assessment for bees for the active substance imidacloprid considering all uses other than seed treatments and granules. EFSA J. 2015;13(8) https://doi.org/10.2903/j.efsa.2015.4211.

Evans MD, Dizdaroglu M, Cooke MS. Oxidative DNA damage and disease: induction, repair and significance. Mutat Res. 2004;567(1):1–61. https://doi.org/10.1016/j.mrrev.2003.11.001.

Fang J, Holmgren A. Inhibition of thioredoxin and thioredoxin reductase by 4-hydroxy-2- nonenal *in vitro* and *in vivo*. J Am Chem Soc. 2006;128(6):1879–85. https://doi.org/10.1021/ja057358l.

Faroon O, Kueberuwa S, Smith L, De Rosa C. ATSDR evaluation of health effects of chemicals. II. Mirex and chlordecone: health effects, toxicokinetics, human exposure, and environmental fate. Toxicol Ind Health. 1995;11(6):1–203. https://doi.org/10.1177/2F074823379501100601.

FDA. FDA Issues Health Advisory Regarding Labeling Changes for Lindane Products. FDA Consum. March 28, 2003; T03-19.

Fenster L, Eskenazi B, Anderson B, Bradman A, Hubbard A, Barr DB. In utero exposure to DDT and performance on the Brazelton neonatal behavioral assessment scale. Neurotoxicology. 2007;28(3):471–7. https://doi.org/10.1016/j.neuro.2006.12.009.

Fernandes LC, Santos AG, Sampaio TB, Sborgi S, Prediger R, Ferro MM, et al. Exposure to para-quat associated with periodontal disease causes motor damage and neurochemical changes in rats. Hum Exp Toxicol. 2021;40(1):81–9. https://doi.org/10.1177/2F0960327120938851.

Ferri A, Duffard R, de Duffard AM. Selective oxidative stress in brain areas of neonate rats exposed to 2,4-Dichlorophenoxyacetic acid through mother's milk. Drug Chem Toxicol. 2007;30(1):17–30. https://doi.org/10.1080/01480540601017629.

Fianko JR, Donkar A, Lowor ST, Yeboah PO. Agrochemicals and the Ghanaian environment: a review. J Environ Prot Sci. 2011;2(3):221–30. https://doi.org/10.4236/jep.2011.23026.

File SE, Lister RG. Do the reductions in social interaction produced by picrotoxin and pentyl-enetetrazole indicate anxiogenic actions? Neuropharmacology. 1984;23(7):793–6. https://doi.org/10.1016/0028-3908(84)90113-8.

Fishman BE, Gianutsos G. CNS biochemical and pharmacological effects of the isomers of hexa-chlorocyclohexane (lindane) in the mouse. Toxicol Appl Pharmacol. 1988;93(1):146–53. https://doi.org/10.1016/0041-008X(88)90034-8.

Fluegge KR, Nishioka M, Wilkins JR 3rd. Effects of simultaneous prenatal exposures to organo-phosphate and synthetic pyrethroid insecticides on infant neurodevelopment at three months of age. J Environ Toxicol Public Health. 2016;1:60–73. https://doi.org/10.5281/zenodo.218417.

Forns J, Lertxundi N, Aranbarri A, Murcia M, Gascon M, Martinez D, et al. Prenatal exposure to organochlorine compounds and neuropsychological development up to two years of life. Environ Int. 2012;45:72–7. https://doi.org/10.1016/j.envint.2012.04.009.

Fruhwirth GO, Loidl A, Hermetter A. Oxidized phospholipids: from molecular properties to disease. Biochim Biophys Acta. 2007;1772(7):718–36. https://doi.org/10.1016/j.bbadis.2007.04.009.

Fujimori K, Benet H, Mehendale HM, Ho IK. *In vivo* and *in vitro* synthesis, release, and uptake of [3-H]-dopamine in mouse striatal slices after *in vivo* exposure to chlordecone. J Biochem Toxicol. 1986;1(4):1–12. https://doi.org/10.1002/jbt.2570010402.

Furlong MA, Barr DB, Wolff MS, Engel SM. Prenatal exposure to pyrethroid pesticides and childhood behavior and executive functioning. Neurotoxicology. 2017;62:231–8. https://doi.org/10.1016/j.neuro.2017.08.005.

Gallegos CE, Bartos M, Bras C, Gumilar F, Antonelli MC, Minetti A. Exposure to a glyphosate-based herbicide during pregnancy and lactation induces neurobehavioral alterations in rat off-spring. Neurotoxicology. 2016;53:20–8. https://doi.org/10.1016/j.neuro.2015.11.015.

Garcia G, Tagliaferro P, Bortolozzi A, Madariaga MJ, Brusco A, Evangelista de Duffard AM, et al. Morphological study of 5-HT neurons and astroglial cells on brain of adult rats perinatal or chronically exposed to 2,4-Dichlorophenoxyacetic acid. Neurotoxicology. 2001;22(6):733–41. https://doi.org/10.1016/S0161-813X(01)00059-6.

Garcia G, Tagliaferro P, Ferri A, Evangelista de Duffard AM, Duffard R, Brusco A. Study of tyrosine hydroxylase immunoreactive neurons in neonate rats lactationally exposed to 2,4-dichlorophenoxyacetic acid. Neurotoxicology. 2004;25:951–7. https://doi.org/10.1016/j.neuro.2004.05.004.

Garry VF, Harkins ME, Erickson LL, Long-Simpson LK, Holland SE, Burroughs BL. Birth defects, season of conception, and sex of children born to pesticide applicators living in the Red River Valley of Minnesota, USA. Environ Health Perspect. 2002;110(Suppl 3):441–9. https://doi.org/10.1289/ehp.02110s3441.

Gasnier C, Dumont C, Benachour N, Clair E, Chagnon MC, Séralini GE. Glyphosate-based herbicides are toxic and endocrine disruptors in human cell lines. Toxicology. 2009;262(3):184–91. https://doi.org/10.1016/j.tox.2009.06.006.

Gerhart JM, Hong JM, Tilson HA. Studies on the mechanism of chlordecone induced tremor in rats. Neurotoxicology. 1985;6(1):211–30.

Gilbert ME, Mack CM. Seizure thresholds in kindled animals are reduced by the pesticides lindane and endosulfan. Neurotoxicol Teratol. 1995;17(2):143–50. https://doi.org/10.1016/0892-0362(94)00065-L.

Gilden RC, Huffling K, Sattler B. Pesticides and health risks. J Obstet Gynecol Neonatal Nurs. 2010;39(1):103–10. https://doi.org/10.1111/j.1552-6909.2009.01092.x.

Goldey ES, Taylor DH. Developmental neurotoxicity following premating maternal exposure to hexachlorobenzene in rats. Neurotoxicol Teratol. 1992;14(1):15–21. https://doi.org/10.1016/0892-0362(92)90024-5.

Golomb BA. Acetylcholinesterase inhibitors and Gulf War illnesses. Proc Natl Acad Sci U S A. 2008;105(11):4295–300. https://doi.org/10.1073/pnas.0711986105.

Gómez-Giménez B, Felipo V, Cabrera-Pastor A, Agustí A, Hernández-Rabaza V, Llansola M. Developmental exposure to pesticides alters motor activity and coordination in rats: sex differences and underlying mechanisms. Neurotox Res. 2018;33(2):247–58. https://doi.org/10.1007/s12640-017-9823-9.

González-Alzaga B, Hernández AF, Rodríguez-Barranco M, Gómez I, Aguilar-Garduño C, López-Flores I, et al. Pre- and postnatal exposures to pesticides and neurodevelopmental effects in children living in agricultural communities from South-Eastern Spain. Environ Int. 2015;85:229–37. https://doi.org/10.1016/j.envint.2015.09.019.

Grandjean P, Harari R, Barr DB, Debes F. Pesticide exposure and stunting as independent predictors of neurobehavioral deficits in Ecuadorian school children. Pediatrics. 2006;117(3):e546–56. https://doi.org/10.1542/peds.2005-1781.

Gupta RC. Brain regional heterogeneity and toxicological mechanisms of organophosphates and carbamates. Toxicol Mech Methods. 2004;14(3):103–43. https://doi.org/10.1080/15376520490429175.

Guzelian PS. The clinical toxicology of chlordecone as an example of toxicological risk assessment for man. Toxicol Lett. 1992;64–65:589–96. https://doi.org/10.1016/0378-4274(92)90236-D.

Handal AJ, Lozoff B, Breilh J, Harlow SD. Effect of community of residence on neurobehavioral development in infants and young children in a flower-growing region of Ecuador. Environ Health Perspect. 2007;115(1):128–33. https://doi.org/10.1289/ehp.9261.

Handal AJ, Harlow SD, Breilh J, Lozoff B. Occupational exposure to pesticides during pregnancy and neurobehavioral development of infants and toddlers. Epidemiology. 2008;19(6):851–9. https://doi.org/10.1097/ede.0b013e318187cc5d.

Harari R, Julvez J, Murata K, Barr D, Bellinger DC, Debes F, et al. Neurobehavioral deficits and increased blood pressure in school-age children prenatally exposed to pesticides. Environ Health Perspect. 2010;118(6):890–6. https://doi.org/10.1289/ehp.0901582.

Hardell L, Lindström G, Van Bavel B. Is DDT exposure during fetal period and breast-feeding associated with neurological impairment? Environ Res. 2002;88(3):141–4. https://doi.org/10.1006/enrs.2002.4337.

Harvey SC. Antiseptics and disinfectants, fungicides, ectoparasiticides. In: Gilman AG, Goodman LS, Gilman A, editors. Pharmacological basis of therapeutics. 6th ed. New York: Macmillan Publishing Co.; 1980. p. 959–79.

Hawkins M. Updated Review of Glyphosate (103601) Incident Reports. Memorandum, EPA Toxicology and Epidemiology Branch. February 26. 2009.

Herst PM, Rowe MR, Carson GM, Berridge MV. Functional mitochondria in health and disease. Front Endocrinol (Lausanne). 2017;8:296. https://doi.org/10.3389/fendo.2017.00296.

Hirano T, Yanai S, Omotehara T, Hashimoto R, Umemura Y, Kubota N, et al. The combined effect of clothianidin and environmental stress on the behavioral and reproductive function in male mice. J Vet Med Sci. 2015;77(10):1207–15. https://doi.org/10.1292/jvms.15-0188.

Holene E, Nafstad I, Skaare JU, Sagvolden T. Behavioural hyperactivity in rats following postnatal exposure to sub-toxic doses of polychlorinated biphenyl congeners 153 and 126. Behav Brain Res. 1998;94(1):213–24. https://doi.org/10.1016/S0166-4328(97)00181-2.

Horner RD, Kamins KG, Feussner JR, Grambow SC, Hoff-Lindquist J, Harati Y, et al. Occurrence of amyotrophic lateral sclerosis among Gulf War veterans. Neurology. 2003;61(6):742–9. https://doi.org/10.1212/01.WNL.0000069922.32557.CA.

Howland MA. Insecticides: chlorinated hydrocarbons, pyrethrins and DEET. In: Goldfrank LR, editor. Goldfrank's toxicologic emergencies. Stanford: Appleton and Lange; 1998. p. 1451–8.

Ismail AA, Rohlman DS, Abdel Rasoul GM, Abou Salem ME, Hendy OM. Clinical and biochemical parameters of children and adolescents applying pesticides. Int J Occup Environ Med. 2010;1(3):132–43.

Ito Y, Tomizawa M, Suzuki K, Shirakawa Y, Ono H, Adachi K, et al. Organophosphate agent induces ADHD-like behaviors via inhibition of brain endocannabinoid-hydrolyzing enzyme(s) in adolescent male rats. J Agric Food Chem. 2020;68(8):2547–53. https://doi.org/10.1021/acs.jafc.9b08195.

Jacobson JL, Jacobson SW. Prenatal exposure to polychlorinated biphenyls and attention at school age. J Pediatr. 2003;143(6):780–8. https://doi.org/10.1067/S0022-3476(03)00577-8.

Jett DA. Chemical toxins that cause seizures. Neurotoxicology. 2012;33(6):1473–145. https://doi.org/10.1016/j.neuro.2012.10.005.

Joiner TE Jr, Brown JS, Wingate LR. The psychology and neurobiology of suicidal behavior. Annu Rev Psychol. 2005;56:287–314. https://doi.org/10.1146/annurev.psych.56.091103.070320.

Joshi JC, Ray A, Gulati K. Effects of morphine on stress induced anxiety in rats: role of nitric oxide and Hsp70. Physiol Behav. 2015;139:393–6. https://doi.org/10.1016/j.physbeh.2014.11.056.

Julvez J, Debes F, Weihe P, Choi AL, Grandjean P. Thyroid dysfunction as a mediator of organochlorine neurotoxicity in preschool children. Environ Health Perspect. 2011;119(10):1429–35. https://doi.org/10.1289/ehp.1003172.

Kamboj SS, Kumar V, Kamboj A, Sandhir R. Mitochondrial oxidative stress and dysfunction in rat brain induced by carbofuran exposure. Cell Mol Neurobiol. 2008;28(7):961–9. https://doi.org/10.1007/s10571-008-9270-5.

Kamel F, Umbach DM, Bedlack RS, Richards M, Watson M, Alavanja MC, et al. Pesticide exposure and amyotrophic lateral sclerosis. Neurotoxicology. 2012;33(3):457–62. https://doi.org/10.1016/j.neuro.2012.04.001.

Kaminski RM, Tochman AM, Dekundy A, Turski WA, Czuczwar SJ. Ethosuximide and valproate display high efficacy against lindane induced seizures in mice. Toxicol Lett. 2004;154(1-2):55–60. https://doi.org/10.1016/j.toxlet.2004.07.002.

Kapoor U, Srivastava MK, Bhardwaj S, Srivastava LP. Effect of imidacloprid on antioxidant enzymes and lipid peroxidation in female rats to derive its No Observed Effect Level (NOEL). J Toxicol Sci. 2010;35(4):577–81. https://doi.org/10.2131/jts.35.577.

Kaur R, Thakur Y. Pesticide induced oxidative stress diseases – a review. JETIR. 2018;5(8)

Kbare SB, Rizvi AG, Shukta OP, Singh RRP, Perkash O, Misra VD. Epidemic outbreak of neuro-ocular manifestations due to chronic BHC poisoning. J Assoc Physicians India. 1977;25(3):215–22.

Keifer M, Rivas F, Moon JD, Checkoway H. Symptoms and cholinesterase activity among rural residents living near cotton fields in Nicaragua. Occup Environ Med. 1996;53(11):726–9. https://doi.org/10.1136/oem.53.11.726.

Keil AP, Daniels JL, Hertz-Picciotto I. Autism spectrum disorder, flea and tick medication, and adjustments for exposure misclassification: the CHARGE (Childhood Autism Risks from Genetics and Environment) case-control study. Environ Health. 2014;13(3) https://doi.org/10.1186/1476-069X-13-3.

Kenet T, Froemke RC, Schreiner CE, Pessah IN, Merzenich MM. Perinatal exposure to a noncoplanar polychlorinated biphenyl alters tonotopy, receptive fields, and plasticity in rat primary auditory cortex. Proc Natl Acad Sci U S A. 2007;104(18):7646–51. https://doi.org/10.1073/pnas.0701944104.

Kim KH, Pessah IN. Perinatal exposure to environmental polychlorinated biphenyls sensitizes hippocampus to excitotoxicity ex vivo. Neurotoxicology. 2011;32(6):981–5. https://doi.org/10.1016/j.neuro.2011.04.004.

Kim KH, Inan SY, Berman RF, Pessah IN. Excitatory and inhibitory synaptic transmission is differentially influenced by two ortho-substituted polychlorinated biphenyls in the hippocampal slice preparation. Toxicol Appl Pharmacol. 2009;237(2):168–77. https://doi.org/10.1016/j.taap.2009.03.002.

Kim S, Lee HS, Park Y. Perinatal exposure to low-dose imidacloprid causes ADHD-like symptoms: evidences from an invertebrate model study. Food Chem Toxicol. 2017;110:402–7. https://doi.org/10.1016/j.fct.2017.10.007.

Kimura-Kuroda J, Komuta Y, Kuroda Y, Hayashi M, Kawano H. Nicotine-like effects of the neonicotinoid insecticides acetamiprid and imidacloprid on cerebellar neurons from neonatal rats. PLoS One. 2012;7(2):e32432. https://doi.org/10.1371/journal.pone.0032432.

Konthonbut P, Kongtip P, Nankongnab N, Tipayamongkholgul M, Yoosook W, Woskie S. Paraquat exposure of pregnant women and neonates in agricultural areas in Thailand. Int J Environ Res Public Health. 2018;15(6):1163. https://doi.org/10.3390/ijerph15061163.

Koureas M, Tsezou A, Tsakalof A, Orfanidou T, Hadjichristodoulou C. Increased levels of oxidative DNA damage in pesticide sprayers in Thessaly Region (Greece). Implications of pesticide exposure. Sci Total Environ. 2014;496:358–64. https://doi.org/10.1016/j.scitotenv.2014.07.062.

Krüger M, Schrödl W, Pedersen IB, Shehata AA. Detection of glyphosate in malformed piglets. J Environ Anal Toxicol. 2014;4(5) https://doi.org/10.4172/2161-0525.1000230.

Kuehn BM. Increased risk of ADHD associated with early exposure to pesticides, PCBs. JAMA. 2010;304(1):27–8. https://doi.org/10.1001/jama.2010.860.

Kühlbrandt W. Structure and function of mitochondrial membrane protein complexes. BMC Biol. 2015;13(89) https://doi.org/10.1186/s12915-015-0201-x.

Kumar J, Monica Lind P, Salihovic S, van Bavel B, Lind L, Ingelsson E. Influence of persistent organic pollutants on oxidative stress in population-based samples. Chemosphere. 2014;114:303–9. https://doi.org/10.1016/j.chemosphere.2014.05.013.

Lan A, Kalimian M, Amram B, Kofman O. Prenatal chlorpyrifos leads to autism-like deficits in C57Bl6/J mice. Environ Health. 2017;16(1) https://doi.org/10.1186/s12940-017-0251-3.

Landrigan PJ, Lambertini L, Birnbaum LS. A research strategy to discover the environmental causes of autism and neurodevelopmental disabilities. Environ Health Perspect. 2012;120(7):a258–60. https://doi.org/10.1289/ehp.1104285.

Lee DH, Jacobs DR, Porta M. Association of serum concentrations of persistent organic pollutants with the prevalence of learning disability and attention deficit disorder. J Epidemiol Community Health. 2007;61(7):591–6. https://doi.org/10.1136/jech.2006.054700.

Lee I, Eriksson P, Fredriksson A, Buratovic S, Viberg H. Developmental neurotoxic effects of two pesticides: behavior and biomolecular studies on chlorpyrifos and carbaryl. Toxicol Appl Pharmacol. 2015;288(3):429–38. https://doi.org/10.1016/j.taap.2015.08.014.

Lekei E, Ngowi AV, London L. Acute Pesticide poisoning in children: hospital review in selected hospitals of Tanzania. J Toxicol. 2017;4208405 https://doi.org/10.1155/2017/4208405.

Lenaz G. The mitochondrial production of reactive oxygen species: mechanisms and implications in human pathology. IUBMB Life. 2001;52(3-5):159–64. https://doi.org/10.1080/15216540152845957.

Leung MCK, Meyer JN. Mitochondria as a target of organophosphate and carbamate pesticides: Revisiting common mechanisms of action with new approach methodologies. Reprod Toxicol. 2019;89:83–92. https://doi.org/10.1016/j.reprotox.2019.07.007.

Li B, He X, Sun Y, Li B. Developmental exposure to paraquat and maneb can impair cognition, learning and memory in Sprague-Dawley rats. Mol BioSyst. 2016;12:3088–97. https://doi.org/10.1039/C6MB00284F.

Liew Z, von Ehrenstein OS, Ling C, Yuan Y, Meng Q, Cui X, et al. Ambient exposure to agricultural pesticides during pregnancy and risk of cerebral palsy: a population-based study in California. Toxics. 2020;8(3):52. https://doi.org/10.3390/toxics8030052.

Lifshitz M, Shahak E, Sofer S. Carbamate and organophosphate poisoning in young children. Pediatr Emerg Care. 1999;15(2):102–3. https://doi.org/10.1097/00006565-199904000-00006.

Lin PC, Lin HJ, Liao YY, Guo HR, Chen KT. Acute poisoning with neonicotinoid insecticides: a case report and literature review. Basic Clin Pharmacol Toxicol. 2013;112(4):282–6. https://doi.org/10.1111/bcpt.12027.

Lindahl T, Barnes DE. Repair of endogenous DNA damage. Cold Spring Harb Symp Quant Biol. 2000;65:127–33. https://doi.org/10.1101/sqb.2000.65.127.

Liu J, Cao S, Chen Z, Raine A, Hanlon A, Ai Y, et al. Cohort profile update: The China Jintan child cohort study. Int J Epidemiol. 2015;44(5):1548. https://doi.org/10.1093/ije/dyv119.

Liu JJ, Guo C, Wang B, Shi MX, Yang Y, Yu Z, et al. Maternal fenvalerate exposure during pregnancy impairs growth and neurobehavioral development in mouse offspring. PLoS ONE. 2018;13(10):e0205403. https://doi.org/10.1371/journal.pone.0205403.

Liu S, Yu M, Xie X, Ru Y, Ru S. Carbofuran induces increased anxiety-like behaviors in female zebrafish (Danio rerio) through disturbing dopaminergic/norepinephrinergic system. Chemosphere. 2020;253 https://doi.org/10.1016/j.chemosphere.2020.126635.

Lizardi PS, O'Rourke MK, Morris RJ. The effects of organophosphate pesticide exposure on Hispanic children's cognitive and behavioral functioning. J Pediatr Psychol. 2008;33(1):91–101. https://doi.org/10.1093/jpepsy/jsm047.

Llorens J, Tusell JM, Suñol C, Rodríguez-Farré E. Effects of lindane on spontaneous behavior of rats analyzed by multivariate statistics. Neurotoxicol Teratol. 1989;11(2):145–51. https://doi.org/10.1016/0892-0362(89)90053-6.

Llorens J, Tusell JM, Suñol C, Rodrífguez-Farré E. On the effects of lindane on the plus-maze model of anxiety. Neurotoxicol Teratol. 1990;12(6):643–7. https://doi.org/10.1016/0892-0362(90)90078-Q.

Lonare M, Kumar M, Raut S, Badgujar P, Doltade S, Telang A. Evaluation of imidacloprid-induced neurotoxicity in male rats: a protective effect of curcumin. Neurochem Int. 2014;78:122–9. https://doi.org/10.1016/j.neuint.2014.09.004.

Lopez-Sandoval J, Sanchez-Enriquez S, Rivera-Leon EA, Bastidas-Ramirez BE, Garcia-Garcia MR, Gonzalez-Hita ME. Cardiovascular risk factors in adolescents: role of insulin resistance and obesity. Acta Endocrinol (Buchar). 2018;14(3):330–7. https://doi.org/10.4183/aeb.2018.330.

Lyall K, Croen LA, Sjödin A, Yoshida CK, Zerbo O, Kharrazi M, et al. Polychlorinated biphenyl and organochlorine pesticide concentrations in maternal mid-pregnancy serum samples: association with autism spectrum disorder and intellectual disability. Environ Health Perspect. 2017;125(3):474–80. https://doi.org/10.1289/EHP277.

Marfo JT, Fujioka K, Ikenaka Y, Nakayama SMM, Mizukawa H, Aoyama Y, et al. Relationship between Urinary N-Desmethyl-Acetamiprid and Typical Symptoms including Neurological Findings: A Prevalence Case-Control Study. PLoS One. 2015;10(11):e0142172. https://doi.org/10.1371/journal.pone.0142172.

Mariussen E, Fonnum F. Neurochemical targets and behavioral effects of organo-halogen compounds: an update. Crit Rev Toxicol. 2006;36(3):253–89. https://doi.org/10.1080/10408440500534164.

Marks AR, Harley K, Bradman A, Kogut K, Barr DB, Johnson C, et al. Organophosphate pesticide exposure and attention in young Mexican-American children: the CHAMACOS study. Environ Health Perspect. 2010;118(12):1768–74. https://doi.org/10.1289/ehp.1002056.

Matin MA, Jaffery FN, Siddiqui RA. A possible neurochemical basis of the central stimulatory effects of pp'DDT. J Neurochem. 1981;36(3):1000–5. https://doi.org/10.1111/j.1471-4159.1981.tb01692.x.

Mehdi SH, Qamar A. Paraquat-induced ultrastructural changes and DNA damage in the nervous system is mediated via oxidative-stress-induced cytotoxicity in Drosophila melanogaster. Toxicol Sci. 2013;134(2):355–65. https://doi.org/10.1093/toxsci/kft116.

Mesnage R, Bernay B, Séralini GE. Ethoxylated adjuvants of glyphosate-based herbicides are active principles of human cell toxicity. Toxicology. 2013;313(2-3):122–8. https://doi.org/10.1016/j.tox.2012.09.006.

Miodovnik A. Environmental neurotoxicants and developing brain. Mt Sinai J Med. 2011;78(1):58–77. https://doi.org/10.1002/msj.20237.

Miranda-Contreras L, Dávila-Ovalles R, Benítez-Díaz P, Peña-Contreras Z, Palacios-Prü E. Effects of prenatal paraquat and mancozeb exposure on amino acid synaptic transmission in developing mouse cerebellar cortex. Brain Res Dev Brain Res. 2005;160(1):19–27. https://doi.org/10.1016/j.devbrainres.2005.08.001.

Mishra D, Tiwari SK, Agarwal S, Sharma VP, Chaturvedi RK. Prenatal carbofuran exposure inhibits hippocampal neurogenesis and causes learning and memory deficits in offspring. Toxicol Sci. 2012;127(1):84–100. https://doi.org/10.1093/toxsci/kfs004.

Mittler R. ROS Are Good. Trends Plant Sci. 2017;22(1):11–9. https://doi.org/10.1016/j.tplants.2016.08.002.

Mohammadi H, Ghassemi-Barghi N, Malakshah O, Ashari S. Pyrethroid exposure and neurotoxicity: a mechanistic approach. Arh Hig Rada Toksikol. 2019;70(2):74–89. https://doi.org/10.2478/aiht-2019-70-3263.

Morales E, Sunyer J, Castro-Giner F, Estivill X, Julvez J, Ribas-Fitó N, et al. Influence of glutathione S-transferase polymorphisms on cognitive functioning effects induced by p,p'-DDT among preschoolers. Environ Health Perspect. 2008;116(11):1581–5. https://doi.org/10.1289/ehp.11303.

Moser VC, Padilla S, Simmons JE, Haber LT, Hertzberg RC. Impact of chemical proportions on the acute neurotoxicity of a mixture of seven carbamates in preweanling and adult rats. Toxicol Sci. 2012;129(1):126–34. https://doi.org/10.1093/toxsci/kfs190.

Muñoz-Quezada MT, Lucero BA, Gutiérrez-Jara JP, Buralli RJ, Zúñiga-Venegas L, Muñoz MP, et al. Longitudinal exposure to pyrethroids (3-PBA and trans-DCCA) and 2,4-D herbicide in rural schoolchildren of Maule region, Chile. Sci Total Environ. 2020;749 https://doi.org/10.1016/j.scitotenv.2020.141512.

Nakagami A, Koyama T, Kawasaki K, Negishi T, Ihara T, Kuroda Y, et al. Maternal plasma polychlorinated biphenyl levels in cynomolgus monkeys (Macaca fascicularis) affect infant social skills in mother-infant interaction. Dev Psychobiol. 2011;53(1):79–88. https://doi.org/10.1002/dev.20493.

Nasuti C, Gabbianelli R, Falcioni ML, Di Stefano A, Sozio P, Cantalamessa F. Dopaminergic system modulation, behavioral changes, and oxidative stress after neonatal administration of pyrethroids. Toxicology. 2007;229(3):194–205. https://doi.org/10.1016/j.tox.2006.10.015.

Negga R, Stuart JA, Machen ML, Salva J, Lizek AJ, Richardson SJ, et al. Exposure to glyphosate- and/or Mn/Zn-ethylene-bis-dithiocarbamate containing pesticides leads to degeneration of gamma-aminobutyric acid and dopamine neurons in Caenorhabditis elegans. Neurotox Res. 2012;21(3):281–90.

Newschaffer CJ, Croen LA, Daniels J, Giarelli E, Grether JK, Levy SE, et al. The epidemiology of autism spectrum disorders. Annu Rev Public Health. 2007;28:235–58. https://doi.org/10.1146/annurev.publhealth.28.021406.144007.

Nicolson GL. The Fluid-Mosaic Model of Membrane Structure: still relevant to understanding the structure, function and dynamics of biological membranes after more than 40 years. Biochim Biophys Acta. 2014;1838(6):1451–66. https://doi.org/10.1016/j.bbamem.2013.10.019.

Niki E. Lipid peroxidation: physiological levels and dual biological effects. Free Radic Biol Med. 2009;47(5):469–84. https://doi.org/10.1016/j.freeradbiomed.2009.05.032.

NIOH. Final Report of the Investigation of Unusual Illnesses Allegedly Produced by Endosulfan Exposure in Padre Village of Kasargod District (N. Kerala). Ahmedabad: National Institute of Occupational Health, Indian Council of Medical Research; 2002.

Nunes MEM, Schimith LE, Costa-Silva DGD, Lopes AR, Leandro LP, Martins IK, et al. Acute exposure to permethrin modulates behavioral functions, redox, and bioenergetics parameters and induces DNA damage and cell death in larval zebrafish. Oxidative Med Cell Longev. 2019;2019:9149203. https://doi.org/10.1155/2019/9149203.

Nyström T. Role of oxidative carbonylation in protein quality control and senescence. EMBO J. 2005;24(7):1311–7. https://doi.org/10.1038/sj.emboj.7600599.

Ojha A, Srivastava N. Redox imbalance in rat tissues exposed with organophosphate pesticides and therapeutic potential of antioxidant vitamins. Ecotoxicol Environ Saf. 2012;75(1):230–41. https://doi.org/10.1016/j.ecoenv.2011.08.013.

Ojha A, Yaduvanshi SK, Pant SC, Lomash V, Srivastava N. Evaluation of DNA damage and cyto-toxicity induced by three commonly used organophosphate pesticides individually and in mixture, in rat tissues. Environ Toxicol. 2013;28(10):543–52. https://doi.org/10.1002/tox.20748.

Ornoy A, Weinstein-Fudim L, Ergaz Z. Prenatal factors associated with autism spectrum disorder (ASD). Reprod Toxicol. 2015;56:155–69. https://doi.org/10.1016/j.reprotox.2015.05.007.

Ostrea EM Jr, Reyes A, Villanueva-Uy E, Pacifico R, Benitez B, Ramos E, et al. Fetal exposure to propoxur and abnormal child neurodevelopment at 2 years of age. Neurotoxicology. 2012;33(4):669–75. https://doi.org/10.1016/j.neuro.2011.11.006.

Özdemir HH, Kara M, Yumrutas O, Uckardes F, Eraslan E, Demir CF, et al. Determination of the effects on learning and memory performance and related gene expressions of clothianidin in rat models. Cogn Neurodyn. 2014;8(5):411–6. https://doi.org/10.1007/s11571-014-9293-1.

Pellegrino D, La Russa D, Marrone A. Oxidative imbalance and kidney damage: new study perspectives from animal models to hospitalized patients. Antioxidants (Basel). 2019;8(12):594. https://doi.org/10.3390/antiox8120594.

Pesticide Action Network. Communities in Peril: global report on health impacts of pesticide use in agriculture. Penang: PAN Asia Pacific; 2010.

Popa-Wagner A, Mitran S, Sivanesan S, Chang E, Buga AM. ROS and brain diseases: the good, the bad, and the ugly. Oxidative Med Cell Longev. 2013;963520 https://doi.org/10.1155/2013/963520.

Porta M, Grimalt JO, Jariod M, Ruiz L, Marco E, López T, Malats N, Puigdomènech E, Zumeta E. PANKRAS II Study Group. The influence of lipid and lifestyle factors upon correlations between highly prevalent organochlorine compounds in patients with exocrine pancreatic cancer. Environ Int. 2007;33(7):946–54. https://doi.org/10.1016/j.envint.2007.05.005.

Pu Y, Yang J, Chang L, Qu Y, Wang S, Zhang K, et al. Maternal glyphosate exposure causes autism-like behaviors in offspring through increased expression of soluble epoxide hydrolase. PNAS. 2020;117(21):11753–9. https://doi.org/10.1073/pnas.1922287117.

Quijano RF. Endosulfan poisoning in Kasargod, Kerala, India: report of a fact-finding mission. Penang: Pesticide Action Network Asia and the Pacific; 2002. http://www.panap.net/sites/default/files/endosulfan_report_Kerala_1.pdf.

Rai DK, Sharma B. Carbofuran-induced oxidative stress in mammalian brain. Mol Biotechnol. 2007;37(1):66–71. https://doi.org/10.1007/s12033-007-0046-9.

Rauh VA, Garfinkel R, Perera FP, Andrews HF, Hoepner L, Barr DB, et al. Impact of prenatal chlorpyrifos exposure on neurodevelopment in the first 3 years of life among inner-city children. Pediatrics. 2006;118(6):e1845–59. https://doi.org/10.1542/peds.2006-0338.

Rauh VA, Perera FP, Horton MK, Whyatt RM, Bansal R, Hao X, Liu J, Barr DB, Slotkin TA, Peterson BS. Brain anomalies in children exposed prenatally to a common organophosphate pesticide. PNAS. 2012;109(20):7871–6.

Ribas-Fitó N, Torrent M, Carrizo D, Muñoz-Ortiz L, Júlvez J, Grimalt JO, et al. In utero exposure to background concentrations of DDT and cognitive functioning among preschoolers. Am J Epidemiol. 2006;164(10):955–62. https://doi.org/10.1093/aje/kwj299.

Rice DC. Parallels between attention deficit hyperactivity disorder and behavioral deficits produced by neurotoxic exposure in monkeys. Environ Health Perspect. 2000;108(Suppl 3):405–8. https://doi.org/10.1289/ehp.00108s3405.

Rice D, Barone S Jr. Critical periods of vulnerability for the developing nervous system: evidence from humans and animal models. Environ Health Perspect. 2000;108(Suppl 3):511–33. https://doi.org/10.1289/ehp.00108s3511.

Richardson JR, Taylor MM, Shalat SL, Guillot TS 3rd, Caudle WM, Hossain MM, et al. Developmental pesticide exposure reproduces features of attention deficit hyperactivity disorder. FASEB J. 2015;29(5):1960–72. https://doi.org/10.1096/fj.14-260901.

Roberts JR, Karr CJ. Council On Environmental Health Pesticide exposure in children. Pediatrics. 2012;130(6):e1765–88. https://doi.org/10.1542/peds.2012-2758.

Roberts EM, English PB, Grether JK, Windham GC, Somberg L, Wolff C. Maternal residence near agricultural pesticide applications and autism spectrum disorders among children in the California Central Valley. Environ Health Perspect. 2007;115(10):1482–9. https://doi.org/10.1289/ehp.10168.

Roberts JR, Dawley EH, Reigart JR. Children's low-level pesticide exposure and associations with autism and ADHD: a review. Pediatr Res. 2019;85(2):234–41. https://doi.org/10.1038/s41390-018-0200-z.

Rodrigues KJ, Santana MB, Do Nascimento JL, Picanço-Diniz DL, Maués LA, Santos SN, et al. Behavioral and biochemical effects of neonicotinoid thiamethoxam on the cholinergic system in rats. Ecotoxicol Environ Saf. 2010;73(1):101–7. https://doi.org/10.1016/j.ecoenv.2009.04.021.

Rogan WJ, Chen A. Health risks and benefits of bis(4-chlorophenyl)-1,1,1-trichloroethane (DDT). Lancet. 2005;366(9487):763–73. https://doi.org/10.1016/S0140-6736(05)67182-6.

Rohlman DS, Arcury TA, Quandt SA, Lasarev M, Rothlein J, Travers R, et al. Neurobehavioral performance in preschool children from agricultural and non-agricultural communities in Oregon and North Carolina. Neurotoxicology. 2005;26(4):589–98. https://doi.org/10.1016/j.neuro.2004.12.002.

Rosso SB, Cáceres AO, de Duffard AM, Duffard RO, Quiroga S. 2,4-Dichlorophenoxyacetic acid disrupts the cytoskeleton and disorganizes the Golgi apparatus of cultured neurons. Toxicol Sci. 2000;56(1):133–40. https://doi.org/10.1093/toxsci/56.1.133.

Rowe C, Gunier R, Bradman A, Harley KG, Kogut K, Parra K, et al. Residential proximity to organophosphate and carbamate pesticide use during pregnancy, poverty during childhood, and cognitive functioning in 10-year-old children. Environ Res. 2016;50:128–37. https://doi.org/10.1016/j.envres.2016.05.048.

Ruckart PZ, Kakolewski K, Bove FJ, Kaye WE. Long-term neurobehavioral health effects of methyl parathion exposure in children in Mississippi and Ohio. Environ Health Perspect. 2004;112(1):46–51. https://doi.org/10.1289/ehp.6430.

Rudyk CA, McNeill J, Prowse N, Dwyer Z, Farmer K, Litteljohn D, et al. Age and chronicity of administration dramatically influenced the impact of low dose paraquat exposure on behavior and hypothalamic-pituitary-adrenal activity. Front Aging Neurosci. 2017;9:222. https://doi.org/10.3389/fnagi.2017.00222.

Saeedi Saravi SS, Dehpour AR. Potential role of organochlorine pesticides in the pathogenesis of neurodevelopmental, neurodegenerative, and neurobehavioral disorders: a review. Life Sci. 2016;145:255–64. https://doi.org/10.1016/j.lfs.2015.11.006.

Sagiv SK, Nugent JK, Brazelton TB, Choi AL, Tolbert PE, Altshul LM, et al. Prenatal organochlorine exposure and measures of behavior in infancy using the Neonatal Behavioral Assessment Scale (NBAS). Environ Health Perspect. 2008;116(5):666–73. https://doi.org/10.1289/ehp.10553.

Sagiv SK, Thurston SW, Bellinger DC, Tolbert PE, Altshul LM, Korrick SA. Prenatal organochlorine exposure and behaviors associated with attention deficit hyperactivity disorder in school-aged children. Am J Epidemiol. 2010;171(5):593–601. https://doi.org/10.1093/aje/kwp427.

Sagiv SK, Thurston SW, Bellinger DC, Altshul LM, Korrick SA. Neuropsychological measures of attention and impulse control among 8-year-old children exposed prenatally to organochlorines. Environ Health Perspect. 2012;120(6):904–9. https://doi.org/10.1289/ehp.1104372.

Sagiv SK, Harris MH, Gunier RB, Kogut KR, Harley KG, Deardorff J, et al. Prenatal organophosphate pesticide exposure and traits related to autism spectrum disorders in a population living in proximity to agriculture. Environ Health Perspect. 2018;126(4):047012. https://doi.org/10.1289/ehp2580.

Samsel A, Seneff S. Glyphosate, pathways to modern diseases II: Celiac sprue and gluten intolerance. Interdiscip Toxicol. 2013;6(4):159–84. https://doi.org/10.2478/intox-2013-0026.

Samsel A, Seneff S. Glyphosate, pathways to modern diseases III: manganese, neurological diseases, and associated pathologies. Surg Neurol Int. 2015;6:45. https://doi.org/10.4103/2152-7806.153876.

Sandström J, Broyer A, Zoia D, Schilt C, Greggio C, Fournier M, et al. Potential mechanisms of development-dependent adverse effects of the herbicide paraquat in 3D rat brain cell cultures. Neurotoxicology. 2017;60:116–24. https://doi.org/10.1016/j.neuro.2017.04.010.

Schmuck G, Röhrdanz E, Tran-Thi QH, Kahl R, Schlüter G. Oxidative stress in rat cortical neurons and astrocytes induced by paraquat in vitro. Neurotox Res. 2002;4:1–13. https://doi.org/10.1080/10298420290007574.

Schuh RA, Richardson JR, Gupta RK, Flaws JA, Fiskum G. Effects of the organochlorine pesticide methoxychlor on dopamine metabolites and transporters in the mouse brain. Neurotoxicology. 2009;30(2):274–80. https://doi.org/10.1016/j.neuro.2008.12.015.

Seneff S, Swanson N, Li C. Aluminum and glyphosate can synergistically induce pineal gland pathology: connection to gut dysbiosis and neurological disease. J Agric Sci. 2015;6:42–70. https://doi.org/10.4236/as.2015.61005.

Sheikhansari G, Soltani-Zangbar MS, Pourmoghadam Z, Kamrani A, Azizi R, Aghebati-Maleki L, et al. Oxidative stress, inflammatory settings, and microRNA regulation in the recurrent implantation failure patients with metabolic syndrome. Am J Reprod Immunol. 2019;82(4):e13170. https://doi.org/10.1111/aji.13170.

Shelton JF, Hertz-Picciotto I, Pessah IN. Tipping the balance of autism risk: potential mechanisms linking pesticides and autism. Environ Health Perspect. 2012;120(7):944–51. https://doi.org/10.1289/ehp.1104553.

Shelton JF, Geraghty EM, Tancredi DJ, Delwiche LD, Schmidt RJ, Ritz B, et al. Neurodevelopmental disorders and prenatal residential proximity to agricultural pesticides: the CHARGE study. Environ Health Perspect. 2014;122(10):1103–9. https://doi.org/10.1289/ehp.1307044.

Shen H, Main KM, Kaleva M, Virtanen H, Haavisto AM, Skakkebaek NE, et al. Prenatal organochlorine pesticides in placentas from Finland: exposure of male infants born during 1997-2001. Placenta. 2005;26(6):512–4. https://doi.org/10.1016/j.placenta.2004.10.001.

Shimada H, Hirai K, Simamura E, Pan J. Mitochondrial NADH-quinone oxidoreductase of the outer membrane is responsible for paraquat cytotoxicity in rat livers. Arch Biochem Biophys. 1998;351(1):75–81. https://doi.org/10.1006/abbi.1997.0557.

Simon-Delso N, Amaral-Rogers V, Belzunces LP, Bonmatin JM, Chagnon M, Downs C, et al. Systemic insecticides (neonicotinoids and fipronil): trends, uses, mode of action and metabolites. Environ Sci Pollut Res Int. 2015;22(1):5–34. https://doi.org/10.1007/s11356-014-3470-y.

Sinaii N, Cleary SD, Ballweg ML, Nieman LK, Stratton P. High rates of autoimmune and endocrine disorders, fibromyalgia, chronic fatigue syndrome and atopic diseases among women with endometriosis: a survey analysis. Hum Reprod. 2002;17(10):2715–24. https://doi.org/10.1093/humrep/17.10.2715.

Sinha C, Agrawal AK, Islam F, Seth K, Chaturvedi RK, Shukla S, et al. Mosquito repellent (pyrethroid-based) induced dysfunction of blood-brain barrier permeability in developing brain. Int J Dev Neurosci. 2004;22(1):31–7. https://doi.org/10.1016/j.ijdevneu.2003.10.005.

Sinha C, Seth K, Islam F, Chaturvedi RK, Shukla S, Mathur N, et al. Behavioral and neurochemical effects induced by pyrethroid-based mosquito repellent exposure in rat offsprings during prenatal and early postnatal period. Neurotoxicol Teratol. 2006;28(4):472–81. https://doi.org/10.1016/j.ntt.2006.03.005.

Slotkin TA, Seidler FJ. Developmental neurotoxicants target neurodifferentiation into the serotonin phenotype: Chlorpyrifos, diazinon, dieldrin and divalent nickel. Toxicol Appl Pharmacol. 2008;233(2):211–9. https://doi.org/10.1016/j.taap.2008.08.020.

Slotkin TA, MacKillop EA, Ryde IT, Tate CA, Seidler FJ. Screening for developmental neurotoxicity using PC12 cells: comparisons of organophosphates with a carbamate, an organochlorine, and divalent nickel. Environ Health Perspect. 2007;115(1):93–101. https://doi.org/10.1289/ehp.9527.

Slotkin TA, Bodwell BE, Levin ED, Seidler FJ. Neonatal exposure to low doses of diazinon: long-term effects on neural cell development and acetylcholine systems. Environ Health Perspect. 2008a;116(3):340–8. https://doi.org/10.1289/ehp.11005.

Slotkin TA, Bodwell BE, Ryde IT, Levin ED, Seidler FJ. Exposure of neonatal rats to parathion elicits sex-selective impairment of acetylcholine systems in brain regions during adolescence and adulthood. Environ Health Perspect. 2008b;116(10):1308–14. https://doi.org/10.1289/ehp.11451.

Smulders CJ, Bueters TJ, Van Kleef RG, Vijverberg HP. Selective effects of carbamate pesticides on rat neuronal nicotinic acetylcholine receptors and rat brain acetylcholinesterase. Toxicol Appl Pharmacol. 2003;193(2):139–46. https://doi.org/10.1016/j.taap.2003.07.011.

Song SB, Xu Y, Zhou BS. Effects of hexachlorobenzene on antioxidant status of liver and brain of common carp (Cyprinus carpio). Chemosphere. 2006;65(4):699–706. https://doi.org/10.1016/j.chemosphere.2006.01.033.

Swanson NL, Leu A, Abrahamson J, Wallet B. Genetically engineered crops, glyphosate and the deterioration of health in the United States of America. J Org Syst. 2014;9(2):6–37.

Taira K. Human neonicotinoids exposure in Japan. Jpn J Clin Ecol. 2014;23(1):14–24.

Taira K, Aoyama Y, Kawakami T, Kamata M, Aoi T. Detection of chloropyridinyl neonicotinoid insecticide metabolite 6-chloronicotinic acid in the urine: six cases with subacute nicotinic symptoms. Chudoku Kenkyu. 2011;24(3):222–30.

Tanaka T. Reproductive and neurobehavioral effects of clothianidin administered to mice in the diet. Birth Defects Res B Dev Reprod Toxicol. 2012;95(2):151–9. https://doi.org/10.1002/bdrb.20349.

Taylor JR. Neurological manifestations in humans exposed to chlordecone and follow-up results. Neurotoxicology. 1982;3(2):9–16.

Thiel NA, Sachett A, Schneider SE, Garbinato C, Decui L, Eichwald T, et al. Exposure to the herbicide 2,4-dichlorophenoxyacetic acid impairs mitochondrial function, oxidative status, and behavior in adult zebrafish. Environ Sci Pollut Res. 2020;27:45874–82. https://doi.org/10.1007/s11356-020-10497-6.

Tochman AM, Kamiński R, Turski WA, Czuczwar SJ. Protection by conventional and new antiepileptic drugs against lindane-induced seizures and lethal effects in mice. Neurotox Res. 2000;2(1):63–70. https://doi.org/10.1007/BF03033328.

Torres-Sánchez L, Rothenberg SJ, Schnaas L, Cebrián ME, Osorio E, Del Carmen Hernández M, et al. In utero p,p'-DDE exposure and infant neurodevelopment: a perinatal cohort in Mexico. Environ Health Perspect. 2007;115(3):435–9. https://doi.org/10.1289/ehp.9566.

Torres-Sánchez L, Schnaas L, Cebrián ME, Hernández Mdel C, Valencia EO, García Hernández RM, et al. Prenatal dichlorodiphenyldichloroethylene (DDE) exposure and neurodevelopment: a follow-up from 12 to 30 months of age. Neurotoxicology. 2009;30(6):1162–5. https://doi.org/10.1016/j.neuro.2009.08.010.

Trachootham D, Lu W, Ogasawara MA, Nilsa RD, Huang P. Redox regulation of cell survival. Antioxid Redox Signal. 2008;10(8):1343–74. https://doi.org/10.1089/ars.2007.1957.

Tuchman R, Cuccaro M. Epilepsy and autism: neurodevelopmental perspective. Curr Neurol Neurosci Rep. 2011;11(4):428–34. https://doi.org/10.1007/s11910-011-0195-x.

United Nations Children's Fund (UNICEF). Annual Report 2018. 2018.

Vester AI, Chen M, Marsit CJ, Caudle WM. A neurodevelopmental model of combined pyrethroid and chronic stress exposure. Toxics. 2019;7(2):24. https://doi.org/10.3390/toxics7020024.

Viel JF, Warembourg C, Le Maner-Idrissi G, Lacroix A, Limon G, Rouget F, et al. Pyrethroid insecticide exposure and cognitive developmental disabilities in children: the PELAGIE mother-child cohort. Environ Int. 2015;82:69–75. https://doi.org/10.1016/j.envint.2015.05.009.

Viel JF, Rouget F, Warembourg C, Monfort C, Limon G, Cordier S, et al. Behavioural disorders in 6-year-old children and pyrethroid insecticide exposure: the PELAGIE mother-child cohort. Occup Environ Med. 2017;74(4):275–81. https://doi.org/10.1136/oemed-2016-104035.

von Ehrenstein OS, Ling C, Cui X, Cockburn M, Park AS, Yu F, et al. Prenatal and infant exposure to ambient pesticides and autism spectrum disorder in children: Population based case-control study. BMJ. 2019;364 https://doi.org/10.1136/bmj.l962.

Voorhees JR, Rohlman DS, Lein PJ, Pieper AA. Neurotoxicity in preclinical models of occupational exposure to organophosphorus compounds. Front Neurosci. 2017;10:590. https://doi.org/10.3389/fnins.2016.00590.

Vreugdenhil HJ, Mulder PG, Emmen HH, Weisglas-Kuperus N. Effects of perinatal exposure to PCBs on neuropsychological functions in the Rotterdam cohort at 9 years of age. Neuropsychology. 2004;18(1):185–93. https://doi.org/10.1037/0894-4105.18.1.185.

Wagner-Schuman M, Richardson JR, Auinger P, Braun JM, Lanphear BP, Epstein JN, et al. Association of pyrethroid pesticide exposure with attention-deficit/hyperactivity disorder in a nationally representative sample of U.S. children. Environ Health. 2015;14:44. https://doi.org/10.1186/s12940-015-0030-y.

Wei YH, Lu CY, Wei CY, Ma YS, Lee HC. Oxidative stress in human aging and mitochondrial disease-consequences of defective mitochondrial respiration and impaired antioxidant enzyme system. Chin J Phys. 2001;44(1):1–11.

Whyatt RM, Barr DB. Measurement of organophosphate metabolites in postpartum meconium as a potential biomarker of prenatal exposure: a validation study. Environ Health Perspect. 2001;109(4):417–20. https://doi.org/10.1289/ehp.01109417.

Whyatt RM, Rauh V, Barr DB, Camann DE, Andrews HF, Garfinkel R, et al. Prenatal insecticide exposures and birth weight and length among an urban minority cohort. Environ Health Perspect. 2004;112(10):1125–32. https://doi.org/10.1289/ehp.6641.

Wolansky MJ, Gennings C, DeVito MJ, Crofton KM. Evidence for dose-additive effects of pyrethroids on motor activity in rats. Environ Health Perspect. 2009;117(10):1563–70. https://doi.org/10.1289/ehp.0900667.

Workman JL, Trainor BC, Finy MS, Nelson RJ. Inhibition of neuronal nitric oxide reduces anxiety-like responses to pair housing. Behav Brain Res. 2008;187(1):109–15. https://doi.org/10.1016/j.bbr.2007.08.033.

World Health Organization (WHO). Principles for evaluating health risks in children associated with exposure to chemicals. Geneva; 2006.

Yamazaki K, Araki A, Nakajima S, Miyashita C, Ikeno T, Itoh S, et al. Association between prenatal exposure to organochlorine pesticides and the mental and psychomotor development of infants at ages 6 and 18 months: the Hokkaido Study on Environment and Children's Health. Neurotoxicology. 2017;69:201–8. https://doi.org/10.1016/j.neuro.2017.11.011.

Yang W, Sun AY. Paraquat-induced free radical reaction in mouse brain microsomes. Neurochem Res. 1998;23(1):47–53. https://doi.org/10.1023/A:1022497319548.

Yang W, Tiffany-Castiglioni E. The bipyridyl herbicide paraquat produces oxidative stress-mediated toxicity in human neuroblastoma SH-SY5Y cells: relevance to the dopaminergic pathogenesis. J Toxicol Environ Health A. 2005;68(22):1939–61. https://doi.org/10.1080/15287390500226987.

Yang W, Carmichael SL, Roberts EM, Kegley SE, Padula AM, English PB, et al. Residential agricultural pesticide exposures and risk of neural tube defects and orofacial clefts among offspring in the San Joaquin Valley of California. Am J Epidemiol. 2014;179(6):740–8. https://doi.org/10.1093/aje/kwt324.

Yolton K, Xu Y, Sucharew H, Succop P, Altaye M, Popelar A, et al. Impact of low-level gestational exposure to organophosphate pesticides on neurobehavior in early infancy: a prospective study. Environ Health. 2013;12(1):79. https://doi.org/10.1186/1476-069x-12-79.

Young JG, Eskenazi B, Gladstone EA, Bradman A, Pedersen L, Johnson C, et al. Association between in utero organophosphate pesticide exposure and abnormal reflexes in neonates. Neurotoxicology. 2005;26(2):199–209. https://doi.org/10.1016/j.neuro.2004.10.004.

Yu CJ, Du JC, Chiou HC, Chung MY, Yang W, Chen YS, et al. Increased risk of attention-deficit/hyperactivity disorder associated with exposure to organophosphate pesticide in Taiwanese children. Andrology. 2016;4(4):695–705. https://doi.org/10.1111/andr.12183.

Zhang Y, Han S, Liang D, Shi X, Wang F, Liu W, et al. Prenatal exposure to organophosphate pesticides and neurobehavioral development of neonates: a birth cohort study in Shenyang, China. PLoS One. 2014;9(2):e8849. https://doi.org/10.1371/journal.pone.0088491.

Zhang J, Guo J, Wu C, Qi X, Jiang S, Zhou T, et al. Early-life carbamate exposure and intelligence quotient of seven-year-old children. Environ Int. 2020;145 https://doi.org/10.1016/j.envint.2020.106105.

Zhu XY, Wu YY, Xia B, Dai MZ, Huang YF, Yang H, et al. Fenobucarb-induced developmental neurotoxicity and mechanisms in zebrafish. Neurotoxicology. 2020;79:11–9. https://doi.org/10.1016/j.neuro.2020.03.013.

Sleep Disturbance–Induced Free Radical Formation in the Gut May Be Blocked by Melatonin

Vaibhav Mishra, Meet Parikh, S. Akanksha, Niraj Kumar Jha, and Kavindra Kumar Kesari

Contents

Abstract All animals—including humans, birds, reptiles, flies, and even worms (*Caenorhabditis elegans*)—not only require sleep but in fact spend nearly half of their lives asleep. This is why sleep attracts scientists and researchers to understand the basic mechanisms behind it. The sleeping brain not only stores collected information but also resets the circadian cycle. Earlier, it was believed that during sleep, the brain does not work, but now scientists realize that the brain actually works more while sleeping. According to literature reports, during sleep, the brain restores all of its data, receives new informative signals from different parts of the body, and

V. Mishra (✉) · M. Parikh
Department of Neurology, School of Medicine, Harry S. Truman Memorial Veterans'
Hospital, University of Missouri, Columbia, MO, USA

S. Akanksha
Department of Microbiology, Banaras Hindu University, Varanasi, Uttar Pradesh, India

N. K. Jha
Department of Biotechnology, School of Engineering & Technology, Sharda University,
Greater Noida, Uttar Pradesh, India

K. K. Kesari (✉)
Department of Applied Physics & Department of Bioproducts & Biosystems,
School of Science, Aalto University, Espoo, Finland
e-mail: kavindra.kesari@aalto.fi

© The Author(s), under exclusive license to Springer Nature Switzerland AG 2021
K. K. Kesari, N. K. Jha (eds.), *Free Radical Biology and Environmental Toxicity*,
Molecular and Integrative Toxicology, https://doi.org/10.1007/978-3-030-83446-3_11

253

sends them for storage. We now know that several pathways and biological agents are involved in sleep, and this has improved our understanding of the function(s) of sleep and sleep-related pathologies. In this chapter, we collate valuable information about all of the pathways that are possibly involved in sleep disruption and consequently influence generation of reactive oxygen species and gut microbes. We also discuss a therapeutic approach to scavenging of reactive oxygen species.

Keywords Sleep/wake cycle · Oxidative stress · Reactive oxygen species · Gut microbiota

1   Introduction

Sleep is a characteristic and periodically repeating condition of latency, described by deficiency of cognizance and diminished responsiveness to external stimuli. Several systematic and descriptive studies have attempted to clarify the mysteries of sleep, yet none have been exhaustive enough to provide all of the relevant information. Sleep is characterized as a rapidly reversible condition of a fixed state and extraordinarily diminished sensory responsiveness (Campbell and Tobler 1984), whereas the waking state is characterized by absence of sleep and is distinguished by cognizance, attention, and performance. In 1929, it was discovered that electrical processes in the cerebrum could be recorded as electrical oscillations and that these rhythms changed between wakefulness and sleep.

Sleep is not a homogeneous state; it is a continuum of different states. In mammalian species, this continuum is divided into two significant states: non-rapid-eye-movement (NREM) sleep and rapid-eye-movement (REM) sleep.

Three different sources of electrophysiological data have been utilized to distinguish the different sleep phases (Datta and MacLean 2007):

1. Electroencephalography (EEG), which analyzes electrical activity in the brain
2. Electro-oculography (EOG) which tracks eye movement
3. Electromyography (EMG), which quantifies electrical activity in muscles

By distinguishing the coordinated movement of cortical neurons and recording voltage variances regarding the amplitude of the subsequent rhythms and their concurrencies, EEG is used to distinguish changes in attention (wakefulness) and sleep stages. For investigation of the conditions of sleep and the waking state, EEG frequencies are gathered into groups.

2   Sleep Physiology and Wake Cycle

Sleep is regulated mainly by two processes: i) a circadian alerting process known as the C process and ii) a sleep-wake Homeostatic, also known as S process. C process is an independent process which may controls the timing of sleep along with coordinating the light-dark cycle of day and night (Achermann 2004). However, S process is an internal biochemical system that decreases during sleep and increases during wakefulness. In normal wakefulness, increasing sleep homeostatic pressure is rewarded by more alerting circadian signals which promote wakefulness in light. Contrary in night circadian and homeostatic signals maintain sleep. Although, lab animals exhibit polyphasic sleep patterns suggesting that process C and S have an important role in the management of the sleep and wake cycle. Disturbance in sleep indicates a higher imbalance of antioxidants and accumulation of reactive oxygen species (ROS) in the body, especially in the gut (Vaccaro et al. 2020). Therefore, in this chapter we attempted to establish the association between sleep deprivation and an accumulation of ROS in the gut with possible different therapeutic options.

3   What Is Sleep Deprivation?

Sleep plays a significant part in human existence by enabling body metabolites and consequently helping to maintain overall and psychological well-being (Alzoubi et al. 2013). In sleep research, the most important question is how and why sleep loss occurs? Ongoing investigations have shown that sleep deprivation leads to accumulation of ROS, affecting different body parts and their function. In the brain, ROS accumulate in several regions and affect their functioning; if ROS accumulate in the hippocampus, memory impairment occurs, which may be brought about by an increase in oxidative pressure in the hippocampus. That pressure may be due to generation of ROS during the wake cycle and may be exhibited by changes in brain-derived neurotrophic factor (BDNF) levels (Alzoubi et al. 2013). According to Vorroco et al. shortened durations of sleep in flies result in increased levels of ROS in the cerebrum.

Therefore, according to Vaccaro et al. (2020), ROS accumulates in the gut when sleep is insufficient, which may lead to cause severe barin related diseases. Such effects may find more promising in the transgenic animal models, where an incraesed level of ROS may affect cerebrum or non-neuronal tissues of the human body. Bhattacharyya et al. (2014) have also reported that the intestinal ROS might be tumorigenic through several other pathways for example, causing inflammation (Aviello and Knaus 2017; Lasry et al. 2016) or DNA damage (Evans et al. 2004; Markowitz and Bertagnolli 2009). The action of noradrenergic neurons during sleep loss has been underestimated. An increase in levels of aryl sulfotransferase (a protein that is important for catabolism of catecholamines) in rodents, which has been observed consistently during lack of sleep, suggests that sleep deprivation may be linked to increased activity of the locus coeruleus. Furthermore, it has been reported that norepinephrine can help to protect insomniac animals from the effects of ROS. Markers of oxidative damage have been observed in rodents exposed to prolonged sleeplessness, followed by specific lesioning in the locus coeruleus. Since most cellular oxidant formation happens in the mitochondria, it is imperative to link these discoveries with those of investigations into metabolic rates during the homeostatic cycle and after sleep loss.

During prolonged lack of sleep, cerebral metabolism may seem to be unchanged or may decline (in comparison with ordinary awakening) in mammals, healthy patients, and sleep-deprived patients. These discoveries have confirmed a connection between ROS-instigated oxidative damage and the lethality of extreme lack of sleep. The brain stem, hypothalamus and thalamus are some of the brain regions and known to be engaged with sleep and its controlling activities (Ikeda et al. 2005). Indeed, basal forebrain and cortex are also associated with sleep, but however, a limited studies are available to explore the possible association with ROS. It is understood that oxidative pressure debilitates electrical activity in the brain. Numerous antioxidants and catalysts that shield cells from oxidative stress display diurnal cycles in their expression or performance levels. Adverse effects of oxidative pressure—for example, those associated with DNA damage, protein degradation, or lipid peroxidation—are coordinated with circadian rhythmicity. Some of these effects coincide with the times of day when they are most highly expressed in humans. Moreover, the peak times of expression of genes and proteins during the circadian period have been recorded for examination. Those whose expression is highest in the morning include glutathione peroxidase (GPx), glutathione reductase (GR), catalase, superoxide dismutase (SOD), uric acid, and peroxiredoxins (PRXs), whereas in the evening, the activities of melatonin, plasma thiols, lipid peroxidation, ascorbic acid, period 1, period 2, and the cryptochromes (CRYs) are highest. In this way, decreases of oxidative stress in the cerebrum, brain stem, and cortex have useful impacts on their electrical activity (Sharma et al. 1993). An increase in brain stem electrical activity may additionally boost cerebral cortex activity. Hence, improvised EEG recording (subsequently physiological movement) generally observed after oxidative stress reduction, which could be an indicator of sleep deprivation in stress.

Here, we are focusing on sleep deprivation. Sleep deficiency is generated by a change in lifestyle and disruption of the day–night cycle. Consequently, sleep deprivation causes oxidative stress, which is the key factor for generation of ROS in individual cell. After ROS are generated, they damage the cell membrane, causing leakage of cell material, and cell death ultimately occurs. ROS accrue in different parts of the body (the gastrointestinal tract (GIT), brain, etc.). Furthermore, accumulation of ROS affects the gut microbiota, and this may play a crucial role in different health-related issues.

4 Sleep Deprivation and Generation of Reactive Oxygen Species

Sleep deprivation may be responsible for generation and accumulation of oxidative stress and ROS (according to the stress model). According to literature reports, sleep deprivation can increase physical activity, which increases the metabolic rate. Consequently, more ROS will be generated and because of their elevated activity, it is very difficult for the normal physiological functions of cells to operate. Furthermore, it is very challenging to measure the quantities of ROS. We are only able to measure the final products of stress and check their reactivity with antioxidant parameters (Ramanathan et al. 2010). ROS generated in the body use up antioxidants obtained from the food we ingest to perform our daily bodily functions Pandey and Kar (2018).

4.1 Reactive Oxygen Species

ROS are produced mainly by natural killer cells, white blood cells (such as macrophages and neutrophils), and phagocytic cells, and are byproducts of different mechanistic pathways (Li and Shah 2003), such as the electron transfer system (oxidative phosphorylation) that occurs in mitochondria to generate adenosine triphosphate (ATP). Moreover, electrons are transferred from different cytochromes to complete the cycle. In that transfer, excess anions (superoxides) are generated by ubiquinone (complex I) (Li and Shah 2003). These anions increase oxidative stress in cells (Feissner et al. 2009). In addition, uncontrolled generation of ROS in cells causes leakage of ions (especially calcium (Ca^{2+}) simultaneously with the release of antioxidant enzymes (Lungato et al. 2013). It also obstructs nucleic acid strands and translation of nucleic acids, causes protein oxidation, and ultimately causes cell death. Different research groups have shown that ROS worsen pathophysiological conditions (Nelson et al. 2006). Furthermore, ROS play important roles in neurodegenerative disorders (Alzheimer's disease, depression, Parkinson's disease, Huntington's disease, stroke, epilepsy, and many others). They affect not just the brain but also other important organs (the gut, lungs, heart, kidneys, etc.). As a result of different stress factors, the production and degradation cycle of ROS disturbs the normal physiological functions of cells and leads to cell death (Nelson et al. 2006).

5 Disturbance of the Gut Microbiota Can Affect Sleep

The ways in which the gut microbiota affect sleep and circadian rhythms are still being researched. According to Yuanyuan Li, the gut microbiota are responsible for the restfulness of sleep in their host. It has been suggested that the main system that regulates the function of the gut microbiota is the central nervous system (CNS); the other relevant system is the hypothalamic–pituitary–adrenal (HPA) axis. To understand how this works, research has focused on neurotransmitters involved in both the brain and the gut, such as serotonin, acetylcholine, and histamine. Some amino acids, such as tryptophan and glutamate, are also present in both regions. Last, but not least, the vagus nerve pathway plays an important role via the enteric nervous system (ENS) (Powley et al. 2008). It is evident from literature reports that the intestinal myenteric plexuses have sensory neurons that are in direct contact with gut microbes. These neurons make connections with motor neurons and control secretion of intestinal juices and peristalsis movements. In addition, a key feature is the connection of the vagus nerve to the intestinal nervous system, forming a neuronal pathway (Powley et al. 2008; Bonaz et al. 2013). Therefore, it is supposed that the vagus may play a mediator role between the gut and the brain, where the gut microbes produce toxic metabolites that may affect sleep, generation of ROS, and normal brain functions via the vagus nerve (Powley et al. 2008; Bonaz et al. 2013).

6 Therapeutics to Reduce Reactive Oxygen Species Levels

It is very interesting that the brain has no defensive mechanisms against oxidative stress. Hence, it is highly susceptible to damage by ROS, which promote malfunctions in cellular and molecular pathways (Gupta et al. 2003). ROS accumulation not only affects the total duration of sleep but also pulls the organism toward death. The most common form of therapeutics aims to control ROS generation to prevent their deposition. This can be achieved by oral administration of antioxidant molecules.

6.1 Melatonin

Melatonin is known as a sleep hormone or pineal hormone, and is released by the pineal gland, which is situated in the hypothalamus. It has very powerful ROS-scavenging properties and very high cell permeability, which helps it to cross the blood–brain barrier and other cellular and subcellular barriers easily. According to the literature (Tarocco et al. 2019), during aging, levels of melatonin decrease, and this has been attributed to neurodegeneration. There is growing evidence highlighting the possible role of melatonin in neurodegenerative disorders (Gupta et al. 2003). It has been found to enhance the activity of antioxidant enzymes and to stimulate genetic expression of GPx, GR, catalase, and SOD. Furthermore, melatonin normalizes ROS such as nitric oxide, hydrogen peroxide, singlet oxygen, superoxide radicals, hypochlorous acid, peroxynitrite anions, peroxyl radicals, and hydroxyl free radicals (Gupta et al. 2003). Melatonin also has a chelating action, which makes it unique. Melatonin reduces the free radical formation and protect against inflammation and oxidative stress, where hydroxyl ion which make melatonin as an aspirant molecule. Reports in the literature support the concept that oral administration of melatonin is safe (Reiter et al. 2016; Tarocco et al. 2019), which makes melatonin a promising candidate molecule for future use.

6.2 Natural Products Such as Antioxidants

Use of natural products is very effective to regulate generation and accumulation of ROS. It is well established that certain plants (*Peganum harmala* (Singh et al. 2013), *Hedranthera barteri* (Onasanwo et al. 2010), *Terminalia chebula* (Mishra et al. 2013a, b), *Xylocarpus granatum*, *Xylocarpus moluccensis* (Lakshmi et al. 2014,

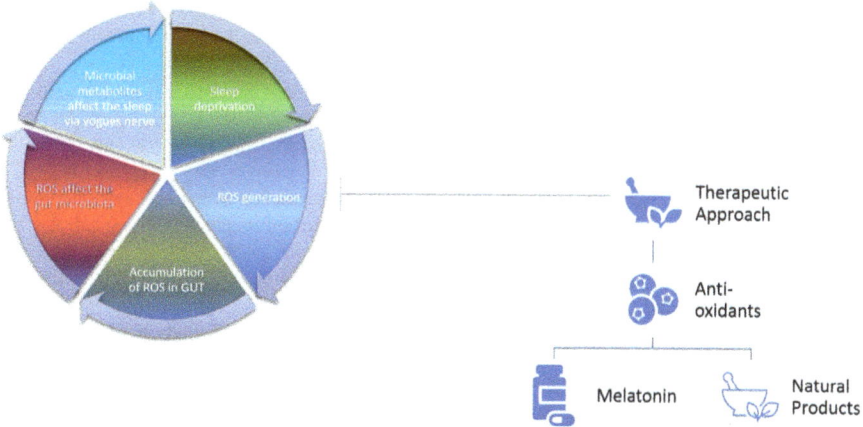

Fig. 1 The cycle of sleep deprivation and generation of ROS affects the microbiome, which could be restored by the use of antioxidants

2015), *Nyctanthes arbortristis* (Pandeti et al. 2013), *Dysoxylum binectariferum* (Mishra et al. 2014), etc.) have antioxidant properties, which may make them effective in decreasing ROS levels (Fig. 1).

7   Conclusion

Sleep deprivation due to an unexpected or stressed lifestyle causes accumulation of ROS and affects gut microbes, thereby disturbing brain function, maybe via vagus nerve stimulation. In this chapter, we have discussed possible therapeutics that could be used to treat the cumulative effects of ROS. We have highlighted melatonin and natural products that may be used in future therapy to decrease ROS-generated effects. Furthermore, we believe that sleep deprivation is a cycle and that if one step is disturbed, the fault is forwarded on to every step; thus, if we aim to normalize the cycle, we have to restore the cycle at multiple intervals. This will rebuild sleep and may normalize the function of the gut microbiota.

References

Alzoubi KH, Khabour OF, Salah HA, Abu Rashid BE. The combined effect of sleep deprivation and Western diet on spatial learning and memory: role of BDNF and oxidative stress. J Mol Neurosci. 2013;50(1):124–33.

Achermann P. The two-process model of sleep regulation revisited. Aviat Space Environ Med. 2004 Mar;75(3 Suppl):A37–43. PMID: 15018264.

Aviello, G., and Knaus, U.G. (2017). ROS in gastrointestinal inflammation: Rescue Or Sabotage? Br. J. Pharmacol. 174, 1704–1718.

Bhattacharyya, A., Chattopadhyay, R., Mitra, S., and Crowe, S.E. (2014). Oxidative stress: an essential factor in the pathogenesis of gastrointestinal mucosal diseases. Physiol. Rev. 94, 329–354.

Bonaz B, Picq C, Sinniger V, Mayol JF, Clarençon D. Vagus nerve stimulation: from epilepsy to the cholinergic anti-inflammatory pathway. Neurogastroenterol Motil. 2013;25:208–21.

Campbell SS, Tobler I. Animal sleep: a review of sleep duration across phylogeny. Neurosci Biobehav Rev. 1984;8:269–300.

Datta S, Maclean RR. Neurobiological mechanisms for the regulation of mammalian sleep–wake behavior: reinterpretation of historical evidence and inclusion of contemporary cellular and molecular evidence. Neurosci Biobehav Rev. 2007; 31 (5): 775–824.

Evans, M.D., Dizdaroglu, M., and Cooke, M.S. (2004). Oxidative DNA damage and disease: induction, repair and significance. Mutat. Res. 567, 1–61.

Feissner RF, Skalska J, Gaum WE, Sheu SS. Crosstalk signaling between mitochondrial Ca^{2+} and ROS. Front Biosci. 2009;14:1197–218.

Gupta YK, Gupta M, Kohli K. Neuroprotective role of melatonin in oxidative stress vulnerable brain. Indian J Physiol Pharmacol. 2003;47(4):373–86.

Ikeda M, Ikeda-Sagara M, Okada T, Clement P, Urade Y, Nagai T, Sugiyama T, Yoshioka T, Honda K, Inoué S. Brain oxidation is an initial process in sleep induction. Neuroscience. 2005;130(4):1029-40.

Lakshmi V, Singh N, Shrivastva S, Mishra SK, Dharmani P, Mishra V, Palit G. Gedunin and photogedunin of *Xylocarpus granatum* show significant antisecretory effects and protect the gastric mucosa of peptic ulcer in rats. Phytomedicine. 2010;17(8–9):569–74. https://doi.org/10.1016/j.phymed.2009.10.016.

Lakshmi V, Mishra V, Palit G. A new gastroprotective effect of limonoid compounds xyloccensins X and Y from *Xylocarpus molluccensis* in rat. Nat Prod Bioprospect. 2014;4(5):277–83.

Lakshmi V, Mishra V, Palit G. In vivo gastroprotective effect of xyloccensin-E and xyloccensin-I from *Xylocarpus molluccensis* in rats. Nat Prod Res. 2015;29:469–73.

Laura Dumitrescu, Iulia Popescu-Olaru, Liviu Cozma, Delia Tulbă, Mihail Eugen Hinescu, Laura Cristina Ceafalan, Mihaela Gherghiceanu, and Bogdan Ovidiu Popescu. Stress and the Microbiota-Gut-Brain Axis. Oxi Med and Cellu Longevity Oxi. 2018;2406594:12. doi.org/10.1155/2018/2406594.

Lasry, A., Zinger, A., and Ben-Neriah, Y. (2016). Inflammatory networks under- lying colorectal cancer. Nat. Immunol. 17, 230–240.

Li JM, Shah AM. ROS generation by nonphagocytic NADPH oxidase: potential relevance in diabetic nephropathy. J Am Soc Nephrol. 2003;14:S221–6.

Lungato L, Marques MS, Pereira VG, Hix S, Gazarini ML, Tufik S and Almeida VD. Sleep Deprivation Alters Gene Expression and Antioxidant Enzyme Activity in Mice Splenocytes. Scand J Immuno. 2013;195-199. https://doi.org/10.1111/sji.12029.

Markowitz, S.D., and Bertagnolli, M.M. (2009). Molecular origins of cancer: Molecular basis of colorectal cancer. N. Engl. J. Med. 361, 2449–2460.

Mishra V, Agrawal M, Onasanwo SA, Madhur G, Rastogi P, Pandey HP, Palit G, Narender T. Antisecretory and cytoprotective effects of chebulinic acid isolated from the fruits of *Terminalia chebula* on gastric ulcers. Phytomedicine. 2013a;20:506–11.

Mishra V, Shukla A, Pandeti S, Barthwal MK, Pandey HP, Palit G, Narender T. Arbortristoside-A and 7-O-trans-cinnamoyl-6β-hydroxyloganin isolated from *Nyctanthes arbortristis* possess anti-ulcerogenic and ulcer-healing properties. Phytomedicine. 2013b;20:1055–63.

Mishra SK, Mishra V, Pandey AK, Srivastava A, Tripathi CKM, Palit G, Mir SS, Mahdi AA, Agrawal SK, Lakshmi V. Docking study of the rohtukine for the prevention of peptic ulcer—a new target. J Phytopharmacol. 2014;3:9–15.

Mishra SK, Tiwari S, Shrivastava S, Sonkar R, Mishra V, Nigam SK, Saxena AK, Bhatia G, Mir SS. Pharmacological evaluation of the efficacy of *Dysoxylum binectariferum* stem bark and its active constituent rohitukine in regulation of dyslipidemia in rats. J Nat Med. 2018;72(4):837–45. https://doi.org/10.1007/s11418-014-0830-3.

Nelson SK, Bose SK, Grunwald GK, Myhill P, McCord JM. The induction of human superoxide dismutase and catalase in vivo: a fundamentally new approach to antioxidant therapy. Free Radic Biol Med. 2006;40:341–7.

Onasanwo SA, Singh N, Olaleye SB, Mishra V, Palit G. Antiulcer and antioxidant activities of *Hedranthera barteri* {(Hook F.) Pichon} with possible involvement of H^+, K^+ ATPase inhibitory activity. Ind J Med Res. 2010;132:442–9.

Pandey A, Kar SK. Rapid eye movement sleep deprivation of rat generates ROS in the hepatocytes and makes them more susceptible to oxidative stress. Sleep Sci. 2018;11(4):245–53.

Powley TL, Wang XY, Fox EA, Phillips RJ, Liu LW, Huizinga JD. Ultrastructural evidence for communication between intramuscular vagal mechanoreceptors and interstitial cells of Cajal in the rat fundus. Neurogastroenterol Motil. 2008;20:69–79.

Ramanathan L, Hu S, Frautschy SA, Siegel JM. Short-term total sleep deprivation in the rat increases antioxidant responses in multiple brain regions without impairing spontaneous alternation behavior. Behav Brain Res. 2010;207:305–9.

Reiter RJ, Mayo JC, Tan DX, Sainz RM, Alatorre-Jimenez M, Qin L. Melatonin as an antioxidant: under promises but over delivers. J Pineal Res. 2016;61(3):253–78. https://doi.org/10.1111/jpi.12360.

Sharma D, Maurya AK and Singh R. Age-related decline in multiple unit action potentials of CA3 regions of rat hippo-campus: correlation with lipid peroxidation and lipofuscin concentration and the effect of centrophenoxine. Neurobiol Aging. 1993;14: 319–330.

Singh R, Pandeti S, Palit G, Barthwal MK, Pandey HP and Narender T. Cytoprotective and antisecretory effects of Azadiradione isolated from the seeds of Azadirachta indica (neem) on gastriculcers in rat models. Phytother Res. 2015; 29:910-916.

Singh VK, Mishra V, Tiwari S, Khaliq T, Barthwal MK, Pandey HP, Palit G, Narender T. Antisecretory and cytoprotective effects of peganine hydrochloride isolated from the seeds of *Peganum harmala* on gastric ulcers. Phytomedicine. 2013;20:1180–5.

Tarocco A, Caroccia N, Morciano G, Wieckowski MR, Ancora G, Garani G, Pinton P. Melatonin as a master regulator of cell death and inflammation: molecular mechanisms and clinical implications for newborn care. Cell Death Dis. 2019;10(4):317.

Vaccaro A, Dor YK, Nambara K, Pollina EA, Lin C, Greenberg ME, Rogulja D. Sleep loss can cause death through accumulation of reactive oxygen species in the gut. Cell. 2020;181:1307–1328.

Yuanyuan L, Yanli H, Fang F and Bin Z. The Role of Microbiome in Insomnia, Circadian Disturbance and Depression. Front in Psych. 2018;9:669.

Initiation of Neurodegenerative Disorders (NDDs) Through Metal Toxicity Generated Oxidative Stress

Vinayak Agarwal, Divya Jindal, Shriya Agarwal, Shalini Mani, and Manisha Singh

Contents

Abstract The appropriate functioning of central nervous system is imperative for its physical integrity and even a slight change in the physicochemical properties leads to permanent neuronal injuries causing cognitive impairments and proteinopathies. Additionally, according to the statistics by WHO, it is expected that the progression of AD is estimated to swell up from 66% to 88% universally by the end of 2050. Epidemiologic studies expose that the foremost cause of neural collapse is the accumulation of heavy metals leading to DNA impairment and distressing the regular functioning of nearly all vital enzymes. Therefore, there exists the urgent prerequisite to alter such oxidative environment which not only limits the impedance of the biochemical processes but also initiates the chronic pathologies. Thus, in this chapter, the authors depict various effects of metal toxicity in human brain and its favorable antidotal strategies in treating the intoxication based on their pharmacological antagonists.

Keywords Neuronal damage · Alzheimer's disease · Reactive oxygen species · NMDA receptor · Antidotal strategy

V. Agarwal · D. Jindal · S. Agarwal · S. Mani · M. Singh (✉)
Department of Biotechnology, Jaypee Institute of Information Technology (JIIT),
Noida, Uttar Pradesh, India

K. K. Kesari, N. K. Jha (eds.), *Free Radical Biology and Environmental Toxicity*,
Molecular and Integrative Toxicology, https://doi.org/10.1007/978-3-030-83446-3_12

1 Introduction

Physiological integrity of the central nervous system (CNS) is essential for the healthy functioning of the brain and any disruptions in its physicochemical conformity results in irreversible neuronal damages like cognitive impairments and proteinopathies. Such irreversible changes also occur in neurodegenerative diseases (NDDs) that are concerned with continuous loss of clustered neuronal cells and are associated with aggregates of protein. The progressive neuronal loss results in the emergence of diseases like Alzheimer's disease (AD), Parkinson's disease (PD), and amyotrophic lateral sclerosis (ALS) (Barnham et al. 2004). Moreover, around 50% people having NDDs are suffering from AD. According to WHO, AD has shown a huge increase in projection of 66% from 2005 to 2030 and approximately 56 million population globally is suffering from AD and is estimated to grow up to 88 million by 2050 (2020 Alzheimer's disease facts and figures 2020). Till now, most prominent elements involved in AD include amyloid precursor protein (APP), tau (τ), amyloid beta (Aβ), and APOE, further causing the plaque formation and then NFTs (neurofibrillary tangles) and the plaques formation usually take place when 40–42 APP polymerizes with interference in amyloidogenic pathway of CNS (Lashuel et al. 2002). AD is reported to be caused by arctic mutation in the APP, and Aβ is thought to be the autosomal dominant form in Alzheimer's. It has been demonstrated that APOE4 plays an important role in the development of AD by causing immune reactivity with Aβ deposits and aggregation of total τ proteins and phosphorylated τ proteins are majorly responsible for NDDs (Masters et al. 2015). Furthermore, out of the many sources responsible for causing oxidative stress and neurodegenerative pathologies, the human system interaction and intoxication with heavy metals is one of the leading causes of degenerative changes in the human brain in today's industrial world. It has been observed in many research based studies that the population is exposed to heavy metal poisoning, that is coming from industrial wastes and pesticides are severely toxic to the human system (Singh et al. 2017). Heavy metals are natural elements on earth and are oldest toxins for humans since thousands of years now and in today's time these heavy metals originate from industrial processes, ground water, contaminated foods, herbal components, etc. The heavy metal forms like iron, mercury, manganese, arsenic, and lead are more toxic to human reproductive, neural, and cardiovascular systems (Sharma et al. 2014). Heavy metals target processes and proteins which are involved in early onset of any disease, such as DNA and genotoxic damage, LPO, protein sulfhydryl's depletion, protein's oxidation of thiol group, cell differentiation, and cell cycle regulation. These effects of heavy metals result in the activation of certain transcription factors such as NF-kB, AP-1, and p53 which are redox-sensitive (Valko et al. 2005). Lead and cadmium are involved in the inhibition of acetylcholinesterase in blood and brain (Vivek Kumar Gupta et al. 2016).

Epidemiological studies have shown that these heavy metals cause degeneration of neurons in various ways via DNA damage, amyloidogenesis, disturbing brain enzymes activities, etc. (Lee et al. 2018). Neurodegenerative disease, i.e., Alzheimer's disease, is mainly caused by heavy accumulation of lead, cadmium, and mercury. Accumulated lead results in developmental defects and hypertension (Flora et al. 2012). Lead commences by causing oxidative DNA damage, which methylates APP genes (hypo-methylated gene). It will result in overexpression of the APP gene and activates TF SP1. With the activation of beta and gamma secretase enzymes, these overexpressed APPS will be cleaved into overexpressed Aβ proteins and their aggregation results in plaque formation. Mercury is one of the risk factors involved in AD. Their increased concentrations have been dominantly seen in hair. Various studies demonstrated that mercury is involved in memory loss and dementia (Zahir et al. 2005; Mutter et al. 2010). Initially, mercury binds with tubulin and frails it. This will lead to inhibition of polymerization tubulin and cause apoptosis of neural cells. This will bring out formation of neurofibrillary tangles. Mercury can also cause oxidative stress which phosphorylates τ proteins (Lee et al. 2018). Third major heavy metal intricate in AD is cadmium. Cadmium has a potential neurotoxicity effect which is shown to have neurobehavioral problems such as memory, attentiveness, and psychomotor speed. Studies have shown that cadmium exposure interferes the normal functioning of amyloid beta (Aβ) and presenilin (PS1). Cadmium decreases the activity of Aβp1-42 channel and blocks it which will result in the formation of Aβp fibrils or oligomers. Cadmium can also block M1 receptors which indicate the overexpression of GSK-3β and AChE-S and downregulate AChE-R. This will increase the concentration of τ and Aβ proteins, ultimately engendering Alzheimer's disease.

1.1 Metal Toxicity Generates Oxidative Stress

Plethora of prior literature elucidating approaches through which human system experiences raised oxidative stress have considered the heavy metal ions mediated oxidative stress to be highly prevalent in the world population. The reactive oxygen species (ROS) are generated by activation of molecular oxygen which serves to be extremely serviceable in the breakdown of regulatory processes. ROS can be formed by various direct pathways by redox active metals and O_2 or N_2 interaction, processed by Haber–Weiss reaction or indirect pathways via metallo-enzyme activation by calcium. Calcium is one of the common metals involved in many signaling pathways and disruption in any of the calcium content will elicit negative regulation of the cellular responses leading to neurodegeneration (Lewén et al. 2000). Certain reactive species tends to interact with several elements & molecules causing neuronal cell death facilitated by an array of different signalling pathways. These pathways have been further found to contribute to the formation of toxic species such as ROS (reactive oxygen species), cholesterol oxide, peroxides, and RNS (reactive nitrogen species). Besides ROS another major reactive species class is reactive nitrogen species (RNS) which exhibits significant deleterious effects on CNS mediated via nitric oxide (NO); furthermore, these free radical species inhibit the antioxidant functioning, mitochondrial dysfunction causing abnormal cell proliferation, apoptosis, and neurodegeneration (Salazar-Flores et al. 2019).

Extensive research has shown that expression of these species is not limited to just hampering biochemical processes but have the potential to initiate chronic pathologies. Under such oxidative environment, unsaturated lipids have been reported several oxidative modifications along with alteration in lipid peroxidation, which occurs due to attack by metals on double bonds of lipid to form reactive radicals which initiate a chain reaction and act on adjacent fatty acid chains. This reaction initiates the formation of NHE (4-hydroxy-2, 3-nonenal) and their increased levels is prominently reported in AD, ALS, and PD (Barnham et al. 2004). All the bases of DNA have affinity toward oxidative damage which involves carbonylation, nitration, etc. (Alam et al. 1997).

1.2   Sources Responsible for Metal Toxicity in Human Brain

Various sources contribute to the accumulation of heavy metals in the human brain causing neurodegenerative diseases (Table 1). Heavy metal may engender from sewage disposals, industrial smoke, natural weathering of earth's crust, coal mining, pesticides, and many more. Arsenic stands out at twentieth position in most abundant elements, and when present in inorganic form becomes toxic (Sauvé 2014). Arsenic is an important heavy metal which is involved in causing disquiet from ecological as well as health standpoints. They may engender from industrial effluents or any natural means, or any unintended source (Hughes et al. 1988). Arsenic can also be derived from mining and ores processing, smelting processes, and meta-sedimentary bed rocks (Matschullat 2000). Another heavy metal is lead, which when present in dry atmosphere appears bright silver metal, otherwise is light blue. Lead is very toxic and results in environmental contamination and various human disorders. Elevated levels of lead will diffuse across blood–brain barrier and disrupt communication between endothelial cells and astrocytes. These are mainly sourced from fertilizers and pesticides, smelting of ores, soil wastes, exhaust from automobiles, smoking, drinking, food, industrial wastes, gasoline, etc. (Jaishankar et al. 2014). Mercury is a heavy d-block element, liquid metal, and odorless upon heating. Mercury in their methylated form is very toxic and bio-accumulative. This heavy metal is found in many anthropogenic activities like municipal waste water discharges, incineration, battery degradation, and seawaters (Ferrara et al. 2000).

Cadmium is zinc's by-product and can be found in mineral fertilizers, smelting, and ores processing. They are classified under group I of carcinogens by IARC (Henson and Chedrese 2004). Chromium is predominantly found in petroleum and coal, industries working on chemical production and metallurgy, water sediments, soil, plants, metal alloys, cement, leather tanning, etc. (Ghani 2011; Sabine Martin 2009). Copper is one of the ductile and malleable elements which are highly conductive. It is commonly used in industries concerned with cables, cookware, pharmaceuticals (birth control pills), swimming pools, and accumulated in plants and soil (Agency for Toxic Substances and Disease Registry (ATSDR) 2004). Aluminum shows much more affinity toward oxygen and is commonly found in drinking water,

Table 1 Types of heavy metals in the ecosystem and their impact on the central nervous system

S. no.	Heavy metals	Sources	Impact on CNS
1.	Iron	Enzymes (catalase), proteins involved in oxygen transportation (hemoglobin and myoglobin), various dietary elements and soil	Iron deposition on basal ganglia results in neural degeneration, neurodegeneration with brain Fe accumulation (Friedreich ataxia, Kufor-Rakeb disease, aceruloplasminemia, and neuroferritinopathy), elevated levels of Fe can be released in brain regions after local hemorrhage caused by brain trauma (Farina et al. 2013).
2.	Manganese	Diet, airborne exposure via fumes, aerosols, or anthropogenic sources, parental nutrition and manganese carrying drugs	Chronic exposure to Mn has been shown to cause the degeneration of nigrostriatal DAergic neurons, aggregate in mitochondria of neurons and astrocytes, and involved in disruption of synthesis of ATP (McDougall et al. 2008; Milatovic et al. 2017).
3.	Mercury	Anthropogenic activities like municipal wastewater discharges, incineration, battery degradation, and seawaters	MeHg-induced neurotoxicity and neurodegeneration (Farina et al. 2011).
4.	Arsenic	Industrial effluents, ore mining, smelting processes, meta-sedimentary bed rocks, drinking water	Alteration in central monoaminogenic system neurotoxicity and cytotoxicity (Mejía et al. 1997), other neurological problems (Huy et al. 2014).
5.	Cadmium	Mineral fertilizers, smelting, and ores processing	Elevation of norepinephrine in hypothalamus and mid brain (Flora and Tandon 1987).
6.	Lead	Fertilizers, pesticides, smelting of ores, soil waste, automobile exhausts, smoking, alcohol, food, gasoline, industrial waste	Anemia (less Hb), hypertension, disruption of nervous systems, brain damage, blocking neurotransmission by inhibiting NMDAR and neurotransmitter release, blocking calcium gated channels (Nihei et al. 2000; Guilarte and McGlothan 2003), intellectual disorders.
7.	Chromium	Petroleum and coal, industries working on chemical production and metallurgy	
8.	Copper	Industries involved with cables, cookware, pharmaceuticals (birth control pills), swimming pools, accumulated in plants and soil	Neurotoxicity in hepatocerebral region (Pal 2014).
9.	Aluminum	Drinking water, Al coated drugs, beverages	Inflammation in brain and induce neurotoxicity when present in higher concentration by NFT formation (Maya et al. 2016).

(continued)

Table 1 (continued)

S. no.	Heavy metals	Sources	Impact on CNS
10.	Nickel	Batteries production, jewelry, cereals, unrefined grains, hydrogenated oils	Induce toxicity by activating CDK5/p25 and GSK-3β (Gorantla et al. 2020).
11.	Zinc	Food as dairy products, cereals, nuts, beans, and whole grains and industrially via smelting of ores and draining of mining operations	Ischemic strokes (Galasso and Dyck 2007).

Al coated drugs, beverages, etc. (Jaishankar et al. 2014). Manganese is not counted as a free element and is generally taken in diet, airborne exposure via fumes, aerosols, anthropogenic sources, parental nutrition, or Mn containing drugs (Farina et al. 2013). Iron is an essential element which is required for body functions, one of the important components in algae and enzymes, i.e., catalase and cytochromes along with proteins helping in transporting oxygen (Valko et al. 2005). Iron is the most abundant element, mainly found in diet and soil. Sporadic accidental or occupational exposure has shown neurotoxicity (Farina et al. 2013). Nickel is a hard, transition element and ductile. Powered nickel has significant chemical activity because of their maximized surface area. Nickel is usually used in batteries production, jewelry, cereals, unrefined grains, hydrogenated oils, etc. (Amirah et al. 2013). Zinc is a universal element, present in a divalent state, which acts as catalyst, regulatory ion for various cellular processes, and acts as cofactor for SOD. It engenders from food dairy products, cereals, nuts, beans, and whole grains and industrially via smelting of ores and draining of mining operations (Mocchegiani et al. 2000).

2   Effects of Toxicity Generated by ROS/RNS

2.1   Alzheimer's Disease

Since ROS and RNS are sensitive species, any modification in the cells can be targeted as a marker for oxidative stress. Alzheimer's disease is characterized by formation of amyloid beta plaques which are cleaved by beta and gamma secretase from APP. Recent studies have suggested that APPs maintain Cu homeostasis. Elevated concentrations of copper, iron, and zinc have been seen in deposits of amyloid from patients affected with AD (Table 1). Zinc induces aggregation and precipitation of amyloid beta peptides while copper and iron only induce aggregation at acidic pH (Huang et al. 1997; Smith et al. 1997). Their excessive generation causes neuronal cell death. Many evidences propose the homeostasis of zinc, iron, and copper, which are altered in AD (Atwood et al. 1999). When Cu^{2+} is present, synthetic amyloid beta becomes toxic and can be inhibited by extracellular catalase, which implicates hydrogen peroxide in the toxic pathway. Redox reaction will occur when iron and copper associate with amyloid beta, and reduces oxidative states of both the metals. This will lead to the production of hydrogen peroxide from oxygen in reduced oxidative states, initiating Fenton chemistry. This will lead to the formation of highly toxic OḢ (hydroxyl) radical. These synthetic Aβ, when interacting with copper, induces oxidative modification of sulfur at Met35 which results in Aβ oligomers and various adducts. These combinations of cellular components are toxic to cells (Barnham et al. 2003).

2.2 Parkinson's Disease

Loss of dopaminergic neurons is a characteristic feature of Parkinson's disease, caused by deposition of α-synuclein. Neuromelanin accumulates iron ions and their loss in cells containing it marks a prominent target (Huang et al. 1999). They contain products of dopamine chemistry. Dopamine coordinates the metal and reduces their oxidation states, as it is a strong chelator and an electron donor. Subsequently, this will lead to the production of H_2O_2 (Table 1), initiating Fenton chemistry. Their association will form synthetic melanin, become pro-melanin at high concentrations, and induce redox-associated toxicity (Double et al. 2003).

Another emerging evidence causes α-synuclein to affect the activity of dopamine by mutation at A35T, which causes accumulation of dopamine in the cell and results in ROS when interacted with iron. This mutation will alter the activity of dihydropteridine reductase (directly associated with synthesis of dopamine) (Baptista et al. 2003). High levels of iron aggravate cell damage by lipid peroxidation. NO synthesized by NOS obstructs enzymes in mitochondrial ETC.

In short, when dopamine is released from vesicles, it starts interacting with iron and initiates redox reaction, which results in the formation of neuromelanin (NM) and this NM associated with iron to form ROS. The equilibrium is maintained by α-synuclein. ROS formation leads to oxidative stress, causing degeneration of neuronal cells.

2.3 Amyotrophic Lateral Sclerosis

Amyotrophic lateral sclerosis is characterized by loss of motor neurons of spinal cord and cerebral cortex. It is distinguished by aggregation of misfolded proteins in neuronal tissues, toxic via SOD. In ALS, mutations in SOD can convert from antioxidants to pro-oxidants, resulting in oxidative stress by forming hydrogen peroxide and peroxynitrite (OONO–) (Uversky et al. 2001). Mitochondrial dysfunction ultimately leads to oxidative stress, which assists ROS/RNS-elevated concentrations.

These reactive species will cause DNA mutation, Ca^{2+} homeostasis, and permeability problems, thus leading to ALS (Table 1).

Methylated mercury causes elevated extracellular glutamate levels by inhibiting astrocytes glutamate uptake and stimulating glutamate excretion from presynaptic terminals. This will over-activate NMDA receptors and hence increases Ca^{2+} influx into the neurons. This increased concentration of Ca^{2+} will cause mitochondrial collapse and activate neuronal NOS (nitric oxide synthase), increasing NO formation. The methylated mercury hinders the ETC which will lead to the formation of ROS and H_2O_2. This will lead to oxidative stress in the postsynaptic terminal of the neuron (Farina et al. 2011).

3   Exposure, Uptake, and Bioaccumulation of Heavy Metals in Human System

Past decade witnessed a significant rise of metal toxin induction in the human system through certain traditional means of origin and distribution such as environmental pollution; however, the neoteric views regarding the same are unlike the prior assumptions. Contemporary world has been marked by a ubiquitous presence of heavy metals in the atmosphere, lithosphere, and hydrosphere; primary source of such omnipresence is achieved due to the elevated levels of copious natural and anthropogenic sources (Keshav Krishna and Rama 2016). There are several metals existing in our ecosystem; however, this category of metals has been classed separately not only due to their high-density profile but also due to the severe adverse effects caused by them to the ecosystem and human body. Some of these heavy metals are antimony, cadmium, lead, mercury, manganese, nickel, copper, cobalt, chromium, arsenic, iron, bismuth, zinc, vanadium, and gallium (Godwill et al. 2019). Detailed exploratory studies have been conducted to investigate their pathophysiological importance in association with the human system as their bioaccumulation has been reported to cause severe impairments to vital organ systems encompassing central nervous, gastrointestinal, and reproductive systems (Kumar et al. 2018).

Recent experimental observations have explored various facets associated with heavy metals exposure to the human system that include the consumption of adulterated or contaminated food (marine fauna), exposure at the workplace. Although the most common source of them all is the inhalation of contaminated fumes or dust particles (Alina et al. 2012; Dondero et al. 2011). The chief facilitator for this chain of contamination to attain isochronic function and exponentially increasing its reach to the human body is biomagnification which enables the contaminants from industries to biosphere and then to the human system. There are numerous routes via which heavy metals can enter our body such as gastrointestinal tracts (utilized by cadmium, manganese, arsenic, and lead) by simply consuming food orally. However, most of these toxic metals enter via inhalation or adsorption at the epidermal layer (skin) which is followed by a rapid redistribution pattern in systemic circulation to tissues throughout the body (Florea and Büsselberg 2006).

Predominantly besides acute toxicity caused by these heavy metals, what is more pressing is the bioaccumulation of such substances in the body which poses the potential to cause severe adverse effects. The ability of heavy metals to bind to distinct proteins such as nucleic acid facilitates the transport of metal bound proteins, further these metal-bound proteins compartmentalize inside the cellular framework of the system and degrades macromolecules which blocks metabolic activities. Various studies have reported central nervous system to be the foremost target of these toxins leading to various classes of mental disorders; however, heavy metals have observed to impair blood constituents which enable them to damage several ancillary targets such as lungs, liver, kidney, and other vital organs along with several other pathologies which is even more fatal than its primary target. Thereby chronic exposure and subsequent accumulation of these heavy metals may lead to stagnation of muscular, physical, and cognitive activities which mimics certain initial attributes of Alzheimer's and Parkinson's disease. Furthermore, extended continuity with these metals may cause permanent damage to DNA leading to mutation and thus eventually causing cancer (Farina et al. 2013).

4 Cellular and Molecular Mechanism of Metal Accumulation in Brain

Several prior literatures provide us with evidence which help us to understand the role of heavy metal accumulation in the motor region of the central nervous system causing neurotoxicity along with associated pathologies (Fig. 1). Amidst various toxins, there are only few such as lead, manganese, and organic mercury which could easily cross the blood–brain barrier (BBB) and reach CNS. Mercury in biological system requires getting associated with methyl group (CH_3) to attain its stable and transportable conformation, Methylmercury ($M_eH_g^+$) (Yin et al. 2011; Yorifuji et al. 2011). It is generally found in marine organisms in miniscule quantities as a contaminant; however, biomagnification elevates it to dangerous levels till it reaches the human system. M_eH_g exhibits extensive absorption (90–95%) in the GI tract, after which M_eH_g forms a complex with cysteine residues which enables it to cross BBB. Complex mimics the structural moiety of methionine which is reported to be a specific amino acid transporter into the CNS. Following its transportation into CNS, the complex undergoes demethylation resulting in inorganic mercury accumulation in the brain. The toxic accumulation in CNS leads to discrepancies in homeostasis of excitatory neurotransmitters along with overexpression of NMDA (N-methyl-D-aspartate) type glutamate receptors which resulted in soaring Ca^{2+} influx into neurons. This large pool of intracellular Ca^{2+} serves as a potent neurotoxin which hinders with the biosynthesis of neurotransmitters, thereby causing a significant imbalance in the development of the brain (Lafon-Cazal et al. 1993).

Besides mercury, a considerable sum of individuals has been observed with manganese (Mn) accumulation in the CNS in the past couple of decades, which prompted extensive research exploring various facets associated with such toxicity. Research

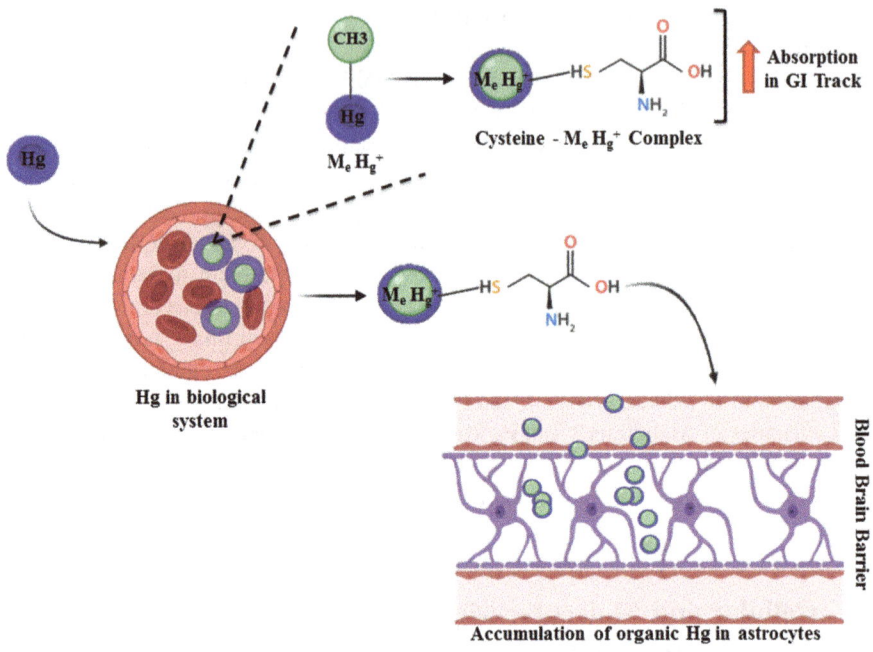

Fig. 1 Accumulation of metal toxin (mercury) in astrocytes after subsequent penetration through BBB

studies suggested that extended exposure to Mn leads to its accumulation in astrocytes, neuronal mitochondria, and oligodendrocyte cells. Amassing such heavy metal (Mn) in neurons emanates distinct factors inhibiting F1/F0 ATP synthase or complex 1 (NADH dehydrogenase) of mitochondrial respiratory chain (Abdel-Megeed 2020; Chen et al. 2001). Binding of these toxins occurs distinctly at either of the two sites; complex II (succinate dehydrogenase) or the glutamate/ aspartate exchanger which depends upon the source of mitochondrial energy; thereby hampering ATP synthesizing biochemical processes (Milatovic et al. 2009). Reduction in intracellular ATP production further paves the way for the generation of free radical species (ROS and RNS) causing cellular toxicity. Elevated free radicals' levels greatly contribute to the deterioration of dopamine transporter (DAT1) which further results in dopaminergic neurotoxicity (Benedetto et al. 2010).

5 Inhibition in Neural Pathways and Signal Transmission

Heavy metals have always been strongly associated with neural toxicity causing discrepancies in regular neuronal functioning and signal transduction. NMDA receptor (NMDAR) is a crucial component which is profoundly involved in enhancing certain facets associated with hippocampus such learning and memory. On exposure to lead (Pb) several animal studies have unveiled cognitive defects leading to learning difficulty which is analogous in case of NMDAR impairment or absence. Researchers

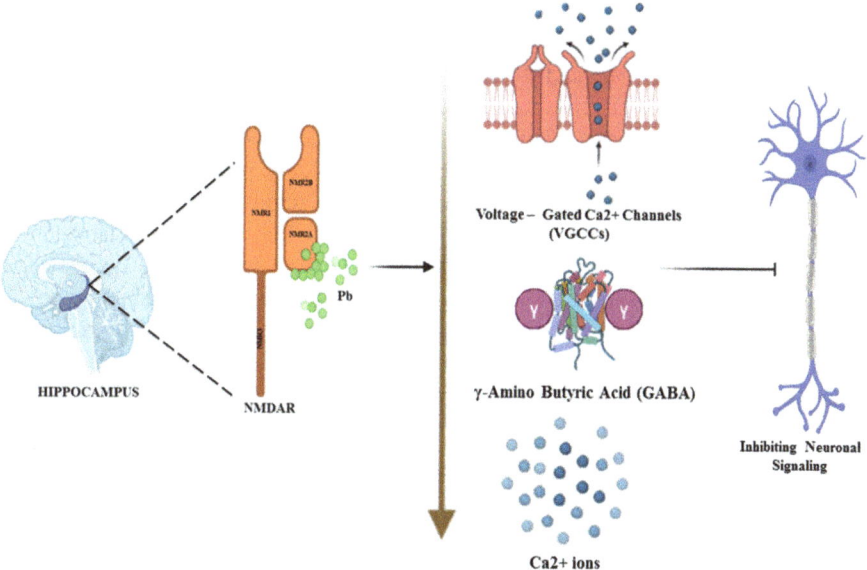

Fig. 2 Schematic representation of lead-mediated inhibition of neuronal signaling at hippocampus in brain

have experimentally verified lead to be a potent noncompetitive antagonist of NMDAR receptor in the hippocampal region which also tends to exhibit higher affinity toward NR2A regulatory subunit of the NMDAR (Fig. 2). Longer receptor (NMDA)–ligand (lead) interaction results in lowering neurotransmission in rat along with systematic reduction in the release of γ-aminobutyric acid (GABA) and Ca^{2+}-dependent glutamate in the hippocampal region, which further deteriorate the condition to dysfunctional presynaptic neuron signaling (Lasley and Gilbert 1996; Xiao et al. 2006). Besides this chronic lead exposure may further damage the postsynaptic signaling as inhibitory postsynaptic currents (IPSCs) and excitatory postsynaptic currents (EPSCs) are modulated by presynaptic glutamate and GABA which are reduced under the influence of lead toxicity (Braga et al. 1999). Reduced levels of Ca^{2+} can also hamper the VGCCs (voltage-gated calcium (Ca^{2+}) channels) functioning which further contributes to dismantling of the neural network in the hippocampal region of the CNS paving the way for copious neural pathologies (Peng et al. 2002).

6   Antidotal Strategies

Various bioactive compounds have been demonstrated to act against mercury toxicity. Some of the agents that have shown beneficial effects are natural products (Farina et al. 2005), vitamins E and K (Shichiri et al. 2007; Sakaue et al. 2011), calcium channel blocking agents, compounds containing thiols (Falluel-Morel et al. 2012), or glutamatergic antagonists (Ramanathan and Atchison 2011). While administering these bioactive compounds, supportive care is needed to sustain the vital functions of the body to stamp out toxic metal. Thiol compounds reduce half-life of blood while chelating agents increase elimination of mercury by using DMPS, D-penicillamine, and N-acetyl-DL-penicillamine. However, these compounds are not much effective but can be used to minimize the concentration (Ruha et al. 2009). DMSA (meso-2,3-dimercaptosuccinic acid) and BAL (2,3- dimercaprol) have shown efficacy in treating arsenic, cadmium, and mercury toxicity, but due to DMSA's lipophobic nature, it may interfere with removal of heavy metal from intracellular sites (Kalia and Flora 2005). In recent studies, it has been reported that combination therapy of thiol chelator and $CaNa_2EDTA$ can be active against chromic Pb toxicity (Flora et al. 2007). To the best of our knowledge, there is no potential evidence of antidotal effects of these compounds against mercury toxicity, but experimental studies have shown that they can protect cells against mercury toxicity and also abate the mercury deposition in tissues (de Freitas et al. 2009).

7 Conclusion

Metals can serve as essential components for the normal well-being and physiology but at the same time carry some serious toxic exposures. The foremost treatment against metal toxicity is the antidotal therapy which permits the exclusion of surplus toxic metals from the system that makes it instantly toxic and diminishes the late effects on normal brain activity leading to complication in the supplementary therapies. Although, various ranges of metal chelators are available for the chelation of toxic metals, however, antidotal therapy can be a significant method against metal poisoning. Such approaches are still in clinical trials and even committing better clinical therapeutic benefits. In any case, despite all the prevailing disadvantages to the metal toxicity, it is more vital to understand the necessity for unraveling more specific and advanced cheating molecules not only to resolve the toxicity but also to improve the clinical recovery. Therefore, the newer antidotal therapeutic strategies like chelating agents, antioxidants, or nutraceuticals should be focused as it could provide better outcomes when explored for inhibiting metal poisoning.

References

2020 Alzheimer's disease facts and figures. Alzheimer's Association Report. 2020;16(3).
Abdel-Megeed RM. Probiotics: a promising generation of heavy metal detoxification. Biol Trace Elem Res. 2020.
Agency for Toxic Substances and Disease Registry (ATSDR). Toxicological profile for Copper Atlanta. US Department of Health and Humans Services, Public Health Service, Centers for Diseases Control; 2004.
Alam ZI, Jenner A, Daniel SE, Lees AJ, Cairns N, Marsden CD, et al. Oxidative DNA damage in the Parkinsonian brain: an apparent selective increase in 8-hydroxyguanine levels in substantia nigra. J Neurochem. 1997;69(3):1196–203.

Alina MAA, Mohd Yunus AS, Mohd Zakiuddin S, Mohd Izuan Effendi H, Muhammad Rizal R. Heavy metals (mercury, arsenic, cadmium, plumbum) in selected marine fish and shellfish along the straits of Malacca. Int Food Res J. 2012;19(1):135–40.

Amirah MN, Faizal WIW, Nurliyana MH, Laili S. Human health risk assessment of metal contamination through consumption of fish. J Environ Pollut Human Health. 2013;1(1):1–5.

Atwood CS, Huang X, Moir RD, Tanzi RE, Bush AI. Role of free radicals and metal ions in the pathogenesis of Alzheimer's disease. Met Ions Biol Syst. 1999;36:309–64.

Baptista MJ, O'Farrell C, Daya S, Ahmad R, Miller DW, Hardy J, et al. Co-ordinate transcriptional regulation of dopamine synthesis genes by alpha-synuclein in human neuroblastoma cell lines. J Neurochem. 2003;85(4):957–68.

Barnham KJ, Ciccotosto GD, Tickler AK, Ali FE, Smith DG, Williamson NA, et al. Neurotoxic, redox-competent Alzheimer's β-amyloid is released from lipid membrane by methionine oxidation. J Biol Chem. 2003;278(44):42959–65.

Barnham KJ, Masters CL, Bush AI. Neurodegenerative diseases and oxidative stress. Nat Rev Drug Discov. 2004;3(3):205–14.

Benedetto AAC, Avila DS, Milatovic D, Aschner M. Extracellular dopamine potentiates mn-induced oxidative stress, lifespan reduction, and dopaminergic neurodegeneration in a BLI-3–dependent manner in Caenorhabditis elegans. PLoS Genet. 2010;6(8)

Braga MFM, Pereira EFR, Albuquerque EX. Nanomolar concentrations of lead inhibit glutamatergic and GABAergic transmission in hippocampal neurons. Brain Res. 1999;826(1):22–34.

Chen J-Y, Tsao GC, Zhao Q, Zheng W. Differential cytotoxicity of Mn(II) and Mn(III): special reference to mitochondrial [Fe-S] containing enzymes. Toxicol Appl Pharmacol. 2001;175(2):160–8.

de Freitas AS, Funck VR, Rotta Mdos S, Bohrer D, Mörschbächer V, Puntel RL, et al. Diphenyl diselenide, a simple organoselenium compound, decreases methylmercury-induced cerebral, hepatic and renal oxidative stress and mercury deposition in adult mice. Brain Res Bull. 2009;79(1):77–84.

Dondero F, Banni M, Negri A, Boatti L, Dagnino A, Viarengo A. Interactions of a pesticide/heavy metal mixture in marine bivalves: a transcriptomic assessment. BMC Genomics. 2011;12:195.

Double KL, Gerlach M, Schünemann V, Trautwein AX, Zecca L, Gallorini M, et al. Iron-binding characteristics of neuromelanin of the human substantia nigra. Biochem Pharmacol. 2003;66(3):489–94.

Falluel-Morel A, Lin L, Sokolowski K, McCandlish E, Buckley B, DiCicco-Bloom E. N-acetyl cysteine treatment reduces mercury-induced neurotoxicity in the developing rat hippocampus. J Neurosci Res. 2012;90(4):743–50.

Farina M, Franco JL, Ribas CM, Meotti FC, Missau FC, Pizzolatti MG, et al. Protective effects of Polygala paniculata extract against methylmercury-induced neurotoxicity in mice. J Pharm Pharmacol. 2005;57(11):1503–8.

Farina M, Aschner M, Rocha JB. Oxidative stress in MeHg-induced neurotoxicity. Toxicol Appl Pharmacol. 2011;256(3):405–17.

Farina M, Avila DS, da Rocha JBT, Aschner M. Metals, oxidative stress and neurodegeneration: a focus on iron, manganese and mercury. Neurochem Int. 2013;62(5):575–94.

Ferrara R, Mazzolai B, Lanzillotta E, Nucaro E, Pirrone N. Temporal trends in gaseous mercury evasion from the Mediterranean seawaters. Sci Total Environ. 2000;259(1–3):183–90.

Flora SJ, Tandon SK. Effect of combined exposure to cadmium and ethanol on regional brain biogenic amine levels in the rat. Biochem Int. 1987;15(4):863–71.

Flora SJS, Saxena G, Mehta A. Reversal of Lead-induced neuronal apoptosis by chelation treatment in rats: role of reactive oxygen species and intracellular Ca^{2+}. J Pharmacol Exp Ther. 2007;322(1):108–16.

Flora G, Gupta D, Tiwari A. Toxicity of lead: a review with recent updates. Interdiscip Toxicol. 2012;5(2):47–58.

Florea A-M, Büsselberg D. Occurrence, use and potential toxic effects of metals and metal compounds. Biometals. 2006;19(4):419–27.

Galasso SL, Dyck RH. The role of zinc in cerebral ischemia. Mol Med (Cambridge, Mass). 2007;13(7–8):380–7.

Ghani AGA. Effect of chromium toxicity on growth, chlorophyll and some mineral nutrients of Brassica juncea L. Egypt Acad J Biol Sci H Bot. 2011;2(1)

Godwill E, Ferdinand P, Nwalo N, Unachukwu M. Mechanism and health effects of heavy metal toxicity in humans; 2019. p. 1–23.

Gorantla NV, Das R, Balaraman E, Chinnathambi S. Transition metal nickel prevents Tau aggregation in Alzheimer's disease. Int J Biol Macromol. 2020;156:1359–65.

Guilarte TR, McGlothan JL. Selective decrease in NR1 subunit splice variant mRNA in the hippocampus of Pb^{2+}-exposed rats: implications for synaptic targeting and cell surface expression of NMDAR complexes. Brain Res Mol Brain Res. 2003;113(1-2):37–43.

Henson MC, Chedrese PJ. Endocrine disruption by cadmium, a common environmental toxicant with paradoxical effects on reproduction. Exp Biol Med (Maywood, NJ). 2004;229(5):383–92.

Huang X, Atwood CS, Moir RD, Hartshorn MA, Vonsattel JP, Tanzi RE, et al. Zinc-induced Alzheimer's Abeta1-40 aggregation is mediated by conformational factors. J Biol Chem. 1997;272(42):26464–70.

Huang X, Atwood CS, Hartshorn MA, Multhaup G, Goldstein LE, Scarpa RC, et al. The A beta peptide of Alzheimer's disease directly produces hydrogen peroxide through metal ion reduction. Biochemistry. 1999;38(24):7609–16.

Hughes JP, Polissar L, van Belle G. Evaluation and synthesis of health effects studies of communities surrounding arsenic producing industries. Int J Epidemiol. 1988;17(2):407–13.

Huy TB, Tuyet-Hanh TT, Johnston R, Nguyen-Viet H. Assessing health risk due to exposure to arsenic in drinking water in Hanam Province, Vietnam. Int J Environ Res Public Health. 2014;11(8):7575–91.

Jaishankar M, Tseten T, Anbalagan N, Mathew BB, Beeregowda KN. Toxicity, mechanism and health effects of some heavy metals. Interdiscip Toxicol. 2014;7(2):60–72.

Kalia K, Flora SJS. Strategies for safe and effective therapeutic measures for chronic arsenic and lead poisoning. J Occup Health. 2005;47(1):1–21.

Keshav Krishna A, Rama Mohan K. Distribution, correlation, ecological and health risk assessment of heavy metal contamination in surface soils around an industrial area, Hyderabad, India. Environ Earth Sci. 2016;75(5):411.

Kumar A, Singh N, Pandey R, Gupta VK, Sharma B. Biochemical and molecular targets of heavy metals and their actions. In: Rai M, Ingle AP, Medici S, editors. Biomedical applications of metals. Cham: Springer International Publishing; 2018. p. 297–319.

Lafon-Cazal M, Pietri S, Culcasi M, Bockaert J. NMDA-dependent superoxide production and neurotoxicity. Nature. 1993;364(6437):535–7.

Lashuel HA, Hartley D, Petre BM, Walz T, Lansbury PT Jr. Neurodegenerative disease: amyloid pores from pathogenic mutations. Nature. 2002;418(6895):291.

Lasley SM, Gilbert ME. Presynaptic glutamatergic function in dentate gyrus in vivo is diminished by chronic exposure to inorganic lead. Brain Res. 1996;736(1):125–34.

Lee HJ, Park MK, Seo YR. Pathogenic mechanisms of heavy metal induced-Alzheimer's disease. Toxicol Environ Health Sci. 2018;10(1):1–10.

Lewén A, Matz P, Chan PH. Free radical pathways in CNS injury. J Neurotrauma. 2000;17(10):871–90.

Masters CL, Bateman R, Blennow K, Rowe CC, Sperling RA, Cummings JL. Alzheimer's disease. Nat Rev Dis Primers. 2015;1(1):15056.

Matschullat J. Arsenic in the geosphere – a review. Sci Total Environ. 2000;249(1–3):297–312.

Maya S, Prakash T, Madhu KD, Goli D. Multifaceted effects of aluminium in neurodegenerative diseases: a review. Biomed Pharmacother. 2016;83:746–54.

McDougall SA, Reichel CM, Farley CM, Flesher MM, Der-Ghazarian T, Cortez AM, et al. Postnatal manganese exposure alters dopamine transporter function in adult rats: potential impact on nonassociative and associative processes. Neuroscience. 2008;154(2):848–60.

Mejía JJ, Díaz-Barriga F, Calderón J, Ríos C, Jiménez-Capdeville ME. Effects of lead-arsenic combined exposure on central monoaminergic systems. Neurotoxicol Teratol. 1997;19(6):489–97.

Milatovic D, Zaja-Milatovic S, Gupta RC, Yu Y, Aschner M. Oxidative damage and neurodegeneration in manganese-induced neurotoxicity. Toxicol Appl Pharmacol. 2009;240(2):219–25.

Milatovic D, Gupta RC, Yin Z, Zaja-Milatovic S, Aschner M. Chapter 32 – Manganese. In: Gupta RC, editor. Reproductive and developmental toxicology. 2nd ed. Academic; 2017. p. 567–81.

Mocchegiani E, Muzzioli M, Giacconi R. Zinc, metallothioneins, immune responses, survival and ageing. Biogerontology. 2000;1(2):133–43.

Mutter J, Curth A, Naumann J, Deth R, Walach H. Does inorganic mercury play a role in Alzheimer's disease? A systematic review and an integrated molecular mechanism. J Alzheimer's Dis. 2010;22(2):357–74.

Nihei MK, Desmond NL, McGlothan JL, Kuhlmann AC, Guilarte TR. N-methyl-D-aspartate receptor subunit changes are associated with lead-induced deficits of long-term potentiation and spatial learning. Neuroscience. 2000;99(2):233–42.

Pal A. Copper toxicity induced hepatocerebral and neurodegenerative diseases: an urgent need for prognostic biomarkers. NeuroToxicology. 2014;40:97–101.

Peng S, Hajela RK, Atchison WD. Characteristics of block by Pb^{2+} of function of human neuronal L-, N-, and R-type Ca^{2+} channels transiently expressed in human embryonic kidney 293 cells. Mol Pharmacol. 2002;62(6):1418–30.

Ramanathan G, Atchison WD. Ca^{2+} entry pathways in mouse spinal motor neurons in culture following in vitro exposure to methylmercury. Neurotoxicology. 2011;32(6):742–50.

Ruha AM, Curry SC, Gerkin RD, Caldwell KL, Osterloh JD, Wax PM. Urine mercury excretion following meso-dimercaptosuccinic acid challenge in fish eaters. Arch Pathol Lab Med. 2009;133(1):87–92.

Sabine Martin WG. Human health effects of heavy metals. Environ Sci Technol Briefs Citizens. 2009;15

Sakaue M, Mori N, Okazaki M, Kadowaki E, Kaneko T, Hemmi N, et al. Vitamin K has the potential to protect neurons from methylmercury-induced cell death in vitro. J Neurosci Res. 2011;89(7):1052–8.

Salazar-Flores JT-JJ, Rojas-Bravo D, Reyna-Villela ZM, Torres-Sánchez ED. Effects of mercury, lead, arsenic and zinc to human renal oxidative stress and functions: a review. J Heavy Metal Toxic Dis. 2019;4

Sauvé S. Time to revisit arsenic regulations: comparing drinking water and rice. BMC Public Health. 2014;14(1):465.

Sharma B, Singh S, Siddiqi NJ. Biomedical implications of heavy metals induced imbalances in redox systems. Biomed Res Int. 2014;2014:640754.

Shichiri M, Takanezawa Y, Uchida K, Tamai H, Arai H. Protection of cerebellar granule cells by tocopherols and tocotrienols against methylmercury toxicity. Brain Res. 2007;1182:106–15.

Singh N, Gupta VK, Kumar A, Sharma B. Synergistic effects of heavy metals and pesticides in living systems. Front Chem. 2017;5(70)

Smith MA, Harris PL, Sayre LM, Perry G. Iron accumulation in Alzheimer disease is a source of redox-generated free radicals. Proc Natl Acad Sci U S A. 1997;94(18):9866–8.

Uversky VN, Li J, Fink AL. Metal-triggered structural transformations, aggregation, and fibrillation of human alpha-synuclein. A possible molecular NK between Parkinson's disease and heavy metal exposure. J Biol Chem. 2001;276(47):44284–96.

Valko M, Morris H, Cronin MT. Metals, toxicity and oxidative stress. Curr Med Chem. 2005;12(10):1161–208.

Vivek Kumar Gupta AK, Siddiqi NJ, Sharma B. Rat brain acetyl cholinesterase as a biomarker of cadmium induced neurotoxicity. Open Access J Toxicol. 2016;1(1)

Xiao C, Gu Y, Zhou C-Y, Wang L, Zhang M-M, Ruan D-Y. Pb^{2+} impairs GABAergic synaptic transmission in rat hippocampal slices: a possible involvement of presynaptic calcium channels. Brain Res. 2006;1088(1):93–100.

Yin Z, Lee E, Ni M, Jiang H, Milatovic D, Rongzhu L, et al. Methylmercury-induced alterations in astrocyte functions are attenuated by ebselen. NeuroToxicology. 2011;32(3):291–9.

Yorifuji T, Debes F, Weihe P, Grandjean P. Prenatal exposure to lead and cognitive deficit in 7- and 14-year-old children in the presence of concomitant exposure to similar molar concentration of methylmercury. Neurotoxicol Teratol. 2011;33(2):205–11.

Zahir F, Rizwi SJ, Haq SK, Khan RH. Low dose mercury toxicity and human health. Environ Toxicol Pharmacol. 2005;20(2):351–60.

Reactive Oxygen Species and Oxidative Stress on the Formation of Diabetic Ulcer

Vini Nagaraj and Suneel Kumar (iD)

Contents

Abstract Reactive oxygen species (ROS) appears to be a key player in mediating molecular insult in a plethora of chronic diseases and majorly contributes to disease pathogenesis. Oxidative stress, which is the consequence of an imbalance between the production of free radicals and the intrinsic counterfeit action by antioxidants is widely considered to be a hallmark of secondary damage in many disease conditions. ROS are considered both beneficial by regulating signaling molecules and harmful by causing debilitating complications under different conditions. One such devastating secondary impediment of the major health issues caused by ROS is pressure ulcers. Pressure ulcers are chronic wounds that pose challenges for patients concerning multiple surgeries and further devastating complications. It is believed

V. Nagaraj
Keck Center for Collaborative Neuroscience and Department of Cell Biology and Neuroscience, Rutgers University, Piscataway, NJ, USA
e-mail: vn149@dls.rutgers.edu

S. Kumar (✉)
Department of Biomedical Engineering, Rutgers, The State University of New Jersey, Piscataway, NJ, USA
e-mail: sk1350@soe.rutgers.edu

© The Author(s), under exclusive license to Springer Nature Switzerland AG 2021
K. K. Kesari, N. K. Jha (eds.), *Free Radical Biology and Environmental Toxicity*,
Molecular and Integrative Toxicology, https://doi.org/10.1007/978-3-030-83446-3_13

279

that the presence of excess amounts of activated macrophages and neutrophils in chronic wounds often result in excess levels of ROS and further delays wound healing. This chapter aims at reviewing and deciphering the role of ROS and oxidative stress in the pathogenesis of pressure ulcer/foot ulcer triggered by diabetes-related immobility and pathogenesis.

Keywords Pressure ulcer · Foot ulcer · Reactive oxygen species · Oxidative stress · Diabetes · Immobility · Antioxidant · Wound healing · Chronic wounds

1 Introduction

The world diabetic patient population has reached 463 million in the year 2020, making this disease condition the most prevalent healthcare concern. The number of cases is predicted to increase to 700 million by 2045 (Ogurtsova et al. 2017). In the United States alone, more than 100 million patients are living with diabetes or prediabetes according to the Centers for Disease Control and Prevention (CDC) (Divers 2020). The World Health Organization prioritizes diabetes as one of the major noncommunicable conditions globally (Global report on diabetes 2021). According to CDC, the highest and the total US healthcare spending on diabetes in 2017 was $327 billion in medical costs and diagnosis, which resulted in a heavy economic burden (Economic Costs of Diabetes in the U.S. in 2017 2021). The secondary complications after several years from the initial diagnosis are devastating to the patient and the immediate family and burdensome to the country's healthcare system. Secondary complications from diabetes include macrovascular complications such as cardiovascular or microvascular complications such as amputations, nephropathy, and neuropathy (Global report on diabetes 2021). The most common wounds caused by diabetes are primarily diabetic foot ulcers and pressure ulcers (PUs). Chronic diabetic wounds are one of the most important and long-term sequelae of diabetes. Among the 463 million diabetic patients, 25% of them will develop foot ulcers in their lifetime (The management of diabetic foot 2021). The morbidity and mortality rate of this devastating health condition falls between 43% to 55%, which is arguably higher than other life-threatening disease conditions such as cancers, which makes it a major healthcare problem (Quaresma et al. 2015). The prevalence of these types of ulcers in the United States is ~2% of its total population (Are You At Risk for Chronic Wounds 2021). The prevalence and impact of chronic wounds are also devastating worldwide. Diabetic foot ulcers and PUs are the most common cause of amputations globally; they are slow-healing wounds that cost billions of dollars toward treatment and place a great burden on any healthcare system (Uccioli et al. 2015). Around 67% of total amputations around the world occur in diabetic patients (Sen et al. 2009). Annually, the United States' healthcare system spends up to $17.8 billion for treating pressure injuries, making it the most expensive measurable medical errors (Van Den Bos et al. 2011). The demand for wound care products has been on the rise in the United States. The advanced wound-dressing markets targeting products for treating PUs are expected to exceed $22 billion by 2024, owing to the rising incidences of chronic wounds

(Global Advanced Wound Care Market 2021). Since the spending expenditure for treating pressure wounds is on the increase, there is an urgent need to understand the pathophysiology of this devastating health condition and develop new preventive treatments to benefit both the patients and relieve the burden on a country's healthcare system. This review has aimed at understanding the precise mechanism elicited by diabetic PUs and the summary of the available treatment options currently under clinical trials.

2   Diabetes and Diabetic Ulcers

Type 2 diabetes (T2D) is a serious multiracial chronic condition, which poses a major impact on the lives and well-being of entire humankind worldwide. Diabetes and its complications have remarkable after-effects on the patient's quality of life. One of the destructive complications of diabetes is the development of chronic wounds and ulcers caused by impaired wound healing. Chronic wounds are wounds that do not follow the standard healing process, but rather persist for a longer duration of time, leading to a decreased standard of living and mobility of patients, which poses a major threat to any healthcare system worldwide. Chronic wounds could be assigned to three major categories such as PUs, diabetic ulcers, venous ulcers, and arterial ischemia (Mustoe et al. 2006). Pressure ulcers are often common in patients who have compromised mobility and sensation due to causal conditions such as paralysis or those in ICUs for an extended period. Prolonged pressure on specific soft tissues in such patients often leads to immense ischemic necrosis and extensive tissue damage (Mustoe et al. 2006; Gebhardt 2002). In a normal healthy person, a physical wound is healed by normal phases of cutaneous healing such as hemostasis, inflammatory phase, proliferative phase, and maturation phase. In patients suffering from PUs, the process of normal orderly phases of healing is inhibited. Hence, the wound in these patients sustain in the inflammatory phase for extended amounts of time and remains unhealed (Singer and Clark 2008). Since wound healing is arrested at the inflammation phase in patients with a PU, it is important to understand the perturbed mechanisms and circumvent this phase to discover clinically relevant treatment options. While several factors cause chronic inflammation in these patients, ROS has gained major focus in the last few decades due to the hypothesis that excessive ROS can contribute to delayed wound healing (Wlaschek and Scharffetter-Kochanek 2005).

2.1   Hyperglycemia and Pressure Ulcer

PUs are also known as decubitus ulcers which are ulcers or sores primarily caused by pressure against the skin and the underlying tissue for extended periods. Excessive pressure causes reduced or no capillary blood flow to the area, which leads to tissue ischemia and ultimately results in the formation of PUs. PU usually develops around bony prominence areas such as elbows, greater trochanter, and tailbones (Pressure ulcers: Current understanding and newer modalities of treatment 2021). Hyperglycemic type 2 diabetic patients have about a 30% chance of developing chronic wounds, such as foot ulcers, venous leg ulcers, and PUs. Moreover, diabetic patients with PUs experience severe forms of ulcers and delayed wound healing. Diabetes-caused wounds are considered chronic if the healing does not occur beyond 8 weeks. The normal pattern of healing is deterred in these patients owing to their impaired immune system triggered by poor glycemic control (Koivukangas et al. 1999).

Diabetic patients with a body mass index (BMI) of over 30 are generally considered obese. Obesity decreases the mobility and overall physical activity of these patients making them prone to complete inactivity and development of ulcers, importantly PUs in various body parts, so much more in ICU patients (Walden et al. 2013; Allman et al. 1995). In diabetic rats, it has been shown that the tolerance of the underlying collagen network in the skin may be compromised or lowered when exposed to external constant pressure and this may delay the healing process and ultimately result in chronic wounds (Huang et al. 2010). Pressure ulcers can be triggered by complex processes with influencing factors such as the (i) exposure and intensity of pressure applied to the area (How Much Time Does it Take to Get a Pressure Ulcer 2021), (ii) vascular permeability and tensile strength of the underlying muscle tissue, (iii) subcutaneous fat and dermis (Berlowitz and Brienza 2007; Vande Berg and Rudolph 1995), and (iv) excessive glycation of hemoglobin (Patel et al. 2019). These factors could dictate the amount of oxygen and nutrients delivered to the tissue, resulting in secondary complications such as necrosis, hypoxia, thrombosis, and free radical generation (Gefen 2008). Hyperglycemia and prolonged immobility are major risk factors for developing PUs in diabetic patients (Understanding Pathophysiology 2021). One may arrive at questioning how hyperglycemia elevates free radical levels and how this, in turn, affects the underlying tissue, which leads to the development of PUs.

3   Roles of Hyperglycemia, Advanced Glycated End Products (AGE), Oxidative Stress, and ROS in the Formation of Ulcer

Prominent target cells or areas affected by hyperglycemia in T2D patients are those that fail to decline their glucose transport in response to the extracellular high glucose levels, leading to high intracellular glucose levels. In diabetic PU, hyperglycemia induces the activation of important independent pathways such as (i) polyol, (ii) protein kinase C, (iii) increased production of AGE, (iv) hexosaminase, and (v) mitochondrial superoxide production. The end products of each of these pathways are NADPH, NAD(P)H oxidase, O^{2-}, hydrogen peroxide (H_2O_2), and ROS (Brownlee 2001; Dhall et al. 2014). These end products culminate and lead to increased oxidative or redox stress, which ultimately leads to significant increases in the ROS levels and reduced antioxidant defenses. This increase crucially impacts the etiology of diabetic-related complications, including PUs (Williamson et al. 1993; Duscher et al. 2015). Increased production of ROS could also impair the normal wound healing process in these patients (Chen and Rogers 2007). The natural oxidative process is a beneficial and important process, which provides the energy necessary for many crucial cellular functions. A serious imbalance in the physiological equilibrium between pro-oxidative radicals and their antioxidant defense levels leads to oxidative stress within a cell, tissue, or organ (Betteridge 2000).

3.1 Oxidative Stress

The radical derivatives from cells that undergo cellular metabolism are intermediates of molecular oxygen such as superoxide anion, hydrogen peroxide; single oxygen molecules that are highly reactive with added electrons are generally termed as ROS (Baynes 1991). At normal physiological concentrations, ROS acts as secondary messengers for various biological processes. However, overproduction or excess levels of ROS can cause irreversible damage to cells by oxidizing proteins, and hence the ROS levels are tightly controlled by counteractive antioxidant scavenger enzymes. An imbalance between ROS production and buffering capacity of antioxidant enzymes primarily leads to oxidative stress. Oxidative stress is considered to play a major role in diabetes-related complications, especially in diabetes-induced ulcers (Oxidative stress and diabetic complications 2021). Several normal pancreatic cell signaling pathways are perturbed by hyperglycemic conditions and pave way for diabetes-related complications by ROS formation. ROS is a byproduct of different ROS-generating systems such as mitochondria and peroxisomes. Oxidation reactions generally yield H_2O_2, which acts as precursors for a free radical generation. NADPH oxidase, AGE, PKC, and DAG, which are byproducts of glycolysis, are also intermediates for ROS production in diabetic wounds (Fig. 1) (Thannickal and Fanburg 2000). Thus, overproduction of ROS under hyperglycemic conditions is observed to be elevated in diabetic wounds and is one of the major inducers of oxidative stress (Oxidative stress and diabetic complications 2021). This section aims at understanding the different pathways that lead to the overproduction of oxidative stress.

3.2 Mitochondrial ROS Production

Elevated glucose levels enter glycolysis and Krebs cycle and oxidation of glucose results in increased ATP/ADP ratio and this leads to depolarization of mitochondrial membrane potential. The membrane depolarization leads to the production of free radical superoxide that is converted to H_2O_2, which is another form of ROS, produced via excess production of NADH and FADH2 (Wu et al. 2016).

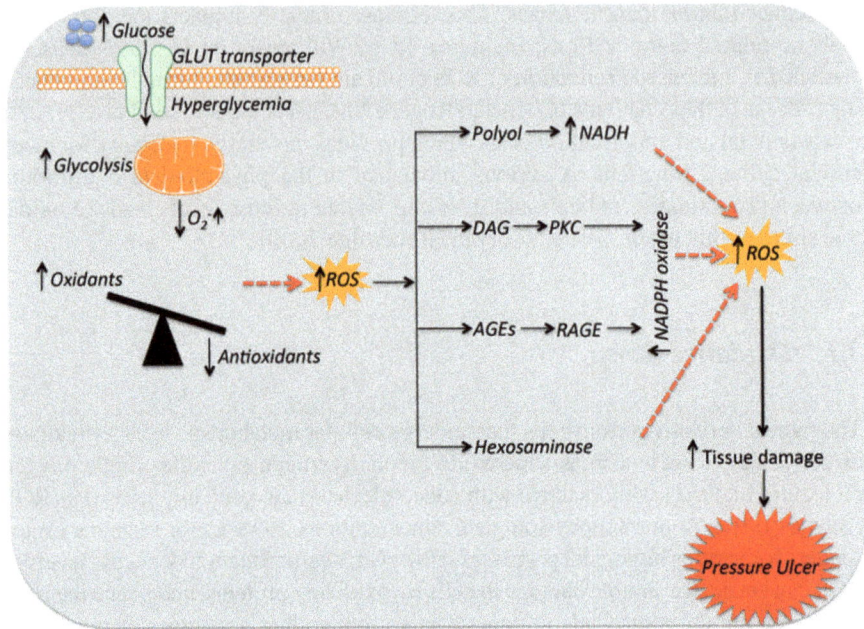

Fig. 1 Schematic of hyperglycemia-induced ROS production via different metabolic pathways

Excessive release of mitochondrial ROS leads to accumulation of upstream glyc-eraldehyde 3-phosphate dehydrogenase (GAPDH) because of the inhibition of GAPDH by poly(ADP-ribose) modification. Increased level of GAPDH acts as a substrate for AGE formation, protein kinase C (PKC), polyol, and hexosaminase pathways (Nishikawa et al. 2000), and these pathways are activated, which results in the depletion of antioxidant resources and further favors even more production of ROS and AGEs. ROS-induced ROS production is an important source for excess ROS generation (Zinkevich and Gutterman 2011). The overproduction of ROS tips off the balance between oxidants and antioxidant ratio leading to excessive oxida-tive stress (Wlaschek and Scharffetter-Kochanek 2005).

3.3   AGE Product Formation

AGEs are formed when highly reactive aldehyde or ketone groups of reducing sug-ars and free amino groups on proteins, lipids, and amino acids are nonenzymatically glycated. It has been shown in diabetic patients that AGEs form and accumulate at a rapid and high rate in circulation owing to their intracellular high glucose levels (Monnier et al. 2005). The binding of AGE to the receptor of AGE produces super-oxide anion (O^{2-}) and hydrogen peroxide (H_2O_2) (RAGE-induced cytosolic ROS promote mitochondrial superoxide generation in diabetes 2021) and aids in the pro-duction of excess levels of AGEs which in turn perpetuating further ROS generation affecting mitochondrial respiratory chain proteins, thus forming an infinite loop that amplifies the levels of AGE molecules (Thomson et al. 2016). The resulting high levels of ROS from this reaction also add to the oxidative stress generated from the mitochondrial oxidative stress (Nowotny et al. 2015).

3.4   Activation of PKC Pathway

Protein kinase C belongs to a serine-threonine protein kinase, which is activated by diacylglycerol (DAG). Hyperglycemia increases the de novo synthesis of DAG and increases several forms of PKC, which then activates other signaling processes that regulate protein functions. In diabetic wounds or ulcers, PKC targets NADPH oxidase, a strong precursor for superoxide ROS production, thus exacerbating the already elevated ROS levels within the wounds (Koya and King 1998).

3.5   Delayed Wound Healing

A decade ago, it was hypothesized that excess production of ROS majorly delays the wound healing capacity of diabetes-induced ulcers. Neutrophils are a major hallmark of the inflammation phase during the wound healing process. In a normal wound healing process, neutrophils release ROS, which then aids in rapid healing. In patients with diabetic ulcers, it was observed that excess neutrophils were present in the chronically infected regions for prolonged periods releasing excess ROS. Besides, diabetic patients have excess production of ROS due to hyperglycemia. The local and overall overproduction of ROS saturates the antioxidant levels, thus tipping off the endogenous oxidant-antioxidant levels in the wounds resulting in oxidative stress. Elevated levels of ROS in the wound region lead to increased inflammation ultimately leading to prolonged tissue damage (Wlaschek and Scharffetter-Kochanek 2005; Nouvong et al. 2016; James et al. 2003). We found only 2 clinical trials related to ROS and oxidative stress on wound healing in diabetes patients where one trial is completed (NCT03649243) testing the role of propolis spray-on diabetic foot ulcers. The trial suggested the beneficial role of spray-on wound healing (Mujica et al. 2019). Another clinical trial (NCT04315909) that is ongoing is testing platelet-rich-plasma along with Vitamin E and C on diabetic wounds and foot ulcers.

4   Conclusion and Future Perspective

Diabetes-induced wounds or PUs are a growing cause of burden for the current healthcare system. Oxidative stress plays a pivotal role in its development. We have aimed at understanding the different pathways responsible for increased oxidation stress under hyperglycemic conditions. We have deciphered the role of ROS, a major player in tissue damage. Pathways affected by hyperglycemia such as AGEs, mitochondrial ROS production, and PKC play an important part in catalyzing wound progression. Understanding the precise mechanism of how PUs trigger the cascade of events leading to disability and depreciated quality of life in patients enables scientists to discover new treatment options and also to relieve the overall burden on the healthcare system. There are very few clinical trials going on in this area of research; therefore, much more research is needed in terms of studying the anti-oxidant therapies to improve the quality of life of these diabetic patients.

References

Allman RM, Goode PS, Patrick MM, Burst N, Bartolucci AA. Pressure ulcer risk factors among hospitalized patients with activity limitation. JAMA. 1995;273(11):865–70.

Are You At Risk for Chronic Wounds? – Future of Personal Health [Internet]. [cited 2021 Feb 22]. Available from: https://www.futureofpersonalhealth.com/skin-health/are-you-at-risk-for-chronic-wounds/#.

Baynes JW. Role of oxidative stress in development of complications in diabetes. Diabetes. 1991;40(4):405–12.

Berlowitz DR, Brienza DM. Are all pressure ulcers the result of deep tissue injury? A review of the literature. Ostomy Wound Manage. 2007;53(10):34–8.

Betteridge DJ. What is oxidative stress? Metabolism. 2000;49(2 Suppl 1):3–8.

Brownlee M. Biochemistry and molecular cell biology of diabetic complications. Nature. 2001;414(6865):813–20.

Chen WYJ, Rogers AA. Recent insights into the causes of chronic leg ulceration in venous diseases and implications on other types of chronic wounds. Wound Repair Regen Off Publ Wound Heal Soc Eur Tissue Repair Soc. 2007;15(4):434–49.

Dhall S, Do DC, Garcia M, Kim J, Mirebrahim SH, Lyubovitsky J, et al. Generating and reversing chronic wounds in diabetic mice by manipulating wound redox parameters [Internet]. J Diab Res. Hindawi. 2014;2014:e562625 [cited 2021 Feb 22]. Available from: https://www.hindawi.com/journals/jdr/2014/562625/.

Divers J. Trends in incidence of type 1 and type 2 diabetes among youths – Selected counties and Indian reservations, United States, 2002–2015. MMWR Morb Mortal Wkly Rep [Internet]. 2020;69 [cited 2021 Feb 22]. Available from: https://www.cdc.gov/mmwr/volumes/69/wr/mm6906a3.htm.

Duscher D, Neofytou E, Wong VW, Maan ZN, Rennert RC, Inayathullah M, et al. Transdermal deferoxamine prevents pressure-induced diabetic ulcers. Proc Natl Acad Sci U S A. 2015;112(1):94–9.

Economic Costs of Diabetes in the U.S. in 2017 I Diabetes Care [Internet]. [cited 2021 Feb 22]. Available from: https://care.diabetesjournals.org/content/early/2018/03/20/dci18-0007.

Gebhardt K. Pressure ulcer prevention. Part 1. Causes of pressure ulcers. Nurs Times. 2002;98(11):41–4.

Gefen A. How much time does it take to get a pressure ulcer? Integrated evidence from human, animal, and in vitro studies. Ostomy Wound Manage. 2008;54(10):26–8, 30–5

Global Advanced Wound Care Market – Analysis and Forecast, 2018–2024: Focus on Advanced Wound Care Dressings, NPWT Devices, Wound Care Biologics, HBOT Devices, Ultrasonic Devices, and Electromagnetic Devices [Internet]. [cited 2021 Feb 22]. Available from: https://bisresearch.com/industry-report/advanced-wound-care-market.html.

Global report on diabetes [Internet]. [cited 2021 Feb 22]. Available from: https://www.who.int/publications-detail-redirect/9789241565257.

How Much Time Does it Take to Get a Pressure Ulcer? Integrated Evidence from Human, Animal, and In Vitro Studies | Wound Management & Prevention [Internet]. [cited 2021 Feb 22]. Available from: https://www.o-wm.com/content/how-much-time-does-it-take-get-a-pressure-ulcer-integrated-evidence-human-animal-and-in-vitr.

Huang L, Nakagami G, Minematsu T, Kinoshita A, Sugama J, Nakatani T, et al. Ulceration and delayed healing following pressure loading in hyperglycemic rats with an immature dermal collagen fiber network. Wounds Compend Clin Res Pract. 2010;22(9):237–44.

James TJ, Hughes MA, Cherry GW, Taylor RP. Evidence of oxidative stress in chronic venous ulcers. Wound Repair Regen Off Publ Wound Heal Soc Eur Tissue Repair Soc. 2003 Jun;11(3):172–6.

Koivukangas V, Annala AP, Salmela PI, Oikarinen A. Delayed restoration of epidermal barrier function after suction blister injury in patients with diabetes mellitus. Diabet Med J Br Diabet Assoc. 1999;16(7):563–7.

Koya D, King GL. Protein kinase C activation and the development of diabetic complications. Diabetes. 1998;47(6):859–66.

Monnier VM, Sell DR, Genuth S. Glycation products as markers and predictors of the progression of diabetic complications. Ann N Y Acad Sci. 2005;1043:567–81.

Mujica V, Orrego R, Fuentealba R, Leiva E, Zúñiga-Hernández J. Propolis as an adjuvant in the healing of human diabetic foot wounds receiving care in the diagnostic and treatment centre from the regional hospital of Talca. J Diabetes Res. 2019;2019:2507578.

Mustoe TA, O'Shaughnessy K, Kloeters O. Chronic wound pathogenesis and current treatment strategies: a unifying hypothesis. Plast Reconstr Surg. 2006;117(7 Suppl):35S–41S.

Nishikawa T, Edelstein D, Du XL, Yamagishi S, Matsumura T, Kaneda Y, et al. Normalizing mitochondrial superoxide production blocks three pathways of hyperglycaemic damage. Nature. 2000;404(6779):787–90.

Nouvong A, Ambrus AM, Zhang ER, Hultman L, Coller HA. Reactive oxygen species and bacterial biofilms in diabetic wound healing. Physiol Genomics. 2016;48(12):889–96.

Nowotny K, Jung T, Höhn A, Weber D, Grune T. Advanced glycation end products and oxidative stress in type 2 diabetes mellitus. Biomolecules. 2015;5(1):194–222.

Ogurtsova K, da Rocha Fernandes JD, Huang Y, Linnenkamp U, Guariguata L, Cho NH, et al. IDF Diabetes Atlas: global estimates for the prevalence of diabetes for 2015 and 2040. Diabetes Res Clin Pract. 2017;128:40–50.

Oxidative stress and diabetic complications – PubMed [Internet]. [cited 2021 Feb 22]. Available from: https://pubmed.ncbi.nlm.nih.gov/21030723/.

Patel S, Srivastava S, Singh MR, Singh D. Mechanistic insight into diabetic wounds: pathogenesis, molecular targets and treatment strategies to pace wound healing. Biomed Pharmacother Biomedecine Pharmacother. 2019;112:108615.

Pressure ulcers: Current understanding and newer modalities of treatment [Internet]. [cited 2021 Feb 22]. Available from: https://www.ncbi.nlm.nih.gov/pmc/articles/PMC4413488/.

Quaresma M, Coleman MP, Rachet B. 40-year trends in an index of survival for all cancers combined and survival adjusted for age and sex for each cancer in England and Wales, 1971–2011: a population-based study. Lancet Lond Engl. 2015;385(9974):1206–18.

RAGE-induced cytosolic ROS promote mitochondrial superoxide generation in diabetes – PubMed [Internet]. [cited 2021 Feb 22]. Available from: https://pubmed.ncbi.nlm.nih.gov/19158353/.

Sen CK, Gordillo GM, Roy S, Kirsner R, Lambert L, Hunt TK, et al. Human skin wounds: a major and snowballing threat to public health and the economy. Wound Repair Regen Off Publ Wound Heal Soc Eur Tissue Repair Soc. 2009;17(6):763–71.

Singer AJ, Clark RAF. Cutaneous wound healing [Internet]. https://doi.org/10.1056/NEJM199909023411006. Massachusetts Medical Society; 2008 [cited 2021 Feb 22]. Available from: https://www.nejm.org/doi/10.1056/NEJM199909023411006.

Thannickal VJ, Fanburg BL. Reactive oxygen species in cell signaling. Am J Physiol Lung Cell Mol Physiol. 2000;279(6):L1005–28.

The management of diabetic foot: A clinical practice guideline by the Society for Vascular Surgery in collaboration with the American Podiatric Medical Association and the Society for Vascular Medicine – PubMed [Internet]. [cited 2021 Feb 22]. Available from: https://pubmed.ncbi.nlm. nih.gov/26804367/.

Thomson M, Al-Qattan KK, Js D, Ali M. Anti-diabetic and anti-oxidant potential of aged garlic extract (AGE) in streptozotocin-induced diabetic rats. BMC Complement Altern Med. 2016;16:17.

Uccioli L, Izzo V, Meloni M, Vainieri E, Ruotolo V, Giurato L. Non-healing foot ulcers in diabetic patients: general and local interfering conditions and management options with advanced wound dressings. J Wound Care. 2015;24(4 Suppl):35–42.

Understanding Pathophysiology, Fifth Edition – Sue Huether.pdf | Infection | Medical [Internet]. Scribd. [cited 2021 Feb 22]. Available from: https://www.scribd.com/document/414429961/ Understanding-Pathophysiology-Fifth-Edition-Sue-Huether-pdf.

Van Den Bos J, Rustagi K, Gray T, Halford M, Ziemkiewicz E, Shreve J. The $17.1 billion problem: the annual cost of measurable medical errors. Health Aff Proj Hope. 2011;30(4):596–603.

Vande Berg JS, Rudolph R. Pressure (decubitus) ulcer: variation in histopathology – a light and electron microscope study. Hum Pathol. 1995;26(2):195–200.

Walden CM, Bankard SB, Cayer B, Floyd WB, Garrison HG, Hickey T, et al. Mobilization of the obese patient and prevention of injury. Ann Surg. 2013;258(4):646–50; discussion 650–651

Williamson JR, Chang K, Frangos M, Hasan KS, Ido Y, Kawamura T, et al. Hyperglycemic pseudohypoxia and diabetic complications. Diabetes. 1993;42(6):801–13.

Wlaschek M, Scharffetter-Kochanek K. Oxidative stress in chronic venous leg ulcers. Wound Repair Regen Off Publ Wound Heal Soc Eur Tissue Repair Soc. 2005;13(5):452–61.

Wu J, Jin Z, Zheng H, Yan L-J. Sources and implications of NADH/NAD(+) redox imbalance in diabetes and its complications. Diabetes Metab Syndr Obes Targets Ther. 2016;9:145–53.

Zinkevich NS, Gutterman DD. ROS-induced ROS release in vascular biology: redox-redox signaling. Am J Physiol Heart Circ Physiol. 2011;301(3):H647–53.

Chronic Oxidative Stress Leads to Genomic Instability in the Pathogenesis of Fanconi Anemia

K. Jagadeesh Chandra Bose, Sarishty Gour, and Jyoti Sarvan

Contents

Abstract Chronic oxidative stress is one of the consequences of cell transformation, which causes DNA damage and adducts culminating into inter-cross-linking of DNA strands. Such aberrant DNAs may lead to cancers. Oxidative stress may directly and indirectly regulate numerous mitochondrial enzymes like peroxidases, superoxide dismutase, catalases, and peroxiredoxins within the cell. Precise understanding of how oxidative stress-mediated DNA damage and DNA inter-cross-link repair system work critically seeks exploiting a genetic disease model. As this, Fanconi anemia cancer-prone DNA damage to aggressively outproportioned oxidative stress. Fanconi anemia (FA) is an infrequent, complex, heterogeneous genetic disorder that is generally apparent in early the days of life and leads to bone marrow failure, acute myeloid leukemia (AML), or solid tumors as aging proceeds. Till date, there are 22 known genes associated with FA proteins functioning to guard the DNA against damage from defective DNA interstrand cross-linkage (ICL). The mutations in any of these critical 22 FA proteins are sensed to create genomic instability, a

K. Jagadeesh Chandra Bose (✉) · S. Gour · J. Sarvan
University Institute of Biotechnology, Chandigarh University, Mohali, Punjab, India
e-mail: jagadeesh.e8748@cumail.in

© The Author(s), under exclusive license to Springer Nature Switzerland AG 2021   289
K. K. Kesari, N. K. Jha (eds.), *Free Radical Biology and Environmental Toxicity*,
Molecular and Integrative Toxicology, https://doi.org/10.1007/978-3-030-83446-3_14

hallmark of cancer. The traditional treatment with DNA cross-linking agents causes hypersensitivity and leads to fatality. Some of these FA proteins reported that they cling to extra nuclear activities also and participate in redox balance, apoptosis, and energy metabolism. FA cells are perceptive to oxidizing agents. Some of the clinical symptoms in FA imply malfunctioned mitochondria, viz., declining membrane potential, low ATP production, impaired oxygen uptake, and deterred morphology. Mitochondria are major targets to reactive oxidative species (ROS). DNA damage by mismanaged damage and repair mechanism is one of the various reasons shaping the pathological conditions in Fanconi anemia cases. Many suggest that bioenergetic pathways are compromised with several types of FA proteins which typically functions in maintaining and controlling the balance of redox reactions, enhancing apoptotic activities of the cell besides energy metabolism. Until now, investigations on respiratory defects in Fanconi anemia cells demonstrate mitochondrial dysfunction (MDF). The previous investigations on FA cells exhibited decreased mitochondrial membrane potential and diminished ATP production along with impaired oxygen uptaking mechanism, thus drastically elevating mitochondrial ROS which further negatively influence morphology of mitochondria. This was also along with the inactivation of enzymes that are essential for energy production. These results suggest that dysregulated mitochondrial activity and their metabolism could to some extent converge with the pathological manifestations of FA. It was also revealed that the human cells of FANCG-mutant proteins exhibit impairments with respect to the mitochondrial functionalities also. Various studies suggest the mitochondrial role in genomic instability. But till date, no human pathogenic mutation has been reported where mitochondrial instability can be correlated with the genomic instability. Fanconi anemia (FA) causes genomic instability which further turns into cancer susceptibility syndrome. FA proteins are involved in nuclear DNA damage repair. Earlier it was shown that Fanconi protein subtype G (FANCG) is involving in mitochondrial oxidative stress metabolism. There are eight FA patients with single nucleotide substitution (C.65G>C) in the mitochondrial targeting peptide signal of FANCG protein; as a result, the mutant protein R22P (p.Arg22Pro) fails to migrate into mitochondria but it can successfully translocate to the nucleus. These R22P-mutant FANCG cells are resistant to DNA damaging agents similar to FANCG-corrected cells, but demonstrated their sensitivity to oxidative stress just like FANCG parental cells. Oxidative stress-mediated mitochondrial damage of R22P cells produces defective Fanconi anemia protein subtype J, a helicase enzyme that requires iron-sulfur clustering, participate in genomic DNA repair. This is the first human disease evidence where mitochondria's role in genomic instability can be argued. Dysfunctional mitochondria pinpoint situations in Fanconi anemia clinical phenotype. Studying MDF would offer deeper insights as well as diagnostic and therapeutic solutions to FA.

Keywords Fanconi anemia · Mitochondria · Oxidative stress · DNA damage repair · ROS · NOS

1   Introduction

Extensive studies with Fanconi anemia (FA) highlight a lost loop in ICL-dependent DNA damage repair in response to oxidative stress. Both the endogenous and exogenous DNA inter-cross-linking agents such as cis-platin, diepoxybutane (DEB), Mitomycin C (MMC), hexavalent chromium compounds, cyclophosphamide, and 8-methoxypsoralen plus near ultraviolet (PUVA) induce DNA damage. The increased sensitivity level of FA cells to the alkylating agents like Mitomycin C (MMC) and diepoxybutane DEB has confirmed with their interference with oxidative stress and the redox cycle in the attention of this genetic disease Fanconi anemia. The ideology of oxidative stress in relation to FA was primarily suggested (Nordenson 1977; Joenje et al. 1987) then other early studies (Schindler et al. 1987; Schindler and Hoehn 1988; Pagano et al. 1997) directed evidence to oxidative stress as a significant and serious phenotypic standpoint in FA. Supportively, FA proteins functions are associated with the regulation of redox states and energy metabolism (Cumming et al. 2001). Clinical inferences point out certain malfunctioning mitochondria. The latter are epicenters for cell energy metabolisms but also presence of reactive oxygen species (ROS). Therefore, alterations in FA mitochondria might possibly relate to the susceptibility of cells to oxidative stimuli. We discuss the investigative purview on these notions in this chapter ahead.

2   Role of Mitochondria in the Regulation of Oxidative Stress

Oxidative stress will be defined as an imbalance between ROS over antioxidant defense enzymes or mechanisms (Borut Poljsak and Milisav 2013). In eukaryotic cellular systems, the mitochondria are associated with the development of energetic metabolism as well as ROS regulations. Due to such proximity of ROS generating cascades, mitochondria also are the main target for ROS. Increased ROS levels are implicated in evading cellular antioxidant defense mechanisms and thus causing damages and impairments in mitochondria. When the intracellular endogenous network of antioxidants and their repair systems are overwhelmed, this kind of impairment is augmented in FA (Fariss et al. 2001). ROS are concerned during a large variety of pathognomonic conditions such as type II diabetes, numerous malignancies, atherosclerosis, chronic inflammatory process ischemia/reperfusion injury, besides other neurodegenerative diseases (Fariss et al. 2001; Pagano et al. 2014). At the same time, increased ROS play a controlling role in metabolic processes of the cell by activation of diverse enzymatic cascades as well as numerous transcription factors (Gupta et al. 1999; Brown et al. 2004). Approximately 1–3% of the molecular oxygen will be consumed during regular physiological respiration process and the levels of ROS in turn through a feedback mechanism assist in stabilizing p53 and hence contributing to its heightened activity (Gupta et al. 1999; Fridman and Lowe 2003).

3   Mitochondrial Imperfections in FA Cells and the Role of Cross-Linking Agents

The prooxidant state is rise of ROS and it is interfering with redox state of the cell (Degan et al. 1995) as well as ATP production (Bogliolo et al. 2002), hypersensitivity to oxidative stress and oxidant stimuli (Saadatzadeh et al. 2004) 2004, (Rani et al. 2008) and build-up of oxidized proteins (Lyakhovich and Surralles 2010) are characterized for several FA subgroups (both in vitro and in vivo). A renowned proof is provided in (Mukhopadhyay et al. 2006; Jagadeesh Chandra Bose et al. 2020b) for mitochondrial dysfunction. In specific, it had been exposed that the FANCG protein localization to mitochondria and its ability to interact with the one of the mitochondrial ROS regulating enzyme, peroxidase, peroxiredoxin 3 (PRDX3), and consequently the G546R genomic mutant expressive R22P-mutant FANCG protein fails in physical interaction with PRDX3 in contrast to FANCG wild type protein. PRDX3 was deregulated in FA cells and they undergo calpain-like cysteine protease-mediated cleavage as well as mislocalization to mitochondria. The human population with the mutant in FANCG mitochondrial localization signal caused impaired mitochondrial localization, thus resulting in brutal levels in their oxidative stress in the mutant cell lines when they are compared with wild cell lines besides the loss of FANCJ helicase actions that actually requires mitochondrial dependent iron-sulfur clustering (Jagadeesh Chandra Bose et al. 2020b).

These FA cells were also characterized by distortions in the mitochondrial structures and these mitochondrial extracts demonstrated significant decrease of thioredoxin-dependent peroxidase activity. The FAG cells that exhibit more sensitivity to OS were dropped by the overexpression of enzyme PRDX3. In turn, downregulation of PRDX3 led to augmented sensitivity toward MMC. Some studies revealed that mitochondria have oxygen-dependent sensitivity and observed a response against and in relevance to 8-methoxypsoralen photoreaction or ultraviolet A (UV-A) irradiation.

It was also observed in FA-mutant cells, unlike FA control cells, when treated with these DNA damaging agents, resulted in depressed levels of ATP and increased apoptotic cellular death. On the other hand, FA cells were very sensitive to 2-deoxy-d-glucose and iodoacetic acid, two inhibitors of the glycolytic metabolism, and therefore these findings suggested that FA cells have accomplished mitochondrial stress during studies on glycolytic metabolism (Bogliolo et al. 2002). Numerous reports on the mitochondrial morphology in FA suggested that there are various potential changes within the structure of FA mitochondria. FANCD2 complementation group affects the mitochondria from similar abnormalities as in groups A and C which include low MMP and oxygen consumption rate (OCR), reduced ATP synthesis, and elevated ROS levels (Kumari et al. 2014). Numerous numbers of distorted mitochondria along with mislocalization of mitochondrial PRDX3 have been also observed in FA-G cells. FA-G fibroblast cell lines demonstrated the recurrent elongation and irregular shapes in mitochondria (Figs. 1 and 2).

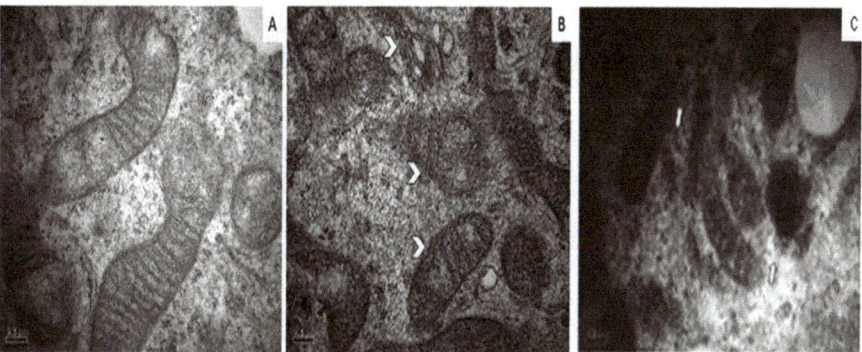

Fig. 1 Mitochondrial abnormalities in FA cells. (Adapted from Mukhopadhay et al. 2006)

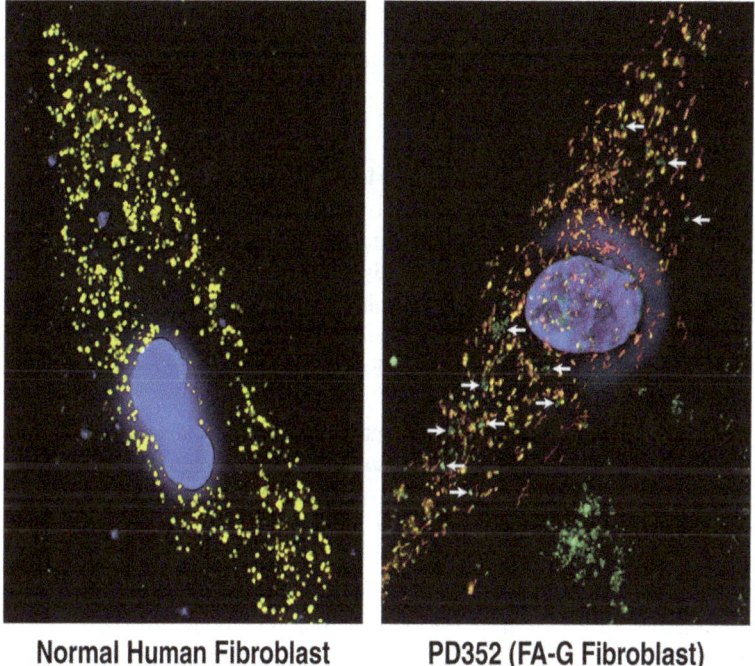

Normal Human Fibroblast   **PD352 (FA-G Fibroblast)**

Fig. 2 Mislocalization of PRDX3 and distorted mitochondrial structures in FA-G–mutant fibro-blasts. Normal primary human fibroblasts, FA-G-mutant primary fibroblasts (PD352) were trans-fected with mitochondrial marker (pDsRed-Mito). The C-terminal PRDX3 antibody and FITC-labeled secondary antibody were used to detect the endogenous PRDX3 protein. Merge of the green and red images. Arrows indicate the PRDX3 (green signals) that does not co-localize with the red mitochondrial marker in FA-G Fibroblasts. (Adapted from Mukhopadhay et al. 2006)

These results are indicative of certain mitochondrial specific enzymes account-able for ROS generation are unresponsive to H_2O_2 in FA cells, signifying the demo-lition of mitochondrial function and their detoxifying machinery in FA cells. These

Fig. 3 Model of peroxiredoxin 3 regulation in Fanconi anemia cells: Schematic model of regulation of PRDX3 activity and the impact of decreased PRDX3 activity on oxidative stress-induced apoptosis and possible DNA damage in Fanconi anemia cells

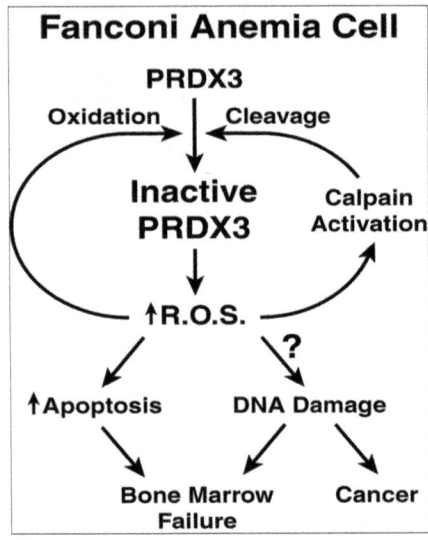

FA cells, when they were treated with H_2O_2, leading to further augmented mitochondrial dysfunctions besides elevated fragmentation. On the contrary, the occurrence of ROS scavenger N-acetylcysteine (NAC) helps to lower ROS levels and NOS levels in siRNA FA-diminished cells and the number of cells were decreased along with induced ROS-defragmented mitochondria. FA-like cells, exhausted from mitochondrial number did not express high sensitivity to MMC suggesting that mitochondria-mediated ROS could be equally important for FA cell sensitivity toward ROS (Lyakhovich 2013).

Quite a lot of incidents have demonstrated that FA cells have impaired mitochondrial membrane potential and disrupted mitochondrial chains, which leads to elevated intracellular ROS and the massive abnormality in the morphological feature of mitochondria that ultimately results in the destruction in ATP production. Excessive resultant with augmented ROS may cause the damages of mitochondria in FA cells and result in incarceration of ATP synthesis besides carbonylation and deactivation of a variety of mitochondrial membrane proteins and decrease of mitochondrial membrane potential. After treating the Fanconi anemia cell with alkylating agents like MMC leads to rise of unfolding in respiration. Since the uncoupled respiration was higher in normal cell lines as compared to FA cells, the apoptotic pathways lead to FA cellular death, which is further mediated by MMC and it was confirmed by using fluorescence microscopy along with ethidium bromide and acridine orange. These results suggest that oxidative stress as well as the cell death due to apoptosis influenced by MMC in FA cells may perhaps relate with the distinct mitochondrial membrane permeability (Jagadeesh Chandra Bose et al. 2020b).

4   FANCGR22P Cells are Sensitive to Oxidative Stress But Resistant to ICL Agents

In order to understand the capacity of the FANCG-mutant R22P protein with respect to DNA repair ability in vivo, Mitomycin drug sensitivity assay was performed. The R22P-mutant cells along with FANCG-corrected and FANCG knockout cells were treated with increasing concentration of Mitomycin (MMC) and cisplatin separately for 2 and 5 days, respectively. The cell survival assay was done with MTT and Trypan blue dye. Surprisingly, even after treating the cells with drugs in increasing concentration for 5 days, the R22P-mutant cells demonstrated resistant to both MMC and cisplatin similar to FANCG-corrected cells (wild type). The R22P stable mutant cells were resistant to formaldehyde treatment for 2 h, when they were compared with FANCG parental cells. All these assay results suggest that FANCG human mutant protein R22P is effective in the formation of FA core complex in the cytoplasm and results in monoubiquitination of FANCD2 which is hallmark of DNA damage repair and these R22P-mutant cells are effective in repairing the nuclear DNA damages, particularly the ICL damage repair. On the contrary, when hydrogen peroxide was used to induce oxidative stress for 2 and 24 h, R22P-mutant cells show much more sensitivity to oxidative stress similar to FANCG knockout cells. This result confirms the observation (Mukhopadhyay et al. 2006) that the role of FANCG protein in mitochondria related with sensitivity toward oxidative stress as a result of diminished peroxidases activity. Taken together all these kinds of results conclude the dual role of FANCG protein in the nuclear DNA damage repair along with oxidative stress regulation in the mitochondria.

5   Correlation Between Mitochondrial Instability and Genomic Instability

Genomic instability is the hallmark of cancer and insufficient damage repair of DNA is the prerequisite of genomic instability. It was observed that despite the nuclear DNA damage repair ability, the R22P FANCG-mutant patients are susceptible to cancer (D'Andrea 2003). Now the question comes whether mitochondria have influence in the genomic DNA repair or not? If the cellular status of the FANCG R22P patients to be considered, then the mitochondria of these FANCG R22P patients are under chronic endogenous oxidative stress right from their birth and the same will be constantly increasing with their age. Their genomic DNA is also susceptible for numerous ICL agents (both exogenous and endogenous). By considering these two facts, an experiment on the cell lines was designed. The R22P-mutant cells, FANCG-corrected cells (wild type), and the FANCG parental (knockout) cells were treated with very mild concentration of H_2O_2 (10 μM) continuously for 14 h to induce chronic oxidative stress, and at an interval of every 2 h, these cells were treated with MMC at low doses around 100 nM for half an hour.

Then the cells were assessed with JC-1 dye for analyzing the loss of mitochondrial membrane potential ($\Delta\Psi$) that was caused by mitochondrial instability. γ-H2AX foci assay for analyzing the nuclear DNA damage is also done for this set of cells (Fig. 9). In FANCG parental and R22P-mutant cells almost similar amount of depolarization in their mitochondria was observed at each time which is significantly minimal when compared to FANCG-corrected cells and depolarization was detected from 2 h of treatment in both cells. In 14 h, nearly 35% and 40% of depolarization mitochondria was observed in R22P-mutant and FANCG parental cells, respectively. When the average numbers of foci were measured in the nucleus of respective cells, it was found that the number of foci in FANCG-corrected cells was significantly less than the other two types of cells at each time point. Upon 14 h of incubation time, the number of the foci was diminished and it strongly suggests the nuclear DNA repair ability of the cells at that time point. In FANCG parental cells, highest numbers of foci were developed immediately after 2 h and their number was increased when compared to R22P-mutant cells. As there is no FANCG protein in parental cell, the cells were unable to protect neither their nuclear DNA nor their mitochondria. In R22P-mutant cells, the number of foci was less than the FANCG parental cell. But after 14 hours of treatments the number of foci was almost equal in both cell types. Hopefully, in R22P-mutant protein is able to repair the DNA at early stages. After that the increased accumulation of depolarized mitochondria crossed the threshold limit and the R22P cell fails to repair the nuclear DNA. These observations strongly suggest the role of mitochondria indirectly on the nuclear DNA damage repair (Jagadeesh Chandra Bose et al. 2020a).

6 Antioxidants Will Regulate the ROS Production by the Mitochondria

Subsequently mitochondria are the chief source for the formation of ROS intracellularly; they are concerned about regular protection from the cytotoxic action of these ROS. Such sorts of ROS protection mechanisms are normally provided by a variety of antioxidants with low molecular weight, along with certain multiple enzymes associated with ROS defense systems (Sies 1997). Among all cellular antioxidants, ubiquinone (Quiles et al. 2004; Marriage et al. 2004) and vitamin E (Moore and Roberts 2nd 1998; Pfluger et al. 2004; Stephens et al. 1996) were predominantly important in ROS regulation. Selective and specific delivery of various antioxidants made synthetically to mitochondria achieved by the attachment of a lipophilic triphenyl phosphonium cation either to α-tocopherol (MitoVit E) or to ubiquinone (MitoQ) (Smith et al. 2003). The total overall positive charge displayed by these synthetic antioxidant derivatives empowers them to get accumulated in mitochondria owing to their negative mitochondrial membrane potential inside these subcellular organelles as well as to provide substantial protection from toxic oxidative stress. Further, recent investigations were proposing that supplementation of vitamin E along with diet involves in enrichment of hepatic mitochondria with elevated levels of α-tocopherol protection in dose dependent manner, may prove the most beneficial effects in the treatment of oxidative stress-mediated liver diseases, mainly when this is from mitochondrial origin (Sukalski et al. 1993; Zhang et al. 2001). Most important for mitochondrial antioxidant protection mechanisms include tripeptide glutathione, GSH (l-γ-glutamyl-l-cysteinylglycine), and multiple GSH-linked enzymatic defense systems (Orrenius et al. 2003). Actually, it was proved that the cellular death is due to toxicity of oxidative stress that was frequently correlated with depleted conditions of the pool of 1 GSH especially in mitochondria than depletion of GSH proteins intracellularly.

7   Antioxidants and Their Effect on Mitochondrial Parameters in FA Cells

The pro-oxidant Fanconi anemia phenotype cells are well studied and characterized by modifications and alterations in their efficient functions like electron transport mechanism in the inner mitochondrial membrane space along with significant loss of electron transfer from Complex I to Complex III. These changes can be restored by administration of the supplements of Coenzyme Q in exogenous route (Ravera et al. 2013). As N-acetylcysteine generally minimizes production of ROS levels it's adminiteration as antioxidant ensures normalized mitochondrial respiration (Zafarullah et al. 2003). N-acetylcysteine has already been proven as moderately effective in limiting the depletion of GSH, predominantly when usage of the drugs those are targeted to mitochondria (Cuccarolo et al. 2012), it is also expected that the treatment of FA cells with N-acetylcysteine may increase oxygen consumption. Investigations on the usage of N-acetylcysteine for the treatment of FA cells were made in (Cuccarolo et al. 2012) as well as the beneficial effects with almost significant restoration of biochemical parameters were demonstrated, whereas the cellular structures did not recover, and in fact it affected mitochondrial structure. The ability of N-acetylcysteine to reestablish Complex I activities is nevertheless inherently associated with increase of ROS levels and their production and consequently results in the deterioration of the already existing typical structural damage of Fanconi anemia-A type A cells. This is also indicative of the presence of a principal damage to mitochondrial membranes in the Fanconi anemia-A phenotype. There are some instructive reports made on lymphoblastoid cell lines which are derived from FA patients who display significant confrontation to oxidative stress induced by actions of H_2O_2 and abundant glutathione (GSH) inhibitors induce the production of ROS, GSH depletion and results in mitochondrial membrane depolarization.

8   Mitochondrial Membrane Depolarization

Numerous information put forwarded that the resistance to the oxidative stress positively alters mitochondrial functionality and their metabolic activities in the FA-mutant cells which might demonstrate the survival policy attained in Fanconi anemia cells were undergoing transformation. The investigations of redox potential and regulation of mitochondria in FA also might be one of supportive evidences in the diagnosis of the FA disease as well as to take care of patients (Cuccarolo et al. 2012). Irrespective of highly heterogeneous nature, the cells collected from all Fanconi anemia patients are demonstrated hypersensitivity to the alkylating agents, particularly to diepoxybutane (DEB), as well as to oxidative stress-mediated DNA damage. Current studies demonstrated that the defective mitochondria in Fanconi anemia cells are closely associated with the overproduction of intracellular ROS and also connected to the depletion of antioxidant defenses like glutathione as a renowned biomarker. The assessment studies revealed that the assumed shielding effect of a mitochondrial specific protective agents like α-lipoic acid (ALA) and N-acetylcysteine (NAC), an immediate antioxidant and recognized as precursor for synthesis of glutathione. These results recommended that the combinatorial studies will be frequently used as a prophylactic method for the postponement of progressive clinical symptoms in Fanconi anemia patients caused by inter-cross-linkage in DNA, which leads to bone marrow failure and early cancer development.

9   Mitochondria and FA Pathology

It has been reported that the cells with mutations either in FA-A or FA-C partly depend on the glycogenolytic metabolism, usually related to transformation process and the resistance of mutant FA-A and FA-C cells.

Though, in pathological condition of FA, accumulation of ROS distresses mitochondrial functions such as production of ATP, detoxification of ROS, and in the maintenance of mitochondrial membrane potential along with oxygen consumption rate. It is like in numerous FA cells, the mitochondria are shifted to the conditions of State 4 semi-resting state, where production of ATP is substandard and the rate of oxygen consumption is diminished. Elevated ROS levels might cause the failure in the functions of detoxifying enzymes in response against ROS resulting in numerous mitochondrial abnormalities. Such abnormalities will be increased along with the mitochondria-based aging theory depicting that ROS generated by the altered electron transport chain in the mitochondria with damaged mitochondrial structure and later leading to further elevated production of ROS and intensifying the damage to the mitochondria. This aging theory even does have certain sorts of limitations. The furthermost understandable point is that augmentation of ROS levels in Fanconi anemia cells with defective mitochondria does not last indefinitely.

This might propose that certain neutralizing mechanism which can aim either to minimize oxidative stress or to prevent the production of ROS levels by mitochondrial specialized enzymes. Certainly, ROS overproduction and/or failure in detoxification of ROS results in pathogenicity.

Hence, therapeutic approaches which are targeted at reducing ROS levels might provide the effective combating against the development of malignancies (Kumar et al. 2008). There are some significant clinical implications for FA-associated cancer patients who require chemotherapy as well as the patients that undergoing hematopoietic somatic cell transplantation. Treatment with chemotherapeutic drugs is essential that leads to decrease in the toxic effects of DNA cross-linkers besides adding antioxidative therapy.

First, palliative care and treatment methods will not correct the definite cause of Fanconi anemia. Secondly, regulation of ROS levels is very much important for proper functions of the metabolism in the cells. Significant decrease in the regulation of ROS levels will be the consequence of the inability of a cell to fight with pathognomonic infections (immune responses based on the lysosomal and peroxisomal degradation of targeted pathogens) or impairment of neurotransmitters in cellular signaling processes.

10 Unique Mitochondrial Localization Signal of Human FANCG

Various in-silico tools were exploited for the identification of the mitochondrial localization signal in human FANCG protein. These tools are mainly focused for basic prediction of two important things such as (i) whether the FA protein contains any mitochondrial localization (MLS)/targeting signal (MTS) or (ii) is the protein has any possibility to localize into mitochondria. The results of the tools robustly predicted (MLS) by considering the first 30 amino acid residues at N-terminal end of human FANCG protein as well as their existence in mitochondria of the human FANCG. At the same time, when the FANCG sequence from other species were analyzed for mitochondrial localization signal, none of them was found to carry the same either at N-terminal end or C-terminal end except human FANCG protein. It may happen that the mitochondrial localization signal region has evolved later in human beings and as a result of possessing the mitochondrial localization signal; human FANCG achieved the ability of taking care of the mitochondria besides nuclear DNA damage repair mechanism. Interestingly, the FANCG knockout mice do not display any severe phenotype. The cells which are derived from FANCG knockout mice are showing mild sensitivity only to the DNA inter-cross-linking agents, but not to oxidative stress (Yang et al. 2001; Parmar et al. 2009; Pulliam-Leath et al. 2010). But expression studies of FANCG of other species are required for confirmation of this information.

Several mutations were identified in the mitochondrial localization signal region of human FANCG protein from the LOVD as well as from the COSMIC (catalogue of somatic mutation in cancer) databases and their mitochondrial localization patterns have been studied. One pathogenic mutation was selected due to its hot-spot position for the mutation, and this mutant was analyzed where the 22nd position, the amino acid residue Arginine was replaced by Proline residue (FANCG-R22P). GFP expression and N-terminal end of this fusion constructs suggested the FANCG protein's inability to migrate into mitochondria. The predicted secondary structure of FANCG protein at this position suggests that helix part is broken due to replacement of Proline in the place of Arginine at 22nd position of the N-terminal end. Numerous mitochondrial investigative studies suggested the significance of positively charged amino acid residues like Arginine/Lysine at the N-terminal end of the proteins for mitochondrial localization. But most interesting feature of this R22P FANCG-mutant protein is that it fails to migrate into the mitochondria but it can successfully translocate to the nucleus. These mutant phenotypes were confirmed by MMC drug sensitivity assay. R22P-mutant stable cells demonstrated resistance to different ICL drugs just similar to FANCG-corrected cells (wild type), but they are exhibiting sensitivity to oxidative stress just similar to FANCG parental cells. The monoubiquitination of FANCD2 in R22P-mutant cells were normal and it also suggests the ability of the mutant FANCG protein to be the part of FA complex during the process of DNA damage repair. So, the phenotypes of R22P pathogenic mutation resolve the long-lasting debate of Fanconi anemia protein's role in the mitochondria. How this result can be concerned with the clinical investigation or diagnosis of FA patients? Some patients are diagnosed as Fanconi anemia by limited assays like phenotypic features, but only drug sensitivity test against ICL agents of their cells suggests negative. In that case combinatorial drug sensitivity test that can be performed in the presence of mild oxidative stress would be a better choice for the confirmation of FA.

11 FANCD2 Mislocalization to Mitochondria

The co-localization studies of FANCD2 wild type along with various truncated deletion mutants at N-terminal end of FANCD2-EGFP in fibroblast cells revealed that FANCD2 localize mitochondria. But truncated deletion mutants of FANCD2 (data not shown) and R05T mutant (Fig. 4) failed to merge with MitoTracker Red indicating that the mitochondrial localization signal (MLS) is crucial for FANCD2 localization into mitochondria. From the above observation, it is confirmed that FANCD2 have some specific functions in the mitochondria. Similar type of results was obtained and confirmed by A R Meetei et al. (Zhang et al. 2017). In their studies also it was confirmed that the loss of FANCD2 protein or FANCA protein in the cells results in dysregulation of mitochondrion genes as well as FANCD2 protein is very essential for the maintenance of the stability of the mitochondrion nucleoid complex. The recent investigations on either Fanconi anemia mice or patient-derived human primary cells were exhibiting their phenotypes such as deregulation in the apoptotic response, elevated levels of hypersensitivity to the oxidative stress along with mitochondrial deficiency in terms of their number and functionalities. The FANCD2 knock-in mouse model suggests that most of the FANCD2-binding proteins are not involved in DNA damage repair. But numerous FANCD2-interactive proteins were found to be mitochondrial specific. It was also shown that FANCD2 is localized in the mitochondria as well as its association with the nucleoid complex protein components such as Atad3, Tufm, and Tfam, all are very much required for biosynthesis of mitochondria. Furthermore, the Atad3/Tufm/Tfam complex was disrupted in FANCD2−/− mice. These findings will strongly highlight the role of FA proteins and their major functions in mitochondria along with oxidative stress regulation.

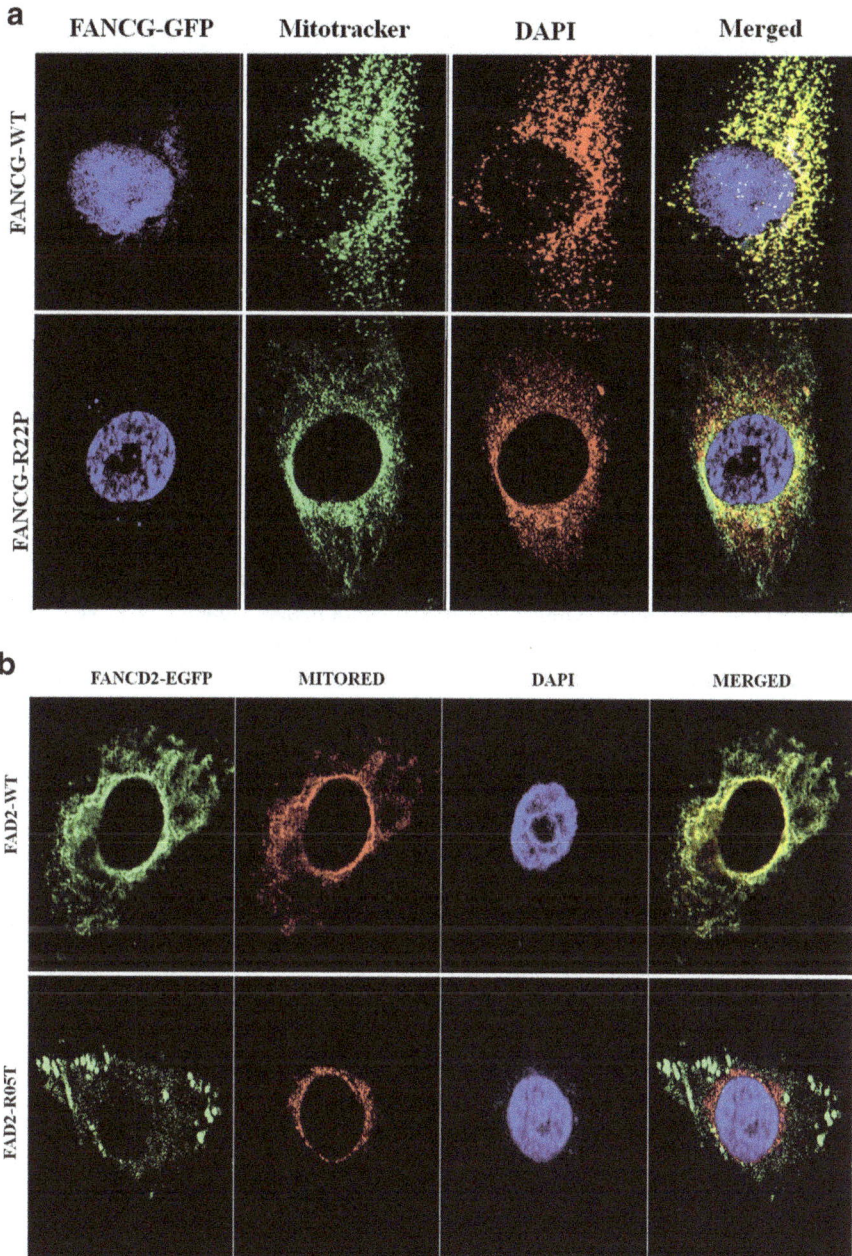

Fig. 4 In vivo co-localization of FANCG and FANCD2 in mitochondria. The above panels that show perfect co-localization of C-terminal GFP tagged FANCG wild type (**a**) and C-terminal EGFP tagged FANCD2 wild type (**b**) describe their existence in mitochondria whereas the R22P-mutant FANCG and R05T-mutant FANCD2 (substitution point mutations which were generated) show mislocalization resulting their loss in mitochondrial translocation

12 Mitochondrial Instability Causes Defective FANCJ

Cancer may be developed from reduced or hampered cellular machinery that is responsible for DNA damage repair. However, recent studies have another direction to this. Cells with FANCG-R22P-mutant protein are considerably more sensitive to oxidative stress endowing a loss of both mitochondrial structure and their membrane potential (Mukhopadhyay et al. 2006). These infer that there is a strong correlation between mitochondrial malfunctions and malfunctioning genomic DNA repair mechanism(s) especially DNA inter-cross-damage repair. Previous literature also supports that both mutations in mtDNA and loss of MMP leading to cancer (Tokarz and Blasiak 2014). Other mechanisms that are resulting in free radicals generation under oxidative stress may render mitochondrial dysfunctions and later they may indirectly cause the impact by destabilizing vital macromolecules involved in DNA damage repair (Nunnari and Suomalainen 2012). However, relatedness of ICL DNA damage into this was never explored. The lead author in his previous work, however, clearly suggested that R22P cells can repair the DNA damages caused by the ICL agents instead of mitochondrial dysfunction within certain limits. When percentage of mitochondrial dysfunction crosses the limit, R22P cells fail to show either protective effect or DNA repair against ICL-linked genomic DNA damages. Proteins with Fe-S complexes essentially play critical roles reportedly in catalysis, nucleic acid synthesis, and repair (Netz et al. 2014). Iron-sulfur clustered proteins majorly rely on mitochondria for providing clusters of iron-sulfur (ISC) to the some essential proteins (Lill and Muhlenhoff 2008) and that their fabrication and catalytic activation requires structurally normal and functionally active mitochondria (Biederbick et al. 2006). Nonetheless, aberrant mtDNA (mitochondrial DNA) or MMP downregulates ISC formation (Lange et al. 1999). Defects in iron metabolism have been reported in yeast system concomitant with a loss of mtDNA (Veatch et al. 2009). A theoretical knowledge gap, however, existed with whether mitochondrial malfunctions caused by oxidative stress could infer defects in iron-sulfur proteins. Studies on R22P cells in animal models instead showed that MMP losses can culminate into substantial deficiency of iron in FANCJ protein that have helicase function during DNA damage repair, although we could not show any helicase activity in the defective FANCJ which could be attributable to either insufficient proteins or limited affinity of immunoprecipitated FANCJ protein to link with DNA. But significant evidences relate impaired helicase activity is due to loss of iron (Wu et al. 2010). Although the study focused on FANCJ, however, factually other cell Fe-S factors known or unknown yet in DNA damage repair might similarly have hampered (Netz et al. 2014). Hence, there is strong evidence that in R22P cells FANCJ is also defective as it might not clustered with significant iron-sulfur cluster due to mitochondrial dysfunctions and is unable to repair the genomic DNA specifically disfigured by ICL. So, future studies with R22P cells would entail elucidating the threshold of dysfunctional mitochondria in cells. Studies above at least confirm the importance of nonrespiratory function of mitochondria in disease progression like cancer which was under speculation for a long time. Therefore, antioxidant therapy in conjunction with modern treatment methodologies may derive more promising avenue for the FA patients. Other than this, as mitochondria-mediated ROS in FA shows a challenging and distinctive strategy, mitochondrial activity and metabolism act as a relative cause for the pathological expression of FA. Resistance to apoptosis can be achieved by cancer cells with proper knowledge of cause and treatment. This can help not only in successful treatment options but also help in developing diagnosis and treatment of the cancer.

References

Biederbick A, et al. Role of human mitochondrial Nfs1 in cytosolic iron-sulfur protein biogenesis and iron regulation. Mol Cell Biol. 2006;26(15):5675–87.

Bogliolo M, et al. Alternative metabolic pathways for energy supply and resistance to apoptosis in Fanconi anaemia. Mutagenesis. 2002;17(1):25–30.

Borut Poljsak DŠ, Milisav I. Achieving the balance between ROS and antioxidants: when to use the synthetic antioxidants. Oxidative Med Cell Longev. 2013;2013(Article ID 956792):11.

Brown DM, et al. Calcium and ROS-mediated activation of transcription factors and TNF-alpha cytokine gene expression in macrophages exposed to ultrafine particles. Am J Phys Lung Cell Mol Phys. 2004;286(2):L344–53.

Cuccarolo P, Viaggi S, Degan P. New insights into redox response modulation in Fanconi's anemia cells by hydrogen peroxide and glutathione depletors. FEBS J. 2012;279(14):2479–94.

Cumming RC, et al. Fanconi anemia group C protein prevents apoptosis in hematopoietic cells through redox regulation of GSTP1. Nat Med. 2001;7(7):814–20.

D'Andrea AD. The Fanconi Anemia/BRCA signaling pathway: disruption in cisplatin-sensitive ovarian cancers. Cell Cycle. 2003;2(4):290–2.

Degan P, et al. In vivo accumulation of 8-hydroxy-2′-deoxyguanosine in DNA correlates with release of reactive oxygen species in Fanconi's anaemia families. Carcinogenesis. 1995;16(4):735–41.

Fariss MW, et al. Enhanced antioxidant and cytoprotective abilities of vitamin E succinate is associated with a rapid uptake advantage in rat hepatocytes and mitochondria. Free Radic Biol Med. 2001;31(4):530–41.

Fridman JS, Lowe SW. Control of apoptosis by p53. Oncogene. 2003;22(56):9030–40.

Gupta A, Rosenberger SF, Bowden GT. Increased ROS levels contribute to elevated transcription factor and MAP kinase activities in malignantly progressed mouse keratinocyte cell lines. Carcinogenesis. 1999,20(11).2063–73.

Jagadeesh Chandra Bose K, Mondal BSKK, Ghosh S, Mokhamatam RB, Manna SK, Mukhopadhyay SS. Despite of DNA repair ability the Fanconi anemia mutant protein FANCGR22P destabilizes mitochondria and leads to genomic instability via FANCJ helicase. bioRxiv. 2020a; https://doi.org/10.1101/2020.01.15.907303.

Jagadeesh Chandra Bose K, et al. Loss of mitochondrial localization of human FANCG causes defective FANCJ helicase. Mol Cell Biol. 2020b;40(23):e00306.

Joenje H, et al. Cytogenetic toxicity of paraquat and streptonigrin in Fanconi's anemia. Cancer Genet Cytogenet. 1987;25(1):37–45.

Kumar B, et al. Oxidative stress is inherent in prostate cancer cells and is required for aggressive phenotype. Cancer Res. 2008;68(6):1777–85.

Kumari U, et al. Evidence of mitochondrial dysfunction and impaired ROS detoxifying machinery in Fanconi anemia cells. Oncogene. 2014;33(2):165–72.

Lange H, Kispal G, Lill R. Mechanism of iron transport to the site of heme synthesis inside yeast mitochondria. J Biol Chem. 1999;274(27):18989–96.

Lill R, Muhlenhoff U. Maturation of iron-sulfur proteins in eukaryotes: mechanisms, connected processes, and diseases. Annu Rev Biochem. 2008;77:669–700.

Lyakhovich A. Damaged mitochondria and overproduction of ROS in Fanconi anemia cells. Rare Dis. 2013;1:e24048.

Lyakhovich A, Surralles J. Constitutive activation of caspase-3 and Poly ADP ribose polymerase cleavage in fanconi anemia cells. Mol Cancer Res. 2010;8(1):46–56.

Marriage BJ, et al. Cofactor treatment improves ATP synthetic capacity in patients with oxidative phosphorylation disorders. Mol Genet Metab. 2004;81(4):263–72.

Moore K, Roberts LJ 2nd. Measurement of lipid peroxidation. Free Radic Res. 1998;28(6):659–71.

Mukhopadhyay SS, et al. Defective mitochondrial peroxiredoxin-3 results in sensitivity to oxidative stress in Fanconi anemia. J Cell Biol. 2006;175(2):225–35.

Netz DJ, et al. Maturation of cytosolic and nuclear iron-sulfur proteins. Trends Cell Biol. 2014;24(5):303–12.

Nordenson I. Effect of superoxide dismutase and catalase on spontaneously occurring chromosome breaks in patients with Fanconi's anemia. Hereditas WILEY Online Library. 1977;86(2):147–50.

Nunnari J, Suomalainen A. Mitochondria: in sickness and in health. Cell. 2012;148(6):1145–59.

Orrenius S, Zhivotovsky B, Nicotera P. Regulation of cell death: the calcium-apoptosis link. Nat Rev Mol Cell Biol. 2003;4(7):552–65.

Pagano G, et al. In vitro hypersensitivity to oxygen of Fanconi anemia (FA) cells is linked to ex vivo evidence for oxidative stress in FA homozygotes and heterozygotes. Blood. 1997;89(3):1111–2.

Pagano G, et al. Damaged mitochondria in Fanconi anemia – an isolated event or a general phenomenon? Onco Targets Ther. 2014;1(4):287–95.

Parmar K, D'Andrea A, Niedernhofer LJ. Mouse models of Fanconi anemia. Mutat Res. 2009;668(1–2):133–40.

Pfluger P, et al. Vitamin E: underestimated as an antioxidant. Redox Rep. 2004;9(5):249–54.

Pulliam-Leath AC, et al. Genetic disruption of both Fancc and Fancg in mice recapitulates the hematopoietic manifestations of Fanconi anemia. Blood. 2010;116(16):2915–20.

Quiles JL, et al. Coenzyme Q supplementation protects from age-related DNA double-strand breaks and increases lifespan in rats fed on a PUFA-rich diet. Exp Gerontol. 2004;39(2):189–94.

Rani R, Li J, Pang Q. Differential p53 engagement in response to oxidative and oncogenic stresses in Fanconi anemia mice. Cancer Res. 2008;68(23):9693–702.

Ravera S, et al. Mitochondrial respiratory chain Complex I defects in Fanconi anemia complementation group A. Biochimie. 2013;95(10):1828–37.

Saadatzadeh MR, et al. Oxidant hypersensitivity of Fanconi anemia type C-deficient cells is dependent on a redox-regulated apoptotic pathway. J Biol Chem. 2004;279(16):16805–12.

Schindler D, Hoehn H. Fanconi anemia mutation causes cellular susceptibility to ambient oxygen. Am J Hum Genet. 1988;43(4):429–35.

Schindler D, et al. Confirmation of Fanconi's anemia and detection of a chromosomal aberration (1Q12-32 triplication) via BrdU/Hoechst flow cytometry. Am J Pediatr Hematol Oncol. 1987;9(2):172–7.

Sies H. Oxidative stress: oxidants and antioxidants. Exp Physiol. 1997;82(2):291–5.

Smith RA, et al. Using mitochondria-targeted molecules to study mitochondrial radical production and its consequences. Biochem Soc Trans. 2003;31(Pt 6):1295–9.

Stephens NG, et al. Randomised controlled trial of vitamin E in patients with coronary disease: Cambridge Heart Antioxidant Study (CHAOS). Lancet. 1996;347(9004):781–6.

Sukalski KA, Pinto KA, Berntson JL. Decreased susceptibility of liver mitochondria from diabetic rats to oxidative damage and associated increase in alpha-tocopherol. Free Radic Biol Med. 1993;14(1):57–65.

Tokarz P, Blasiak J. Role of mitochondria in carcinogenesis. Acta Biochim Pol. 2014;61(4):671–8.

Veatch JR, et al. Mitochondrial dysfunction leads to nuclear genome instability via an iron-sulfur cluster defect. Cell. 2009;137(7):1247–58.

Wu Y, et al. Fanconi anemia group J mutation abolishes its DNA repair function by uncoupling DNA translocation from helicase activity or disruption of protein-DNA complexes. Blood. 2010;116(19):3780–91.

Yang Y, et al. Targeted disruption of the murine Fanconi anemia gene, Fancg/Xrcc9. Blood. 2001;98(12):3435–40.

Zafarullah M, et al. Molecular mechanisms of N-acetylcysteine actions. Cell Mol Life Sci. 2003;60(1):6–20.

Zhang JG, et al. Vitamin E succinate protects hepatocytes against the toxic effect of reactive oxygen species generated at mitochondrial complexes I and III by alkylating agents. Chem Biol Interact. 2001;138(3):267–84.

Zhang T, et al. Fancd2 in vivo interaction network reveals a non-canonical role in mitochondrial function. Sci Rep. 2017;7:45626.

Toxicity with Waste-Generated Ionizing Radiations: Blunders Behind the Scenes

Anirudh Sharma, Kartar Chand, Gajendra B. Singh, and Gaurav Mudgal

Contents

Abstract Besides natural causes, human-led contamination of the planet is significant and impacts the health of all living beings in the planet ascribed to population-borne challenges we are encountering in the face of either global warming, or pollution, or scarcity of water, as well to an increasing extent that of ionizing radiations. Understandably, studies on the effects of ionizing radiations on life forms are important from the scope of ecology; macroscopic and microscopic lifestyles; plant, human, and animal diseases; and also considering the emerging new power and energy promises from the existing and upcoming nuclear power plants, and other inevitable introductions of radiations from the space and graving situations with wastes that emit radiations as well as those that lead to holes in the ozone blanket, which protects the biosphere from harmful galactic rays. It is imperative to note that

A. Sharma (✉) · K. Chand · G. B. Singh · G. Mudgal
University Institute of Biotechnology, Chandigarh University, Mohali, Punjab, India

© The Author(s), under exclusive license to Springer Nature Switzerland AG 2021
K. K. Kesari, N. K. Jha (eds.), *Free Radical Biology and Environmental Toxicity*,
Molecular and Integrative Toxicology, https://doi.org/10.1007/978-3-030-83446-3_15

technologies are now necessary to curb the growing pressures from the evolution of distress from ionizing radiation, which even at low but long-term doses may drastically impact the biosphere for its life. Answers may lie within the many forms of lives which stay uninfluenced. In this chapter, we focus on various sources of ionizing radiations that affect human and other life forms by means of free radical-induced ailments as well as lay detrimental impact on the environment health stature. We also provide a narrative over studies depicting radiation tolerance in some extremophilic life kinds and how this may show biotechnological implications at managing the impeding adversities with ionizing radiations.

Keywords UV · radiation · ionizing · contamination · biotechnology · tardigrade

1 Introduction

Radiation is a process in which energy travels through a medium or space to be absorbed by another body. Non-physicists often associate the word ionizing radiation (e.g., as occurring in nuclear weapons, nuclear reactors, and radioactive substances), but it can also refer to electromagnetic radiation (i.e., radio waves, infrared light, visible light, ultraviolet light, and X-rays), acoustic radiation, or radiations emitted from other more ambiguous processes. Ionizing radiation is everywhere as it arrives from outer space as cosmic rays. It is in the air as emissions from radioactive radon and its progeny. Naturally occurring radioactive isotopes enter and remain in all living things. It is inescapable, indeed, all species on this planet evolved in the presence of ionizing radiation. While humans exposed to small doses of radiation may not immediately show any apparent biological effects, but undoubtedly ionizing radiations, when dosed beyond a tolerated threshold causes mild to severe harm. These effects are well known both in kind and in degree (Block 1978).

While ionizing radiation can cause harm, it also has many beneficial uses. Radioactive uranium generates electricity in nuclear power plants in many countries. In medicine, X-rays produce radiographs for the diagnosis of internal injuries and diseases. Nuclear medicine physicians use radioactive material as tracers to form detailed images of internal structures and to study metabolism. Biophysicists and molecular biologists involved in protein crystallography use X-ray beam photons to solve structural and mechanistic riddles of biomolecules. Therapeutic radiopharmaceuticals are available to treat disorders such as hyperthyroidism and cancer. Radiotherapy physicians specifically use gamma rays, pion beams, electron beams, neutrons, and other types of radiations to treat cancer. Engineers use radioactive material in oil well logging operations and in soil moisture density gauges. Industrial radiographers use X-rays in quality control to look at internal structures of manufactured devices. Exit signs in buildings and aircraft contain radioactive tritium to make them glow in the dark in the event of a power failure. Many smoke detectors in homes and commercial buildings contain radioactive americium. These many uses of ionizing radiation and radioactive materials enhance the quality of life and help society in many ways. The benefits of each use, however, must always consider the long-term health and environmental risks associated with the use and fabrication of materials based on ionizing radiations. The risks may be to workers directly involved in applying the radiation or radioactive material, to the public, to future generations, and to the environment or to any combination of these. Albeit this, on account of political and economic considerations, benefits have always outweighed risks when ionizing radiation is involved (Goodhead 1988).

2 Ionizing Radiation

Ionizing radiation consists of particles, including photons, which cause the separation of electrons from atoms and molecules. However, some types of radiations of relatively low energy, such as ultraviolet light, can also cause ionization under certain circumstances. To distinguish these types of radiation from radiation that always causes ionization, an arbitrary lower energy limit for ionizing radiation usually is set around 10 kilo electron volts (keV). Directly ionizing radiation consists of charged particles. Such particles include energetic electrons (sometimes called negatrons), positrons, protons, alpha particles, charged mesons, muons, and heavy ions (ionized atoms). This type of ionizing radiation interacts with matter primarily through the Coulomb force, repelling or attracting electrons from atoms and molecules by virtue of their charges (Hall 1994).

A photon that can transfer enough energy to one electron to remove it from its atomic orbital will cause the ionization of an atom and the formation of an ion pair. To form an ion pair, the removed electron constitutes the negative ion, the anion; and the remaining atom is the positive ion or cation. Radiation with sufficient energy to ionize matter is called ionizing radiation, and X-rays and gamma rays are the only known electromagnetic ionizing radiations. The only difference between X-rays and gamma rays understandably exists is their origin. X-rays are emitted from the electron cloud of a stimulated atom, whereas gamma rays come from the nucleus of a radioactive atom. At the same energy level, however, there is no other way to differentiate one from the other, and the same applies to electron and beta particles. Beta particles are different from electrons due to their origin within the nucleus. The energy is directly proportional to the frequency and inversely proportional to the wavelength of the radiation. Besides X-ray and gamma ray electromagnetic radiation, some particles with high kinetic energies (particles traveling at very high speeds) are also capable of inducing ionization. This includes alpha and beta particle radiation, which can cause ionization when emitted (Haynie and Olsher 1981).

2.1 Sources of Ionizing Radiation

We are living in a world in which radiation including ionizing radiation has always been omnipresent. Radiation is part of our natural environment, and it is only very recently, since about 100 years ago, that other sources of man-made radiation were

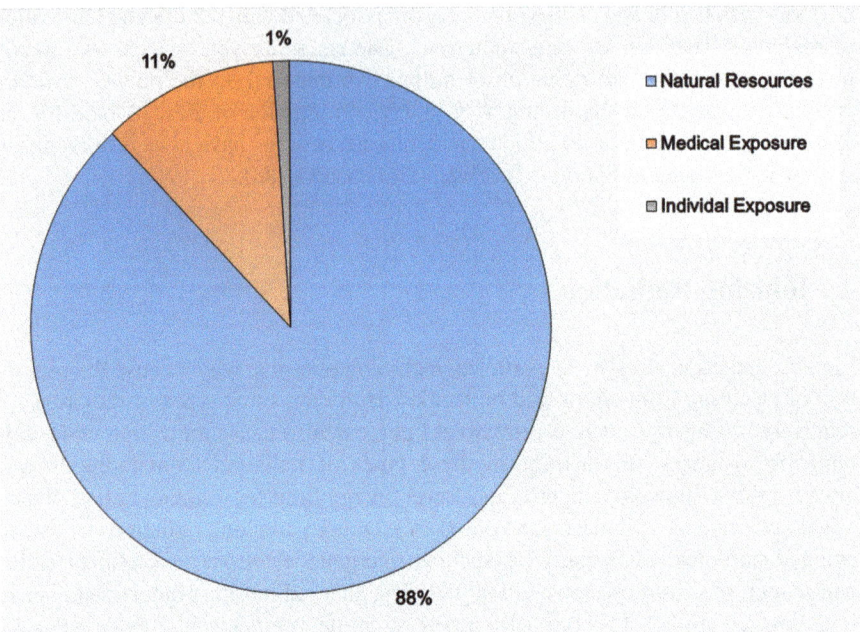

Fig. 1 Sources of impactful ionizing radiations

progressively developed and added to the natural background radiation. For an individual, all radiation sources, both natural and man-made, contribute to a total average annual radiation dose of about 2 mSv. About 88% of this total comes from natural sources, and of the 12% left, the largest part, 11%, is due to medical exposures (Fig. 1). All other sources contribute about 1% of individual radiation exposure altogether. However, these values are average and need to be adjusted according to individual lifestyle and residence location. For example, healthy people who are not undergoing medical care that involves the use of ionizing radiation will receive 99% of their annual dose from natural sources. People living in the mountains or traveling frequently on jet airplanes cruising at high altitudes are more exposed to cosmic radiation than others. Data obtained from radiotherapy facilities indicate that high radiation doses contribute significantly to the average annual radiation exposure of the population, but in medical cases, the benefit that the patient gets from radiation exposure always inevitably exceeds the possible radiation ill-dosage risks, much often this is witnessed in cases of recurrence of diseases and relevant radiation therapy(s) (de Saint-Georges 2009).

2.2   Radiation in Environment

Radiation in environment, also called radiological contamination, is the deposition or presence of radioactive substances on surfaces or within solids, liquids, or gases (including the human body), where their presence is unintended or undesirable (International Atomic Energy Agency [IAEA]). Such contamination presents a hazard because of the radioactive decay of the contaminants, which produces such harmful effects as ionizing radiation (namely α, β, and γ rays) and free neutrons. The degree of hazard is determined by the concentration of the contaminants, the energy of the radiation being emitted, the type of radiation, and the proximity of the contamination to organs of the body. It is important to be clear that the contamination gives rise to the radiation hazard, and the terms "radiation" and "contamination" are not interchangeable (Delves 2007).

The sources of radioactive pollution can be classified into two groups: natural and man-made. Following an atmospheric nuclear weapon discharge or a nuclear reactor containment breach, the air, soil, people, plants, and animals in the vicinity will become contaminated by nuclear fuel and fission products. A spilled vial of radioactive material like uranyl nitrate may contaminate the floor, and many rags are used to wipe up the spill. Cases of widespread radioactive contamination include the Bikini Atoll, the Rocky Flats Plant in Colorado, the Fukushima Daiichi nuclear disaster, the Chernobyl disaster, and the area around the Mayak facility in Russia.

2.3   Sources of Contamination

The sources of radioactive pollution can be natural or man-made. Radioactive contamination can be due to a variety of causes. It may occur due to the release of radioactive gases, liquids, or particles. For example, if a radionuclide used in nuclear medicine is spilled (accidentally or, as in the case of the Goiânia accident, through ignorance), the material could be spread by people as they walk around.

Radioactive contamination may also be an inevitable result of certain processes, such as the release of radioactive xenon in nuclear fuel reprocessing. In cases that radioactive material cannot be contained, it may be diluted to safe concentrations. Nuclear fallout is the distribution of radioactive contamination by the 520 atmospheric nuclear explosions that took place from the 1950s to the 1980s.

In nuclear accidents, a measure of the type and amount of radioactivity released, such as from a reactor containment failure is known as the **source term**. The United States Nuclear Regulatory Commission defines this as, *types and amounts of radioactive or hazardous material released to the environment following an accident* (Wilson 2017).

Contamination does not include residual radioactive material remaining at a site after the completion of decommissioning. Therefore, radioactive material in sealed and designated containers is not properly referred to as contamination, although the units of measurement might be the same.

3   Radiation Contamination Hazards

3.1   Low-Level Contamination

The hazards to people and the environment from radioactive contamination depend on the nature of the radioactive contaminant, the level of contamination, and the extent of the spread of contamination. Low levels of radioactive contamination pose little risk, but can still be detected by radiation instrumentation. If a survey or map is made of a contaminated area, random sampling locations may be labeled with their activity in becquerels or curies on contact. Low levels may be reported in counts per minute using a scintillation counter.

In the case of low-level contamination by isotopes with a short half-life, the best course of action may be to simply allow the material to naturally decay. Longer-lived isotopes should be cleaned up and properly disposed of because even a very low level of radiation can be life-threatening when in long exposure to it. Facilities and physical locations that are deemed to be contaminated may be cordoned off by a health physicist and labeled "Contaminated area." Persons coming near such an area would typically require anti-contamination clothing ("anti-Cs") (International Atomic Energy Agency 2005).

3.2   High-Level Contamination

High levels of contamination may pose major risks to people and the environment. People can be exposed to potentially lethal radiation levels, both externally and internally, from the spread of contamination following an accident (or a deliberate initiation) involving large quantities of radioactive material. The biological effects of external exposure to radioactive contamination are generally the same as those from an external radiation source not involving radioactive materials, such as X-ray diagnostic machines, and are dependent on analyzing signals from the absorbed dose.

When radioactive contamination is being measured or mapped *in situ*, any location that appears to be a point source of radiation is likely to be heavily contaminated. A highly contaminated location is colloquially referred to as a "hot spot." On a map of a contaminated place, hot spots may be labeled with their "on contact" dose rate in mSv/h. In a contaminated facility, hot spots may be marked with a sign, shielded with bags of lead shot, or cordoned off with warning tape containing the radioactive trefoil symbol. The hazard from contamination is the emission of ionizing radiation. The principal radiations that will be encountered are alpha, beta, and gamma; but these have quite different characteristics. They have widely differing penetrating powers and radiation effects, and the (Fig. 2) shows the penetration of these radiations in simple terms. For an understanding of the different ionizing effects of these radiations and the weight factors applied. Radiation monitoring involves the measurement of radiation dose or radionuclide contamination for reasons related to the assessment or control of exposure to radiation or radioactive substances, and the interpretation of the results (International Atomic Energy Agency 2005).

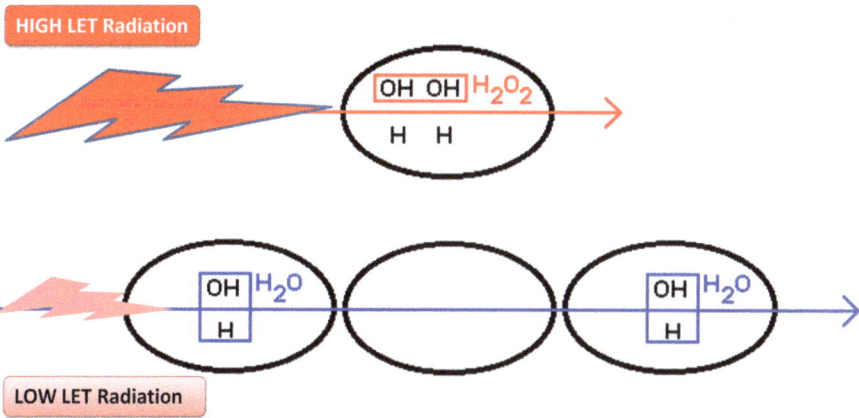

Fig. 2 Low LET and high LET ionizing radiations

4 Ionization of Biological Material

The transfer of energy to a medium by either electromagnetic or particulate radiation may be sufficient to overcome the binding energy of an electron, and the electron may be ejected from the atom. This process is called ionization, and these events cause approximately 33 eV of energy to be deposited into the absorbing medium. Comparing with other types of radiation, ionizing radiation deposits a relatively large amount of energy into a small area. In fact, the 33 eV from one ionization is more than enough energy to disrupt the chemical bond between 2-carbon atoms. The mechanism by which incident radiation interacts with the medium to cause ionization may be direct or indirect. Electromagnetic radiations (X-rays and gamma photons) are indirectly ionizing, that is, they give up their energy in various interactions, and energy is then utilized to produce fast-moving charged particles, such as an electron. It is the electron that then secondarily may react with target molecules, such as oxygen and water. These particulate radiations are directly ionizing radiation which includes alpha and beta particles. Indirectly ionizing radiation includes X-rays, gamma rays, and neutrons and are almost always more penetrating than directly ionizing particulate radiation. The amount of energy deposited into the tissue can be measured as a function of distance along the track of radiations. Various types of ionizing radiations are divided into high-LET and low-LET radiation (LET, linear energy transfer) or the amount of energy deposited in a unit of track length (Fig. 2). Scientifically, it would be preferable to specify the LET of each radiation utilization in experiments. Unfortunately, only qualitative parameters are routinely used in experimentation. Overall comparison of the experiment is extremely difficult. A more general term, the relative biologic effectiveness (RBE), is normally reported (Prasad 1995; Mettler et al. 1992).

The biologic effectiveness of ionizing radiation is usually due to the rather localized deposition of energy, which may affect critical structures such as genetic material (e.g., DNA). The total amount of energy delivered in a lethal dose of radiation is, in fact, extremely small but is effectively utilized. A total body dose of 7 Gray (700 rads) represents an absorption of only 1 cal in a 70-kg man. In terms of temperature rise, this would represent an increase of less than 0.002 degree Celsius.

5 Radiation-Induced Free Radicals Formation in Biological System

Free radicals are neutral atoms or molecules having an unpaired number of electrons. When X-rays interact with water two types of free radicals are formed:

$$X-rays + H_2O \rightarrow \underset{(Hydrogen)}{H^*} + \underset{(Hydroxy)}{OH^*}$$

The recombination of free radicals yields the following:

$$H^* + H^* \rightarrow H_2$$

$$OH^* + OH* \rightarrow H_2O_2$$

The presence of an excess of oxygen during irradiation of cells allows the formation of additional free radicals:

$$H^* + O_2 \rightarrow HO_2^* \,(\text{hydroperoxy free radical})$$

$$HO_2^* + HO_2^* \rightarrow H_2O_2 + O_2$$

When organic molecules (RH) combine with hydroxy free radicals, the organic free radical (R*) is formed.

$$RH + OH^* \rightarrow R^* + H_2O$$
$$(\text{organic})$$

The organic free radicals (R*) then combine with O_2 to form peroxy (RO$_2$*) free radicals:

$$R^* + O_2 \rightarrow RO_2^* \,(\text{peroxy free radical})$$

Thus, the presence of excess oxygen allows the formation of two additional free radicals, hydroperoxy (HO$_2$*) and peroxy (RO$_2$*). This may, in part, account for the increased radiation damage in the presence of excess of oxygen (Prasad 1995).

Because free radicals contain unpaired electrons, they are very reactive and can oxidize or reduce the biological molecules within the cell. The free radicals OH* and HO$_2$* are oxidizing agents, whereas H* is a reducing agent. Free radicals can damage the molecules such as DNA, RNA, and protein as well as membranes. Free radicals have been implicated in the etiology of cancer as well as in neurodegenerative diseases. Free radicals are highly reactive and have a life span of 10-9 to 10-11. They are constantly generated endogenously through several metabolic processes in a cellular system, and their level in the cell is maintained by homeostasis mechanisms. Ionizing radiation is the major exogenous source of free-radical production, and their exposure to the cellular system leads to imbalances in the homeostasis system; as a result, several malfunctions and deleterious effects are observed in the cellular system. Free radicals are generally categorized into two groups, viz, reactive oxygen species and reactive nitrogen species. Both the species are highly reactive and can damage the macromolecules of the cells leading to several types of diseases or cell death. When cells are irradiated, the damage is pronounced primarily by ionization and free radicals. It has been estimated that about two-thirds of biological damage by low LET radiation is due to an indirect effect. Biological damage by high LET is primarily by ionization (Fig. 2). The extent of radiation damage of a given cell type depends upon the total dose, dose rate, type of radiation, mode of radiation delivery, and the environmental condition of the medium. Radiation effects may be early or late and depend upon the type of damage. Precisely many have reviewed the effect of radiation-induced free radicals in body systems, and the figure (Fig. 3) we adapted is from the Wei group's review (Wei et al. 2019). It highlights that radiation-induced tissue damage elicits both ROS and RNS free radicals. They can upregulate various enzymes that follow epigenetic regulation of radiation-induced damage.

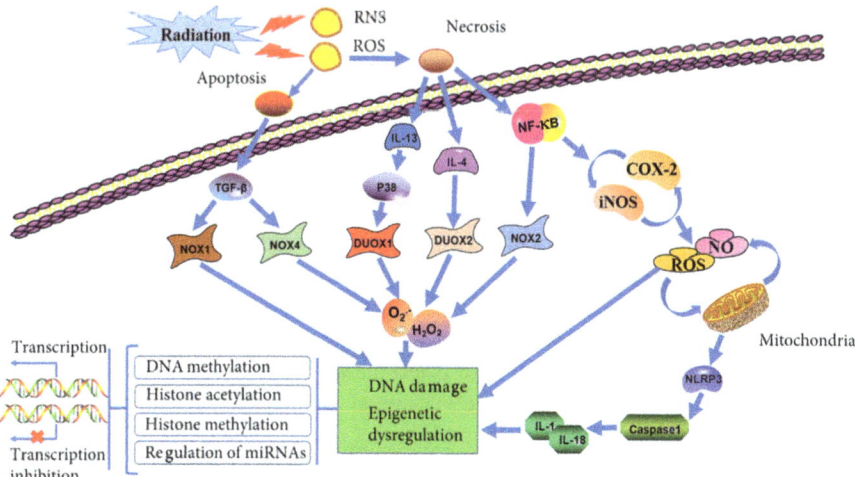

Fig. 3 Radiation induced in epigenetically regulated events that culminate into tissue damage. (Figure as available from Wei's group (Wei et al. 2019))

6   Biological Effects of Radiation

Radiation produces its effect in two ways – (a) Direct effect: radiation interacts directly to macromolecules of the cell; (b) Indirect effect: radiation produces its effect through the generation of free radicals by interacting with water molecules within the cell.

6.1   Effect of Radiation on Proteins

It is increasingly evident that protein molecules are critical targets of free-radical attack both intra- and extracellular because many of them play an important role as enzymes catalyzing diverse important reactions and transportation of different molecules present across membranes (Beal 2002). These physical changes in the proteins after interaction with free radicals are characterized into three groups: (1) fragmentation, (2) aggregation, and (3) susceptibility to proteolysis degradation (Yu 1994). Oxidation of the critical -SH group (sulfhydryl) present in the calcium transport system located in the endoplasmic reticulum, mitochondria, and plasma membrane may promote an increase in Ca^{2+} levels in the cytosol. The elevated calcium levels have been shown to have a number of deleterious effects including induction of cell death (Franklin and Johnson Jr 1992).

6.2   Effects of Radiation on Membrane Lipids

Several studies have supported the idea that membrane damage induced by radiation is a critical event (Agrawal and Kale 2001). Cregan and coworkers have suggested that damage to membrane organization is the initial step in triggering cell death (Cregan et al. 1999). A correlation has been observed between unrepaired membrane damage and loss of clonogenicity in cells. Lipid peroxidation is a highly destructive process and brings about change in structure, fluidity, and permeability of membranes, ultimately altering the structure and function of the cellular membrane, loss of -SH groups, and inactivation of a number of membrane-bound enzymes and receptors; induces swelling and alterations of respiratory functions; mediates DNA damage; and alters RNA transport from the nucleus to cytoplasm (Agrawal and Kale 2001). Spontaneous oxidation of lipid molecules in the membrane by oxygen at room temperature is termed as lipid per oxidation. Among different components of the membrane system, the phospholipid component of the cellular membrane is a highly vulnerable target of LPO due to its polyunsaturated fatty acid side chains. Lipid per oxidation is a free radical-mediated chain reaction and involves three distinct steps, i.e., initiation, propagation, and termination.

6.3 Effect of Radiation on Genetic Material

Among different biomolecules, damage to DNA has been shown to be most important and contribute maximally to cell death (Sutherland et al. 2000; Olinski et al. 2002). Studies made on DNA irradiated *in vitro* in solution, in the dry state or *in vivo* in the biological system have revealed that radiation causes a spectrum of damages to DNA (Reisz et al. 2014).

6.4 External Irradiation

This is due to radiation from contamination located outside the human body. The source can be in the vicinity of the body or can be on the skin surface. The level of health risk is dependent on the duration and the type and strength of irradiation. Penetrating radiation such as gamma rays, X-rays, neutrons, and beta particles pose the greatest risk from an external source. Low penetrating radiation such as alpha particles have a low external risk due to the shielding effect of the top layers of the skin (Johnson and Thaul 1997).

6.5 Internal Irradiation

Radioactive contamination can be ingested into the human body if it is airborne or is taken in as contamination of food or drink, and will irradiate the body internally. The art and science of assessing internally generated radiation dose is **internal dosimetry**.

The biological effects of ingested radionuclides depend greatly on the activity, the bio-distribution, and the removal rates of the radionuclide, which in turn depends on its chemical form, particle size, and route of entry. Effects may also depend on the chemical toxicity of the deposited material, independent of its radioactivity. Some radionuclides may be generally distributed throughout the body and rapidly removed, as is the case with tritiated water.

Some elements show affinitive binding and absorptivity to certain organs, and given this, their radionuclide variants would interpret longer stays over these organs. This action may lead to much lower removal rates. For instance, the thyroid gland takes up a large percentage of any iodine that enters the body. Large quantities of inhaled or ingested radioactive iodine may impair or destroy the thyroid, while other tissues are affected to a lesser extent. Radioactive iodine-131 is a common fission product; it was a major component of the radioactivity released from the Chernobyl disaster, leading to nine fatal cases of pediatric thyroid cancer and hypothyroidism. On the other hand, radioactive iodine is used in the diagnosis and treatment of many diseases of the thyroid precisely because of the thyroid's selective uptake of iodine (Mettler et al. 1992).

The radiation risk proposed by the International Commission on Radiological Protection (ICRP) predicts that an effective dose of one sievert (100 rem) carries a 5.5% chance of developing cancer. Such a risk is the sum of both internal and external radiation doses (Protection 2007).

The ICRP states, "Radionuclides incorporated in the human body irradiate the tissues over time periods determined by their physical half-life and their biological retention within the body" (Paquet et al. 2015). Thus, they may give rise to extended dosage to body tissues for many months or years after the intake. The need to regulate exposures to radionuclides and the accumulation of radiation dose over extended periods of time has led to the definition of "committed dose quantities." The ICRP further states, "For internal exposure, committed effective doses are generally determined from an assessment of the intakes of radionuclides from bioassay measurements or other quantities (e.g., activity retained in the body or in daily excreta). The radiation dose is determined from the intake using recommended dose coefficients" (Protection 2007; Upton 2010).

7 Social and Psychological Effects

A 2015 report in Lancet explained that serious impacts of nuclear accidents were often not directly attributable to radiation exposure, but rather social and psychological effects (Hasegawa et al. 2015). The consequences of low-level radiation are often more psychological than radiological. Because damage from very-low-level radiation cannot be detected, people exposed to it are left in anguished uncertainty about the future outcome. Many believe that they have been fundamentally contaminated for life and may refuse to have children for fear of birth defects. They may be shunned by others in their community who fear a sort of mysterious contagion (Revkin 2011).

Forced evacuation from a radiological or nuclear accident may lead to social isolation, anxiety, depression, psychosomatic medical problems, reckless behavior, and even suicide. Such was the outcome of the 1986 Chernobyl nuclear disaster in Ukraine. A comprehensive 2005 study concluded that "the mental health impact of Chernobyl is the largest public health problem unleashed by the accident to date" (Revkin 2011). Frank N. von Hippel, a US scientist, commented on the 2011 Fukushima nuclear disaster, saying that "fear of ionizing radiation could have long-term psychological effects on a large portion of the population in the contaminated areas" (Von Hippel 2011). Evacuation and long-term displacement of affected populations create problems for many people, especially the elderly and hospital patients (Hasegawa et al. 2015).

Such graver psychological danger does not accompany other materials that put people at risk of cancer and other deadly illness. Visceral fear is not widely aroused by, for example, the daily emissions from coal burning, although, as a National Academy of Sciences study found, this causes 10,000 premature deaths a year in the US population of 317,413,000. Medical errors leading to death in US hospitals are estimated to be between 44,000 and 98,000. It is "only nuclear radiation that bears a huge psychological burden – for it carries a unique historical legacy" (Revkin 2011).

8 Radiation Safety

The objective of radiation safety is to eliminate or minimize the harmful effects of ionizing radiation and radioactive material on workers, the public, and the environment while allowing their beneficial uses (Council CoHEoEtRNR 1994). Most radiation safety programs will not have to implement every one of the elements described below. The design of a radiation safety program depends on the types of ionizing radiation sources involved and how they are used.

8.1 Radiation Safety Principles

The International Commission on Radiological Protection (ICRP) has proposed that the following principles should guide the use of ionizing radiation and the application of radiation safety standards:

I. No practice involving exposures to radiation should be adopted unless it produces sufficient benefit to the exposed individuals or to society to offset the radiation detriment it causes.
II. In relation to any particular source within a practice, the magnitude of individual doses, the number of people exposed, and the likelihood of incurring exposures where these are not certain to be received should all be kept as low as reasonably achievable (ALARA), economic and social factors being taken into account. This procedure should be constrained by restrictions on the doses to individuals (dose constraints), so as to limit the inequity likely to result from the inherent economic and social judgments (the optimization of protection).
III. The exposure of individuals resulting from the combination of all the relevant practices should be subject to dose limits, or to some control of risk in the case of potential exposures. These are aimed at ensuring that no individual is exposed to radiation risks that are judged to be unacceptable from these practices in any normal circumstances. Not all sources are susceptible to control by action at the source and it is necessary to specify the sources to be included as relevant before selecting a dose limit (individual dose and risk limits) (Council CoHEoEtRNR 1994).

8.2 Dosimetry

Dosimetry is used to indicate the dose equivalents that workers receive from external radiation fields to which they may be exposed. Dosimeters are characterized by the type of device, the type of radiation they measure, and the portion of the body for which the absorbed dose is to be indicated. Three main types of dosimeters that are most commonly employed are thermoluminescent dosimeters, film dosimeters, and ionization chambers (Parwaie et al. 2018). Other types of dosimeters include fission foils, track-etch devices, and plastic "bubble" dosimeters (Turner 2008).

Thermoluminescent dosimeters are the most commonly used type of personnel dosimeter. They take advantage of the principle that when some materials absorb energy from ionizing radiation, they store it such that later it can be recovered in the form of light when the materials are heated. To a high degree, the amount of light released is directly proportional to the energy absorbed from the ionizing radiation and hence to the absorbed dose the material received. This proportionality is valid over a very wide range of ionizing radiation energy and absorbed dose rates.

Film was the most popular material for personnel dosimetry before thermoluminescent dosimetry became common. The degree of film darkening depends on the energy absorbed from the ionizing radiation, but the relationship is not linear. The dependence of film response on the total absorbed dose, absorbed dose rate, and radiation energy is greater than that for thermoluminescent dosimeters and can limit film's range of applicability. However, film has the advantage of providing a permanent record of the absorbed dose to which it was exposed.

Self-reading, small ionization chambers, also called pocket chambers, are used to obtain immediate dosimetry information. Their use is often required when personnel must enter high or very high radiation areas, where personnel could receive a large absorbed dose in a short period of time. Pocket chambers often are calibrated locally, and they are very sensitive to shock. Consequently, they should always be supplemented by thermoluminescent or film dosimeters, which are more accurate and dependable but do not provide immediate results.

Dosimetry is required for a worker when he or she has a reasonable probability of accumulating a certain percentage, usually 5 or 10%, of the maximum permissible dose equivalent for the whole-body or certain parts of the body (Radiation Protection Procedures 1973).

8.3   Bioassay

Bioassay (also called radiobioassay) means the determination of kinds, quantities, or concentrations, and, in some cases, the locations of radioactive material in the human body, whether by direct measurement (in vivo counting) or by analysis and evaluation of materials excreted or removed from the human body. Bioassay is usually used to assess worker dose equivalent due to radioactive material taken into the body. It also can provide an indication of the effectiveness of active measures taken to prevent such intake. More rarely it may be used to estimate the dose a worker received from massive external radiation exposure. Bioassay must be performed when a reasonable possibility exists that a worker may receive into the body more than a certain percentage (usually 5 or 10%) of the all for a radionuclide. The chemical and physical form of the radionuclide sought in the body determines the type of bioassay necessary to detect it (Radiation Protection Procedures 1973; World Health Organization 1969).

Bioassay can consist of analyzing samples taken from the body (e.g., urine, feces, blood, or hair) for radioactive isotopes. In this case, the amount of radioactivity in the sample can be related to the radioactivity in the person's body and subsequently to the radiation dose that the person's body or certain organs have received or are committed to receive. Urine bioassay for tritium is an example of this type of bioassay. Bioassay can be performed in-house or samples or personnel can be sent to a facility or organization that specializes in the bioassay to be performed. In either case, proper calibration of equipment and accreditation of laboratory procedures is essential to ensure accurate, precise, and defensible bioassay results (Doll et al. 1994).

8.4 Protective Clothing

Protective clothing is supplied by the employer to the worker to reduce the possibility of radioactive contamination of the worker or his or her clothing or to partially shield the worker from beta, X-ray, or gamma radiation. Examples of the former are anticontamination clothing, gloves, hoods, and boots. Examples of the latter are leaded aprons, gloves, and eyeglasses (World Health Organization 1969).

8.5 Environmental Monitoring

Environmental monitoring refers to collecting and measuring environmental samples for radioactive materials and monitoring areas outside the environs of the workplace for radiation levels (Mukhopadhyay et al. 2016). Purposes of environmental monitoring include estimating consequences to humans resulting from the release of radionuclides to the biosphere, detecting releases of radioactive material to the environment before they become serious, and demonstrating compliance with regulations.

9 Lessons from Radiation-Tolerant Life-Forms: Biotechnology Would Help at Mitigating and Managing Radiation Damage-Led Detriments?

Various studies in different life forms have also show evidences with absorption, tolerance, bioconversion of radiation, and/or radiation-induced damages. A peculiar example is shown by the water bears or the tardigrades, one of the microscopic metazoans famed to be known as the most radiation-tolerant life forms on the planet. They are known to withstand both low (X-rays and gamma rays) (Beltrán-Pardo et al. 2015; Ingemar Jönsson et al. 2005; Horikawa et al. 2006; May 1964) and high extremes of LET (alpha particles, protons) (Horikawa et al. 2006; Charlotta Nilsson et al. 2010), and UV ionizing radiations (Altiero et al. 2011; Horikawa et al. 2013).

In fact, they have been taken to the vacuum of the open space and have evidenced survival to combined pressures from comic and UV radiations, happily tolerating numerously thousand Grays of dose (Jönsson et al. 2008). A dosage of 5 grays of such exposures even is life-threatening on the other hand for humans (Jönsson et al. 2013). Tardigrades are known to thrive in mosses and are playful in thin films of water. Under gradual desiccation in the niche they thrive in, they are amenable to exhibit a cryptobiotic lifestyle by removing almost 99% of body water turning into rock-like *tun* structures wherein all their metabolism halts and the metabolic machinery with components preserved variously by secretion of trehalose, intrinsi-cally disordered proteins (IDPs), etc. Cryptobiosis helps a tardigrade variously to counter stressful extremes of temperature, desiccation, organic solvents, freezing, pressures, etc. (Kinchin 1994; Wright 2001). Other studies elaborate the presence of multiple heat shock proteins, more than mere duplications of the inherent hox gene clusters, and surprisingly latter , which also introduced horizontally from other extremophiles (Boothby et al. 2015). A group from Japan reported that mitochon-dria-targeted heat shock proteins of *Ramazzottius varieornatus* can improve osmo-tolerance in human cells (Tanaka et al. 2015). This group also reported that the occurrence of a DNA-binding protein in *R. varionatus* and more surprisingly that these proteins could suppress X-ray–induced DNA damage in cultured human cells by 40% (Hashimoto et al. 2016). Tardigrades can survive lethal dosage of radiations in both active and tun states. The bases to radiation tolerance are known to be adapted from those related to their desiccation tolerance feature. For example, in one study, hydrated and active *Echiniscoides sigismundi* survived up to 4000 Gy of gamma radiation for 7 days of exposure (Jönsson et al. 2016). Tolerance, however, as seen in tardigrades also comes following a slower rejuvenation in context to developmental and maturation delays which has been reasoned as required for cer-tain unique damage repair machinery in them. Very recently, a group from India has reported that a eutardigrade, *Paramacrobiotus sp.,* exhibits natural fluorescence that shields it from UV radiation (Suma et al. 2020). Moreover, they state that these tardigrades were comparatively more UV tolerant (surviving 30 days with just 15 minutes of exposure) than the sensitive *Hypsibius exemplaris* radiation (which died within 24 hrs of the same UV dosage). Surprisingly they also showed that the pho-tobleached fluorescent extract from this UV tolerant tardigrade can confer similar protective benefits to the UV-sensitive species. This can be supported by some stud-ies that have inferred the attributes of certain radiation-absorbing primary and sec-ondary metabolites such as the scytinemin, shinorine, playthine, biopterin, porphyra-334, mycrosporin, phlorotannin, etc. found in radiation-tolerant extremo-philic microbes. Other studies in tardigrades confirm the presence of certain DNA-associating proteins (Dsup), which are capable of suppressing radiation-induced DNA breaks in cultured human cells.(Hashimoto et al. 2016)

Other than the tardigrades, the microbial world offers many radiation-tolerant entities as well. An extremophilic bacterium, *Deinococcus radiodurans,* also falls under the category of the most radiation-tolerant organism. Like some of the tardi-grades, it can be considered as polyextremophile and also has been listed as the world's toughest bacterium in the Guinness Book of world records (DeWeerdt

2002). As studied, it can repair both single- and double-stranded DNA breaks (Madigan et al. 2008); can survive for three years in outer space (Ott et al. 2020). No doubt that this microbe can withstand oxidative damage from reactive oxygen species as per another study (Slade and Radman 2011), which supports the protection from radiation-induced detriments. It can withstand an acute dosage of 5000 Gy of ionizing radiation without any viability loss. In fact, somehow they are even more robust than the tardigrades in that they can withstand 15,000 Gy dose with just 37% viability. Some researchers report that manganese complexation within this bacterium protects its proteins from being oxidized due to radiation damages (Pearson 2004). Antioxidant systems in such microbes can be used to study the effect of radiation-triggered generation of ROS in human cells, as well as how certain tumors become radiation resistant (Vishambra et al. 2019; Rew 2003). Radiation tolerance in microbial entities may follow various defensive mechanisms like altering intracellular ionic strength, DNA repair, and use of antioxidants and radiation-absorbing secondary metabolites.

10   Bioremediation of Radionuclides

The above examples of life forms reportedly showcase overwhelming tolerance and/or resistance to ionizing radiations. However, complex mechanisms that drive such responses remain yet unresolved. Sizable literature in these contexts indeed exists for the microorganisms, thus vouching their applicability in the bioremediation of toxic pollutants from soil and water. These microorganisms are well known to degrade or detoxify a wide range of toxic pollutants from the environment, like polycyclic aromatic hydrocarbons, toxic heavy metals, pesticides, and other xenobiotics (Singh et al. 2013; Singh et al. 2011; Verma and Sharma 2017; Al-Tohamy et al. 2020; Parte et al. 2017). Nonetheless, microorganisms have also proven themselves as good candidates for the bioremediation of radioactive wastes (Shukla et al. 2017). Various biological approaches used by microorganisms to remediate radionuclides have been depicted in (Fig. 4). Biosorption, intracellular accumulation, biotransformation, biosolubilization, and biomineralization are the fundamental mechanisms used by microorganisms to variously mitigate radionuclides. Biosorption here embraces both the process of adsorption on the cell surface and absorption (internalization). Bioaccumulation of toxic pollutants by microorganisms depends on the adsorption of radioactive waste. However, compared to biosorption, microbial accumulation is a slow process. Biotransformation (redox), another important phenomenon, results in the change in solubility of radionuclides. Oxidized forms of radionuclides are hydrophilic (soluble) while reduction of nuclides results in precipitation.

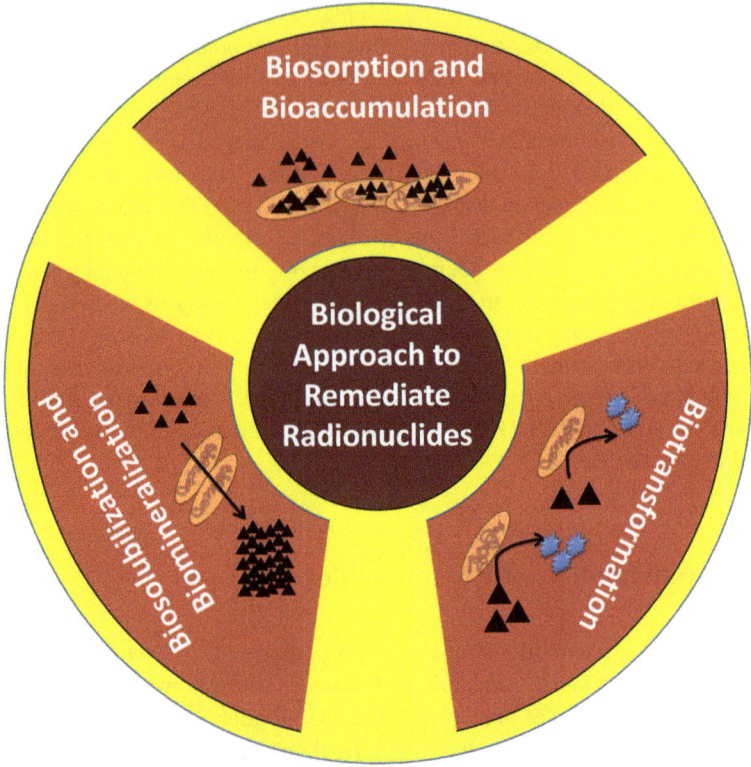

Fig. 4 Various biological approaches toward bioremediating radionuclides

11 Conclusive Comments

Dosage of ionizing radiations has been considered while assessing the effects on life forms and the environment. The need for the applications of such radiations in various industrial, medical, and energy sectors has underrated the effects of long-term exposure to trace dosages. However, as supported by many, such continued exposures may pose critical health impacts on humans, plants, and other entities of the environment. There have been many physical and chemical ways outlined above to provide protective measures at managing and controlling such impacts. Biotechnology seeks at finding additional and more consistent solutions at this as well. Studies are underway on how some organisms are able to tolerate, resist, and convert radiation entities and their effects and to bioprospect these surprises into favorable technological solutions for the needful. The potential usage of such microbes in biotech industries has been recently reviewed (Jung et al. 2017; Gabani and Singh 2013). The probable mechanisms by which the protection is offered to the tardigrade, and/ or especially the *D. radiodurans* from its genetic or metabolic arsenal is yet unknown. But these may carry potential biotechnological prospects for humans toward acquiring protection from harmful UV and other ionizing radiations.

Acknowledgments The authors thank the University Institute of Biotechnology (UIBT) and the University Center of Research and Development (UCRD) at Chandigarh University for academic support. AS carries expertise acquired from works involving radioprotective drugs at INMAS-DRDO labs, as well as he had associated with a group working on a gamma-irradiated mutant fungus for enhanced transesterification, at BARC and Thapar University. KC is working as a JRF with the principal investigator, GM, awarded with a grant (EMR/2016/008022/AS) from the Department of Science and Technology (DST-SERB) for a project that deals with the biodiversity of North-Western Tardigrades in India. GBS works in the theme projects dealing with "bioremediation for bettering environment" at UIBT. Views highlighted here are of authors and do not relate to any mandated activity at Chandigarh University till this manuscript was being shaped. The authors declare not having any conflict of interest.

References

Agrawal A, Kale R. Radiation induced peroxidative damage: mechanism and significance. 2001.

Altiero T, Guidetti R, Caselli V, Cesari M, Rebecchi L. Ultraviolet radiation tolerance in hydrated and desiccated eutardigrades. J Zool Syst Evol Res. 2011;49:104–10.

Al-Tohamy R, Sun J, Fareed MF, Kenawy E-R, Ali SS. Ecofriendly biodegradation of Reactive Black 5 by newly isolated Sterigmatomyces halophilus SSA1575, valued for textile azo dye wastewater processing and detoxification. Sci Rep. 2020;10(1):1–16.

Beal MF. Oxidatively modified proteins in aging and disease. Free Radic Biol Med. 2002;32(9):797–803.

Beltrán-Pardo E, Jönsson KI, Harms-Ringdahl M, Haghdoost S, Wojcik A. Tolerance to gamma radiation in the tardigrade Hypsibius dujardini from embryo to adult correlate inversely with cellular proliferation. PLoS One. 2015;10(7):e0133658.

Block S. Radiation safety for x-ray diffraction and fluorescence analysis equipment. American National Standard: American National Standards Inst. 1978.

Boothby TC, Tenlen JR, Smith FW, Wang JR, Patanella KA, Osborne Nishimura E, et al. Evidence for extensive horizontal gene transfer from the draft genome of a tardigrade. Proc Natl Acad Sci. 2015;112(52):15976–81. https://doi.org/10.1073/pnas.1510461112.

Charlotta Nilsson E, Ingemar Jönsson K, Pallon J. Tolerance to proton irradiation in the eutardigrade Richtersius coronifer–a nuclear microprobe study. Int J Radiat Biol. 2010;86(5):420–7.

Council CoHEoEtRNR. Health effects of exposure to radon: time for reassessment? National Academy Press; 1994.

Cregan SP, MacLaurin JG, Craig CG, Robertson GS, Nicholson DW, Park DS, et al. Bax-dependent caspase-3 activation is a key determinant in p53-induced apoptosis in neurons. J Neurosci. 1999;19(18):7860–9.

de Saint-Georges L. Environmental ionizing radiation. Hazardous Waste Manag. 2009;9:237.

Delves D. International Atomic Energy Agency. IAEA safety glossary: terminology used in nuclear safety and radiation protection. Vienna: International Atomic Energy Agency; 2007. p. 227.

DeWeerdt S. The world's toughest bacterium: deinococcus radiodurans may be a tool for cleaning up toxic waste and more. Genome News. 2002;

Doll R, Evans H, Darby S. Paternal exposure not to blame. Nature. 1994;367(6465):678–80.

Franklin JL, Johnson EM Jr. Suppression of programmed neuronal death by sustained elevation of cytoplasmic calcium. Trends Neurosci. 1992;15(12):501–8.

Gabani P, Singh OV. Radiation-resistant extremophiles and their potential in biotechnology and therapeutics. Appl Microbiol Biotechnol. 2013;97(3):993–1004.

Goodhead D. Spatial and temporal distribution of energy. Health Phys. 1988;55(2):231–40.

Hall E. Predictive assay. Radiobiology for the radiologist. Philadelphia: JB Lippincott Company; 1994. p. 245–56.

Hasegawa A, Tanigawa K, Ohtsuru A, Yabe H, Maeda M, Shigemura J, et al. Health effects of radiation and other health problems in the aftermath of nuclear accidents, with an emphasis on Fukushima. Lancet. 2015;386(9992):479–88.

Hashimoto T, Horikawa DD, Saito Y, Kuwahara H, Kozuka-Hata H, Shin T, et al. Extremotolerant tardigrade genome and improved radiotolerance of human cultured cells by tardigrade-unique protein. Nat Commun. 2016;7

Haynie J, Olsher R. A summary of X-ray machine exposure accidents at the Los Alamos National Laboratory. Los Alamos Unclassified Publications (LAUP). 1981.

Horikawa DD, Sakashita T, Katagiri C, Watanabe M, Kikawada T, Nakahara Y, et al. Radiation tolerance in the tardigrade Milnesium tardigradum. Int J Radiat Biol. 2006;82(12):843–8.

Horikawa DD, Cumbers J, Sakakibara I, Rogoff D, Leuko S, Harnoto R, et al. Analysis of DNA repair and protection in the Tardigrade Ramazzottius varieornatus and Hypsibius dujardini after exposure to UVC radiation. PLoS One. 2013;8(6):e64793.

Ingemar Jönsson K, Harms-Ringdahl M, Torudd J. Radiation tolerance in the eutardigrade Richtersius coronifer. Int J Radiat Biol. 2005;81(9):649–56.

International Atomic Energy Agency. Environmental and source monitoring for purposes of radiation protection. Safety Guide RSG1. 8. International Atomic Energy Agency; 2005.

Johnson JC, Thaul S. An evaluation of radiation exposure guidance for military operations: interim report. 1997.

Jönsson KI, Rabbow E, Schill RO, Harms-Ringdahl M, Rettberg P. Tardigrades survive exposure to space in low Earth orbit. Curr Biol. 2008;18(17):R729–R31.

Jönsson I, Beltran-Pardo E, Haghdoost S, Wojcik A, Bermúdez-Cruz RM, Bernal Villegas JE, et al. Tolerance to gamma-irradiation in eggs of the tardigrade Richtersius coronifer depends on stage of development. J Limnol. 2013;72(s1):73–9.

Jönsson KI, Hygum TL, Andersen KN, Clausen LKB, Møbjerg N. Tolerance to gamma radiation in the marine heterotardigrade, Echiniscoides sigismundi. PLoS One. 2016;11(12):e0168884. https://doi.org/10.1371/journal.pone.0168884.

Jung K-W, Lim S, Bahn Y-S. Microbial radiation-resistance mechanisms. J Microbiol. 2017;55(7):499–507.

Kinchin IM. The biology of tardigrades. Portland Press; 1994.

Madigan MT, Martinko JM, Dunlap PV, Clark DP. Brock biology of microorganisms 12th edn. Int Microbiol. 2008;11:65–73.

May R. Action différentielle des rayons x et ultraviolets sur le tardigrade Macrobiotus areolatus, a l'état actif et desséché. Bull Biol France Belgique. 1964;98:349–67.

Mettler FA, Williamson MR, Royal HD, Hurley JR, Khafagi F, Sheppard MC, et al. Thyroid nodules in the population living around Chernobyl. JAMA. 1992;268(5):616–9.

Mukhopadhyay S, Halligan J, Hastak M. Assessment of major causes: nuclear power plant disasters since 1950. Int J Disaster Resil Built Environ. 2016;

Olinski R, Gackowski D, Foksinski M, Rozalski R, Roszkowski K, Jaruga P. Oxidative DNA damage: assessment of the role in carcinogenesis, atherosclerosis, and acquired immunodeficiency syndrome. Free Radic Biol Med. 2002;33(2):192–200.

Ott E, Kawaguchi Y, Kölbl D, Rabbow E, Rettberg P, Mora M, et al. Molecular repertoire of deinococcus radiodurans after 1 year of exposure outside the international space station within the tanpopo mission. Microbiome. 2020;8(1):1–16.

Paquet F, Etherington G, Bailey MR, Leggett RW, Lipsztein J, Bolch W, et al. ICRP publication 130: occupational intakes of radionuclides: Part 1. Ann ICRP. 2015;44(2):5–188. https://doi.org/10.1177/0146645315577539.

Parte SG, Mohekar AD, Kharat AS. Microbial degradation of pesticide: a review. Afr J Microbiol Res. 2017;11(24):992–1012.

Parwaie W, Refahi S, Ardekani MA, Farhood B. Different dosimeters/detectors used in small-field dosimetry: pros and cons. J Med Signals Sens. 2018;8(3):195–203. https://doi.org/10.4103/jmss.JMSS_3_18.

Pearson H. Secret of radiation-proof bugs proposed. Nature Publishing Group; 2004.

Prasad KN. Handbook of radiobiology. CRC Press; 1995.

Protection R. ICRP publication 103. Ann ICRP. 2007;37(2.4):2.

Radiation Protection Procedures. Safety Series No. 38. IAEA, Vienna. 1973.

Reisz JA, Bansal N, Qian J, Zhao W, Furdui CM. Effects of ionizing radiation on biological molecules--mechanisms of damage and emerging methods of detection. Antioxid Redox Signal. 2014;21(2):260–92. https://doi.org/10.1089/ars.2013.5489.

Revkin AC. Nuclear risk and fear, from hiroshima to fukushima. New York Times; 2011.

Rew D. Deinococcus radiodurans. Eur J Surg Oncol. 2003;29(6):557–8.

Shukla A, Parmar P, Saraf M. Radiation, radionuclides and bacteria: An in-perspective review. J Environ Radioact. 2017;180:27–35.

Singh GB, Srivastava A, Saigal A, Aggarwal S, Bisht S, Gupta S, et al. Biodegradation of carbazole and dibenzothiophene by bacteria isolated from petroleum-contaminated sites. Bioremed J. 2011;15(4):189–95.

Singh GB, Gupta S, Gupta N. Carbazole degradation and biosurfactant production by newly isolated Pseudomonas sp. strain GBS. 5. Int Biodeterior Biodegradation. 2013;84:35–43.

Slade D, Radman M. Oxidative stress resistance in Deinococcus radiodurans. Microbiol Mol Biol Rev. 2011;75(1):133–91.

Suma HR, Prakash S, Eswarappa SM. Naturally occurring fluorescence protects the eutardigrade *Paramacrobiotus* sp. from ultraviolet radiation. Biol Lett. 2020;16(10):20200391. https://doi.org/10.1098/rsbl.2020.0391.

Sutherland BM, Bennett PV, Sidorkina O, Laval J. Clustered DNA damages induced in isolated DNA and in human cells by low doses of ionizing radiation. Proc Natl Acad Sci. 2000;97(1):103–8.

Tanaka S, Tanaka J, Miwa Y, Horikawa DD, Katayama T, Arakawa K, et al. Novel Mitochondria-Targeted Heat-Soluble Proteins Identified in the Anhydrobiotic Tardigrade Improve Osmotic Tolerance of Human Cells. PLoS One. 2015;10(2):e0118272. https://doi.org/10.1371/journal.pone.0118272.

Turner JE. Atoms, radiation, and radiation protection. Wiley; 2008.

Upton AC. Carcinogenic effects of ionising radiation. Mechanisms of oncogenesis. Springer; 2010. p. 43-61.

Verma N, Sharma R. Bioremediation of toxic heavy metals: a patent review. Recent Pat Biotechnol. 2017;11(3):171–87.

Vishambra D, Jaiswalb V, Priyac V, Pair Wise Comparison of D. Radiodurans with radioresistant bacteria and non radioresistant bacteria on habitat basis. J Crit Rev. 2019;6(5):71–80.

Von Hippel FN. The radiological and psychological consequences of the Fukushima Daiichi accident. Bull At Sci. 2011;67(5):27–36.

Wei J, Wang B, Wang H, Meng L, Zhao Q, Li X, et al. Radiation-induced normal tissue damage: oxidative stress and epigenetic mechanisms. Oxidative Med Cell Longev. 2019;2019:3010342. https://doi.org/10.1155/2019/3010342.

Wilson GA. Nuclear regulatory commission. 2017.

World Health Organization. Handling of radiation accidents. Proceedings of a symposium on the handling of radiation accidents; 1969.

Wright JC. Cryptobiosis 300 years on from van Leuwenhoek: what have we learned about tardigrades? Zool Anz. 2001;240(3-4):563–82. https://doi.org/10.1078/0044-5231-00068.

Yu BP. Cellular defenses against damage from reactive oxygen species. Physiol Rev. 1994;74(1):139–62.

The Antioxidant Arsenal Against COVID-19

Gaurav Mudgal, Jaspreet Kaur, Kartar Chand, and Gajendra B. Singh

Contents

Abstract The world is foreseeing an ever-increasing human health crisis by virtue of threats from viral disease outbreaks recently showcased by COVID-19 and also those in the previous two decades by other coronaviruses. Many lives have been lost and various world economies have been constrained with the effect from the current pandemic. Corona-warriors, as the name is floating in the air, have also emerged in the shape of the medical and healthcare practitioners, fundraising charity trusts, and sanitation workforces, who are variously catering to the cause. On this mission, most scientific communities and pharmaceutical giants are also investing their best efforts toward studying the molecular biology of these pathogens and discovering and/or designing potential vaccine candidates against the notorious killers. The scenarios, however, have been felt graver with new mutant SARS-2 CoV strains hitting with sporadic emergence, complicating development and contradicting efficacies of these putative vaccines and standard therapies up close to dissemination into the public. Many viral and other pathogenic entities have been found to correlate their bases of pathogenesis and disease severity to, besides other factors, the development of pronounced oxidative stress. Abnormally generated and ill-regulated oxidative stress profiles are reported to effectuate severity, as seen in COVID-19 cases that

G. Mudgal (✉) · J. Kaur · K. Chand · G. B. Singh
University Institute of Biotechnology, Chandigarh University, Mohali, Punjab, India
e-mail: gaurav.mudgal@cumail.in

© The Author(s), under exclusive license to Springer Nature Switzerland AG 2021   327
K. K. Kesari, N. K. Jha (eds.), *Free Radical Biology and Environmental Toxicity*,
Molecular and Integrative Toxicology, https://doi.org/10.1007/978-3-030-83446-3_16

may trigger cytokine/chemokine storms, ill-orchestrated NETosis, pyroptosis, and others that culminate into symptoms like colitis, ARD/DIC, endothelial dysfunction, coagulopathies, and graver chances of sepsis. Here, we narrate important studies with some of the potent antioxidant molecules which have surfaced with their promises at directly and/or indirectly evading the chances of COVID-19 and possibly other viral diseases. We present a compelling view of the various mechanistic networks within which CoV-led oxidative stress could detriment the homeostatic harmony within the cell/tissue systems by interfering with inflammatory and immunological stimuli and how the antioxidants emerge as a highly sought therapy.

Keywords COVID-19 · Antioxidants · Free radical · ROS · Severity · Sepsis

1   Introduction

Homeostasis of a life-form is found over several factors and is a resultant of the checks and balances dealt with the physiological processes, metabolic turnovers, and to a sizeable extent vests with the mercy and discretion of Mother Nature. Pathogens have evolved distinct mechanistic modes to cause diseases in life forms, and similarly hosts' immune defenses to some extent exhibit co-evolution with them at either resisting, tolerating pathogenesis, or also abdicating the pathogens and/or their detrimental effects over the body systems. As with sustenance of body processes that eventually warrant its homeostasis stature, several processes or events contribute to shaping defense to foreign intrusions and imbalances that confer various inflammatory responses, disease symptoms, and/or syndromes. Among other disease-causing agents, most if not all viruses, have been regarded as champions at intruding into living systems and recruiting them as production factories scaling up their infection, spread, and very significantly their host range. The examples are many and to name a few among the deadliest human disease-causing viruses include the coronaviruses (CoV), Ebola virus, human immunodeficiency virus (HIV), Marburg virus, rotavirus, and others like dengue, and flu viruses, etc. Many of which have reported enormous death tolls globally, devastating economies and human health stature.

Advances in science and technology have unraveled subsequent insights into understanding disease biology and also have enabled orthodoxical approaches to treat body systems of one or more disease conditions. Molecular biology continuingly enables us to gain deeper insights into various mechanisms by virtue of which both the pathogens' and hosts' inherent systems showcase their abilities to surpass each other leading to either the disease or defense, respectively, and without doubts evolution into more fit biological entities. Modern medicine and treatment regimens require these pathobiological insights without which drug discoveries, innovation, delivery systems, as well as small scale *in vivo* and *in vitro* studies, and clinical trials would be for a namesake and treatment development like the one which is highly sought for the COVID-19 would get outrageously tricky. Infection by viruses of the host cell (internalization) renders its drug hunt more cumbersome as most potential targets would probably reside internal to cell sol except a few like entry receptors.

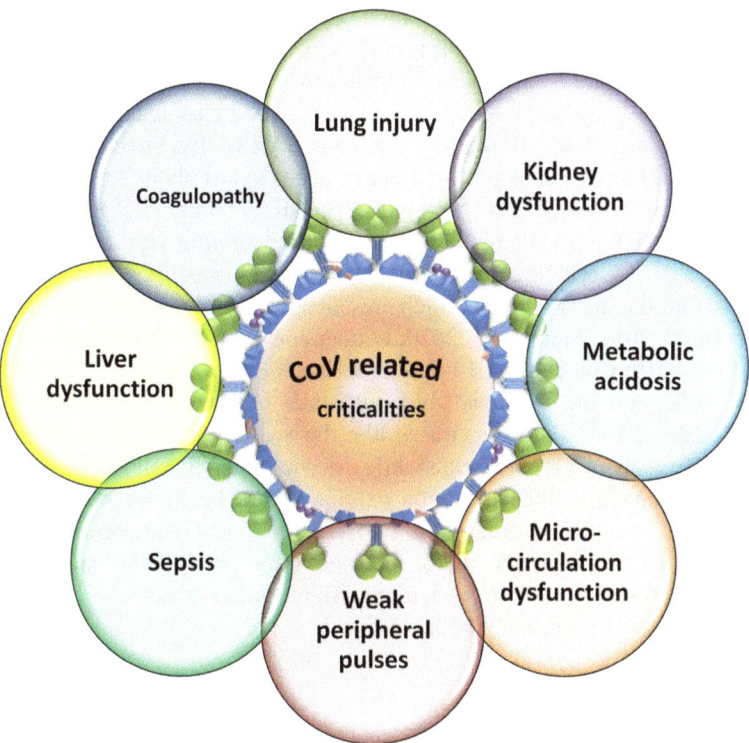

Fig. 1 COVID-19-associated health detriments

Infection by pathogens such as viruses as well as the host system defense responses as inflammation and immunity, both involve multistep and multifactorial reactions. Many from the pathogens' end are known to generate free radicals which in turn lead to reversible and/or irreversible reaction imbalances or disturbance to the housekeeping reactions and/or reaction cascades.

This chapter would focus on the role-play of free radicals relevant in mechanisms causing disease and severity post-coronavirus infection and that of promising antioxidant compounds, many of which carry potential efficacies and implications supporting the emergence of newer health wellness possibilities against CoV-related critical health issues (Fig. 1).

2 Coronavirus Outbreaks: Mild to Severe Health Detriments and Death Tolls

Coronaviruses or CoVs are enveloped, nonsegmented, single-stranded RNA genome viruses with the largest genomes of about 32 Kbs and are the largest group of viruses. CoV-related diseases are not new words of mouth but had been regarded as virology backwaters. They cause numerous pulmonary and gastroenteric infections in animals, including mammals, birds, and even fishes. Past global incidences have been encountered in the outbreaks of SARS-1 in 2003 (Ksiazek et al. 2003), MERS 2013 (CDC 2019), and the currently notorious SARS-2 CoV (causing the COVD19 disease) pandemic (Baric 2020). Especially the SARS-2 has infected more than 70 million global human subjects with a heavy death toll of about 1.6 million, which reminds a similar pandemic of 1918 with Spanish flu (Baric 2020). SARS-2 CoV-triggered COVID-19 pandemic has showcased devastating viral disease outbreak following a decade after the SARS of 2003. CoV outbreaks have followed zoonotic transmissions hailing from bats, which present as a natural reservoir (Mudgal 2014; Reguera et al. 2014; Munir et al. 2020; Leitner and Kumar 2020; Olival et al. 2020; Li and Du 2019; Chu et al. 2014; Lee and Hsueh 2020) However, comparative to that of 2003 and the previous mild outbreaks of not so nasty common cold (HCoV-229E, NL-63) and flu viruses like the swine flu, etc., the SARS-2 CoV majorly affects the adult population while the young population has reported lesser damage. Coronavirus-infected adults exhibit severe health problems posed by inflammations resulting in organ (respiratory, cardiac) and multiple-organ malfunctions, and also that of septic shocks corroborating the mortality status of about 7–15% (WHO 2020). Many conditions which could attribute to severity post COVID-19 causation are highlighted in Fig. 2.

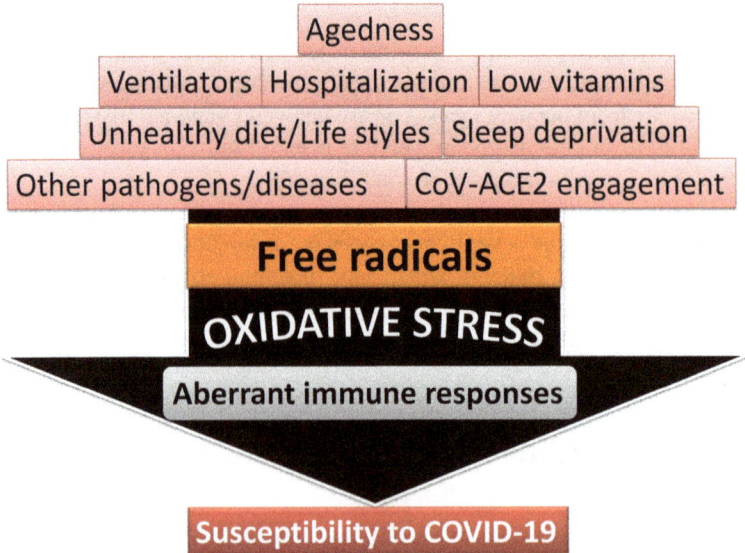

Fig. 2 Reasons for oxidation-stressed body systems adding to susceptibility to COVID-19

3   The Vaccine Candidates and Their Issues

CoVID-19 treatment regimens are yet to be standardized and validated; however, a hunt for potential new drugs is on. Specifically the vaccine research communities are focusing on solution that cater to either (i) restrict viral entry into host cells, (ii) derailing or hampering its replication in the cell, or (iii) more suitably neutralizing the infection by discovering or devising potent antibodies. Other approaches target the host receptor (ACE-2) which binds the CoV entry proteins (spike proteins) for example one by exogenously supplementing soluble (spike affinitive full and/or truncated) versions of the receptor.

Generally higher mutation rates have been reported in RNA viruses than those with DNA genomes (Duffy 2018; Lauring and Andino 2010). Mutations can vary amino acid sequences in the viral surface proteins leading to changes in their post-translational processing such as glycosylation (Li et al. 2020a). This and other changes may deliver, besides detriments also, benefits to the viruses in sense of a heightened infectivity (Diehl et al. 2016; Urbanowicz et al. 2016), transmissibility (Tsetsarkin et al. 2007; Herfst et al. 2012), as well as a protection from neutralizing antibodies (Mudgal 2014; Reguera et al. 2014; Reguera et al. 2012; Ning et al. 2019; Petrie and Lauring 2019). In fact, there have been frequent reports on various versions of the SARS-2 CoV infecting the masses (Baric 2020; Li et al. 2020a; Callaway 2020). Until May 2020, around 329 mutants have been recorded from the affected cases, and among these a more prevalent one was a D614G spike mutant with heightened infectivity (Becerra-Flores and Cardozo 2020).

At treating COVID-19 cases, until the mid of the outbreak some success was assured with convalescent plasma therapy approach (Zeng et al. 2020; Duan et al. 2020); however, it could not catch up that attention due to the nonspecificity, lower levels of neutralizing antibodies, and besides, a recent trial from India finds it ineffective on various other grounds including no reduction in severity and mortality (Agarwal et al. 2020). In yet more recent study, a group of researchers from the USA reported a SARS-2 CoV escape mutant which escapes antibody neutralization and reinfects the COVID-19 convalescent subjects.

Scientific communities around the globe are investigating the possible vaccine candidates with some of the pharma giants with their clinically approved versions against the original SARS-2 CoV. Some commonly followed drug therapies include antiviral drugs in cocktail formulations such as lopinavir/ritonavir (Lim et al. 2020), ribavirin plus CoV-RNAP inhibitory oligos (Elfiky 2020), or former in combination with remdesivir (Ko et al. 2020), more word of mouth and less toxic hydroxyl form of chloroquine in combination with azithromycin (Gautret et al. 2020), and/or mythylprednisolone (Zhou et al. 2020a). Among the above, specifically the lopinavir/ritonavir combination, however, could not impressively benefit at treating HIV cases which is reasoned with the dose-limiting toxicity of this formulation (Cao et al. 2020). However, emerging CoV mutants as above may compel a rethinking of newer vaccine design approaches than conventional. In these times, Complementary and Alternate Medicine (CAM) therapy market has also received global attention.

By these times, many new and known molecules showing implications at solving the anti-CoV-drug/vaccine crises have surfaced, some of which are even under clinical trials. Most of these entities have pronounced credentials with anti-inflammatory, antioxidant, and immunomodulatory efficacies (Castillo et al. 2020). These credentials may have a vital role in the pathophysiology of acute respiratory distress (ARD), which is reported a common scene with SARS-2 CoV-led casualties. Hence drug repurposing trials should also be a norm going hand in hand with vaccine designs.

4 Oxidative Stress and Its Relevance in CoV Immune-Pathology: Deep Insights

Many viral and other pathogenic entities have been found to correlate their bases of pathogenesis and disease severity to, besides other factors, the development of pronounced oxidative stress. For others, instead of a known receptor, receptor determinants are only elucidated. Binding to a receptor marks the specificity of a particular virus to a typical host or tissue(s) within the host based on the bioavailability of such receptor molecule(s). Many of these receptor molecules have intricate housekeeping roles ranging from maintenance of tissue architecture, enzymatic catalysis of important metabolic reactions, cell signaling, and defense of the many to name a few. This is true for the receptors to coronaviruses such as the HCoV-229E, MERS-CoV, HCoV-NL63, and the two SARS-CoVs. Following receptor binding, CoVs are known to deliver their RNA genome into the host cytosol wherein which they take over the cell machinery as a mandate and use it as a production factory to manufacture and release more newer virions which reinfect and thereby spread to other tissues.

Numerous systematic reviews and meta-data have evidenced a positive association of COVID-19 severity with diseases that follow an infective cycle of the viruses in humans (Gheblawi et al. 2020; Jung et al. 2020; Zhang et al. 2020a). Of the many, oxidative stress could be linked multidimensionally to the disease progression caused by SARS viruses (Bosch et al. 2004; Cecchini and Cecchini 2020). During the current COVID-19 and the previous SARS of 2003, both the causative CoVs employ the same receptor, angiotensin-converting enzyme, (ACE2) as also known for another human CoV-NL63 (HCoV-NL63). Still, the SARS-COV-2 is comparatively more affinitive to ACE-2 (Lei et al. 2020). Nonetheless, CoV binding to ACE-2 results in its lower levels rendering its dysregulated housekeeping. ACE-II routinely cleaves an octapeptide vasoconstrictor, Angiotensin II, by virtue of which it would induce vasodilation upon the by-production of Ang1-7. Hence, downregulation of ACE-II housekeeping function is witnessed into COVID-19 which may trigger sequentially deteriorating health stature of the patients (Gan et al. 2020). How this relates variously to surmounting the conditions of oxidative stress in tissue/organ systems would be dealt with further detail (Fig. 3).

Upon CoV binding to ACE-II, the bioavailability of latter on cells is drastically withheld and so is the AngII processing. Uncleaved AngII binds with Angiotensin type 1 (AT1R) which in turn activates the enzyme NADPH oxidase (NOX) (Zablocki and Sadoshima 2013; Rincon et al. 2015; Valente et al. 2012) which signals the generation of reactive oxygen species ROS, and effectuates the state of oxidative stress and followed by inflammatory outcomes adding to the severity of the COVID-19 (Oudit et al. 2007; Sawalha et al. 2020). This is further supported by the finding that NOX-2 is found upregulated in hospitalized subjects (Violi et al. 2020) and also that blocking NOX-2 may reduce the oxidative stress, thereby subduing the severe symptoms (Wu et al. 2020a).

Other studies indicated that endothelial cells stimulated by proinflammatory cytokines, however, can drive NOX proteins and can add to oxidative stress which could lead to endothelial dysfunction (Pennathur and Heinecke 2007; Libby and Lüscher 2020). Endothelial dysfunction may further worsen due to the reduced availability of nitric oxide (caused by ROS), followed by vasoconstriction, redox imbalances, and inflammation. Therefore, CoV-attenuated ACE-2 housekeeping shapes up the renin-angiotensin-aldosterone system (RAAS) to be predominantly oxidizing the vascular environment (Rabelo et al. 2011).

SARS-2 CoV may inevitably remove iron (FeIII) from heme groups in RBC in the bloodstream and therefore trigger Fenton and Haber-Weiss reactions to deliver

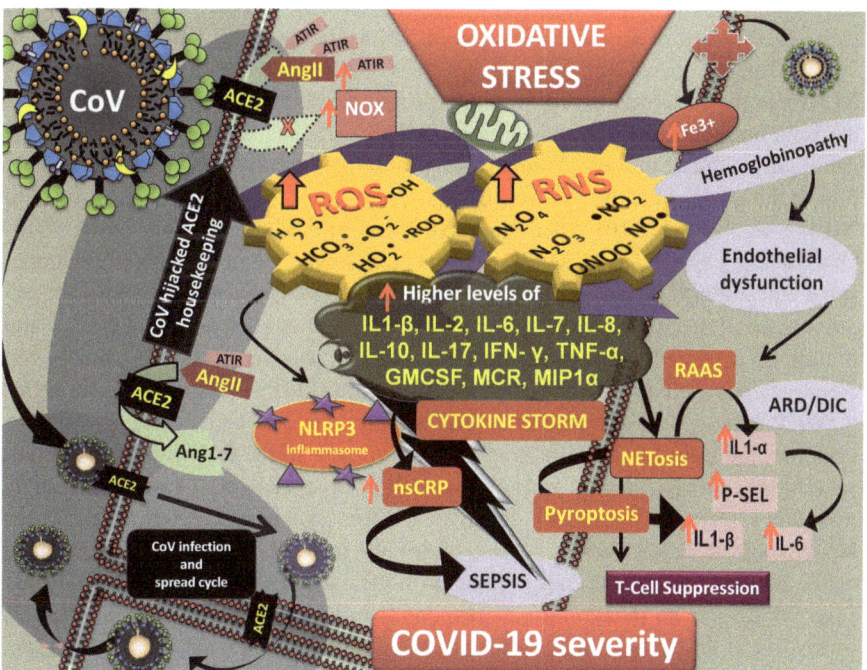

Fig. 3 Mechanistics of oxidative stress-led events contributing to immune imbalances which in turn amount disease severity in COVID-19 and possibly other viral diseases

oxidative stress. Thus hemoglobinopathy may contribute to clinical distress seen as ferroptosis, lipid peroxidation, and mitochondrial damage, etc. (Cavezzi et al. 2020).

Another remarkable outcome of both the SARS-causing CoVs is the heightened levels of proinflammatory cytokines in the patients which is described as a 'cytokine storm' and is confirmed by higher levels of inflammatory markers (Yang et al. 2020a; Qin et al. 2020; Gong et al. 2020; Chen et al. 2020a; Huang et al. 2020) particularly the one which is used in diagnosing the onset of sepsis, the nonspecific C-reactive protein (nsCRP) marker. Not surprising that most COVID-19 casualties have detected elevated levels of both inflammatory cytokines and chemokines (Yang et al. 2020a; Qin et al. 2020; Gong et al. 2020; Chen et al. 2020a; Huang et al. 2020). A plethora of inflammatory entities level up in the blood plasma such as the inter-leukins (IL-1β, -2, -6, -7, -8,-10, -17), interferon γ, IFNγ-inducible protein 10, granulocyte-macrophage colony stimulating factor (GM-CSF), monocyte-chemoattractant protein-1 (MCP-1), TNFα, and macrophage inflammatory protein 1α (MIP1α) among many (Huang et al. 2020; Zhang et al. 2020b; Wu and Yang 2020; Wu et al. 2020b; Ruan et al. 2020; Mehta et al. 2020). Besides, a number of ROS are generated by the action of the macrophages and neutrophils alone (Wang et al. 2020), which oxidatively detriment the immune cells and inflammatory signals (Nagar et al. 2018; Galley 2011).

Oxidative stress may also act as a trigger activating an NLRP3 inflammasome which further may predispose to the cytokine storm as has been evidenced in COVID-19 patients (Mueller et al. 2020; Wu et al. 2020b; Ruan et al. 2020; Mehta et al. 2020; Wang et al. 2020; Nagar et al. 2018; Galley 2011; Yang 2020; Li et al. 2020b; Yang et al. 2020b; Yang et al. 2014; Abais et al. 2015; Martinon 2010). NLRP3 or the NOD-like receptor family protein domain 3, is a major component of the inflammasome and its activation by ROS is effectuated via either or both the deubiquitination of thioredoxin interacting/inhibiting protein (TXNIP) or the activa-tion of the nuclear factor-kB (NF-kB) (Martinon 2010; Zhou et al. 2010; Luo et al. 2014; Donath et al. 2010; Qiao et al. 2012; Boaru et al. 2015; Abais et al. 2015). Other authors vouch for TRMP2 as a factor that links oxidative stress to inflamma-some activation (Zhong et al. 2013). Other than these, 'pyroptosis', a form of pro-grammed cell death which also involves inflammasomes such as the NLRP3, which in turn are activated upon recognition of certain molecular signals called pathogen/damage-associated molecular patterns (PAMPs or DAMPS) (Broz and Dixit 2016).

Well, a great deal of research is still being pursued to understand how inflam-masomes are linked to both facilitating the mandates of the CoV infection to cause severity and spread (Stewart and Cookson 2016; Gram et al. 2012) and/or that of the immune system to halt viral damages to host systems (Lupfer et al. 2015). Nonetheless, oxidative stress is also a driving factor of yet another mechanism called NETosis that is critically linked with the cause of sepsis in COVID-19 dis-ease. NETosis is an innate immune response against pathogens which involves the formation of an extracellular fibrillar network (NET) composed of chromatin and granular proteins released by neutrophils, and this NET can wrap around and atten-uate pathogens like viruses, bacteria, and fungi (Schönrich and Raftery 2016). The downside, however, is that NET formation and release may involve ROS formation by NOX proteins (Almyroudis et al. 2013; Stoiber et al. 2015; Mikacenic et al. 2018) which inherently initiates the NET formation as well. Oxidative stress backed NET formation may suppress T-cell expression while may promote the release of IL-1β (Schönrich et al. 2020). Graver COVID-19 outcomes are thought to be encountered when together with pyroptosis-led release of IL-1β from monocytes and macrophages deliver a positive feedback on exacerbated production of NET entities and that of the noncanonical pyroptotic interleukin IL-1α. This positive feedback may lead to DIC and aberrant immune signaling responses (Barnes et al. 2020). Elevated NETs elevate IL-1α, IL-1β, IL-6, and P-SEL (cell-cell adhesion molecule, P-SELECTIN) (Barbu et al. 2020). Also, IL-1α itself can induce IL-6 (Schönrich and Raftery 2016). This revelation initially shaped a center of attraction to researchers hunting for drugs that block IL-6 to treat COVID-19 maladies (Mehta et al. 2020), but so far could infer a mix bag of pros and cons.

Factors that relate to the severity of COVID-19 are not fully unveiled but for sure are not those related alone to viral infectivity and spread (Hadjadj et al. 2020). Heightened sequential organ failure assessment (SOFA) scores and elevated fibrin degradation derivatives, such as D-dimer, DIC, represent the blueprints for septic organs in COVID-19 cases (Tang et al. 2020a; Tang et al. 2020b; Zhou et al. 2020b). More so, the mechanistic orchestration of inflammatory responses (high levels of IL-6, and other cytokines, dysregulated immunity, as well as microthrombosis) and its interplay with oxidative stress strongly holds water at elucidating the severity (in context to the extent of sepsis) on this zoonotic disease (Beltrán-García et al. 2020; Barnes et al. 2020). More pronounced is the trigger from the under-processed AngII substrate of ACE2, enzyme that is busy as a SARS-CoV entry essential. To talk further, ACE-2 has other housekeeping homeostatic functions as well: i) vascular tone and hypertension control, and that which controls the graveness of acute pulmonary; ii) in arterial and venous endothelial cells, ACE2 takes a protective anti-inflammatory role (Gu et al. 2016; Imai et al. 2005; Burrell et al. 2004). Some authors have showcased COVID-19 as an endothelial disease in context to oxidative stress and inflammation-mediated damages in endothelial harmony (Hamming et al. 2004; Chen et al. 2020b; Hotchkiss et al. 2016; Iba et al. 2020). Oxidative stress further may ensue fibrinogen degradation and clotting whereby it corroborates coagulopathy in COVID-19 cases (White et al. 2016).

5   Do Antioxidants Hold Implications as Anti-virals Against COVID-19?

How the SARS-CoV-2 spreads to other tissues is not largely understood, so is the case with its varying virulence. Even after a year-long agony, any specific anti-SARS-2 CoV vaccine has not shown successful translation into a standardized therapeutic regimen and that other orthodoxical approaches such as the use of oxygen and ventilators are known to be administered under graver outcomes of COVID-19 (Carter et al. 2020). The only hope had been with an antiviral drug remdesivir which works as an inhibitor for the polymerase of SARS-2 CoV and had been in use in many COVID-19 cases (Beigel et al. 2020). Inevitably, in some cases throughout some 30 locations around the globe, that remdesivir has proved a failure (Martinot et al. 2020). With the current severity state in many patients it is being felt challenging to ascertain some personalized treatment approach for patients at heterogeneously varied stages of disease progression or even recurrence, age groups, ethnicities, life style stature, etc. Still, commonly strategized therapies involve those acting on viral multiplication, inflammation control agents such as cytokine blockers, and anticoagulant strategies (Grein et al. 2020; Beltrán-García et al. 2020).

Oxidative stress management could be a game changer as it may holistically manage the severity of COVID-19 and that of other CoV variants currently being aired globally. Antioxidant molecules may subside and/or counteract one or many ROS-triggering mediators talked in the above section that deter immunobalances, and cause damages witnessed as pulmonary dysfunction (ARDs), cytokine storm, and sepsis in subsequent phases post infection (Derouiche 2020; Li et al. 2020c; Chan et al. 2019). Many antioxidant agents are under scrutiny with clinical trials for their antisepsis efficacy and some have been proposed for use as an adjuvant therapy in standard hospital-administered regimens, however, to clarify that there exists no standardized antioxidant therapy yet with regulatory approvals (Prauchner 2017). Accordingly, many antioxidants proposed for use in COVID-19 treatment are being clinically tested, some of which are Vitamin-A, D, C, melatonin, reduced glutathione, resvertrol, N-acetylcysteine, quercitin, curcumin, Boswellia, hesperidin, artemisinin, and silymarin. Others with antioxidant effects include statins, colchicines, and amiodarone (Sanchis-Gomar et al. 2020a; El-Missiry and El-Missiry 2020; Deftereos et al. 2020; Sanchis-Gomar et al. 2020b).

Considering the severity of COVID-19 and the paucity of orthodox treatment regimes, antioxidant supplementation finds its place in the complementary and alternate medicinal arsenal. A somewhat promising notion surrounds at exploiting a critical transcription factor which drives the enzymatic antioxidant defense, the nuclear erythroid-related factor 2 (Nrf-2), and Nrf-2/antioxidant-related entities (ARE), which may need to test repurposing efficacy with the above antioxidants at activating Nrf-2 in order to counteract chances of sepsis prevalent in COVID-19 cases (Cullinan and Diehl 2004; Song et al. 2009; Ungvari et al. 2010; Nakai et al. 2014; Ahmadi and Ashrafizadeh 2020; Marinella 2020). In the sections ahead, we discuss these and other benefits offered by some antioxidants, especially those which have been studied either directly in COVID-19, or the SARS of 2003, or at least in relevance to concurrent symptoms shown by other viral diseases.

6   Promising Antioxidants as Putative Anti-CoVs

Though there are numerous antioxidants that may have been reported and reviewed to find implications in treating many types of viral disease in general and also COVID-19, we include in the below narratives only some to exhibit the various mechanistic modes via which these antioxidants sequester management of oxidative stress and their role in pre- and post-viral pathogenesis, especially considering the severity of COVID-19 disease.

6.1 Melatonin

Chemically N-acetyl-5-methoxytryptamine, melatonin was first known in 1958 (Lerner et al. 1958) as a hormone secreted by the pineal gland in the brain, which promotes dark signals (causing sleeping) and so known to maintain the circadian rhythms (Cardinali and Pévet 1998). It has been reported for its effective immuno-modulatory, apoptotic, antioxidant, antiageing, angiogenic, and anticancer properties (Saha et al. 2019). By virtue of its well-studied direct free radical scavenging property, it offers remarkably reduced oxidative stress by acting against ROS, and RNS both in pathological and natural conditions (Tan 1993; Reiter et al. 2016) and further by activating many antioxidant enzymes (through Nrf2) and by upholding pro-oxidant effects of other enzymes (Pablos et al. 1995; Pablos et al. 1998; Fischer et al. 2013; Shi et al. 2015; Zatta et al. 2003). Specifically for use in treating COVID-19, its antioxidant, immunomodulatory, and anti-inflammatory attributes can be considered which are well appraised for attenuating severe viral infections in general and specially Ebola virus cases (Reiter et al. 2020).

COVID-19 affects majorly the old age population, and it is quite imperative that in this age segment melatonin levels in the blood are lower (10-12 pg/ml) compared to the young people below 20 years of age (20-80 pg/ml) (Iguchi et al. 1982). Bats have been known as a natural reservoir host for SARS-2 CoV, and surprisingly, they are known to have a much higher level of melatonin (90-500 pg/ml), which could reasonably relate to either no or less frequent showcase of mild COVID-19 disease symptoms (Heideman and Bhatnagar 1996). SARS-CoV-2 shares 96% sequence with existing BAT-SARSr-CoV RaTG13 (Zhou et al. 2020c).

The high antioxidant potential of melatonin can be considered as it may bind to 10 free radicals, which is comparatively more impressive than the other known antioxidant compounds like Vitamin C and E, which bind to just one free radical (Tan et al. 2007). Melatonin possesses high bioavailability, by virtue of which it can easily penetrate the blood-brain barrier and placenta (Reppert et al. 1979). Other than this, melatonin may restrict the regulation of cytokines known to deliver proinflammatory and profibrotic effects (Heideman and Bhatnagar 1996). This attribute may implicate its application in COVID-19 treatment as SARS-2 CoV suffering patients usually exhibit the incidences of lung fibrosis (Grillo et al. 2020; Hu et al. 2020; Schwensen et al. 2020).

The use of melatonin as an antiviral has already been supported for infections caused by the encephalitis virus and respiratory syncytial virus (Huang et al. 2010). Moreover, the antioxidant efficacy of melatonin is also known to be linked to activities of enzymes known in oxidative stress and inflammations such as glutathione peroxidase, reductase, catalase, and superoxide dismutase (Reiter et al. 2020; Emerit et al. 2004; Reiter 2000; Reiter et al. 1997).

There are many possible ways in which melatonin may positively influence to overcome pathosymptoms effectuated by coronaviruses. Nonetheless, most respiratory viruses render stress conditions in humans particularly seen in elevated titers of free radicals mostly ROS and RNS (reactive nitrogen species) (Waldhauser et al. 1988). Oxidative stress–resulted lung injury by viruses may also exhibit following a positive feedback loop (Vijay et al. 2015). SARS-CoV endowed oxidative stress is specifically also known to trigger the expression of a phospholipase (PLA2G2D) expression. Higher titers of this enzyme negatively influence the host immune responses to the CoV. This could render the viral infection more lethal. It is noteworthy here that PLA2G2D expression is positively correlated with ageing. Studies with bat models deprived of their antioxidant titers increased their susceptibility to coronavirus, and on the other hand, other models with deleted ROS-pathway candidates resulted in heightened resistance to respiratory viruses (Imai et al. 2008).

Research has shown that COVID-19-linked pulmonary detriments are resulted from what is conferred as a severe inflammation following programmed cell death called pyroptosis (Cookson and Brennan 2001; Yang 2020). This on macrophages and other immunocytes can cause lymphopenia (Panesar 2003) and can obstruct host immune responses to the viral invasion and spread. Melatonin can however serve prospectively at hampering pyroptosis as it can inhibit NLRP3, an inflammasome in the hosts (Ma et al. 2018). Under CoV infection, the viral ORF8b-encoded protein directly interacts with this inflammasome resulting in cell membrane disruption and exposure of cytosolic contents to the extracellular vicinity (Shi et al. 2017). These events potentially may also signal proinflammatory cytokines (Man et al. 2017). Thus, melatonin delivery particularly in the lungs may prove effective at subsiding the inflammatory responses deemed from the CoV-invaded cells. That melatonin-inhibits pyroptosis has been strongly supported with mouse models of bacterial pneumonia (Zhang et al. 2016) besides many other studies (Wang et al. 2019; Arioz et al. 2019; NaveenKumar et al. 2019; Onk et al. 2018; Chen et al. 2018; Liu et al. 2017).

Other than the above prospects, melatonin can also cater to endoplasmic reticulum stress developed from viral infections and can activate unfolded protein response (Chen et al. 2018).

Melatonin can also work at alleviating the toxicity from the standard drugs (discussed earlier). An effective but higher dose of one of the combinations of antiviral drugs lopinavir/ritonavir causes kidney damage due to oxidative stress. Similarly, chloroquine and hydroxchloroquine are believed to be promising drugs against COVID-19 (Colson et al. 2020; Gautret et al. 2020; Yao et al. 2020; Kapoor and Kapoor 2020) but at higher doses may impose toxicity. These side effects can be reduced by using melatonin as an adjuvant therapy (Adikwu et al. 2019) as synergistic effect of melatoin might also help in reducing toxic drug dosage (Michaelides et al. 2011). Also, melatonin can act as an immunostimulant, for example, by increasing the anti-influenza effect of ribavirin (Huang et al. 2019) while this adjuvant combination has also been shown more inhibitory to the replication of RSV (Han et al. 2007).

Oxidative stress can lead to the onset of fibrosis (Su et al. 2019), which has been more reported for both SARS-1 and SARS-2 CoV outbreaks in patients and aftereffects of such condition in the lungs can lead to hypertensions and cardiac complications as is being witnessed as a frequent scene during the current pandemic. Melatonin is reported to restrict the possibility of pulmonary fibrosis following the Hippo/YAP mechanism (Zhao et al. 2018). Castillo and coworkers suggest the use of melatonin in adjuvant therapy for treating CoOVID-19 pneumonia cases (Castillo et al. 2020). In their retrospective hospital trial with 10 CoV-positive cases who were administered a high dose (36–72 mg/day) of melatonin along with a standard therapy, all the patients could be discharged within 7.3–8.6 days of treatment compared to many other CoV-positive hospital entrants who accounted with a longer hospital stay and/or mortality.

A greater dosage of melatonin is not harmful due to the minimal acute toxicity profile known for it (Malhotra et al. 2004; Nordlund and Lerner 1977; Papavasiliou et al. 1972) and as high as 1–6.6 g/day for more than a month even did not report any toxicity. At least the workingly promising therapeutic dose of 75 mg every night as tested on 1400 women under a phase II clinical study did not showcase any toxicity issues.

It sounds imperative that melatonin administration may reduce the severity of the current COVID-19 pandemic. As it is the only known blocking agent to major innate immunity pathways, namely the NG-kV and NLRP3 inflammasome, and also due to the other anti-programmed cell death, anti-ferroptotic and immunomodulatory properties ascribed to it as evaluated in sepsis (Acuña-Castroviejo et al. 2020). Moreover, it is known to be cytoprotective against hypoxia and coagulopathy-associated hemoglobin deterioration (Tesoriere et al. 2001). As highlighted, its use in adjuvant therapeutics may cater to subside problems associated with age (Scheer et al. 2004; Zisapel 2018) as well as side effects following standard therapy. Adding to these, animal studies with exogenous supplementation of melatonin showcased reduced plasma production of proinflammatory cytokines IL-1β (known in pyroptosis) and an appreciated increase in anti-inflammatory cytokine IL-4 (Carrasco et al. 2013).

All in all, melatonin represents a commonly available drug with a high safety profile, which can be self-administered, and so also is more cost-friendly alternative to other antiviral drugs currently being sought (Castillo et al. 2020; Reiter et al. 2020). Other than direct relevance into combatting viral invasion, melatonin also serves as a wellness drug. It is commonly known that sleep-deprived people easily contract infectious diseases. Hence melatonin may simply restore public health by restoring normal sleep and strengthening the immunity stature of the body.

6.2 Vitamin D

About 48 clinical trials testing vitamin-D for various therapeutic attributes are on the go, with almost 10 in advanced level (phase 3-4). Low levels of vitamin D have been associated with increasing death toll among aged males (Raharusun 2020); the study however further requires randomized clinical studies to study if this vitamin's supplementation can therapeutically subside or completely evade the COVID-19.

Vitamin D is known to deliver a vasoprotective effect (Kim et al. 2020) and is dually prospective by counteracting oxidative stress as well as optimizing T-cell immune responses (Von Essen et al. 2010; Sigmundsdottir et al. 2007). These housekeeping roles are made through by its ability to regulate cell signals involving Ca^{2+} and ROS, whereby it also plays a vital character in mechanisms taking care of phosphorus homeostasis (Kutuzova and DeLuca 2007). By regulating oxidative stress and mitochondrial respiration, it has well-pronounced roles in controlling systemic inflammations (Wimalawansa 2019). Nonetheless, it also plays anti-inflammatory roles by controlling adaptive immune responses (Chambers and Hawrylowicz 2011). Its active forms such as calcitrol and hydroxyderivatives can restrict the release of proinflammatory cytokines (that substantiate in the cytokine storm) by downregulating NF-kB and also by inverse agonism on RORY (Slominski et al. 2020). Moreover, Vitamin D can also combat oxidative damage following activation of NRF-2 and p53 dependent mechanisms (Slominski et al. 2020). Its immediate clinical application in practice, however, is limited by FDA regulations. Daily recommended dose is 10000 U/day, but is inefficient at subsiding the cytokine storm and oxidative detriments in severe cases of COVID-19. Mega doses of D3 was found effective and not cytotoxic at attenuating proinflammatory effects alternatively induced by solar radiation (Scott et al. 2017). Except in India and Europe, regulations in the USA don't allow intravenous or intramuscular administrations of D3 and only oral delivery is the option for subjects.

6.3   Vitamin C

As an important antioxidant for the immune system, Vitamin C, when administered intravenously has demonstrated significance at relieving both septic shock and probable sepsis, which are possibly attributed to its immune-suppressive effects (Erol 2020). In one study, Vitamin C (20 mM) was shown to inhibit levels of IL-6 and TNF-α produced by LPS-induced monocytes (Härtel et al. 2004). In yet another study by Nieman's group, 1500 mg week's prior dosage of Vitamin C could reduce the levels of IL-6, IL-10, IL-IRA, and IL-8 in a 90-km marathon athletes than others without the dosage (Nieman et al. 2000).

6.4   Resveratrol

Trans-3, 4′, 5-trihydroxystilbene or Resveratrol, a natural antioxidant on the class of phytoalexins available in grapes, blueberries, and peanuts, besides other plants (Nunes et al. 2018), works by down-regulating the NADPH oxidases, one of the ROS-generating enzymes (Nunes et al. 2018; Xia et al. 2010). In rat models with obstructive lung disease, it has been previously found prospective at enhancing SOD activity and leveling down MDA which culminates into suppressed lipid peroxidation (Wang et al. 2017a). Resveratrol may exhibit immune-regulatory roles by interfering with proinflammatory cytokines, cell regulation, and their gene expression which have thoroughly been reviewed earlier (Malaguarnera 2019). It precisely upregulates NK-cells and macrophages, whereas suppresses B-cell, T17 helper cells, M2-like cells among others (Malaguarnera 2019). It has been shown to block herpesvirus infection (Docherty et al. 1999; Annunziata et al. 2018) as also at suppressing replication of viruses causing dengue, Influenza, and Zika (Mohd et al. 2019; Paemanee et al. 2018). Studies conducted with MERS-CoV indicate its anti-replicative efficacy against this CoV which also prolongs host survival (Lin et al. 2017). Recent studies by Yang's group have reported that resveratrol can both inhibit the SARS-CoV-2 replication and block its entry into Vero cells (Yang et al. 2020c).

6.5 Curcumin

Curcumin is a traditional Indian ayurvedic medicinal agent, extracted from *Curcuma longa* and is used variously for its antipyretic, antipathogenic, and immune booster properties. It contributes to 2-8% of turmeric phytochemicals, and chemically known as diferuloymethane (Modaresi et al. 2017; Aggarwal et al. 2007; Chattopadhyay et al. 2004). It has remarkably lower toxicity supported by its enormous use in Indian ethnic groups as a major component applied variously to the body in rituals and ceremonies. It has a pronounced antioxidant effect; however, this effect has not yet been fully understood to correlate with its immunomodulatory and anti-inflammatory activities (Sahu et al. 2008) nor it has yet been implicated as an antiviral or anti-CoV entity. As a traditional medicament, it has been administered in many inflammatory disease conditions (Joe and Lokesh 2000). Nonetheless, curcumin is known to regulate a plethora of intermediates related to the cell signaling pathways involved in inflammation viz., IBB, NF-kB, ERK1,2, AP-1, TGF-β, TXNIP, STAT3, PPARγ, JAK2-STAT3, NLRP3, p38MAPK, Nrf2, Notch-1, AMPK, TLR-4, and MyD-88 (Saeedi-Boroujeni et al. 2021). That it can find prospects as an anti-COVID-19, especially to subside severe phases, can be supported by its effect especially on inflammasomes, cytotoxic T-cell functions, and adaptive immunity. Its anti-inflammatory property is linked to inhibiting and suppressing prostaglandins, COX2, LOX, iNOS, and that of the cytokine productions. More important to the regulation of antioxidant activity of this compound is its inhibition of cellular NF-kB activity by downregulating its gene expression as well as its productive transfer to the cell nucleus (Singh and Aggarwal 1995; Plummer et al. 1999). Other than this, curcumin suppresses STAT-3 in turn suppressing IL-6 activity (Kamat et al. 2007), inhibits TGF-β, and levels down pro-inflammatory TNF-α and MCP-1 (Liu et al. 2016; Panahi et al. 2016). TGF-β has been considered as an important drug target in COVID-19 treatment strategies (Al-helfawi 2020; Chen 2020). TGF-β has been found upregulated in SARS patients and is particularly known to be over-expressed due to the activity of viral nucleocapsid protein and papain-like proteases (PLpro) (Wang et al. 2017b). In vitro infectivity assays supplemented with OT-101 (intended for TGF-β inhibition) exhibited hampered SARS-CoV1 and -2 replication (Pittet et al. 2001). Hence curcumin finds its relevance at this too. Other than this, curcumin inhibits NLRP3 inflammasomes following epigenetic remodeling events (Ding et al. 2018) due to which it variously impacts the health stature in different disease conditions such as inflammatory, neurological, renal, pulmonary, and other diseases some of which find relevant concurrence with the COVID-19. COVID-19 also registers a state of severe lymphopenia exhibited from the T-cell exhaustion and more importantly culminating into cytotoxic T-cell depletion which is also seen commonly in cancers and other viral pathogenesis (Moon 2020; Diao et al. 2020; Sullivan et al. 2020). All the above notions hint the promising attributes of curcumin for use in targeting critical check-points in viral entry, replication, cell metabolism takeover, and immunopathological and physiological symptoms in diseases caused by CoVs. In fact a recent, though nonrandomized, clinical trial uses an oral nano-curcumin formulation (Sinacurcumin) administered to a group of progressing COVID-19 patients which showed fastly resolved tachypnea, myalgua, chills, and coughing which eventually bettered in view of short the hospital stays, requirement of oxygen support, and perhaps a lesser deteriorating infection extent (Saber-Moghaddam et al. 2021). Nonetheless, before one jumps to conclusions, these studies need to reproduce similar results followed with a randomized placebo trial.

6.6 Ergothioneine

Vitamin C and curcumin have lower bioavailability and hence their dosage will need to be elevated than usually recommended dietary intakes. Ergothioneine (ET) is a more bioavailable dietary amino acid which can accumulate in cells ascribed to its transport roles attributed to an organic cation transporter novel type 1 (OCTN1) (Halliwell et al. 2018; Gründemann et al. 2005; Borodina et al. 2020; Tucker et al. 2019). It is ubiquitously prevalent in animals, bacteria, and fungi. In humans, it shows low catabolism and higher sustenance in body hinting its physiological roles (Halliwell et al. 2018; Gründemann et al. 2005; Tang et al. 2018; Cheah et al. 2017; Melville et al. 1954; Kaneko et al. 1980). ET has multiple unique properties conferring a variety of its cytoprotective roles mediated by its ability of ROS/RNS scavenging (Akanmu et al. 1991; Aruoma et al. 1997; Stoffels et al. 2017; Rougee et al. 1988). How it finds relevance to its potential applications in COVID-19 can be surmised by its modulatory roles on inflammation, metal chelation of iron and copper, radiation damage control, inhibition of vascular adhesion and myelo-peroxidase activity, phagocyte respiratory bursts, upregulation of HSPs, neuronal differentiation, and managing pulmonary and hepatic fibrosis reviewed recently by Cheah and Halliwel (Cheah and Halliwell 2020). Not to elaborate further, as these authors have precisely reviewed numerous studies that vouch for the prospects of ET in treating COVID-19 maladies many of which are exhibited and related to conditions with dementia, depression, atherosclerosis, heart diseases, fatty liver, ischemic injuries, latter more specifically related with ARD (Cheah and Halliwell 2020).

As several age-related disorders make it more probable for contacting COVID-19 and possibly other viral diseases, and ET may serve as a supplemental therapeutic to curb the risks. This is also supported by the lower ET levels in blood of the elderly populations succumbed to age-related disorders such as frailty, neurodegeneration, and heart diseases among many (Cheah et al. 2016; Smith et al. 2020; Kameda et al. 2020; Hatano et al. 2016; Sotgia et al. 2014). Also reviewed is that though specific studies with this compound are yet not progressed on context to account for its effectiveness at evading viral entry and replication, studies using ergotheoniene-rich extracts respectively from fungi and mushrooms on (i) HIV are worth noting when it could dose dependently registered marked inhibition of pro-fs protein helpful in HIV transcription in host cells (Xiao et al. 2006); (ii) on Hepatitis C virus (HCV) where it almost completely inhibits NS3/4A protease and HCV replication (Gallego et al. 2019; Grinde et al. 2006). In to its protective mechanisms against ARDs, its antioxidant properties specifically that culminate with inhibition of active NF-kB and that of the IL-8 expression is more evident (Repine and Elkins 2012; Rahman et al. 2003; Song et al. 2017). Also based on the studies that highlight ergothioneine as an inhibitor of MPO activity and scavenger of MPO-derivatized HOCl (Akanmu et al. 1991; Asahi et al. 2016), it is thought that it may prospectively interfere and subdue the risks related to neutrophil driven NETosis (Cheah and Halliwell 2020).

6.7 Pentoxifylline

Pentoxifylline is a nonspecific inhibitor of phosphodiestrases which has found its extensive application at treating vascular diseases of the brain and peripheral systems by virtue of its blood-thinning rheological effects. Its antioxidant property is witnessed in the suppression of neutrophil activation, the latter which actually generates ROS, and hence pentoxifylline qualifies in subsiding undesired tissue damage (Crouch and Fletcher 1992). Other than this, it has been reported to suppress both *in vitro* and *in vivo* release of IL-6, 1b, 8, TNF-α, and C-reactive protein (CRP) (Neuner et al. 1994; González-Espinoza et al. 2012; Shaw et al. 2009) while it specifically increases the production of anti-inflammatory cytokine IL-10 (Fernandes et al. 2008). It also inhibits cytotoxicity from natural killer cells and T-lymphocytes (Heinkelein et al. 1995; Reed and Degowin 1992) as well as the cell surface expression of adhesion molecule (ICAM-1) and the release of IL-8 and monocyte chemoattractant protein-1 under the effect of cytokine-activated pulmonary epithelia (Neuner et al. 1997; Krakauer 2000). ARD is a common scene with COVID-19 cases as also the frequent instances requiring ventilators. This has been linked to the hyper-inflammation effect following the dysregulation cytokines/chemokines and drastically altered lush immunopathologies, what is called a cytokine storm (Ye et al. 2020). Under this (as seen also in COVID-19) patients exhibit elevated levels of IL-6, 8, 1β, and TNF-α, ROS, chemokines (like CCL-2, -5, and -3, IFN-γ induced protein 10), and that of macrophage colony stimulating factor. Two thirds COVID-19 mortalities in the hospitals exhibit disseminated intravascular coagulation (DIC) which is pronounced as coagulopathy (Tang et al. 2020a) and is well surmised by elevated levels of D-dimer, fibrin and fibrin degradation byproducts. These effectuate vascular thrombosis and vascular occlusions (Li et al. 2020d) and in turn lead to DIC and multiple organ dysfunctions. Pentoxifylline may sequester its anticoagulant effect which can help halt thrombogenesis, its antifibrogenic property can also effectively treat fibrosis (Mandi et al. 1995; Adel et al. 2010; Windmeier and Gressner 1997).

7 Conclusions

One of the notions, mostly supported by our own beliefs and if one would agree, is that surprisingly many countries far off China, where the COVID-19 pandemic emerged, witnessed a fast-paced spread and infectivity than even those neighboring China. In the latter example, India reported the starting infection cases much later to the global November 2019 outbreak dates, and actual lockdowns started in March 2020. Still the number of reported cases in India as for many other Asian neighbors to China were less pronounced at that initial instance compared to massive death tolls reported in the western world including the USA and European nations. Two routes of thought emerged from these observations. Firstly, the migration frequencies to and fro China to these severely affected countries, and secondly, the ethnicity and culinary practices followed at these locations. One would agree to the kind of common household food habits in India and as for many Asian countries, where vegan lifestyles are followed and/or which use numerous and variously proportioned spices in cuisines are known to be embedded in the ancient and traditionally followed indigenous medical systems like the Ayurveda and homeopathy, etc. The use of spices has solid grounding into culinary practices in these countries, many of which are a rich source of antioxidants. Moreover, at these times nowadays when people have inclined more to the wellness products and specially the use of more antioxidants which has been the taglines of the many brands of foods. These could be one of the reasons that initial COVID-19 cases were less pronounced in the countries neighboring China. No doubt a sincere thanks and applause goes to the health and sanitization policies and the seriousness of many governments at enforcing stronger regulations that catered to barricading the spread of COVID-19.

Holistically healthy immune system can be assured with the variously known antioxidant-containing immunity booster foods and many other wellness products that include fruits, probiotics, and prebiotics which may constructively recondition gut and immune system health (Vinayak et al. 2021; Arshad et al. 2020; Olaimat et al. 2020). The use of such agents is the support of growing COVID-19 studies and associated cases of drastic dysbiosis in probiotic microorganisms that habituate the intestine (Xu et al. 2020) as well as others that relate gut microflora, gut infections with COVID-19 (Gu et al. 2020; Yeo et al. 2020; Gao et al. 2020). Probiotic metabolites such as bacteriocins are known with antiviral effects as well. No doubt, a great deal of expectations is with the vaccine candidates from pharmaceutical giants country-wise on their effective administration to the affected masses globally and governments are also making tremendous efforts in these directions. Experts however speculate that successful eradication of SARS-CoV-2 might not be soon possible (Cheah and Halliwell 2020). As we are currently witnessing the airing news on mutated variants of the SARS-CoV-2 and that predictions of it being more impactful over the course of time at adding fuel to the COVID-19 outbreak and other viral outbreaks that might follow, it is quite imperative to focus on holistic wellness of the body systems and equally hunting for antioxidant drug repurposing supported with rigorous clinical trials hand in hand to that with strain-specific vaccine candidates. Perhaps we foresee a highly sought wellness plan too with the futuristic events of similar outbreaks from current or a varying strain or a new viral entity altogether, the outcomes otherwise of which would deem countless lives and economies at stake.

Acknowledgments The authors thank the fraternity of the University Institute of Biotechnology and also the University Center for Research and Development (UCRD) at Chandigarh University (CU) for support. All authors have contributed equally to the manuscript. GM framed the various sections and wrote the whole manuscript; is a PhD in molecular biosciences with previous experience in thrust area themes involved in human and pig coronaviruses, acquired at the National Center of Biotechnology, Spain. JK and KC carried out the literature survey and data collection on various antioxidants, where JK is a research scholar involved with investigating bioprospecting of phytochemicals and their scale-up and KC is a research fellow supported by the Indian government's DST-SERB core research grant (EMR/2016/008022) awarded to the corresponding author and is involved with animal biodiversity research. GBS is a PhD in Environmental Biotechnology who prepared the introduction and conclusion words' section. All authors equally revised the manuscript. The institute to which authors are currently affiliated has no role in shaping the manuscript. All the authors declare no conflict of interest.

References

Abais JM, Xia M, Zhang Y, Boini KM, Li P-L. Redox regulation of NLRP3 inflammasomes: ROS as trigger or effector? Antioxid Redox Signal. 2015;22(13):1111–29.

Acuña-Castroviejo D, Escames G, Figueira JC, de la Oliva P, Borobia AM, Acuña-Fernández C. Clinical trial to test the efficacy of melatonin in COVID-19. J Pineal Res. 2020;69(3):e12683.

Adel M, Awad H, Abdel-Naim A, Al-Azizi M. Effects of pentoxifylline on coagulation profile and disseminated intravascular coagulation incidence in Egyptian septic neonates. J Clin Pharm Ther. 2010;35(3):257–65.

Adikwu E, Brambaifa N, Obianime WA. Melatonin and alpha lipoic acid restore electrolytes and kidney morphology of lopinavir/ritonavir-treated rats. J Nephropharmacol. 2019;9(1):e06-e.

Agarwal A, Mukherjee A, Kumar G, Chatterjee P, Bhatnagar T, Malhotra P. Convalescent plasma in the management of moderate covid-19 in adults in India: open label phase II multicentre randomised controlled trial (PLACID Trial). BMJ. 2020;371:m3939. https://doi.org/10.1136/bmj.m3939.

Aggarwal BB, Surh Y-J, Shishodia S. The molecular targets and therapeutic uses of curcumin in health and disease. Springer; 2007.

Ahmadi Z, Ashrafizadeh M. Melatonin as a potential modulator of Nrf2. Fundam Clin Pharmacol. 2020;34(1):11–9.

Akanmu D, Cecchini R, Aruoma OI, Halliwell B. The antioxidant action of ergothioneine. Arch Biochem Biophys. 1991;288(1):10–6.

Al-helfawi MA. Potential approach for fighting against corona virus disease. Am Sci Res J Eng Technol Sci (ASRJETS). 2020;66(1):127–44.

Almyroudis N, Grimm M, Davidson B, Rohm M, Urban C, Segal B. NETosis and NADPH oxidase: at the intersection of host defense, inflammation, and injury. Front Immunol. 2013;4:45.

Annunziata G, Maisto M, Schisano C, Ciampaglia R, Narciso V, Tenore GC, et al. Resveratrol as a novel anti-herpes simplex virus nutraceutical agent: an overview. Viruses. 2018;10(9):473.

Arioz BI, Tastan B, Tarakcioglu E, Tufekci KU, Olcum M, Ersoy N, et al. Melatonin attenuates LPS-induced acute depressive-like behaviors and microglial NLRP3 inflammasome activation through the SIRT1/Nrf2 pathway. Front Immunol. 2019;10:1511.

Arshad MS, Khan U, Sadiq A, Khalid W, Hussain M, Yasmeen A, et al. Coronavirus disease (COVID-19) and immunity booster green foods: a mini review. Food Sci Nutr. 2020;8(8):3971–6.

Aruoma OI, Whiteman M, England TG, Halliwell B. Antioxidant action of ergothioneine: assessment of its ability to scavenge peroxynitrite. Biochem Biophys Res Commun. 1997;231(2):389–91.

Asahi T, Wu X, Shimoda H, Hisaka S, Harada E, Kanno T, et al. A mushroom-derived amino acid, ergothioneine, is a potential inhibitor of inflammation-related DNA halogenation. Biosci Biotechnol Biochem. 2016;80(2):313–7.

Barbu EA, Mendelsohn L, Samsel L, Thein SL. Pro-inflammatory cytokines associate with NETosis during sickle cell vaso-occlusive crises. Cytokine. 2020;127:154933.

Baric RS. Emergence of a highly fit SARS-CoV-2 variant. N Engl J Med. 2020.

Barnes BJ, Adrover JM, Baxter-Stoltzfus A, Borczuk A, Cools-Lartigue J, Crawford JM, et al. Targeting potential drivers of COVID-19: neutrophil extracellular traps. J Exp Med. 2020;217(6)

Becerra-Flores M, Cardozo T. SARS-CoV-2 viral spike G614 mutation exhibits higher case fatality rate. Int J Clin Pract. 2020.

Beigel JH, Tomashek KM, Dodd LE, Mehta AK, Zingman BS, Kalil AC et al. Remdesivir for the treatment of Covid-19 – preliminary report. N Engl J Med. 2020.

Beltrán-García J, Osca-Verdegal R, Pallardó FV, Ferreres J, Rodríguez M, Mulet S et al. Sepsis and coronavirus disease 2019: common features and anti-inflammatory therapeutic approaches. Crit Care Med. 2020.

Boaru SG, Borkham-Kamphorst E, Van de Leur E, Lehnen E, Liedtke C, Weiskirchen R. NLRP3 inflammasome expression is driven by NF-κB in cultured hepatocytes. Biochem Biophys Res Commun. 2015;458(3):700–6.

Borodina I, Kenny LC, McCarthy CM, Paramasivan K, Pretorius E, Roberts TJ, et al. The biology of ergothioneine, an antioxidant nutraceutical. Nutr Res Rev. 2020;33(2):190–217.

Bosch BJ, Martina BE, Van Der Zee R, Lepault J, Haijema BJ, Versluis C, et al. Severe acute respiratory syndrome coronavirus (SARS-CoV) infection inhibition using spike protein heptad repeat-derived peptides. Proc Natl Acad Sci. 2004;101(22):8455–60.

Broz P, Dixit VM. Inflammasomes: mechanism of assembly, regulation and signalling. Nat Rev Immunol. 2016;16(7):407–20.

Burrell LM, Johnston CI, Tikellis C, Cooper ME. ACE2, a new regulator of the renin–angiotensin system. Trends Endocrinol Metab. 2004;15(4):166–9.

Callaway E. The coronavirus is mutating-does it matter? Nature. 2020;585(7824):174–7.

Cao B, Wang Y, Wen D, Liu W, Wang J, Fan G et al. A trial of lopinavir–ritonavir in adults hospitalized with severe Covid-19. N Engl J Med. 2020.

Cardinali DP, Pévet P. Basic aspects of melatonin action. Sleep Med Rev. 1998;2(3):175–90.

Carrasco C, Marchena AM, Holguín-Arévalo MS, Martín-Partido G, Rodríguez AB, Paredes SD, et al. Anti-inflammatory effects of melatonin in a rat model of caerulein-induced acute pancreatitis. Cell Biochem Funct. 2013;31(7):585–90.

Carter C, Osbourne M, Agagah G, Aedy H, Notter J. COVID-19 Disease: invasive ventilation. Clin Integr Care. 2020:100004.

Castillo RR, Quizon GRA, Juco MJM, Roman ADE, de Leon DG, Punzalan FER, et al. Melatonin as adjuvant treatment for coronavirus disease 2019 pneumonia patients requiring hospitalization (MAC-19 PRO): a case series. Melatonin Res. 2020;3(3):297–310.

Cavezzi A, Troiani E, Corrao S. COVID-19: hemoglobin, iron, and hypoxia beyond inflammation. A narrative review. Clin Pract. 2020;10(2)

CDC. About MERS. National Center for Immunization and Respiratory Diseases (NCIRD), Division of Viral Diseases. 2019. https://www.cdc.gov/coronavirus/mers/about/index.html.

Cecchini R, Cecchini AL. SARS-CoV-2 infection pathogenesis is related to oxidative stress as a response to aggression. Med Hypotheses. 2020;143:110102.

Chambers ES, Hawrylowicz CM. The impact of vitamin D on regulatory T cells. Curr Allergy Asthma Rep. 2011;11(1):29–36.

Chan SM, Selemidis S, Bozinovski S, Vlahos R. Pathobiological mechanisms underlying metabolic syndrome (MetS) in chronic obstructive pulmonary disease (COPD): clinical significance and therapeutic strategies. Pharmacol Ther. 2019;198:160–88.

Chattopadhyay I, Biswas K, Bandyopadhyay U, Banerjee RK. Turmeric and curcumin: biological actions and medicinal applications. Curr Sci Bangalore. 2004;87:44–53.

Cheah IK, Halliwell B. Could ergothioneine aid in the treatment of coronavirus patients? Antioxidants. 2020;9(7):595.

Cheah IK, Feng L, Tang RM, Lim KH, Halliwell B. Ergothioneine levels in an elderly population decrease with age and incidence of cognitive decline; a risk factor for neurodegeneration? Biochem Biophys Res Commun. 2016;478(1):162–7.

Cheah IK, Tang RM, Yew TS, Lim KH, Halliwell B. Administration of pure ergothioneine to healthy human subjects: uptake, metabolism, and effects on biomarkers of oxidative damage and inflammation. Antioxid Redox Signal. 2017;26(5):193–206.

Chen W. A potential treatment of COVID-19 with TGF-β blockade. Int J Biol Sci. 2020;16(11):1954.

Chen Y, Zhao Q, Sun Y, Jin Y, Zhang J, Wu J. Melatonin induces anti-inflammatory effects via endoplasmic reticulum stress in RAW264. 7 macrophages. Mol Med Rep. 2018;17(4):6122–9.

Chen G, Wu D, Guo W, Cao Y, Huang D, Wang H, et al. Clinical and immunological features of severe and moderate coronavirus disease 2019. J Clin Invest. 2020a;130(5)

Chen L, Li X, Chen M, Feng Y, Xiong C. The ACE2 expression in human heart indicates new potential mechanism of heart injury among patients infected with SARS-CoV-2. Cardiovasc Res. 2020b;116(6):1097–100.

Chu DK, Poon LL, Gomaa MM, Shehata MM, Perera RA, Zeid DA, et al. MERS coronaviruses in dromedary camels, Egypt. Emerg Infect Dis. 2014;20(6):1049.

Colson P, Rolain J-M, Lagier J-C, Brouqui P, Raoult D. Chloroquine and hydroxychloroquine as available weapons to fight COVID-19. Int J Antimicrob Agents. 2020:105932.

Cookson B, Brennan M. Pro-inflammatory programmed cell death. Trends Microbiol. 2001;113–114(9):113–4.

Crouch S, Fletcher J. Effect of ingested pentoxifylline on neutrophil superoxide anion production. Infect Immun. 1992;60(11):4504–9.

Cullinan SB, Diehl JA. PERK-dependent activation of Nrf2 contributes to redox homeostasis and cell survival following endoplasmic reticulum stress. J Biol Chem. 2004;279(19):20108–17.

Deftereos SG, Giannopoulos G, Vrachatis DA, Siasos GD, Giotaki SG, Gargalianos P, et al. Effect of colchicine vs standard care on cardiac and inflammatory biomarkers and clinical outcomes in patients hospitalized with coronavirus disease 2019: the GRECCO-19 randomized clinical trial. JAMA Netw Open. 2020;3(6):e2013136-e.

Derouiche S. Oxidative stress associated with SARS-Cov-2 (COVID-19) increases the severity of the lung disease-a systematic review. J Infect Dis Epidemiol. 2020;6:121.

Diao B, Wang C, Tan Y, Chen X, Liu Y, Ning L, et al. Reduction and functional exhaustion of T cells in patients with coronavirus disease 2019 (COVID-19). Front Immunol. 2020;11:827.

Diehl WE, Lin AE, Grubaugh ND, Carvalho LM, Kim K, Kyawe PP, et al. Ebola virus glycoprotein with increased infectivity dominated the 2013–2016 epidemic. Cell. 2016;167(4):1088-98.e6.

Ding X-Q, Wu W-Y, Jiao R-Q, Gu T-T, Xu Q, Pan Y, et al. Curcumin and allopurinol ameliorate fructose-induced hepatic inflammation in rats via miR-200a-mediated TXNIP/NLRP3 inflammasome inhibition. Pharmacol Res. 2018;137:64–75.

Docherty JJ, Fu MMH, Stiffler BS, Limperos RJ, Pokabla CM, DeLucia AL. Resveratrol inhibition of herpes simplex virus replication. Antivir Res. 1999;43(3):145–55.

Donath MY, Böni-Schnetzler M, Ellingsgaard H, Halban PA, Ehses JA. Cytokine production by islets in health and diabetes: cellular origin, regulation and function. Trends Endocrinol Metab. 2010;21(5):261–7.

Duan K, Liu B, Li C, Zhang H, Yu T, Qu J, et al. Effectiveness of convalescent plasma therapy in severe COVID-19 patients. Proc Natl Acad Sci. 2020;117(17):9490–6. https://doi.org/10.1073/pnas.2004168117.

Duffy S. Why are RNA virus mutation rates so damn high? PLoS Biol. 2018;16(8):e3000003.

Elfiky AA. Anti-HCV, nucleotide inhibitors, repurposing against COVID-19. Life Sci. 2020:117477.

El-Missiry MA, El-Missiry ZM. Melatonin is a potential adjuvant to improve clinical outcomes in individuals with obesity and diabetes with coexistence of Covid-19. Eur J Pharmacol. 2020:173329.

Emerit I, Filipe P, Freitas J, Vassy J. Protective effect of superoxide dismutase against hair graying in a mouse model. Photochem Photobiol. 2004;80(3):579–82.

Erol A. High-dose intravenous vitamin C treatment for COVID-19. 2020.

Fernandes JL, de Oliveira RTD, Mamoni RL, Coelho OR, Nicolau JC, Blotta MHS, et al. Pentoxifylline reduces pro-inflammatory and increases anti-inflammatory activity in patients with coronary artery disease – a randomized placebo-controlled study. Atherosclerosis. 2008;196(1):434–42.

Fischer TW, Kleszczyński K, Hardkop LH, Kruse N, Zillikens D. Melatonin enhances antioxidative enzyme gene expression (CAT, GPx, SOD), prevents their UVR-induced depletion, and protects against the formation of DNA damage (8-hydroxy-2'-deoxyguanosine) in ex vivo human skin. J Pineal Res. 2013;54(3):303–12.

Gallego P, Rojas Á, Falcón G, Carbonero P, García-Lozano MR, Gil A, et al. Water-soluble extracts from edible mushrooms (Agaricus bisporus) as inhibitors of hepatitis C viral replication. Food Funct. 2019;10(6):3758–67.

Galley HF. Oxidative stress and mitochondrial dysfunction in sepsis. Br J Anaesth. 2011;107(1):57–64.

Gan R, Rosoman NP, Henshaw DJ, Noble EP, Georgius P, Sommerfeld N. COVID-19 as a viral functional ACE2 deficiency disorder with ACE2 related multi-organ disease. Med Hypotheses. 2020;144:110024.

Gao QY, Chen YX, Fang JY. 2019 novel coronavirus infection and gastrointestinal tract. J Dig Dis. 2020;21(3):125–6.

Gautret P, Lagier J-C, Parola P, Meddeb L, Mailhe M, Doudier B, et al. Hydroxychloroquine and azithromycin as a treatment of COVID-19: results of an open-label non-randomized clinical trial. Int J Antimicrob Agents. 2020:105949.

Gheblawi M, Wang K, Viveiros A, Nguyen Q, Zhong J-C, Turner AJ, et al. Angiotensin-converting enzyme 2: SARS-CoV-2 receptor and regulator of the renin-angiotensin system: celebrating the 20th anniversary of the discovery of ACE2. Circ Res. 2020;126(10):1456–74.

Gong J, Dong H, Xia SQ, Huang YZ, Wang D, Zhao Y et al. Correlation analysis between disease severity and inflammation-related parameters in patients with COVID-19 pneumonia. MedRxiv. 2020.

González-Espinoza L, Rojas-Campos E, Medina-Pérez M, Pena-Quintero P, Gómez-Navarro B, Cueto-Manzano AM. Pentoxifylline decreases serum levels of tumor necrosis factor alpha, interleukin 6 and C-reactive protein in hemodialysis patients: results of a randomized double-blind, controlled clinical trial. Nephrol Dial Transplant. 2012;27(5):2023–8.

Gram AM, Frenkel J, Ressing ME. Inflammasomes and viruses: cellular defence versus viral offence. J Gen Virol. 2012;93(10):2063–75.

Grein J, Ohmagari N, Shin D, Diaz G, Asperges E, Castagna A, et al. Compassionate use of remdesivir for patients with severe Covid-19. N Engl J Med. 2020;382(24):2327–36.

Grillo F, Barisione E, Ball L, Mastracci L, Fiocca R. Lung fibrosis: an undervalued finding in COVID-19 pathological series. Lancet Infect Dis. 2020.

Grinde B, Hetland G, Johnson E. Effects on gene expression and viral load of a medicinal extract from Agaricus blazei in patients with chronic hepatitis C infection. Int Immunopharmacol. 2006;6(8):1311–4.

Gründemann D, Harlfinger S, Golz S, Geerts A, Lazar A, Berkels R, et al. Discovery of the ergo-thioneine transporter. Proc Natl Acad Sci. 2005;102(14):5256–61.

Gu H, Xie Z, Li T, Zhang S, Lai C, Zhu P, et al. Angiotensin-converting enzyme 2 inhibits lung injury induced by respiratory syncytial virus. Sci Rep. 2016;6:19840.

Gu S, Chen Y, Wu Z, Chen Y, Gao H, Lv L et al. Alterations of the gut microbiota in patients with COVID-19 or H1N1 influenza. Clin Infect Dis. 2020.

Hadjadj J, Yatim N, Barnabei L, Corneau A, Boussier J, Pere H et al. Impaired type I interferon activity and exacerbated inflammatory responses in severe Covid-19 patients. MedRxiv. 2020.

Halliwell B, Cheah IK, Tang RM. Ergothioneine–a diet-derived antioxidant with therapeutic potential. FEBS Lett. 2018;592(20):3357–66.

Hamming I, Timens W, Bulthuis M, Lely A. Navis Gv, van Goor H. Tissue distribution of ACE2 protein, the functional receptor for SARS coronavirus. A first step in understanding SARS pathogenesis. J Pathol. 2004;203(2):631–7.

Han Z, Lu W, Huang S. Synergy effect of melatonin on anti-respiratory syscytical virus activity of Ribavirin in vitro. Chin Pharm. 2007;7:10.

Härtel C, Strunk T, Bucsky P, Schultz C. Effects of vitamin C on intracytoplasmic cytokine pro-duction in human whole blood monocytes and lymphocytes. Cytokine. 2004;27(4–5):101–6.

Hatano T, Saiki S, Okuzumi A, Mohney RP, Hattori N. Identification of novel biomark-ers for Parkinson's disease by metabolomic technologies. J Neurol Neurosurg Psychiatry. 2016;87(3):295–301.

Heideman P, Bhatnagar K. Tropical bat, Anoura geoffroyi, that does not use photoperiod to regu-late seasonal reproduction. 1996.

Heinkelein M, Schneider-Schaulies J, Walker BD, Jassoy C. Inhibition of cytotoxicity and cytokine release of CD8+ HIV-specific cytotoxic T lymphocytes by pentoxifylline. J Acquir Immune Defic Syndr Hum Retrovirol. 1995;10(4):417–24.

Herfst S, Schrauwen EJ, Linster M, Chutinimitkul S, de Wit E, Munster VJ, et al. Airborne trans-mission of influenza A/H5N1 virus between ferrets. Science. 2012;336(6088):1534–41.

Hotchkiss RS, Moldawer LL, Opal SM, Reinhart K, Turnbull IR, Vincent J-L. Sepsis and septic shock. Nat Rev Dis Primers. 2016;2(1):1–21.

Hu Z-J, Xu J, Yin J-M, Li L, Hou W, Zhang L-L, et al. Lower circulating interferon-gamma is a risk factor for lung fibrosis in COVID-19 patients. Front Immunol. 2020;11:2348.

Huang SH, Cao XJ, Liu W, Shi XY, Wei W. Inhibitory effect of melatonin on lung oxidative stress induced by respiratory syncytial virus infection in mice. J Pineal Res. 2010;48(2):109–16.

Huang S-H, Liao C-L, Chen S-J, Shi L-G, Lin L, Chen Y-W, et al. Melatonin possesses an anti-influenza potential through its immune modulatory effect. J Funct Foods. 2019;58:189–98.

Huang C, Wang Y, Li X, Ren L, Zhao J, Hu Y, et al. Clinical features of patients infected with 2019 novel coronavirus in Wuhan, China. Lancet. 2020;395(10223):497–506.

Iba T, Levy JH, Levi M, Connors JM, Thachil J. Coagulopathy of coronavirus disease 2019. Crit Care Med. 2020.

Iguchi H, Kato K-I, Ibayashi H. Age-dependent reduction in serum melatonin concentrations in healthy human subjects. J Clin Endocrinol Metab. 1982;55(1):27–9.

Imai Y, Kuba K, Rao S, Huan Y, Guo F, Guan B, et al. Angiotensin-converting enzyme 2 protects from severe acute lung failure. Nature. 2005;436(7047):112–6.

Imai Y, Kuba K, Neely GG, Yaghubian-Malhami R, Perkmann T, van Loo G, et al. Identification of oxidative stress and Toll-like receptor 4 signaling as a key pathway of acute lung injury. Cell. 2008;133(2):235–49.

Joe B, Lokesh B. Dietary n-3 fatty acids, curcumin and capsaicin lower the release of lyso-somal enzymes and eicosanoids in rat peritoneal macrophages. Mol Cell Biochem. 2000;203(1-2):153–61.

Jung S-Y, Choi JC, You S-H, Kim W-Y. Association of renin-angiotensin-aldosterone system inhibitors with COVID-19-related outcomes in Korea: a nationwide population-based cohort study. Clin Infect Dis. 2020.

Kamat AM, Sethi G, Aggarwal BB. Curcumin potentiates the apoptotic effects of chemotherapeutic agents and cytokines through down-regulation of nuclear factor-κB and nuclear factor-κB–regulated gene products in IFN-α–sensitive and IFN-α–resistant human bladder cancer cells. Mol Cancer Ther. 2007;6(3):1022–30.

Kameda M, Teruya T, Yanagida M, Kondoh H. Frailty markers comprise blood metabolites involved in antioxidation, cognition, and mobility. Proc Natl Acad Sci. 2020;117(17):9483–9.

Kaneko I, Takeuchi Y, Yamaoka Y, Tanaka Y, Fukuda T, Fukumori Y, et al. Quantitative determination of ergothioneine in plasma and tissues by TLC-densitometry. Chem Pharm Bull. 1980;28(10):3093–7.

Kapoor KM, Kapoor A. Role of chloroquine and hydroxychloroquine in the treatment of COVID-19 infection-A systematic literature review. MedRxiv. 2020.

Kim D-H, Meza CA, Clarke H, Kim J-S, Hickner RC. Vitamin D and endothelial function. Nutrients. 2020;12(2):575.

Ko W-C, Rolain J-M, Lee N-Y, Chen P-L, Huang C-T, Lee P-I et al. Arguments in favour of remdesivir for treating SARS-CoV-2 infections. Int J Antimicrob Agents. 2020.

Krakauer T. Pentoxifylline inhibits ICAM-1 expression and chemokine production induced by proinflammatory cytokines in human pulmonary epithelial cells. Immunopharmacology. 2000;46(3):253–61.

Ksiazek TG, Erdman D, Goldsmith CS, Zaki SR, Peret T, Emery S, et al. A novel coronavirus associated with severe acute respiratory syndrome. N Engl J Med. 2003;348(20):1953–66.

Kutuzova GD, DeLuca HF. 1, 25-Dihydroxyvitamin D3 regulates genes responsible for detoxification in intestine. Toxicol Appl Pharmacol. 2007;218(1):37–44.

Lauring AS, Andino R. Quasispecies theory and the behavior of RNA viruses. PLoS Pathog. 2010;6(7):e1001005.

Lee P-I, Hsueh P-R. Emerging threats from zoonotic coronaviruses-from SARS and MERS to 2019-nCoV. J Microbiol Immunol Infect. 2020.

Lei C, Qian K, Li T, Zhang S, Fu W, Ding M, et al. Neutralization of SARS-CoV-2 spike pseudo-typed virus by recombinant ACE2-Ig. Nat Commun. 2020;11(1):1–5.

Leitner T, Kumar S. Where did SARS-CoV-2 come from? Mol Biol Evol. 2020;37(9):2463–4.

Lerner AB, Case JD, Takahashi Y, Lee TH, Mori W. Isolation of melatonin, the pineal gland factor that lightens melanocyteS1. J Am Chem Soc. 1958;80(10):2587.

Li F, Du L. MERS coronavirus: an emerging zoonotic virus. Multidisciplinary Digital Publishing Institute; 2019.

Li Q, Wu J, Nie J, Zhang L, Hao H, Liu S, et al. The impact of mutations in SARS-CoV-2 spike on viral infectivity and antigenicity. Cell. 2020a;182(5):1284-94.e9.

Li S, Jiang L, Li X, Lin F, Wang Y, Li B et al. Clinical and pathological investigation of severe COVID-19 patients. JCI Insight. 2020b.

Li X, Geng M, Peng Y, Meng L, Lu S. Molecular immune pathogenesis and diagnosis of COVID-19. J Pharm Anal. 2020c.

Li T, Lu H, Zhang W. Clinical observation and management of COVID-19 patients. Emerg Microb Infect. 2020d;9(1):687–90.

Libby P, Lüscher T. COVID-19 is, in the end, an endothelial disease. Eur Heart J. 2020;41(32):3038–44.

Lim J, Jeon S, Shin H-Y, Kim MJ, Seong YM, Lee WJ, et al. Case of the index patient who caused tertiary transmission of COVID-19 infection in Korea: the application of lopinavir/ritonavir for the treatment of COVID-19 infected pneumonia monitored by quantitative RT-PCR. J Korean Med Sci. 2020;35(6)

Lin S-C, Ho C-T, Chuo W-H, Li S, Wang TT, Lin C-C. Effective inhibition of MERS-CoV infection by resveratrol. BMC Infect Dis. 2017;17(1):1–10.

Liu H, Liu A, Shi C, Li B. Curcumin suppresses transforming growth factor-β1-induced cardiac fibroblast differentiation via inhibition of Smad-2 and p38 MAPK signaling pathways. Exp Ther Med. 2016;11(3):998–1004.

Liu Z, Gan L, Xu Y, Luo D, Ren Q, Wu S, et al. Melatonin alleviates inflammasome-induced pyroptosis through inhibiting NF-κB/GSDMD signal in mice adipose tissue. J Pineal Res. 2017;63(1):e12414.

Luo B, Li B, Wang W, Liu X, Xia Y, Zhang C, et al. NLRP3 gene silencing ameliorates diabetic cardiomyopathy in a type 2 diabetes rat model. PLoS One. 2014;9(8):e104771.

Lupfer C, Malik A, Kanneganti T-D. Inflammasome control of viral infection. Curr Opin Virol. 2015;12:38–46.

Ma S, Chen J, Feng J, Zhang R, Fan M, Han D, et al. Melatonin ameliorates the progression of atherosclerosis via mitophagy activation and NLRP3 inflammasome inhibition. Oxidative Med Cell Longev. 2018;2018

Malaguarnera L. Influence of resveratrol on the immune response. Nutrients. 2019;11(5):946.

Malhotra S, Sawhney G, Pandhi P. The therapeutic potential of melatonin: a review of the science. Medscape Gen Med. 2004;6(2)

Man SM, Karki R, Kanneganti TD. Molecular mechanisms and functions of pyroptosis, inflammatory caspases and inflammasomes in infectious diseases. Immunol Rev. 2017;277(1):61–75.

Mandi Y, Farkas G, Ocsovszky I. Effects of pentoxifyllin and PentaglobinO on TNF and IL-6 production in septic patients. Acta Microbiol Immunol Hung. 1995;42(3):301–8.

Marinella MA. Indomethacin and resveratrol as potential treatment adjuncts for SARS-CoV-2/COVID-19. Int J Clin Pract. 2020:e13535.

Martinon F. Signaling by ROS drives inflammasome activation. Eur J Immunol. 2010;40(3):616–9.

Martinot M, Jary A, Fafi-Kremer S, Leducq V, Delagreverie H, Garnier M et al. Remdesivir failure with SARS-CoV-2 RNA-dependent RNA-polymerase mutation in a B-cell immunodeficient patient with protracted Covid-19. Clin Infect Dis. 2020.

Mehta P, McAuley DF, Brown M, Sanchez E, Tattersall RS, Manson JJ, et al. COVID-19: consider cytokine storm syndromes and immunosuppression. Lancet (London, England). 2020;395(10229):1033.

Melville DB, Horner WH, Lubschez R. Tissue ergothioneine. J Biol Chem. 1954;206(1):221–8.

Michaelides M, Stover NB, Francis PJ, Weleber RG. Retinal toxicity associated with hydroxychloroquine and chloroquine: risk factors, screening, and progression despite cessation of therapy. Arch Ophthalmol. 2011;129(1):30–9.

Mikacenic C, Moore R, Dmyterko V, West TE, Altemeier WA, Liles WC, et al. Neutrophil extracellular traps (NETs) are increased in the alveolar spaces of patients with ventilator-associated pneumonia. Crit Care. 2018;22(1):358.

Modaresi M, HarfBol M, Ahmadi F. A review on pharmacological effects and therapeutic properties of curcumin. J Med Plants. 2017;2(62):1–17.

Mohd A, Zainal N, Tan K-K, AbuBakar S. Resveratrol affects Zika virus replication in vitro. Sci Rep. 2019;9(1):1–11.

Moon C. Fighting COVID-19 exhausts T cells. Nat Rev Immunol. 2020;20(5):277.

Mudgal G. Structural insights into coronavirus binding to host aminopeptidase N and interaction dynamics. 2014.

Mueller AL, McNamara MS, Sinclair DA. Why does COVID-19 disproportionately affect the elderly? 2020.

Munir K, Ashraf S, Munir I, Khalid H, Muneer MA, Mukhtar N, et al. Zoonotic and reverse zoonotic events of SARS-CoV-2 and their impact on global health. Emerg Microb Infect. 2020;9(1):2222–35.

Nagar H, Piao S, Kim C-S. Role of mitochondrial oxidative stress in sepsis. Acute Critic Care. 2018;33(2):65.

Nakai K, Fujii H, Kono K, Goto S, Kitazawa R, Kitazawa S, et al. Vitamin D activates the Nrf2-Keap1 antioxidant pathway and ameliorates nephropathy in diabetic rats. Am J Hypertens. 2014;27(4):586–95.

NaveenKumar SK, Hemshekhar M, Kemparaju K, Girish KS. Hemin-induced platelet activation and ferroptosis is mediated through ROS-driven proteasomal activity and inflammasome activation: protection by melatonin. Biochim Biophys Acta (BBA) Mol Basis Dis. 2019;1865(9):2303–16.

Neuner P, Klosner G, Schauer E, Pourmojib M, Macheiner W, Grünwald C, et al. Pentoxifylline in vivo down-regulates the release of IL-1 beta, IL-6, IL-8 and tumour necrosis factor-alpha by human peripheral blood mononuclear cells. Immunology. 1994;83(2):262.

Neuner P, Klosner G, Pourmojib M, Knobler R, Schwarz T. Pentoxifylline in vivo and in vitro down-regulates the expression of the intercellular adhesion molecule-1 in monocytes. Immunology. 1997;90(3):435–9.

Nieman DC, Peters EM, Henson DA, Nevines EI, Thompson MM. Influence of vitamin C supplementation on cytokine changes following an ultramarathon. J Interf Cytokine Res. 2000;20(11):1029–35.

Ning T, Nie J, Huang W, Li C, Li X, Liu Q, et al. Antigenic drift of influenza A (H7N9) virus hemagglutinin. J Infect Dis. 2019;219(1):19–25.

Nordlund JJ, Lerner AB. The effects of oral melatonin on skin color and on the release of pituitary hormones. J Clin Endocrinol Metab. 1977;45(4):768–74.

Nunes S, Danesi F, Del Rio D, Silva P. Resveratrol and inflammatory bowel disease: the evidence so far. Nutr Res Rev. 2018;31(1):85–97.

Olaimat AN, Aolymat I, Al-Holy M, Ayyash M, Ghoush MA, Al-Nabulsi AA, et al. The potential application of probiotics and prebiotics for the prevention and treatment of COVID-19. NPJ Sci Food. 2020;4(1):1–7.

Olival KJ, Cryan PM, Amman BR, Baric RS, Blehert DS, Brook CE, et al. Possibility for reverse zoonotic transmission of SARS-CoV-2 to free-ranging wildlife: a case study of bats. PLoS Pathog. 2020;16(9):e1008758.

Onk D, Onk OA, Erol HS, Özkaraca M, Çomaklı S, Ayazoğlu TA, et al. Effect of melatonin on antioxidant capacity, inflammation and apoptotic cell death in lung tissue of diabetic rats. Acta Cir Bras. 2018;33(4):375–85.

Oudit GY, Kassiri Z, Patel MP, Chappell M, Butany J, Backx PH, et al. Angiotensin II-mediated oxidative stress and inflammation mediate the age-dependent cardiomyopathy in ACE2 null mice. Cardiovasc Res. 2007;75(1):29–39.

Pablos MI, Agapito MT, Gutierrez R, Recio JM, Reiter RJ, Barlow-Walden L, et al. Melatonin stimulates the activity of the detoxifying enzyme glutathione peroxidase in several tissues of chicks. J Pineal Res. 1995;19(3):111–5.

Pablos MI, Reiter RJ, Ortiz GG, Guerrero JM, Agapito MT, Chuang J-I, et al. Rhythms of glutathione peroxidase and glutathione reductase in brain of chick and their inhibition by light. Neurochem Int. 1998;32(1):69–75.

Paemanee A, Hitakarun A, Roytrakul S, Smith DR. Screening of melatonin, α-tocopherol, folic acid, acetyl-L-carnitine and resveratrol for anti-dengue 2 virus activity. BMC Res Notes. 2018;11(1):1–7.

Panahi Y, Hosseini MS, Khalili N, Naimi E, Simental-Mendía LE, Majeed M, et al. Effects of curcumin on serum cytokine concentrations in subjects with metabolic syndrome: A post-hoc analysis of a randomized controlled trial. Biomed Pharmacother. 2016;82:578–82.

Panesar NS. Lymphopenia in SARS. Lancet. 2003;361(9373):1985.

Papavasiliou PS, Cotzias GC, Düby SE, Steck AJ, Bell M, Lawrence WH. Melatonin and parkinsonism. JAMA. 1972;221(1):88–9.

Pennathur S, Heinecke JW. Oxidative stress and endothelial dysfunction in vascular disease. Curr Diab Rep. 2007;7(4):257–64.

Petrie JG, Lauring AS. Influenza A (H7N9) virus evolution: which genetic mutations are antigenically important? New York: Oxford University Press; 2019.

Pittet J-F, Griffiths MJ, Geiser T, Kaminski N, Dalton SL, Huang X, et al. TGF-β is a critical mediator of acute lung injury. J Clin Invest. 2001;107(12):1537–44.

Plummer SM, Holloway KA, Manson MM, Munks RJ, Kaptein A, Farrow S, et al. Inhibition of cyclo-oxygenase 2 expression in colon cells by the chemopreventive agent curcumin involves inhibition of NF-κB activation via the NIK/IKK signalling complex. Oncogene. 1999;18(44):6013–20.

Prauchner CA. Oxidative stress in sepsis: pathophysiological implications justifying antioxidant co-therapy. Burns. 2017;43(3):471–85.

Qiao Y, Wang P, Qi J, Zhang L, Gao C. TLR-induced NF-κB activation regulates NLRP3 expression in murine macrophages. FEBS Lett. 2012;586(7):1022–6.

Qin C, Zhou L, Hu Z, Zhang S, Yang S, Tao Y et al. Dysregulation of immune response in patients with COVID-19 in Wuhan, China. Clin Infect Dis. 2020.

Rabelo LA, Alenina N, Bader M. ACE2–angiotensin-(1–7)–Mas axis and oxidative stress in cardiovascular disease. Hypertens Res. 2011;34(2):154–60.

Raharusun P. Patterns of COVID-19 mortality and vitamin D: an Indonesian study. Available at SSRN 3585561. 2020.

Rahman I, Gilmour PS, Jimenez LA, Biswas SK, Antonicelli F, Aruoma OI. Ergothioneine inhibits oxidative stress-and TNF-α-induced NF-κB activation and interleukin-8 release in alveolar epithelial cells. Biochem Biophys Res Commun. 2003;302(4):860–4.

Reed WR, Degowin RL. Suppressive effects of pentoxifylline on natural killer cell activity. J Lab Clin Med. 1992;119(6):763–71.

Reguera J, Santiago C, Mudgal G, Ordono D, Enjuanes L, Casasnovas JM. Structural bases of coronavirus attachment to host aminopeptidase N and its inhibition by neutralizing antibodies. PLoS Pathog. 2012;8(8):e1002859.

Reguera J, Mudgal G, Santiago C, Casasnovas JM. A structural view of coronavirus–receptor interactions. Virus Res. 2014;194:3–15.

Reiter RJ. Melatonin: lowering the high price of free radicals. Physiology. 2000;15(5):246–50.

Reiter R, Tang L, Garcia JJ, Muñoz-Hoyos A. Pharmacological actions of melatonin in oxygen radical pathophysiology. Life Sci. 1997;60(25):2255–71.

Reiter RJ, Mayo JC, Tan DX, Sainz RM, Alatorre-Jimenez M, Qin L. Melatonin as an antioxidant: under promises but over delivers. J Pineal Res. 2016;61(3):253–78.

Reiter RJ, Ma Q, Sharma R. Treatment of Ebola and other infectious diseases: melatonin "goes viral". Melatonin Res. 2020;3(1):43–57.

Repine JE, Elkins ND. Effect of ergothioneine on acute lung injury and inflammation in cytokine insufflated rats. Prev Med. 2012;54:S79–82.

Reppert SM, Chez RA, Anderson A, Klein DC. Maternal-fetal transfer of melatonin in the non-human primate. Pediatr Res. 1979;13(6):788–91.

Rincon J, Correia D, Arcaya J, Finol E, Fernández A, Pérez M, et al. Role of Angiotensin II type 1 receptor on renal NAD (P) H oxidase, oxidative stress and inflammation in nitric oxide inhibition induced-hypertension. Life Sci. 2015;124:81–90.

Rougee M, Bensasson R, Land EJ, Pariente R. Deactivation of singlet molecular oxygen by thiols and related compounds, possible protectors against skin photosensitivity. Photochem Photobiol. 1988;47(4):485–9.

Ruan Q, Yang K, Wang W, Jiang L, Song J. Clinical predictors of mortality due to COVID-19 based on an analysis of data of 150 patients from Wuhan, China. Intensive Care Med. 2020;46(5):846–8.

Saber-Moghaddam N, Salari S, Hejazi S, Amini M, Taherzadeh Z, Eslami S, et al. Oral nano-curcumin formulation efficacy in management of mild to moderate hospitalized coronavirus disease-19 patients: An open label nonrandomized clinical trial. Phytother Res. 2021; https://doi.org/10.1002/ptr.7004.

Saeedi-Boroujeni A, Mahmoudian-Sani M-R, Bahadoram M, Alghasi A. COVID-19: a case for inhibiting NLRP3 inflammasome, suppression of inflammation with curcumin? Basic Clin Pharmacol Toxicol. 2021;128(1):37–45. https://doi.org/10.1111/bcpt.13503.

Saha S, Singh KM, Gupta BBP. Melatonin synthesis and clock gene regulation in the pineal organ of teleost fish compared to mammals: Similarities and differences. Gen Comp Endocrinol. 2019;279:27–34.

Sahu A, Kasoju N, Bora U. Fluorescence study of the curcumin – casein micelle complexation and its application as a drug nanocarrier to cancer cells. Biomacromolecules. 2008;9(10):2905–12.

Sanchis-Gomar F, Perez-Quilis C, Favaloro EJ, Lippi G. Statins and other drugs: facing COVID-19 as a vascular disease. Pharmacol Res. 2020a;159:105033.

Sanchis-Gomar F, Lavie CJ, Morin DP, Perez-Quilis C, Laukkanen JA, Perez MV. Amiodarone in the COVID-19 era: treatment for symptomatic patients only, or drug to prevent infection? Am J Cardiovasc Drugs. 2020b;20(5):413–8.

Sawalha AH, Zhao M, Coit P, Lu Q. Epigenetic dysregulation of ACE2 and interferon-regulated genes might suggest increased COVID-19 susceptibility and severity in lupus patients. Clin Immunol. 2020:108410.

Scheer FA, Van Montfrans GA, van Someren EJ, Mairuhu G, Buijs RM. Daily nighttime melatonin reduces blood pressure in male patients with essential hypertension. Hypertension. 2004;43(2):192–7.

Schönrich G, Raftery MJ. Neutrophil extracellular traps go viral. Front Immunol. 2016;7:366.

Schönrich G, Raftery MJ, Samstag Y. Devilishly radical NETwork in COVID-19: oxidative stress, neutrophil extracellular traps (NETs), and T cell suppression. Adv Biol Regul. 2020;77:100741.

Schwensen HF, Borreschmidt LK, Storgaard M, Redsted S, Christensen S, Madsen LB. Fatal pulmonary fibrosis: a post-COVID-19 autopsy case. J Clin Pathol. 2020.

Scott JF, Das LM, Ahsanuddin S, Qiu Y, Binko AM, Traylor ZP, et al. Oral vitamin D rapidly attenuates inflammation from sunburn: an interventional study. J Investig Dermatol. 2017;137(10):2078–86.

Shaw SM, Shah MK, Williams SG, Fildes JE. Immunological mechanisms of pentoxifylline in chronic heart failure. Eur J Heart Fail. 2009;11(2):113–8.

Shi H, Wang X, Tan DX, Reiter RJ, Chan Z. Comparative physiological and proteomic analyses reveal the actions of melatonin in the reduction of oxidative stress in Bermuda grass (Cynodon dactylon (L). Pers.). J Pineal Res. 2015;59(1):120–31.

Shi J, Gao W, Shao F. Pyroptosis: gasdermin-mediated programmed necrotic cell death. Trends Biochem Sci. 2017;42(4):245–54.

Sigmundsdottir H, Pan J, Debes GF, Alt C, Habtezion A, Soler D, et al. DCs metabolize sunlight-induced vitamin D3 to 'program' T cell attraction to the epidermal chemokine CCL27. Nat Immunol. 2007;8(3):285–93.

Singh S, Aggarwal BB. Activation of transcription factor NF-κB is suppressed by curcumin (diferuloylmethane). J Biol Chem. 1995;270(42):24995–5000.

Slominski AT, Chaiprasongsuk A, Janjetovic Z, Kim T-K, Stefan J, Slominski RM, et al. Photoprotective properties of vitamin D and lumisterol hydroxyderivatives. Cell Biochem Biophys. 2020:1–16.

Smith E, Ottosson F, Hellstrand S, Ericson U, Orho-Melander M, Fernandez C, et al. Ergothioneine is associated with reduced mortality and decreased risk of cardiovascular disease. Heart. 2020;106(9):691–7.

Song M-Y, Kim E-K, Moon W-S, Park J-W, Kim H-J, So H-S, et al. Sulforaphane protects against cytokine-and streptozotocin-induced β-cell damage by suppressing the NF-κB pathway. Toxicol Appl Pharmacol. 2009;235(1):57–67.

Song T-Y, Yang N-C, Chen C-L, Thi TLV. Protective effects and possible mechanisms of ergothioneine and hispidin against methylglyoxal-induced injuries in rat pheochromocytoma cells. Oxidative Med Cell Longev. 2017;2017

Sotgia S, Zinellu A, Mangoni AA, Pintus G, Attia J, Carru C, et al. Clinical and biochemical correlates of serum L-ergothioneine concentrations in community-dwelling middle-aged and older adults. PLoS One. 2014;9(1):e84918.

Stewart MK, Cookson BT. Evasion and interference: intracellular pathogens modulate caspase-dependent inflammatory responses. Nat Rev Microbiol. 2016;14(6):346.

Stoffels C, Oumari M, Perrou A, Termath A, Schlundt W, Schmalz H-G, et al. Ergothioneine stands out from hercynine in the reaction with singlet oxygen: resistance to glutathione and

TRIS in the generation of specific products indicates high reactivity. Free Radic Biol Med. 2017;113:385–94.

Stoiber W, Obermayer A, Steinbacher P, Krautgartner W-D. The role of reactive oxygen species (ROS) in the formation of extracellular traps (ETs) in humans. Biomolecules. 2015;5(2):702–23.

Su H, Wan C, Song A, Qiu Y, Xiong W, Zhang C. Oxidative stress and renal fibrosis: mechanisms and therapies. In: Renal fibrosis: mechanisms and therapies. Springer; 2019. p. 585–604.

Sullivan RJ, Johnson DB, Rini BI, Neilan TG, Lovly CM, Moslehi JJ, et al. COVID-19 and immune checkpoint inhibitors: initial considerations. J Immunother Cancer. 2020;8(1)

Tan D-X. Melatonin: a potent, endogenous hydroxyl radical scavenger. Endocr J. 1993;1:57–60.

Tan DX, Manchester LC, Terron MP, Flores LJ, Reiter RJ. One molecule, many derivatives: a never-ending interaction of melatonin with reactive oxygen and nitrogen species? J Pineal Res. 2007;42(1):28–42.

Tang RMY, Cheah IK-M, Yew TSK, Halliwell B. Distribution and accumulation of dietary ergothioneine and its metabolites in mouse tissues. Sci Rep. 2018;8(1):1–15.

Tang N, Li D, Wang X, Sun Z. Abnormal coagulation parameters are associated with poor prognosis in patients with novel coronavirus pneumonia. J Thromb Haemost. 2020a;18(4):844–7.

Tang N, Bai H, Chen X, Gong J, Li D, Sun Z. Anticoagulant treatment is associated with decreased mortality in severe coronavirus disease 2019 patients with coagulopathy. J Thromb Haemost. 2020b;18(5):1094–9.

Tesoriere L, Allegra M, D'Arpa D, Butera D, Livrea M. Reaction of melatonin with hemoglobin-derived oxoferryl radicals and inhibition of the hydroperoxide-induced hemoglobin denaturation in red blood cells. J Pineal Res. 2001;31(2):114–9.

Tsetsarkin KA, Vanlandingham DL, McGee CE, Higgs S. A single mutation in chikungunya virus affects vector specificity and epidemic potential. PLoS Pathog. 2007;3(12):e201.

Tucker RA, Cheah IK, Halliwell B. Specificity of the ergothioneine transporter natively expressed in HeLa cells. Biochem Biophys Res Commun. 2019;513(1):22–7.

Ungvari Z, Bagi Z, Feher A, Recchia FA, Sonntag WE, Pearson K, et al. Resveratrol confers endothelial protection via activation of the antioxidant transcription factor Nrf2. Am J Phys Heart Circ Phys. 2010;299(1):H18–24.

Urbanowicz RA, McClure CP, Sakuntabhai A, Sall AA, Kobinger G, Müller MA, et al. Human adaptation of Ebola virus during the West African outbreak. Cell. 2016;167(4):1079-87.e5.

Valente AJ, Yoshida T, Murthy SN, Sakamuri SSVP, Katsuyama M, Clark RA, et al. Angiotensin II enhances AT1-Nox1 binding and stimulates arterial smooth muscle cell migration and proliferation through AT1, Nox1, and interleukin-18. Am J Phys Heart Circ Phys. 2012;303(3):H282–H96. https://doi.org/10.1152/ajpheart.00231.2012.

Vijay R, Hua X, Meyerholz DK, Miki Y, Yamamoto K, Gelb M, et al. Critical role of phospholipase A2 group IID in age-related susceptibility to severe acute respiratory syndrome–CoV infection. J Exp Med. 2015;212(11):1851–68.

Vinayak A, Mudgal G, Sharma S, Singh GB. Prebiotics for probiotics. In: Advances in probiotics for sustainable food and medicine. Springer; 2021. p. 63–82.

Violi F, Oliva A, Cangemi R, Ceccarelli G, Pignatelli P, Carnevale R, et al. Nox2 activation in Covid-19. Redox Biol. 2020;36:101655.

Von Essen MR, Kongsbak M, Schjerling P, Olgaard K, Ødum N, Geisler C. Vitamin D controls T cell antigen receptor signaling and activation of human T cells. Nat Immunol. 2010;11(4):344–9.

Waldhauser F, Weiszenbacher G, Tatzer E, Gisinger B, Waldhauser M, Schemper M, et al. Alterations in nocturnal serum melatonin levels in humans with growth and aging. J Clin Endocrinol Metab. 1988;66(3):648–52.

Wang X-L, Li T, Li J-H, Miao S-Y, Xiao X-Z. The effects of resveratrol on inflammation and oxidative stress in a rat model of chronic obstructive pulmonary disease. Molecules. 2017a;22(9):1529.

Wang C-Y, Lu C-Y, Li S-W, Lai C-C, Hua C-H, Huang S-H, et al. SARS coronavirus papain-like protease up-regulates the collagen expression through non-Samd TGF-β1 signaling. Virus Res. 2017b;235:58–66.

Wang X, Bian Y, Zhang R, Liu X, Ni L, Ma B, et al. Melatonin alleviates cigarette smoke-induced endothelial cell pyroptosis through inhibiting ROS/NLRP3 axis. Biochem Biophys Res Commun. 2019;519(2):402–8.

Wang J-Z, Zhang R-Y, Bai J. An anti-oxidative therapy for ameliorating cardiac injuries of critically ill COVID-19-infected patients. Int J Cardiol. 2020.

White NJ, Wang Y, Fu X, Cardenas JC, Martin EJ, Brophy DF, et al. Post-translational oxidative modification of fibrinogen is associated with coagulopathy after traumatic injury. Free Radic Biol Med. 2016;96:181–9.

WHO. World Health Organization coronavirus disease (COVID-19) dashboard. 2020.

Wimalawansa SJ. Vitamin D deficiency: effects on oxidative stress, epigenetics, gene regulation, and aging. Biology. 2019;8(2):30.

Windmeier C, Gressner A. Pharmacological aspects of pentoxifylline with emphasis on its inhibitory actions on hepatic fibrogenesis. Gen Pharmacol Vasc Syst. 1997;29(2):181–96.

Wu D, Yang XO. TH17 responses in cytokine storm of COVID-19: an emerging target of JAK2 inhibitor Fedratinib. J Microbiol Immunol Infect. 2020;

Wu H, Wang Y, Zhang Y, Xu F, Chen J, Duan L, et al. Breaking the vicious loop between inflammation, oxidative stress and coagulation, a novel anti-thrombus insight of nattokinase by inhibiting LPS-induced inflammation and oxidative stress. Redox Biol. 2020a:101500.

Wu C, Chen X, Cai Y, Zhou X, Xu S, Huang H et al. Risk factors associated with acute respiratory distress syndrome and death in patients with coronavirus disease 2019 pneumonia in Wuhan, China. JAMA Intern Med. 2020b.

Xia N, Daiber A, Habermeier A, Closs EI, Thum T, Spanier G, et al. Resveratrol reverses endothelial nitric-oxide synthase uncoupling in apolipoprotein E knockout mice. J Pharmacol Exp Ther. 2010;335(1):149–54.

Xiao L, Zhao L, Li T, Hartle DK, Aruoma OI, Taylor EW. Activity of the dietary antioxidant ergothioneine in a virus gene-based assay for inhibitors of HIV transcription. Biofactors. 2006;27(1-4):157–65.

Xu Z, Shi L, Wang Y, Zhang J, Huang L, Zhang C, et al. Pathological findings of COVID-19 associated with acute respiratory distress syndrome. Lancet Respir Med. 2020;8(4):420–2.

Yang M. Cell pyroptosis, a potential pathogenic mechanism of 2019-nCoV infection. Available at SSRN 3527420. 2020.

Yang F, Wang Z, Wei X, Han H, Meng X, Zhang Y, et al. NLRP3 deficiency ameliorates neurovascular damage in experimental ischemic stroke. J Cereb Blood Flow Metab. 2014;34(4):660–7.

Yang Y, Shen C, Li J, Yuan J, Yang M, Wang F et al. Exuberant elevation of IP-10, MCP-3 and IL-1ra during SARS-CoV-2 infection is associated with disease severity and fatal outcome. MedRxiv. 2020a.

Yang Y, Peng F, Wang R, Guan K, Jiang T, Xu G, et al. The deadly coronaviruses: the 2003 SARS pandemic and the 2020 novel coronavirus epidemic in China. J Autoimmun. 2020b:102434.

Yang M, Wei J, Huang T, Lei L, Shen C, Lai J, et al. Resveratrol inhibits the replication of severe acute respiratory syndrome coronavirus 2 (SARS-CoV-2) in cultured Vero cells. Phytother Res. 2020c; https://doi.org/10.1002/ptr.6916.

Yao X, Ye F, Zhang M, Cui C, Huang B, Niu P et al. In vitro antiviral activity and projection of optimized dosing design of hydroxychloroquine for the treatment of severe acute respiratory syndrome coronavirus 2 (SARS-CoV-2). Clin Infect Dis. 2020.

Ye Q, Wang B, Mao J. The pathogenesis and treatment of theCytokine Storm in COVID-19. J Infect. 2020;80(6):607–13.

Yeo C, Kaushal S, Yeo D. Enteric involvement of coronaviruses: is faecal–oral transmission of SARS-CoV-2 possible? Lancet Gastroenterol Hepatol. 2020;5(4):335–7.

Zablocki D, Sadoshima J. Angiotensin II and oxidative stress in the failing heart. Antioxid Redox Signal. 2013;19(10):1095–109.

Zatta P, Tognon G, Carampin P. Melatonin prevents free radical formation due to the interaction between β-amyloid peptides and metal ions [Al (III), Zn (II), Cu (II), Mn (II), Fe (II)]. J Pineal Res. 2003;35(2):98–103.

Zeng H, Wang D, Nie J, Liang H, Gu J, Zhao A, et al. The efficacy assessment of convalescent plasma therapy for COVID-19 patients: a multi-center case series. Signal Transduct Target Ther. 2020;5(1):219. https://doi.org/10.1038/s41392-020-00329-x.

Zhang Y, Li X, Grailer JJ, Wang N, Wang M, Yao J, et al. Melatonin alleviates acute lung injury through inhibiting the NLRP3 inflammasome. J Pineal Res. 2016;60(4):405–14.

Zhang H, Penninger JM, Li Y, Zhong N, Slutsky AS. Angiotensin-converting enzyme 2 (ACE2) as a SARS-CoV-2 receptor: molecular mechanisms and potential therapeutic target. Intensive Care Med. 2020a;46(4):586–90.

Zhang W, Zhao Y, Zhang F, Wang Q, Li T, Liu Z, et al. The use of anti-inflammatory drugs in the treatment of people with severe coronavirus disease 2019 (COVID-19): the experience of clinical immunologists from China. Clin Immunol. 2020b:108393.

Zhao X, Sun J, Su W, Shan H, Zhang B, Wang Y, et al. Melatonin protects against lung fibrosis by regulating the Hippo/YAP pathway. Int J Mol Sci. 2018;19(4):1118.

Zhong Z, Zhai Y, Liang S, Mori Y, Han R, Sutterwala FS, et al. TRPM2 links oxidative stress to NLRP3 inflammasome activation. Nat Commun. 2013;4(1):1–11.

Zhou R, Tardivel A, Thorens B, Choi I, Tschopp J. Thioredoxin-interacting protein links oxidative stress to inflammasome activation. Nat Immunol. 2010;11(2):136–40.

Zhou Y-H, Qin Y-Y, Lu Y-Q, Sun F, Yang S, Harypursat V, et al. Effectiveness of glucocorticoid therapy in patients with severe novel coronavirus pneumonia: protocol of a randomized controlled trial. Chin Med J. 2020a;10

Zhou F, Yu T, Du R, Fan G, Liu Y, Liu Z et al. Clinical course and risk factors for mortality of adult inpatients with COVID-19 in Wuhan, China: a retrospective cohort study. Lancet. 2020b.

Zhou P, Yang X-L, Wang X-G, Hu B, Zhang L, Zhang W, et al. A pneumonia outbreak associated with a new coronavirus of probable bat origin. Nature. 2020c;579(7798):270–3.

Zisapel N. New perspectives on the role of melatonin in human sleep, circadian rhythms and their regulation. Br J Pharmacol. 2018;175(16):3190–9.

Synergistic Effects of Heavy Water in Health Prospects

Jyoti Verma

Contents

Abstract This chapter provides an insight into heavy water/deuterium use in the medicinal chemistry and biotechnology, due to its physicochemical characteristics that are reported to have potential applications in the applied research. The synergistic effects of heavy water in health prospects are vividly sketched to understand its explorable properties. A few case studies, citing the examples of various pharmaceuticals employing the active isotope in the drugs, that are under clinical trials have been discussed. An outline of a few reported patents regarding the same is also presented to get a gist of their commercial significance in the biomedical industry. The biomolecule in question is not only versatile but also exhibits pseudo-effect in the biological system and expected to have a boom in the heavily invested pharma industry. The potential conflicts that may arise regarding the patent rights with respect to the molecule are also covered in the closing perspectives.

J. Verma (✉)
ONGC Energy Centre, Delhi, India
e-mail: verma_jyoti@ongc.co.in

© The Author(s), under exclusive license to Springer Nature Switzerland AG 2021   359
K. K. Kesari, N. K. Jha (eds.), *Free Radical Biology and Environmental Toxicity*,
Molecular and Integrative Toxicology, https://doi.org/10.1007/978-3-030-83446-3_17

Keywords Heavy water · deuteration · deuterium-labeled stimulated Raman
scattering · hydrogen isotope · deuterated drugs · biological disorders

Abbreviations

BNCT	Boron neutron capture therapy
CM	Chylomicron
CMR	Chylomicron remnant
CSDS	Chronic social defeat stress
DIE	Deuterium isotope effect
DNL	De novo lipogenesis
FA	Fatty acid
FDA	Food and Drug Administration
HA	Hyaluronic acid
HWB	Heavy Water Board
LC-MS	Liquid chromatography–mass spectrometry
LNS	Lipid-based nutrient supplements
LPL	Lipoprotein lipase
MID	Mass isotopologue distribution
MPS	Muscle protein synthesis
MRI	Magnetic resonance imaging
NAFLD	Nonalcoholic fatty liver disease
NASH	Nonalcoholic hepatosteatosis
NMDAR	N-Methyl-D-aspartate receptor
NMR	Nuclear magnetic resonance spectroscopy
OPV	Oral polio vaccine
PHWR	Pressurized heavy water reactor
PMMA	Polymethyl methacrylate
QTOF-FI-MS	Quadrupole-time-of-flight mass spectrometry
SRS	Stimulated Raman scattering
TAG	Triacylglycerol
TBW	Total body water
VLDL	Very low-density lipoprotein

1 Introduction

A nobel laureate and chemistry professor at Columbia, Dr. Harold C. Urey in 1932
discovered a versatile heavy water (D_2O) molecule by separating deuterium from
liquid hydrogen that revolutionized tremendous interest in biochemical application

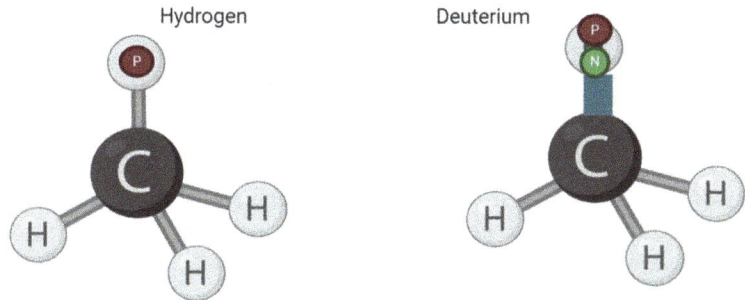

Fig. 1 Hydrogen-carbon structure and deuterium-carbon structure (heavy isotopes make comparatively stronger bonds)

of nuclear power production. The viscosity and melting and boiling points of D_2O are relatively higher than H_2O, while also possessing about 2600 times neutron-moderating ability due to which it has been used as an optimized source for moderator in pressurized heavy water reactor (PHWR). However, major interest in D_2O's biophysical properties resurged during the last two decades regarding its usage in medicinal chemistry. The isotope effects give rise to difference in the zero-point energy and bond energies resulting in composite molecular structure and behavior, thereby exhibiting different physicochemical properties as compared to normal water. Figure 1 shows the bond strength of carbon with hydrogen and deuterium, respectively. Deuterium is also known to track changes in ocean circulation, star formation, and nuclear power generation studies.

Deuterium constitutes about 0.015% of hydrogen atoms in natural water (International Atomic Energy Agency 2011). Initially, chemical plants meant to produce hydrogen produced deuterium as a by-product. Nuclear weapons used heavy water as feed material which was discouraged later. D_2O has been used in analytical studies as solvent for dissolution.

Categorically, D_2O has following effects on the living systems:

1. "Solvent Isotope Effect" based on the molecular properties of D_2O on the structure of water and biological macromolecules.
2. "Deuterium Isotope Effect" (DIE), based on the ability of D_2O to replace H with D in biological molecules. The C-D bond is manyfold stronger than the C-H bond (Fig. 1) and therefore, more resistant to enzymatic and chemical cleavage.

Mammals can tolerate up to 30% deuterium isotope fraction in drinking water ($fD_{\text{drinking water}} = 0.3$), leading to a steady-state blood serum deuterium isotope fraction of $\sim 21\%$ ($fB_{\text{blood serum}} = 0.21$). The life expectancy of mammals remains unaltered within this limit with completely reversible side effects upon restoration of the normal fD (= 0.000 15) value. fB levels up to 0.23 over short periods are harmless to humans, thus enabling viable clinical employment of D_2O (Blagojevic et al. 1994).

2 Attributes of Heavy Water Making It an Ideal Candidate for Studying Biological System

Since long, the medicinal usage of D_2O in lower concentrations due to its non-radioactive nature and instant permeability to all body compartments, tissues and cell types, has been confined to use as a tracer molecule in clinical nutrition and for studying parameters of human physiology in infants, pregnant and lactating mothers as well as healthy adults. These investigations are conducted using NMR (nuclear magnetic resonance) and spectrophotometry.

Preliminary studies have reported the anti-genotoxic and radioprotective effects of D_2O in mice and reduced salt- and ethanol-induced hypertension in rats. However, serious development regarding the idea of replacing normal hydrogen by deuteriums in a drug molecule has been underway only since last decade. Deuterium, due to the presence of one neutron in it, behaves like the "heavy hydrogen" isotope of regular hydrogen (having no neutron), thereby doubling its weight. Needless to say, conversion of hydrogen to tritium, the two-neutron isotope, is the highest weight change percentage, but tritium is radioactive and therefore its use is discouraged in such kind of biological studies. The idea of replacement of hydrogen with deuterium in a drug is being vividly researched because this weight difference makes a huge impact as a higher bond dissociation energy is required to break the bond with the heavier isotope deuterium. Eventually, this can slow down the metabolism of such a drug, preventing its early clearance in the biological system when they encounter the liver, as compared to the drug having regular hydrogen.

Another attribute of D_2O is relevant to thermo-stabilization of biological macro-molecules, cells, and tissues to preserve their biological activity, such as in the case of certain vaccines like oral polio vaccine (OPV) or other macromolecules requiring freezing temperatures.

Few studies have reported the therapeutic use of D_2O in controlling the invasive-ness of malignant cells in biological systems, including those of pancreas which are difficult to treat. Although D_2O is more toxic to malignant cells as compared to the normal animal cells still, new modalities are needed to explore the optimized con-centration requirements to establish their therapeutic efficacy. Deuteration of anti-cancer drugs is known to lower the side effects without affecting its efficacy. The therapeutic effect of photosensitizers can also be modulated by D_2O.

3 Chemistry of Deuteration Reactions

Although basic H-D exchange is a redox-neutral transformation, a few studies have reported using elements like zinc as reducing agent in such processes. Different kinds of additives via NMR spectroscopic analysis have shown remarkable varia-tions in the observed deuteration (Pascal et al. 2019). For example, while using zinc (additive), D_2 gas production can be experimentally proven. The research group suggested that by using systematic variation of ligands, the selectivity trends can be amplified to further improvise the methods for deuteration studies. Pascal et al. (2019) reported a preparative simple selective deuteration method for polyfunctional organic molecules under mild conditions. Based on the additive used, the pre-catalyst $[RuCl_2(PPh_3)_3]$ was converted into various defined Ru complexes by employing either CuI (cat.)/Zn, or KOD (cat.) or KOD (cat.)/Zn. When D_2O was used as a deuterium source, different complex chemo-selectivities were observed in H–D exchange reactions, individually.

The significance of developing such kind of protocols is to facilitate the studies regarding isotopic exchange of hydrogen atoms for deuterium or tritium, to understand biosynthesis and metabolism pathways (Atzrodt et al. 2007, 2018; Sajiki et al. 2004). Also, rapid oxidative metabolic degradation of reactive C-H bonds can be controlled by position-selective deuteration (Pirali et al. 2019; Russak and Bednarczyk 2019). Exploitation of the kinetic isotope effect causes reactivity moderation and is likely to enhance the bioavailability and stability of a pharmacophore in living systems, while also enabling understanding of mechanistic issues (Howland 2015; Andrieu et al. 2006). Therefore, global efforts for improving chemo-selectivity of deuteration methods are underway. The use of D_2O (or T_2O) is found to be more feasible for deuteration, in contrast to D_2 or T_2 gases that are hazardous (Yu et al. 2016; Loh et al. 2017; Garreau et al. 2019).

4 Deuteration Used by Pharmaceuticals

The 1950s–1960s era started shifting its focus on evaluating the bio-therapeutic potential of D_2O in health-care system. Many biological studies suggest that the therapeutic efficacy of pharmaceutical drugs can be enhanced via its deuteration resulting in slowing down of their metabolic pathway(s) to increase their pharmacological effects and/or lowering of toxic manifestations. Inserting deuterium at strategic sites in the compound molecular structure leads to the formation of deuterium–carbon bonds that are ten-fold unassailable as compared to hydrogen–carbon bonds. One such example of FDA-approved deuterated drug is deutetrabenazine (targets the vesicular monoamine transporter type 2, a transport protein on presynaptic neurons) from Teva pharmaceuticals (an American Israeli pharmaceutical company), used for the treatment of Huntington's chorea. The unmodified version of tetrabenazine has 2 methoxy groups on its aryl ring which is targeted by the metabolizing enzymes resulting in demethylation to phenol/catechol. These molecules are further disposed of by kidneys by attaching some other group (like a glucuronic acid) on the phenolic OH. The treatment mechanism involves deutetrabenazine depleting the monoamines dopamine and serotonin and other monoamine neurotransmitters for reducing the involuntary writhing linked with the disorder. Administering lower Austedo's placebo-controlled dose during phase 3 clinical trial as compared to non-deuterated drug for attaining effective and desirable concentrations also suppressed the disease symptoms. Therefore, reduced side effects (sleepiness, depression, and anxiety) were observed.

While deuteration works for a couple of drugs used in clinical practice approved by Canadian and US regulatory agencies, others have not significantly shown enhanced performances. The other example is deuterated drug CTP-499 for the treatment of diabetic nephropathy, which was studied for its pharmacokinetic profile by CoNCERT Pharmaceuticals Inc. In March 2017, CoNCERT Pharmaceuticals (Lexington, Massachusetts) sold its deuterated Kalydeco (ivacaftor) for treating cystic fibrosis (patients with G551D mutations) to Vertex Pharmaceuticals (Boston). CTP-656 (deuterated Kalydeco) can be taken in low doses and, unlike Kalydeco, does not require a high calorie meal for optimal absorption.

CoNCERT's AVP-786 deuterated drug that was the modified version of the cough suppressant dextromethorphan targeted dextrorphan (its major metabolite). The Food and Drug Administration (FDA) approved N-methyl-d-aspartate (NMDA) receptor antagonist (undeuterated dextromethorphan), in combination with the drug quinidine for treating involuntary laughing spells/crying in patients with neurological disorder. Although quinidine has cardiovascular side effects at higher doses, it promotes the accumulation of dextromethorphan by lowering its breakdown rate to attain desirable therapeutic concentration. To troubleshoot this issue, Avanir Pharmaceuticals (California) collaborated with CoNCERT for the production of amalgamated quinidine with more stable deuterated AVP-786 (Nuedexta) to administer controlled quinidine doses in Alzheimer's patients and also patients witnessing depression and schizophrenia. Table 1 shows the examples of a few deuterated drugs manufactured by pharmaceuticals (Schmidt 2017).

5   A Few Reported Case Studies of Heavy Water in Animal and Plant Biotechnology

1. Boron neutron capture therapy (BNCT) is a nonsurgical therapeutic modality, used for the treatment of locally invasive tumorous cancers, which involves selectively concentrating boron compounds into tumor cells, followed by subjecting epithermal neutron beam radiations to such tumor cells. D_2O has shown to enhance the neutron penetration to boron compounds bound to malignant cells while limiting healthy tissue radiation dose deposition during the therapy.

BNCT mechanism employs (1) administration cum accumulation of boronated compound in high concentrations at the tumor area and (2) targeting thermalized neutron flux to the boronated site. The incoming neutrons are captured by boron (^{10}B) atoms at the tumor locus resulting in the formation of an intermediate species that can undergo α-decay to ^7Li, $^{10}B(n, \alpha)^7Li$, depositing 2.31 MeV within 10–15 μm of tissue parenchyma. The only technology limitation is that the incoming neutrons are excessively moderated by elastic scattering with large thermal neutron capture cross section of protons that lowers the neutron count reaching the boronated target site, while also enhancing the recoil γ-radiation dose deposition from the proton capture reaction. Moreover, when the target area is partially deuterated, this

Table 1 Deuterated drugs manufactured by pharmaceutica.s (Schmidt 2017)

S. no.	Deuterated drug	Name	Target/site	Disorder	Pharmaceutical
1.	Apremilast	CT-730	PDE4	Inflammatory	Celgene/Concert
2.	Dextromethorphan	AVP-786	NMDA glutamate receptor	Alzheimer's disease, agitation symptoms, depression/schizophrenia	Avanir/Concert
3.	Doxorubicin	VX-984	DNA-dependent protein kinase	Solid tumors	Vertex
4.	Kalydeco (ivacaftor)	CTP-656	CF transmembrane conductor regulator	Cystic fibrosis	Vertex
5.	Linoleic acid	RT001	Cell membrane	Freidreich's ataxia	Retrotope
6.	Pioglitazone	DRX-065	(PPAR-Y)	Nonalcoholic steatohepatitis/adrenoleukodystrophy	DeuteRx
7.	Rosuvastatin	BMS-986165	Tyrosine kinases	Psoriasis	BMS
8.	Ruxolitinib	CTP-543	JAK/STAT	Alopecia areata (hair loss)	Concert

increments the neutron flux homogeneity and minimizes the local "hot-spots" formation facilitating effective delivery of concentrated radiation doses to tumor cell site, additionally safeguarding the normal cell integrity of healthy tissue (Sakurai 2004).

Daniel et al. (2005) studied the implications for boron neutron capture therapy of malignant brain tumors via pharmaco-thermodynamic effects of deuterium-induced edema in adult male rat brain via ^1H_2O MRI. Confoundingly, deuterium being used in clinical application for having antitumor (antimitotic) activity for conferring protection against γ-irradiation at high doses, deuterium induces brain edema that is inimical to neutron capture therapy. The study reported uniform bio-distribution of both deuterium and deuterium-induced brain edema in brain. During steady-state blood fluid deuteration, mammals consuming highly deuterated water may show neurological and behavioral abnormalities followed by compression of brain structures, such as the basal ganglia, the cerebellum, and the brain stem. It was also revealed that edema negatively impacts thermal neutron flux penetrance and effective dose reduction by $\sim 10\%$ due to increase in the brain volume.

2. Deuterium can also stabilize drug enantiomers (flipping between mirror images, with different properties) into required orientation. For example, thalidomide (mixture of the right-handed R and the left-handed S enantiomer), approved in Europe a few years ago for treating morning sickness in pregnant women. It was reported that while the R enantiomer was therapeutic, the S enantiomer caused teratogenic effects resulting in babies with severe birth defects.

Companies like Retrotope (Los Altos, California) use deuterium to stabilize the therapeutic enantiomers in a single orientation. Clinical trials are underway for treating Freidreich's ataxia, which is an inherited progressive disorder of the central nervous system, and is being treated by a deuterated form of linoleic fatty acid to protect cell and mitochondrial membranes against oxidative damage.

DeuteRx (Andover, Massachusetts) is known to interchange the deuterium component for hydrogen in the chiral center for stabilization of the desired enantiomer drug. Its lead compound, DRX-065, is the R enantiomer of pioglitazone that has been stabilized by deuterium, which is profoundly being used as a generic diabetes drug and for the treatment of nonalcoholic hepatosteatosis (NASH). The pioglitazone S enantiomer induces side effects like weight gain and edema, whereas the R enantiomer does not cause these side effects and works via different mitochondrial pathway. DRX-065 is under clinical trials for treating adrenomyeloneuropathy and NASH.

Zhang et al. (2018) examined the effects of deuterium substitution at the C6 position by determining the levels of (R)-ketamine (N-methyl-D-aspartate receptor (NMDAR) antagonist) and its two main metabolites, (R)-norketamine and (2R,6R)-HNK, in the plasma and brain after a single dose of (R)-ketamine or (R)-D2-ketamine in mice and found that no influence of deuterium isotope in the antidepressant effects of (R)-ketamine in a chronic social defeat stress (CSDS) model was observed in terms of rapid and long-lasting duration. (R)-ketamine is more potent than (S)-ketamine for unknown reasons, although both of them together manifest swift and enduring antidepressant effects and antisuicidal ideation in patients witnessing depression. It was concluded that (2R,6R)-HNK does not contribute to the antidepressant effects of (R)-ketamine.

3. A study by Stanford University (Ryan et al. 2018) revealed that D_2O labeling can determine the sites of active secondary metabolism in plants of medicinal importance over relatively short time scales. Such a mechanism can serve as a prominent method for evaluating plant-derived pharmaceuticals in clinical trials to investigate the biosynthetic pathways in slow-growing or recalcitrant plants to understand the underlying genetic/enzymatic machinery. In lieu of genomic information, metabolic engineering strategies can be used in host heterologous expression to unravel the puzzles of medicinal chemistry. Various quantitative and qualitative assessment of metabolic activities of bioactive molecules can be conducted using heavy water (D_2O) labeling in plants, which is not only a simple but also a cost-effective experimental technique that can potentially be performed on any species. Plants are hydrotropic in nature, which means that essentially all the hydrogen atoms required for synthesis of biochemical molecules in the organic biomass are derived from water during photosynthesis. Hydrogen/deuterium is incorporated by photosystem II for production of bioactive molecules by conversion to plastoquinol-bound hydrogen/deuterium (Vinyard et al. 2013), which are further fixed to a relatively stable carbon-bound form by conversion to NADPH (and NADH), from which all carbon-bound hydrogen atoms are produced. A difference in mass isotopologue distribution (MID) can be calculated using mass spectrometry to quantitatively measure the metabolite turnover and net flux (Baran et al. 2017).

A few case studies reported by the Stanford research group have shown that D_2O labeling can be used in conjunction with advanced LC-MS-based metabolomics that can identify trace amounts of fixed deuterium in various metabolites with high sensitivity, therefore ensuring a short turn-around time, particularly in slow-growing or recalcitrant plants. Pharmacokinetic studies for the incorporated levels of deuterium into specific metabolites at isotopic steady-state conditions may further enable plethora of knowledge generation in future to have a gist of relative as well as absolute biosynthetic capabilities in biological systems. However, there is a requirement to optimize the concentrations in which heavy water use in design of experiments and its analysis can be carried out in labeling studies due to its reported toxicity in eukaryotes at higher doses (Lewis 1934; Kushner et al. 1999). For example, D_2O volume concentrations beyond 37.5% showed attenuation of growth and arrest of germination in barley (Bhandarkar et al. 1971). While excised *Menispermum* leaves have been reported to have shown tolerance up to 100% D_2O for 15 days, with normal metabolism and label incorporation. Logically, deuterium should be incorporated at much higher rates at higher doses, but eventually, the overall duration of viable metabolism is expected to be decreased due to stress conditions.

The only drawback of using deuterium as a tracer molecule in such isotopic studies is that the general assumption of tracer being inert, i.e., it is being detected by only the mass spectrometer, no longer holds true. Actually, the biocatalysts or target cells can detect the difference between deuterium (^{2}H) and protium (^{1}H) that slightly complicates the situation which may lead to stress (Hayes 2001). Efforts are also required to differentiate between "exchangeable" and "non-exchangeable" positions during label incorporation because deuterium atoms linked to heteroatoms (nitrogen or oxygen) get exchanged instantly with the hydrogen in protic solvents, whereas carbon-bound hydrogens do not exhibit this behavior, thereby leading to loss of the label from the exchangeable positions during plant metabolite extraction (with unlabeled methanol and water) and LC-MS analysis (with unlabeled water and acetonitrile) (Fischer et al. 2013). Few studies suggest that, during enzyme-catalyzed biosynthesis, isotopic exchange of even some carbon-linked hydrogen can occur. Needless to mention, solvent exchange is nonsignificant and enzymatic fractionation is minimal in the case of ^{13}C in $^{13}CO_2$, which is also nontoxic to plants but having a disadvantage of being a gas, therefore requiring complex systems for efficient label insertion.

6 Reported Biotechnological Applications of Deuterium (Fig. 2)

1. Deuterium labeling has been successfully used by Pinnick et al. (2019) to measure the human lipid metabolism within the liver in vivo and in vitro using hepatocyte cellular models. The role of hepatic de novo lipogenesis (DNL) specifically in the fatty acid synthesis (e.g., palmitate) in cellular dysfunction in the progression of nonalcoholic fatty liver disease (NAFLD) has been vividly demonstrated (Fig. 3).
2. Decreased cell growth has been reported for cells cultured in D_2O due to decline in DNA replication rate, in contrast to an increase in vesicle transport in leukemia cells cultured in D_2O (Kalkur et al. 2014).
3. Hyaluronic acid (HA), a high-molecular-weight polysaccharide, is one of the main constituents of synovial fluids which is involved in wound healing, tumor progression, and joint lubrication. Study to analyze the association between synovial fluid and lipid membranes has been done using an oligolamellar stack of lipid membranes with a solution of hyaluronic acid in D_2O (Kreuzer et al. 2012).
4. Optimized dose of D_2O leads to control of hypertension, which is a risk factor for the development of myocardial infarction, stroke, and kidney failure as mentioned in US Patent No. 5223269 (Liepins 1993).
5. D_2O has been used to study the pathogenic role of adipose tissue in liver steatosis condition induced by alcohol consumption, which showed that adipose triglycerides formed by hyperlipolysis are transported back and accumulate in the liver (Zhong et al. 2012).
6. D_2O is also used to study the selective brain cooling which is a countercurrent heat exchange process for cooling down of the arterial blood going to the brain (Strauss et al. 2015).

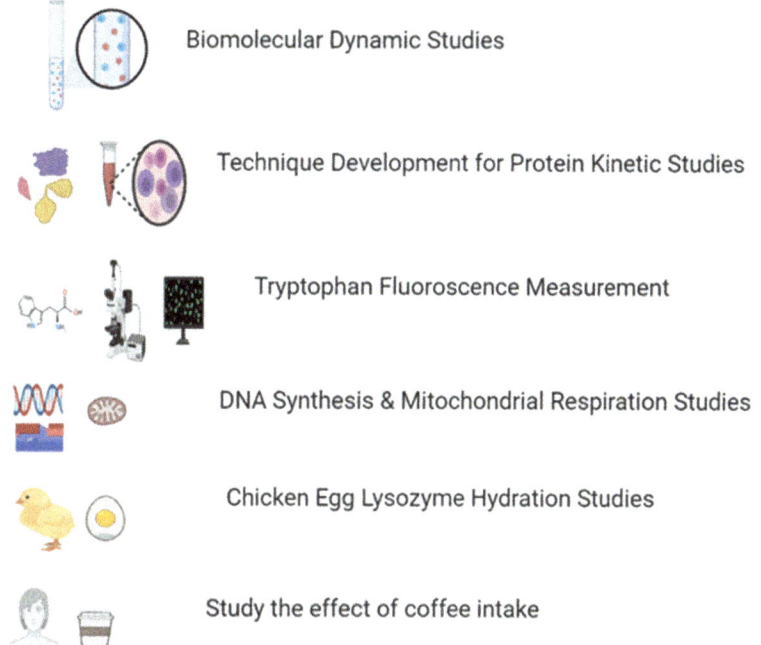

Biomolecular Dynamic Studies

Technique Development for Protein Kinetic Studies

Tryptophan Fluoroscence Measurement

DNA Synthesis & Mitochondrial Respiration Studies

Chicken Egg Lysozyme Hydration Studies

Study the effect of coffee intake

Fig. 2 Applications of heavy water in health science

7. As demonstrated in US Patent No. 8709496, Bayerl studied the viral inhibitory properties of D_2O for restricting its multiplication in the host cell (Bayerl 2014).

8. As described in Patent No. 8609147, D_2O has sought its use in the cure of viral skin diseases due to lower possibility of side effects and enhanced bioavailability (Bayerl 2013). Due to viral replication in the host cell, the infected cell possesses higher metabolic rate as compared to the healthy cells leading to higher water absorption. D_2O, when administered in optimized doses to the host cell, effectively limits the enzymatic reactions (DNA polymerase activity) occurring in the host cell. The uninfected cells do not absorb the same amount of D_2O as compared to the infected cells due to reduced metabolism.

9. Metabolic D_2O-labeling methodology is used to analyze the in vivo relative protein concentrations during calorie restriction regime, which implies that mitochondrial protein turnover is reduced during calorie restriction by analyzing relative protein degradation and synthesis along with relative mitochondrial autophagy and biogenesis (Price et al. 2012).

10. D_2O-labeling method is less invasive, restrictive, and more cost-effective than conventional amino acid tracer techniques, also quantifies acute changes in muscle protein synthesis (MPS) (Wilkinson et al. 2015).

11. Patients with advanced chronic kidney disease, when administered with optimized dose of deuterated water, can provide an estimation of trace amounts of volatile metabolites present in exhaled respiratory gas (Davies et al. 2014).

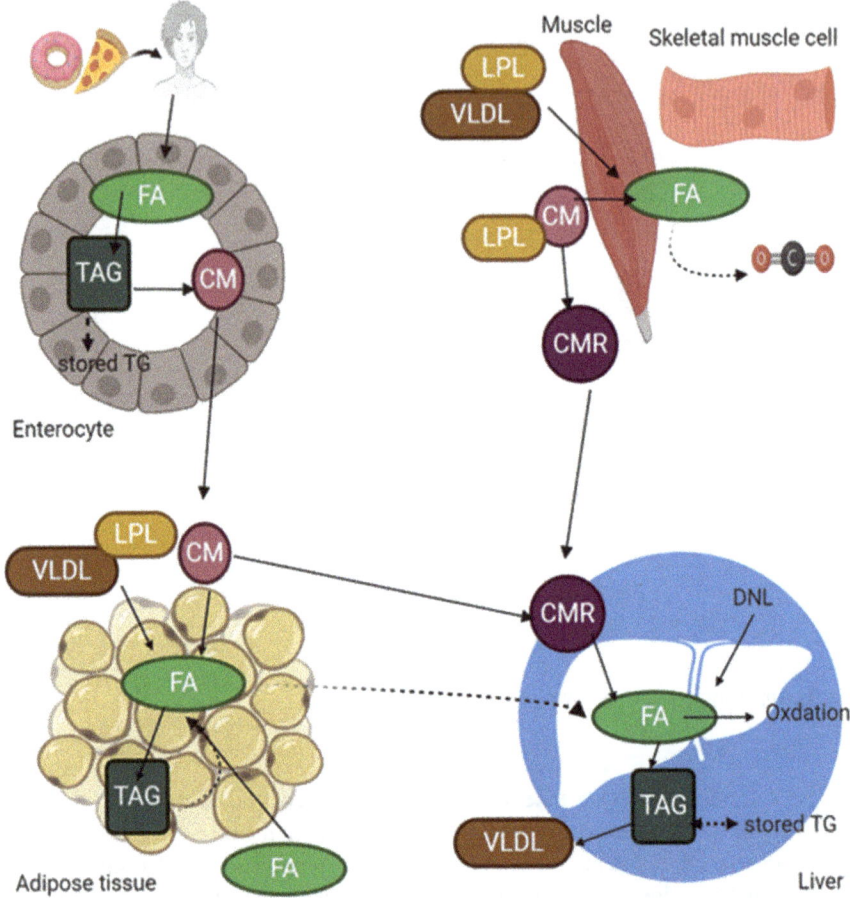

Fig. 3 Process flow profile of lipid metabolism in humans that can be studied by protocol employing D_2O. *FA fatty acid, TAG triacylglycerol, CM chylomicron, VLDL very low-density lipoprotein, LPL lipoprotein lipase, CMR chylomicron remnant, DNL de novo lipogenesis

12. The role of dietary supplementation can be examined by administration of low dose lipid-based nutrient supplements (LNS) for accelerating growth and development post 6 months in infants due to self-regulation of energy consumption. Intake of LNS may affect the uptake of the breast milk negatively as stated in pediatrics (Kumwenda et al. 2014). D_2O dual-energy X-ray absorptiometry, air displacement plethysmography, or quantitative NMR has also been studied to assess the body fat mass in pediatrics (Crook et al. 2012).

13. The in vivo drug/xenobiotic elimination from a biological system is usually initiated by hydroxilation that may be delayed by deuteration, while the principal pharmacological effect remains unchanged (Krumbiegel 2011).

14. D_2O is also used as a solvent in ^1H NMR spectroscopy to nullify the solvent signal interference in metabolomics, where nuclei in a magnetic field absorbs

and reemits electromagnetic radiations (Guy 2015). D_2O being a diffusible perfusion tracer is widely used in magnetic resonance imaging (MRI) for the analysis of tissue perfusion in contrast to gadolinium-based agents having low permeability into blood vessels (Wang et al. 2013).

15. Defoiche et al. (2009) examined that deuterium-labeled glucose can inspect the RNA turnover, DNA replication, and cell reproduction in clinical studies, by culturing a lymphocyte cell line (PM1) in vitro.

16. D_2O also enabled the study of the solvent–solute interactions with acetate (Banno et al. 2012), D-fructose, and tautomers (Barclaya et al. 2012), surveying biophysical properties of phospholipids bilayers (Beranová et al. 2012), transaldolase exchange in gluconeogenesis in various physiological conditions (Browning and Burgess 2012).

17. D_2O can measure the zinc concentration in breast milk by analyzing the plasma zinc concentrations in mothers and infants, which is a crucial mineral for the child's development. Islam and Brown (2014) reported that infant birth-weight category and milk zinc concentration or transfer between mother and child are independent entities.

18. The kinetic effects of halogen substituents of L-tryptophan (L-Trp), which is a metabolic biosynthetic precursor of vital compounds such as niacin, tryptamine, melatonin, plant hormones, and serotonin during biotransformation catalyzed by TPase, has been studied using deuterium (Winnicka et al. 2016).

19. The performance of H/D exchange experimental studies involving D_2O and other substances can be evaluated by flow injection hybrid quadrupole-time-of-flight mass spectrometry (QTOF-FI-MS) (Broecker and Pragst 2012).

20. Deuterium-labeled stimulated Raman scattering (SRS) imaging has been used to study the myelination process in brain cells of developing mice that might help in diagnosing head injuries and monitoring the development of multiple sclerosis (Min lab/Columbia). Once heavy water is metabolized by the body cells, it is inserted into newly synthesized biomolecules like DNA, proteins, and lipids, forming chemical association bonds with carbon, which when incident with light, vibrate with different signature frequencies that could be tracked in the target organ (Shi et al. 2018).

21. Fat production has been studied using deuterium-labeled SRS imaging in ageing roundworms during reproductive cycle. Fat helps in egg maturation and post-reproduction; its synthesis is slowed down.

22. The other nontrivial applications of D_2O include determination of total body water (TBW) (Bila et al. 2016), hypertension treatment (Liepins 1993), skin treatment for herpes virus-based diseases (Bayerl 2013), and respiratory tract treatment for virus-based diseases (Vareille et al. 2011).

23. Peptide and protein formation with thiol radicals (glutathione in human metabolism) (Hofstetter et al. 2012).

24. Investigation of charge state of polyamidoamine dendrimers (Hong et al. 2012).

25. Investigation of protein balance studies (Holm et al. 2013).

26. Temporal quantification of cationic conformational equilibrium (Hatano et al. 2012) and protein side-chain exchange positions (Hansen et al. 2012).

27. In situ microbial studies (Berry et al. 2014).
28. Temperature-dependent structural studies of heavy water (Kamitakahara et al. 2012).
29. Characterization studies of piperazinium salts that are degradation products of nitrogen mustards by NMR (nuclear magnetic resonance spectroscopy) and LC-MS (liquid chromatography–mass spectrometry) (Lee et al. 2012).
30. Analysis of vibrational spectrum of monosaccharides (Jin et al. 2012) and heme iron-dependent process (Mielcarek et al. 2015).
31. Mutagenesis and protein engineering (Minamoto et al. 2012), formation of glutathione-thiol radical, cysteine and transfer reactions of intramolecular hydrogen atoms (Mozziconacci et al. 2012).
32. Stabilization of the native protein structure (Reddy et al. 2012), porphyrins (Rich and Mchale, 2012).
33. Study of urinary metabolism and leptin signaling pathway (Saadat et al. 2012).
34. Study of lifetime of dissolved organic matter, polyelectrolyte solutions, solvent and salt concentration-dependent alignment properties of acrylamide copolymer gels (Sharpless 2012; Shew et al. 2012; Trigo-Mouriño et al. 2012).

7 Discussion

In India, Heavy Water Board (HWB) is a leading producer of heavy water and therefore collaborative R&D studies may be carried out to assess the potential of heavy water in overcoming challenges in biomedical applications. Using biotechnological applications to study the stress and adaptive responses in microbial organisms, such as bacteria and algae that can adapt to grow in almost 100% D_2O, can be evaluated for industrial and medical use.

A comprehensive research program to explore the bio-therapeutic potential of heavy water/deuterium can pave the way for enormous avenues in the biomedical area and deserves to be considered on utmost priority in dealing with critical biological disorders via drug development and therapy. As presently, regulatory obligations may be relatively less stringent, thermo-stabilization of the vaccines used in animal husbandry could be the areas of choice. Likewise, shelf-life extension of the industrial enzymes, preservation of cells, organs is another area of profound interest worth exploring. Extensive research studies for synthesis of deuterated drugs to enhance efficacy, profiling of their pharmacokinetic parameters in understanding mammalian physiology and its health implications can be very rewarding. Likewise, a detailed pharmacological evaluation of D_2O to control the proliferation of malignant cells and photodynamic therapy in cancer treatment could be promising. Based on scientific revelations for physiological deceleration either through caloric curtailment or by central nervous system control, D_2O at lower doses is being projected as a health promoter by numerous North American companies.

Not only in health sector, but D_2O also finds industrial applications, for example, a salt-tolerant photosynthetic microalga *Dunaliella salina* is widely reported to produce up to 50% of its biomass as deuterated glycerol at high NaCl concentrations. Deuterated glycerol can be used as a precursor of transparent plastics and other commercially important compounds used in medical field. Based on the molecular differences between hydrogen and deuterium, the applications include but are not limited to optical fibers with improved intensity and better transmission properties of deuterated PMMA (polymethyl methacrylate) compared to the conventional one, optical recording systems with improved storage features because of sharper and uniform optical recording density with deuterated polymeric substance and improved quality of semiconductors with lower stress induced leakage currents across the metal oxide gates. Lubricant additives composed of deuterated compounds exhibit interesting physicochemical properties and have illustrated remarkable performance vis-à-vis normal lubricants.

8 Challenges and Opportunities

Over 200 patents have already been filed or granted on biotechnological applications of D_2O in industry and health-care system (Table 2). Regarding the isotopic drug form and the deuterated drug form, there could be a debate regarding the implementation of "Doctrine of Equivalents" under patent laws, which is practiced differently in different countries, based on performing the same function the same way and yielding the same result. Also, identifying which particular hydrogen in a drug molecule could be replaced with deuterium, in such a way that prolongs drug's

Table 2 Pharmaceutical deuterated drug patents filed (Timmins 2014)

S. no.	Patent title	Year	Pharmaceutical
1.	Process for the preparation of sulfamide derivatives	2013	Johnson and Johnson® (NJ, USA)
2.	9h-pyrrolo [2,3-b:5,4-c'] dipyridine azacarboline derivatives, preparation thereof, and therapeutic use thereof	2012	Sanofi-Aventis® (FRA)
3.	Substituted 5-fluoro-1H-pyrazolopyridines and their use	2012	Bayer® (GER)
4.	Modified bovine granulocyte-colony stimulating factor polypeptides (G-CSF) and their uses	2010	Eli Lilly® (IN, USA)
5.	Compounds useful as inhibitors of protein kinases	2010	Abbott® (IL, USA)
6.	Compounds for the treatment of hepatitis C	2010	Bristol-Myers Squib® (NY, USA)
7.	Thiazolopyrimidine p13k inhibitor compounds and methods of use	2009	Roche (SWE)
8.	Organic compounds	2009	Novartis® (SWE)
9.	Assays for allosteric modulators of G-protein coupled receptors (GPCRs)	2009	Astra Zeneca® (UK)
10.	41-Methoxy isotope labeled rapamycin 42-ester	2007	Wyeth® (NJ, USA)

G-CSF granulocyte-colony stimulating factor, *GPCRs* G-protein coupled receptors half-life without modifying drug efficacy, is another potent challenge. The slowing metabolic effect affects different systems disproportionally long before the satiation conditions are achieved, therefore optimizing the drug concentrations would be another challenge as at high concentrations, it ends up being lethal to biological cells. Deuteration approach is most likely suitable for increasing the exposure of a drug that is primarily cleared by a process involving C-H bond cleavage and not for other drugs that are removed by other mechanisms, such as Paracetamol (cleared by sulfation). Going into the deep insights of medicinal chemistry would help to understand the actual pros and cons of such deuterated drug discoveries.

Theoretically, assimilating "heavy hydrogen" into smaller molecules makes a drug long-lasting and improvises its toxicity profile. Still, the clinical advantages are questionable in case of an IP (intellectual property) dispute. Seldom, the deuterated compounds at few instances show anomaly behavior. They might direct the generation of toxic metabolites obliging rigorous risk management and therefore necessitating the understanding of action plan for predicting the metabolic enzymes (CYP-450, CYP2D6) activity for processing the drug.

D_2O may affect membrane function at the cellular level. Protozoans can tolerate up to 70% D_2O. D_2O is known to enhance the heat stability of macromolecules, while due to the inhibition of chaperonin formation, the cellular heat stability may be reduced. High D_2O levels in rats can lower salt- and ethanol-induced hypertension. It is also reported that the genotoxicity of the anticancer drug tamoxifen can be reduced by deuteration. The effectiveness of long-chain fatty acids and fluoro-D-phenylalanine can be enhanced by deuteration via breakdown inhibition by target microbes. Deuteration can be studied to troubleshoot the problem of insect resistance to insecticides via cytochrome P450 system pathway.

A pharmaceutical drug can be metabolized via a number of pathways, and C–H breakage is one of them. Therefore, identification of a drug's metabolic pathway, followed by site detection of where to insert the deuterium, is quite a complex mechanism. This can prove to be very advantageous or may have a neutral effect. Needless to mention, an incorrect deuterium insertion may lead to contradictory effects in a biological system.

It is anticipated that novel biosynthetic enzymes will be identified using the deuteration technique, which will facilitate the discovery of advanced approaches regarding transcriptomics and understanding of active metabolism. In biotechnology, it would be worthwhile to examine the distinct morphological features linked to microbial adaptation employing deuterated polymers that provide an excellent model for such kind of illustrations. The discussed technology has enormous potential in the industries based on fundamental and applied research. The unique avenues of D_2O can be substantially exploited for probing biological phenomenon for delineating the underlying enigma to generate an array of classified information in biotic systems.

Acknowledgment The author declares no conflict of interest. The figures were created and exported using paid subscription from BioRender.com. The author is thankful to Dr. Nishant Joshi, from the School of Natural Sciences, Shiv Nadar University, Greater Noida, India for his support in the study.

References

Andrieu J, Camus JM, Balan C, Poli R. Amino-phosphanes in RhI-catalyzed hydroformylation: new mechanistic insights using D_2O as deuterium-labeling agent. Eur J Inorg Chem. 2006;1:62–8. https://doi.org/10.3928/02793695-20150821-55.

Atzrodt J, Derdau V, Fey T, Zimmermann J. The renaissance of H/D exchange. Angew Chem Int Ed. 2007;46(41):7744–65. https://doi.org/10.1002/anie.200700039.

Atzrodt J, Derdau V, Kerr WJ, Reid M. C-H functionalisation for hydrogen isotope exchange. Angew Chem Int Ed. 2018;57(12):3022–47. https://doi.org/10.1002/anie.201708903.

Banno M, Ohtaab K, Tominaga K. Vibrational dynamics of acetate in D_2O studied by infrared pump–probe spectroscopy. Phys Chem Chem Phys. 2012;14:6359–66.

Baran R, Lau R, Bowen BP, Diamond S, Jose N, Garcia-Pichel F, Northen TR. Extensive turnover of compatible solutes in cyanobacteria revealed by deuterium oxide (D_2O) stable isotope probing. ACS Chem Biol. 2017;12:674–81. [PubMed: 28068058]

Barclaya T, Ginic-Markovica M, Johnstona MR, et al. Observation of the keto tautomer of D-fructose in D_2O using ^1H NMR spectroscopy. Carbohydr Res. 2012;347:136–41.

Bayer TM, Inventor; D_2 Bioscience Group Ltd., assignee. Use of deuterium oxide for treatment of herpes virus-based diseases of the skin. United States patent 8,609,147. 2013.

Bayerl TM, Inventor; D_2 Bioscience Group Ltd., assignee. Use of deuterium oxide for treatment of herpes virus-based diseases of the skin. United States patent 8,609,147. 2013.

Bayerl T, Inventor; D_2 Bioscience Group Ltd., assignee. Use of deuterium oxide for the treatment of virus-based diseases of the respiratory tract. United States Patent. 8,709,496. 2014.

Beranová L, Humpolıckova J, Sikora J, et al. Effect of heavy water on phospholipid membranes: experimental confirmation of molecular dynamics simulations. Phys Chem Chem Phys. 2012;14:14516–22.

Berry D, Mader E, Lee TK, et al. Tracking heavy water (D_2O) incorporation for identifying and sorting active microbial cells. Proc Natl Acad Sci U S A. 2014;112:194–203.

Bhandarkar MK, Bhattacharya S, Gaur BK. Physiological studies with heavy water. II. Germination and growth inhibition of barley by D_2O. Physiol Plant. 1971;24:517–21.

Bila W, De Freitas A, Galdino A, et al. Deuterium oxide dilution and body composition in overweight and obese schoolchildren aged 6–9 years. J Pediatr. 2016;92:46–52.

Blagojevic N, Storr G, Allen BJ, Hatanaka H, Nakagawa H. Role of heavy water in boron neutron capture therapy. In: Zamenhof R, Solares G, Harling O, editors. Topics in dosimetry and treatment planning for neutron capture therapy. Madison: Advanced Medical Publishing; 1994. p. 125–34.

Broecker S, Pragst F. Isomerization of cannabidiol and Δ9-tetrahydrocannabinol during positive electrospray ionization. In-source hydrogen/ deuterium exchange experiments by flow injection hybrid quadrupole-time-of-flight mass spectrometry. Rapid Commun Mass Spectrom. 2012;26:1407–14.

Browning JB, Burgess SC. Use of $2H_2O$ for estimating rates of gluconeogenesis: determination and correction of error due to transaldolase exchange. Am J Physiol Endocrinol Metab. 2012;303:E1304–12.

Crook TA, Armbya N, Cleves MA, et al. Air displacement plethysmography, dual-energy X-ray absorptiometry, and total body water to evaluate body composition in preschool-age children. J Acad Nutr Diet. 2012;112:1993–8.

Daniel CM, Xin L, Charles SSJ. Pharmaco-thermodynamics of deuterium-induced oedema in living rat brain via ^1H_2O MRI: implications for boron neutron capture therapy of malignant brain tumours. Phys Med Biol. 2005;50:2127–39.

Davies S, Spanel P, Smith D. Breath analysis of ammonia, volatile organic compounds and deuterated water vapor in chronic kidney disease and during dialysis. Bioanalysis. 2014;6:843–57.

Defoiche J, Zhang Y, Lagneaux L, et al. Measurement of ribosomal RNA turnover in vivo by use of deuterium-labeled glucose. Clin Chem. 2009;55:1824–33.

Fischer CR, Bowen BP, Pan C, Northen TR, Banfield JF. Stable-isotope probing reveals that hydrogen isotope fractionation in proteins and lipids in a microbial community are different and species-specific. ACS Chem Biol. 2013;8:1755–63. [PubMed: 23713674]

Garreau AL, Zhou H, Young MC. A protocol for the *Ortho*-deuteration of acidic aromatic compounds in D_2O catalyzed by cationic Rh^{III}. Org Lett. 2019;21(17):7044–8. https://doi.org/10.1021/acs.orglett.9b02618.

Guy A. Machine perfusion in kidney transplantation: Clinical application & metabolomic analysis [thesis]. Birmingham: University of Birmingham; 2015.

Hansen AL, Lundström P, Velyvis A, et al. Quantifying millisecond exchange dynamics in proteins by CPMG relaxation dispersion NMR using side-chain ^1H probes. J Am Chem Soc. 2012;134:3178–89.

Hatano NN, Watanabe M, Takekiyo T, et al. Anomalous conformational change in 1-butyl-3-methylimidazolium tetrafluoroborate–D_2O mixtures. J Phys Chem A. 2012;116:1208–12.

Hayes JM. Fractionation of carbon and hydrogen isotopes in biosynthetic processes. Rev Mineral Geochem. 2001;43:225–77.

Hofstetter D, Thalmann B, Nauser T, et al. Hydrogen exchange equilibria in thiols. Chem Res Toxicol. 2012;25:1862–7.

Holm L, O'Rourke B, Ebenstein D, et al. Determination of steady-state protein breakdown rate in vivo by the disappearance of protein-bound tracer-labeled amino acids: a method applicable in humans. Am J Physiol Endocrinol Metab. 2013;304:E895–907.

Hong K, Liu Y, Porcar L, et al. Structural response of polyelectrolyte dendrimer towards molecular protonation: the inconsistency revealed by SANS and NMR. J Phys Condens Matter. 2012;24:1–7.

Howland RH. Deuterated drugs. J Psychosoc Nurs. 2015;53(9):13–6. https://doi.org/10.1177/1060028018797110.

International Atomic Energy Agency. Introduction to body composition assessment using the deuterium dilution technique with analysis of urine samples by isotope ration mass spectrometry. Human Health Series [Internet]. 2011. Available from: http://www-pub.iaea.org/books/iaea-books/8369/Introduction-to-Body-Composition-Assessment-Using-the-Deuterium-Dilution-Technique-with-Analysis-of-Saliva-Samples-by-Fourier-Transform-Infrared-Spectrometry.

Islam MM, Brown KH. Zinc transferred through breast milk does not differ between appropriate- and small-for-gestational-age, predominantly breast-fed Bangladeshi infants. J Nutr. 2014;144:771–6.

Jin L, Simons JP, Gerber RB. Monosaccharide-water complexes: vibrational spectroscopy and anharmonic potentials. J Phys Chem A. 2012;116:11088–94.

Kalkur RS, Ballast AC, Triplett AR, et al. Effects of deuterium oxide on cell growth and vesicle speed in RBL-2H3 cells. Peer J. 2014;2:1–13.

Kamitakahara W, Faraone A, Liu K, et al. Temperature dependence of structure and density for D_2O confined in MCM-41-S. J Phys Condens Matter. 2012;24:1–7.

Kreuzer M, Strobl M, Reinhardt M, et al. Impact of a model synovial fluid on supported lipid membranes. Biochim Biophys Acta. 2012;1818:2648–59.

Krumbiegel P. Large deuterium isotope effects and their use: a historical review. Isot Environ Health Stud. 2011;47:1–17.

Kumwenda C, Dewey KG, Hemsworth J, et al. Lipid-based nutrient supplements do not decrease breast milk intake of Malawian infants. Am J Clin Nutr. 2014;99:617–23.

Kushner DJ, Baker A, Dunstall TG. Pharmacological uses and perspectives of heavy water and deuterated compounds. Can J Physiol Pharmacol. 1999;77:79–88. [PubMed: 10535697]

Lee JY, Lee YH, Byun YG. Characterization and study of piperazinium salts, degradation products of nitrogen mustards by nuclear magnetic resonance spectroscopy and liquid chromatography–mass spectrometry. J Chromatogr A. 2012;1227:163–73. [PubMed: 17788137]

Lewis GN. Biology of heavy water. Science. 1934;79:151–3. [PubMed: 17788137]

Liepins A. Methods and composition for the treatment of hypertension [Internet]. 1993. Available from: http://patft.uspto.gov/netacgi/nph-Parser?Sect2=PTO1&Sect2=HITOFF&p=1&u=/netahtml/PTO/search-bool.html&r=1&f=G&l=50&d=PALL&RefSrch=yes&Query=PN/5223269.

Loh YY, Nagao K, Hoover AJ, Hesk D, Rivera NR, Colletti SL, Davies IW, MacMillan DWC. Photoredox-catalyzed deuteration and tritiation of pharmaceutical compounds. Science. 2017;358(6367):1182–7. https://doi.org/10.1126/science.aap9674.

Mielcarek A, Blauenburg B, Miethke M, et al. Molecular insights into frataxin-mediated iron supply for heme biosynthesis in *Bacillus subtilis*. PLoS One. 2015;10:e0122538.

Minamoto T, Wadab E, Shimizu I. A new method for random mutagenesis by error-prone polymerase chain reaction using heavy water. J Biotechnol. 2012;157:71–4.

Mozziconacci O, Williams TD, Schöneich C. Intramolecular hydrogen transfer reactions of thiyl radicals from glutathione: formation of carbon-centered radical at Glu, Cys, and Gly. Chem Res Toxicol. 2012;25:1842–61.

Pascal E, Franziska U, Sven S, Bernd P. Mild, selective Ru-catalyzed deuteration using D_2O as a deuterium source. Chem Eur J. 2019;25(72):16550–4. https://doi.org/10.1002/chem.201904927.

Pinnick KE, Gunn PJ, Hodson L. Measuring human lipid metabolism using deuterium labeling: in vivo and in vitro protocols. In: Fendt SM, Lunt SY, editors. Metabolic signaling: methods and protocols, methods in molecular biology, vol. 1862; 2019. https://doi.org/10.1007/978-1-4939-8769-6_6.

Pirali T, Serafini M, Cargnin S, Genazzani AA. Applications of deuterium in medicinal chemistry. J Med Chem. 2019;62(11):5276–97.

Price JC, Khambatta CF, Li KW, et al. The effect of long term calorie restriction on in vivo hepatic proteostatis: a novel combination of dynamic and quantitative proteomics. Mol Cell Proteomics. 2012;1:1801–14.

Reddy PM, Taha M, Venkatesu P, et al. Destruction of hydrogen bonds of poly (N-isopropylacrylamide) aqueous solution by trimethylamine N-oxide. J Chem Phys. 2012;136:1–10.

Rich CC, Mchale JL. Influence of hydrogen bonding on excitonic coupling and hierarchal structure of a light-harvesting porphyrin aggregate. Phys Chem Chem Phys. 2012;14:2362–74.

Russak EM, Bednarczyk EM. Impact of deuterium substitution on the pharmacokinetics of pharmaceuticals. Ann Pharmacother. 2019;53(2):211–6.

Ryan SN, Xin G, Smith K, Faust AM, Sattely ES, Fischer CR. D_2O labeling to measure active biosynthesis of natural products in medicinal plants. AIChE J. 2018;64(12):4319–30. https://doi.org/10.1002/aic.16413.

Saadat N, Iglayreger HB, Myers MGJ, et al. Differences in metabolomic profiles of male db/db and s/s, leptin receptor mutant mice. Physiol Genomics. 2012;44:374–81.

Sajlki H, Aoki F, Esaki H, Maegawa T, Hirota K. Efficient C-H/C-D exchange reaction on the alkyl side chain of aromatic compounds using heterogeneous Pd/C in D_2O. Org Lett. 2004;6(9):1485–7.

Sakurai Y. Studies on depth-dose-distribution controls by deuteration and void formation in boron neutron capture therapy. Phys Med Biol. 2004;49:3367–78.

Schmidt C. First deuterated drug approved. Nat Biotechnol. 2017;35:6.

Sharpless CM. Lifetimes of triplet dissolved natural organic matter (DOM) and the effect of $NaBH_4$ reduction on singlet oxygen quantum yields: implications for DOM photophysics. Environ Sci Technol. 2012;46:3912–20.

Shew CY, Do C, Hong K, et al. Conformational effect on small angle neutron scattering behavior of interacting polyelectrolyte solutions: a perspective of integral equation theory. J Chem Phys. 2012;137:1–9.

Shi L, Zheng C, Shen Y, Chen ZS, Silveira ES, Zhang L, Wei M, Liu C, Tomas CS, Targoff K, Min W. Optical imaging of metabolic dynamics in animals. Nat Commun. 2018;9(1) https://doi.org/10.1038/s41467-018-05401-3.

Strauss WM, Hetem RS, Mitchell D, et al. Selective brain cooling reduces water turnover in dehydrated sheep. PLoS One. 2015;12:e0115514.

Timmins GS. Deuterated drugs; where are we now? Expert Opin Ther Pat. 2014;24:1067–75.

Trigo-Mouriño P, Navarro-Vázquez A, Sánchez-Pedregal VM. Influence of solvent and salt concentration on the alignment properties of acrylamide copolymer gels for the measurement of RDC. Magn Reson Chem. 2012;50:S29–37.

Vareille M, Kieninger E, Edwards M, et al. The airway epithelium: soldier in the fight against respiratory viruses. Clin Microbiol Rev. 2011;24:210–29.

Vinyard DJ, Ananyev GM, Dismukes GC. Photosystem II: the reaction center of oxygenic photosynthesis. Annu Rev Biochem. 2013;82:577–606. [PubMed: 23527694]

Wang F, Peng S, Lu C, et al. Water signal attenuation by D_2O infusion as a novel contrast mechanism for ^1H perfusion MRI. NMR Biomed. 2013;26:692–8.

Wilkinson DJ, Cegielski J, Phillips BE, et al. Internal comparison between deuterium oxide (D_2O) and L-[ring-$^{13}C_6$] phenylalanine for acute measurement of muscle protein synthesis in humans. Phys Rep. 2015;3:1–9.

Winnicka E, Szymanska J, Kanska M. Synthesis of deuterium-labelled halogen derivates of ljtryptophan catalysed by tryptophanase. Isot Environ Health Stud. 2016;52:231–8.

Yu RP, Hesk D, Rivera N, Pelczer I, Chirik PJ. Iron-catalysed tritiation of pharmaceuticals. Nature. 2016;529(7585):195–9. https://doi.org/10.1038/nature16464.

Zhang K, Toki H, Fujita Y, Ma M, Chang L, Qu Y, Harada S, Nemoto T, Yasuhira AM, Yamaguchi J, Chaki S, Hashimoto K. Lack of deuterium isotope effects in the antidepressant effects of (R)-ketamine in a chronic social defeat stress model. Psychopharmacology. 2018; https://doi.org/10.1007/s00213-018-5017-2.

Zhong W, Zhao Y, Tang Y, et al. Chronic alcohol exposure stimulates adipose tissue lipolysis in mice-role of reverse triglyceride transport in the pathogenesis of alcoholic steatosis. Am J Pathol. 2012;180:998–1007.

Index

© The Author(s), under exclusive license to Springer Nature Switzerland AG 2021   379
K. K. Kesari, N. K. Jha (eds.), *Free Radical Biology and Environmental Toxicity*,
Molecular and Integrative Toxicology, https://doi.org/10.1007/978-3-030-83446-3

Ingram Content Group UK Ltd.
Milton Keynes UK
UKHW022133060423
419683UK00001B/15

9 783030 834456